Sustainable Development Goals (SDG) – Umsetzung in Praxis, Lehre und Entscheidungsprozessen

Diese Buchreihe beleuchtet die UN-Nachhaltigkeitsziele hinsichtlich ihrer wissenschaftlichen Aspekte, disziplinären Verknüpfungen und Bedeutung für die Lehre sowie Transfer in die Gesellschaft.
Kerngerüst bilden die Ziele „Gesundheit und Wohlergehen", „Sauberes Wasser und Sanitäreinrichtungen", „Bezahlbare und saubere Energie", „Industrie, Innovation und Infrastruktur", „Nachhaltige Städte und Gemeinden", „Maßnahmen zum Klimaschutz". Die sechs Ziele bilden das Fundament, deren Betrachtung im Vordergrund steht und den jeweiligen Ausgangspunkt für die weiteren Überlegungen bildet. Klar ist, dass diese sechs, aber insgesamt alle 17 Ziele, nicht isoliert betrachtet werden können. Insgesamt sind die Ziele vor dem Hintergrund einer Reaktion allgemeiner globaler Veränderungen zu sehen. Viele Ziele sind in Europa mit eigenen Themen behaftet, was die technischen Voraussetzungen oder eine internationale Koordination betrifft, und in den Ländern der EU durch den Klimawandel mit eigenen Vorzeichen versehen. Daher will diese Reihe vorausdenken, Probleme frühzeitig erkennen und benennen. Lösungen sollen konkretisiert werden und auf Ihre Praxistauglichkeit überprüft werden.

Jörn Birkmann · Elena-Maria Klopries · Holger Schüttrumpf ·
Alessa Trüdinger · Stefanie Stenger-Wolf
Hrsg.

Die Flutkatastrophe 2021

Vom Wiederaufbau zur Klimaresilienz

Springer

Hrsg.
Jörn Birkmann
Institut für Raumordnung und Entwicklungsplanung
Universität Stuttgart
Stuttgart, Deutschland

Alessa Trüdinger
Institut für Raumordnung und Entwicklungsplanung
Universität Stuttgart
Stuttgart, Deutschland

Elena-Maria Klopries
Lehrstuhl und Institut für Wasserbau und Wasserwirtschaft
RWTH Aachen
Aachen, Deutschland

Stefanie Stenger-Wolf
Lehrstuhl und Institut für Wasserbau und Wasserwirtschaft
RWTH Aachen
Aachen, Deutschland

Holger Schüttrumpf
Lehrstuhl und Institut für Wasserbau und Wasserwirtschaft
RWTH Aachen
Aachen, Deutschland

ISSN 2731-9083 ISSN 2731-9091 (electronic)
Sustainable Development Goals (SDG) – Umsetzung in Praxis, Lehre und Entscheidungsprozessen
ISBN 978-3-658-48687-7 ISBN 978-3-658-48688-4 (eBook)
https://doi.org/10.1007/978-3-658-48688-4

Die Deutsche Nationalbibliothek verzeichnet diese Publikation in der Deutschen Nationalbibliografie; detaillierte bibliografische Daten sind im Internet über https://dnb.d-nb.de abrufbar.

Dieses Werk wurde gefördert durch RWTH Aachen University.

Planung: Dr. Daniel Fröhlich

Springer Vieweg ist ein Imprint der eingetragenen Gesellschaft Springer Fachmedien Wiesbaden GmbH und ist ein Teil von Springer Nature.
Die Anschrift der Gesellschaft ist: Abraham-Lincoln-Str. 46, 65189 Wiesbaden, Germany

Editorial

Liebe Leserin, lieber Leser,

Extremereignisse werden im Kontext des Klimawandels in Zukunft zunehmen und zudem führt der Klimawandel zur Veränderung bekannter Phänomene. Die Flutkatastrophe im Sommer 2021, die allein in Deutschland knapp 200 Todesopfer gefordert hat, war ein solches Extremereignis und erfordert neue Ansätze und eine kritische Reflexion bisheriger Strategien, Methoden und Instrumente zum Umgang mit Hochwasser und Starkregen. Dabei ist Hochwasser für Deutschland keine Neuheit. Das Hochwasser von Mitte Juli 2021 in Westeuropa reiht sich in die Kette historisch bedeutender Hochwasserereignisse in Deutschland ein, wie beispielsweise die Rheinhochwasser von 1993 und 1995, das Oderhochwasser von 1997 und die Elbehochwasser von 2002 und 2013. Insbesondere die hohe Anzahl der Todesopfer und Betroffenen sowie die hohen wirtschaftlichen Schäden erschütterten im Sommer 2021 das Land und erregten internationales Interesse. So wurde bereits im November 2021 das vom Bundesministerium für Bildung und Forschung geförderte Projekt „KAHR – Wissenschaftliche Begleitung der Wiederaufbauprozesse nach der Flutkatastrophe in Rheinland-Pfalz und Nordrhein-Westfalen: Impulse für Resilienz und Klimaanpassung“ ins Leben gerufen. Ein zentrales Ziel des KAHR Projekts war dabei der Erkenntnisgewinn durch die wissenschaftliche Aufarbeitung des Hochwasserereignisses mit einem interdisziplinären und überregionalen Team von Wissenschaftlerinnen und Wissenschaftlern sowie Institutionen an der Wissenschafts-Praxis-Schnittstelle, um Schadensmuster besser zu verstehen und Ansatzpunkte zur Stärkung der Resilienz zu identifizieren. KAHR steht hierbei für **K**lima**A**npassung, **H**ochwasser und **R**esilienz. Das Projekt hatte unter anderem die Weiterentwicklung von Hochwasser- und Schadensmodellen, die Sensibilisierung für Risiken im Kontext von Hochwasser und Starkregen, die Weiterentwicklung von Bewertungskriterien für einzelne Hochwasserschutz- und -vorsorgemaßnahmen sowie die Verbesserung von Resilienz und Hochwasserrisikovorsorge im Rahmen des Wieder- und Neuaufbaus zum Ziel. Ein weiterer zentraler Baustein war die flexible, zeitnahe und unkomplizierte Reaktion auf wissenschaftliche Bedarfe in den flutbetroffenen Regionen während der Wiederaufbauphase. Dies war nicht ganz einfach, da jede Forschungsförderung eigentlich zuvor klar definierte Ziele und Produkte verlangt, aber bei KAHR die Produkte erst im Laufe der wissenschaftlichen Begleitung stetig konkretisiert und verfeinert wurden. Zudem war die Forschung in den flutbetroffenen Regionen wie die Arbeit an einem lebenden Patienten, da sich Problemlagen und Akteure über die verschiedenen Phasen des Wiederaufbaus deutlich veränderten bzw. weiterentwickelten. Beispielsweise waren am Anfang des Projekts die Wiederaufbaugesellschaften noch nicht vorhanden, die aber später ein wichtiger Adressat und Nutzer der wissenschaftlichen Empfehlungen des KAHR Projekts wurden.

Neben der hohen Flexibilität und großen Praxisnähe und der Herausforderung in stark zerstörten und im „Sondermodus“ arbeitenden Städten und Kreisen zu forschen, war auch der transdisziplinäre Aufbau des Konsortiums mit 13 Verbundpartnern eine Besonderheit. Dieses umfasste das Institut für Raumordnung und Entwicklungsplanung der Universität Stuttgart, das Institut für Wasserbau und Wasserwirtschaft der RWTH Aachen University, das Institut für Umweltwissenschaften und Geographie, AG Geographie und Naturrisikenforschung der Universität Potsdam, den Lehrstuhl für Regionalentwicklung und Risikomanagement, Fakultät Raumplanung der Technischen Universität Dortmund, bauen-kunst-werkstoffe der Hochschule Koblenz, das Fachgebiet Wasserbau und Wasserwirtschaft der Rheinland-Pfälzischen Technischen Universität Kaiserslautern-Landau, das Institut für qualifizierende Innovationsforschung und -beratung in Bad Neuenahr-Ahrweiler, das Helmholtz-Zentrum für Umweltforschung, das Helmholtz-Zentrum für Geoforschung Potsdam, das Deutsche Institut für Urbanistik und das HochwasserKompetenzCentrum e. V. sowie den Landkreis Ahrweiler und den Wasserverband Eifel-Rur als Praxispartner.

Gemeinsam wurden das Hochwasser von Mitte Juli 2021 in Nordrhein-Westfalen und Rheinland-Pfalz aufgearbeitet und Schadensmuster erfasst sowie Herausforderungen beim Wiederaufbau systematisch, u. a. durch Expertengespräche und Haushaltsbefragungen, ermittelt. Der Wiederaufbau wurde begleitet, wobei insbesondere die Weiterentwicklung von Hochwasser- und Starkregenrisikokonzepten im Fokus stand. Dabei wurden die Expertisen aus der Wasserwirtschaft, der Raumplanung und der Risiko- und sozialwissenschaftlichen Forschung systematisch zusammengeführt. Es stand u. a. die Frage im Raum, welche Schutzziele bei der Anpassung und Stärkung der Resilienz gegenüber zukünftigen Ereignissen im Rahmen des Wieder- und Neuaufbaus zu empfehlen sind und welche Standards gesetzlich und im Rahmen der Wiederaufbauförderung vorgeschrieben sind. Gerade beim Wiederaufbau sog. sensibler und kritischer Infrastrukturen zeigten sich hier deutliche Schwächen in der bisherigen Praxis. Auch die Rolle der Brücken wurde durch die wissenschaftlichen Befunde verdeutlicht und es wurden Ansatzpunkte zur Stärkung der Hochwasserresilienz formuliert. Des Weiteren wurden Hemmnisse und Chancen für den resilienten Wiederaufbau in verschiedenen Phasen des Wiederaufbaus ermittelt und Empfehlungen abgeleitet und in zahlreichen Veranstaltungen und Fachgesprächen vor Ort und in den betroffenen Ländern kommuniziert. Auch die Zusammenarbeit zwischen verschiedenen Akteuren und Institutionen sowie Entscheidungsprozesse im Wiederaufbau wurden beleuchtet. Dabei zeigt sich u. a., dass die Wiederaufbaugesellschaften eine sehr unterschiedliche Form und Personalstärke je nach Stadt oder Ortschaft entwickelt haben. Darüber hinaus wurden die Betroffenen und Entscheidungsträger zu ausgewählten Themen direkt befragt und beraten. Dabei konnten neue und innovative Konzepte und Lösungen erprobt und direkt für die Wiederaufbauentscheidungen bereitgestellt werden. Durch frühzeitig formulierte „10 Empfehlungen des KAHR Projekts für einen zukunftsfähigen und klimaresilienten Wiederaufbau" konnte das Projekt Akteure auf verschiedenen Ebenen für die Belange eines resilienten Wieder- und Neuaufbaus sensibilisieren. Die 10 Empfehlungen des KAHR Projekts wurden von der Praxis in konkreten Förderentscheidungen aufgegriffen, beispielsweise für die Begründung höherer Schutzstandards gegenüber zukünftigen Hochwasserereignissen bei ausgewählten Projekten.

Das vorliegende Buch gibt einen Überblick und Einblick in die facettenreiche Forschung des KAHR Projektes und bietet zudem eine Synthese wichtiger Erkenntnisse und Erfahrungen. Aufgrund des transdisziplinären Charakters des Projekts und seiner Zielsetzung der wissenschaftlichen Begleitung und Beratung ist das Projekt aber nicht auf ein Produkt oder einen spezifischen Bereich beschränkt. Vielmehr zeigt das Buch die Bandbreite der zu behandelnden Themen im Wiederaufbau auf, die das Konsortium aus seiner Expertise heraus aufgreifen konnte.

An dem vorliegenden Buch hat das gesamte Projektkonsortium zusammen mit weiteren assoziierten Partnern und Forschungspartnern mitgewirkt. Es erwarten Sie interessante und spannende inter- und transdisziplinäre Beiträge, die sich mit der Aufarbeitung der Flutkatastrophe 2021 und dem anschließenden Wieder- und Neuaufbau befassen. Dabei werden die Auswirkungen und Schadensmuster der Katastrophe in verschiedenen Beiträgen beleuchtet. Ebenso wird sich zentralen Fragen und Konzepten, wie im Rahmen des Wieder- und Neuaufbaus mehr Resilienz und wirksame Hochwasservorsorge erreicht werden können, gewidmet. Ein besonderer Schwerpunkt liegt auf Perspektiven aus der Wasserwirtschaft und Raumplanung, ergänzt durch weitere Beiträge, z. B. zum operativen und baulichen Hochwasserschutz sowie zur sozialwissenschaftlichen Aufbereitung des Wiederaufbaus.

Das Buch hat nicht nur die Sicherung der vielfältigen Projektergebnisse zum Ziel, sondern möchte einen Transfer der Erkenntnisse in weitere (Mittelgebirgs-)Regionen ermöglichen. Allerdings kann das Buch nicht alle Problemstellungen oder Fragen aufgreifen oder gar beantworten – es bietet aber mit seinen umfangreichen Perspektiven sowie Befunden einen wichtigen Beitrag für die weitere Diskussion in Wissenschaft und Praxis.

Vier Jahre nach der Flutkatastrophe 2021 ist der Wiederaufbau noch lange nicht abgeschlossen und der Weg zur klimaresilienten Entwicklung trotz eines sehr hohen Engagements aller Akteure vor Ort noch weit. Zudem wird ersichtlich, dass auch nach Abschluss des KAHR Projekts (sprich nach 3 Jahren intensiver Begleitforschung) Bedarf an weiterer wissenschaftlicher Begleitung besteht, da weder der Wieder- und Neuaufbau abgeschlossen noch die Schutzmaßnahmen vollständig implementiert sind. Die Beiträge können daher auch in Teilen einen Anstoß für weitere Forschungen bieten oder skizzieren.

Wenn die aufgezeigten Befunde und Empfehlungen für Sie interessant sind oder Fragen aufwerfen, sind Sie herzlich eingeladen mit uns Kontakt aufzunehmen. Zuerst einmal wünschen wir Ihnen aber viel Freude und viele neue Erkenntnisse und Gedanken beim Lesen der Beiträge des Buches!

Die Editorinnen und Editoren

Stefanie Stenger-Wolf
Alessa Trüdinger
Jörn Birkmann
Elena-Maria Klopries
Holger Schüttrumpf

Inhaltsverzeichnis

Abbildungsverzeichnis

Tabellenverzeichnis

Book Summary

Jörn Birkmann, Elena-Maria Klopries, Holger Schüttrumpf, Alessa Trüdinger und Stefanie Stenger-Wolf

J. Birkmann et al. (Hrsg.), *Die Flutkatastrophe 2021*, Sustainable Development Goals (SDG) – Umsetzung in Praxis, Lehre und Entscheidungsprozessen,
https://doi.org/10.1007/978-3-658-48688-4_1

The research project KAHR (KlimaAnpassung, Hochwasser und Resilienz, transl.: ClimateAdaptation, Flood and Resilience) scientifically accompanied communities and regions affected by the flood disaster of July 2021 in Rhineland-Palatinate (RP) and North Rhine-Westphalia (NRW) for three years during their reconstruction efforts. The accompanying research, along with its findings and recommendations, is based on work related to various research subjects, questions, and requirements that have evolved throughout the project. Research activities all address the overarching question of flood- and climate-resilient reconstruction and new construction. This encompasses everything from capturing and analyzing damage patterns to examining the role of bridges, as well as aspects of social vulnerability and critical and sensitive infrastructures. Within the framework of the research project, solutions relating to various Sustainable Development Goals (SDGs) were discussed and developed. Notably, this includes goals related to health and well-being, strengthening industry, innovation, and infrastructure, reducing inequalities, as well as promoting sustainable cities and communities. The different chapters of this volume thus also illustrate the range of investigations and scientific consultations conducted by 13 partners in the consortium.

A distinctive feature of the KAHR project is that the scientific accompanying research took center stage rather than focusing on a single product or technical system. The research within KAHR encompasses a wide range of topics and objects of study, with all approaches and products sharing the commonality that they do not merely engage in theoretical discussions about reconstruction, climate resilience, and flood preparedness, but also address concrete problems, questions, and processes in flood-affected regions. At its core, the research functioned like the treatment of a "living patient" – a subject still experiencing open wounds, yet simultaneously requiring answers to urgent and specific questions about reconstruction and recovery. Research subjects that were not prominently featured on the scientific radar immediately following the flood disaster include questions concerning the roles of bridges as well as critical and sensitive infrastructures (SDG 9 – "Industry, Innovation, and Infrastructure"). For example, scientific studies within KAHR have examined in detail how bridges influence flood events, and recommendations for improving bridge standards have been derived from these findings. Additionally, KAHR's research on critical and sensitive infrastructures not only addressed their general significance, but also intensely discussed specific challenges associated with rebuilding on sites at risk, like the old location of the Levana School (▶ Chap. 16). Statements arising from the KAHR project significantly contributed to decisions made regarding relocating this sensitive infrastructure to a new site. It became evident that scientific findings were crucial not only for local communities or districts, but also at the state level to secure funding for such relocations.

Scientific findings and recommendations encountered an environment characterized by tensions and conflicts, as well as differing solution ideas. The scientific data and methods were helpful in objectifying the discussion and also provided the opportunity to bring stakeholders together for neutral consultations based on scientific facts and recommendations. A central aim of the research activities was to identify strategies and approaches for improving flood risk preparedness and strengthening climate resilience within reconstruction and rebuilding efforts. The term "reconstruction" has already been critically discussed, as the prevailing view was originally that the reconstruction process equated to compensation for damages. In contrast, the KAHR project, as reflected in the contributions to this book, takes the view that the reconstruction and rebuilding process must include significant improvements and new standards for enhanced flood protection and greater climate resilience so that the region is better prepared for future events and the substantial funds allocated for reconstruction are invested in safer structures. Moreover, the contributions in this book clearly show that KAHR's research findings are not only aimed at the municipalities affected, but also contain recommendations for higher levels of government, such as the state or federal level. Additionally, despite its focus on the 2021 flood disaster in Germany, the KAHR project also addresses international concepts and strategies like the "Building Back Better" approach from the Sendai Framework (see UNDRR 2015).

This involved contributions to the specification of resilience approaches and the concept of "Building Back Better", as well as the identification of different phases of reconstruction and various stakeholders, since the entry points for corresponding strategies and measures also depend on these phases and actors. For example, it became evident that the reconstruction process, particularly in the second to fourth phases and thus after disaster response, i.e. when it comes to temporary infrastructure and, in particular, the restoration of landscapes and the rebuilding of houses and infrastructure, can take into account the requirements of climate-resilient settlement development. This means that different phases can be distinguished within the reconstruction process, which may overlap to some extent but nevertheless offer different options for action in terms of flood risk prevention and climate resilience (see Birkmann et al. 2023a). Immediately after the disaster, central challenges included questions related to disaster management, providing temporary shelters, or ensuring electricity and lighting (SDG 13 – "Climate Action"). The removal and storage of large quantities of waste and debris from the affected areas and destroyed houses also posed considerable problems. In a subsequent phase, discussions focused on the reconstruction of private homes and public infrastructures such as schools and kindergartens. This discussion is still ongoing (as of January 2025). In the third year after the disaster, more advanced concepts for flood protection and strengthening resilience were also discussed, such as locations for larger flood retention basins (► Sect. 11.4); however, implementation of these measures is often still far off. An example demonstrating that concepts can be quickly established after disasters while measures are developed and their potential assessed, and subsequently implemented, is the master plan "Flood-resilient Urban and Regional Development for the Inde and Vicht Catchment Area," developed by Wasserverband Eifel-Rur (waterboard Eifel-Rur) in collaboration with RWTH Aachen University's Institute for Hydraulic Engineering and Water Resources Management. This plan identifies potential measures that enhance flood protection and strengthen regional flood resilience to minimize future flooding risks (see ► Chap. 12). The master plan focuses on a holistic approach that creates sustainable and resilient structures beyond those mandated by law under HQ_{100} standards (HQ_{100} refers to a flood event with a statistical reoccurrence period of 1 time in 100 years).

Overall, the reconstruction process, particularly in parts of the Ahr valley, but also, for example, in areas like the Erft and Inde catchments, takes significantly longer than initially assumed after the disaster. It becomes clear that the rebuilding of individual districts as well as larger infrastructures, such as schools, bridges, and railway lines, requires much more time than one or two years. Furthermore, data and planning bases for reconstruction are changing; for instance, there are more accurate or revised flood hazard maps being utilized along with innovative approaches to flood modeling (see also ► Chap. 4), necessitating adjustments in the reconstruction process. An important finding is that scientific accompanying research is not only significant and impactful in the first year following a disaster, but the scientific insights and recommendations can also be taken into account during the ongoing reconstruction process in subsequent years. For example, scientific recommendations have been developed regarding the relocation of critical and sensitive infrastructures or flood-adapted bridges that are being applied (today) three years after the event in concrete planning and project execution. For instance, the school administration has now decided to relocate Levana School, which accommodates nearly 100 children with intellectual and some physical disabilities, from the Ahrweiler district (see Bonner General-Anzeiger 2024 [newspaper]), allowing for new construction at a different location in the coming years.

A general conclusion from the scientific accompanying research is that new data on various topics, such as flood and damage modeling (see ▶ Chaps. 4 and 7), mental strain and trauma among individuals (see ▶ Chap. 9), the specific challenges of rebuilding in socially disadvantaged neighborhoods (see ▶ Chap. 18), the analysis of bridge clogging (see ▶ Chap. 10), and findings regarding precautionary strategies of affected households (see ▶ Chap. 19) as well as businesses (see ▶ Chap. 8), can be integrated into ongoing reconstruction efforts and the associated discussion and decision-making processes. In some cases, these findings and recommendations have also had direct effects in practice.

Despite numerous successes of the scientific accompanying research, it should be noted that the scientific recommendations of the KAHR project have also faced criticism from some stakeholders. For example, criticism regarding the interpretation of reconstruction funding was raised by certain ministries, leading to discussions directly with those ministries. The KAHR project was also able to contribute through its scientific research on-site with statements made in contexts such as inquiry commissions in state parliaments (see Heidbreder 2023; Schüttrumpf 2022; Thieken 2024) as well as in the construction committee of the Bundestag (see Birkmann 2022).

The high media attention directed towards research on reconstruction has additionally contributed to a stronger recognition of the voice of science in public discourse and politics. Accordingly, relevant press conferences were held each time in connection with five science-practice dialogues organized by the KAHR project as well as during the synthesis conference. Reports in local, regional, and national media led to a greater visibility for scientific assessments within politics and administration, thus indirectly achieving more impact. Consequently, one of our conclusions is that communicating scientific findings and recommendations through media plays an important role in raising awareness and objectifying discussions around reconstruction efforts.

Particularly after major disasters, such as the floods in the summer of 2021, where enormous damage must be repaired (see Mohor et al. 2024) and, at the same time, infrastructure must be rebuilt and social and health consequences must be managed over long periods of time (see Zenker et al. 2024), important topics for further research and starting points for measures to strengthen preparedness against future risks arise. Until now, these phases have not been sufficiently taken into account in traditional risk management cycles for building resilience (see Lumbroso 2007; ◘ Fig. 1.1).

The accompanying research in KAHR shows very clearly that the approaches and measures in the response and recovery phase contain important steps for prevention and risk reduction. Approaches to preventing risks and strengthening resilience should therefore also be actively strengthened in the response and recovery phases. This concerns, among other things, funding instruments for reconstruction or temporary infrastructures, but also operational flood protection, which should not be limited to the phase during and right after an event, but must be expanded to include pre- and post-event care (see also ▶ Chap. 13).

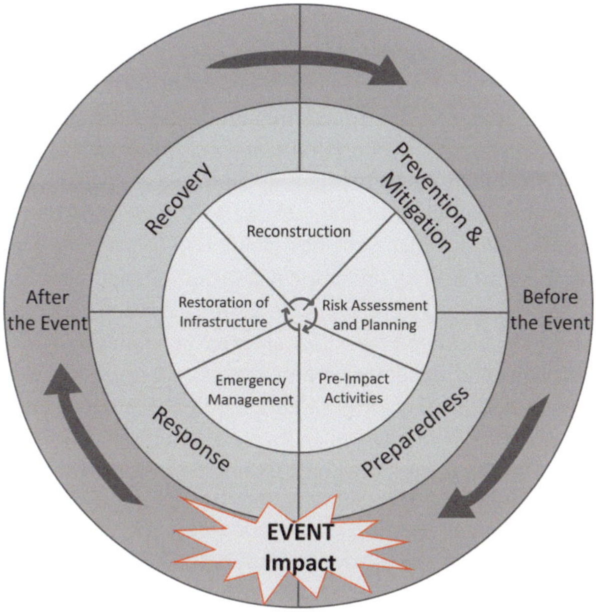

Fig. 1.1 The disaster risk management cycle. (Own figure based on Lumbroso (2007) and Vanneuville et al. (2011))

As early as in the first year of scientific monitoring and support, the KAHR project, in cooperation with other experts from various fields, jointly formulated "10 recommendations from a scientific perspective on the topic of reconstruction and sustainability of flood-affected regions" (see Birkmann et al. 2022), which have already been frequently referred to.

1.1 Recommendations for climate- and flood-resilient reconstruction and new construction

The 10 KAHR recommendations, which were compiled early on in the project, are outlined below. In the following, selected recommendations are discussed in terms of their impact. The recommendations also reflect the interdisciplinary composition of the research team in the KAHR project.

- **[1] Reconstruction after the 2021 flood event also offers an opportunity to initiate a strategic transformation process and strengthen disaster resilience.**

Reconstruction after the flood disaster highlights the need for change and is therefore also an opportunity to concretize, promote and implement resilient, flood- and climate-adapted spatial and settlement development concepts as well as management concepts for rivers. Synergies with other strategic goals such as climate protection should also be utilized. This includes, for example, moving away from oil heating systems, which both significantly increase the level of damage in the event of flooding and require a change in terms of climate protection. For this targeted transformation, reconstruction funds should be strategically combined with other subsidies. Simply rebuilding the old structures is not sufficient (SDG 11 – "Sustainable cities and communities", SDG 13 – "Climate action").

- **[2] All potentials of flood modeling and risk analysis should be fully utilized for planning protection strategies as well as for the preparation and warning of those affected.**

New methods of flood modeling are being successfully employed in science, with detail levels varying between kilometers and millimeters and computation times ranging from fractions of a second to hours. These methods should be critically examined for their practical applicability and purposefully integrated into the creation and revision of flood hazard and risk maps, the designation of flood-prone areas, flood forecasting and warning systems, as well as operational disaster management. Extreme events can become catastrophic particularly when populations, decision-makers, and disaster response teams are surprised and unprepared for an extreme situation. Risk analyses should help avoid such surprises and integrate preparedness early on into the strategic development of villages, towns, and infrastructures. This requires a broader approach to risk modeling that does not limit itself to a few scenarios but also includes worst-case scenarios beyond extreme flood levels. It should also be estimated in advance where particularly unfavorable situations could arise that might lead to loss of life (SDG 13 – "Climate Action").

- **[3] More space for rivers is important; however, this does not only mean retreating settlements, but also adapted land uses. Rivers need space.**

If this space is not available, residential areas, commercial zones, and industrial sites will be flooded. Settlement retreat is especially advisable during the reconstruction phase where structures are already damaged or destroyed. Besides considering the intensity of hazards (flooding/heavy rainfall) at affected locations, it is also important to look at the vulnerability of the affected structures. Facilities housing particularly vulnerable individuals (e.g., hospitals, schools, kindergartens, nursing homes) should be treated differently than "normal" residential uses or commercial sites. Businesses that could cause secondary effects in case of damage must also be given special consideration. Therefore, different vulnerabilities need to come more into focus. Additionally, creating multifunctional land uses offers opportunities for urban development concerning waterways. For instance, parks and sports fields can serve as floodplain areas during floods or heavy rainfall without incurring significant damage while remaining available to the public most of the time. Furthermore, protective goals should be formulated for selected land uses since risks and damages from flooding and heavy rainfall depend on varying degrees of vulnerability and preparedness. These aspects must become integral components of adaptive land-use concepts. Reconstruction should go hand in hand with promoting model approaches (SDG 15 – "Life on Land").

- **[4] Bridges must be given greater consideration in future assessments of flood hazards.**

Bridges can significantly increase flooding risks during flood events. During the July 2021 floods, over 50 % of existing municipal, state-owned, federal bridges as well as railway bridges were damaged or destroyed in some river regions because they were not designed to withstand enormous hydraulic pressures caused by large amounts of sedimentation and debris flow. In future planning and evaluations regarding bridge constructions over flowing waters existing design criteria regarding hydraulic performance requirements, such as necessary foundation forms/depths, as well as static load cases must be expanded to accommodate increased loads from transported sediments and debris flow adequately. Simultaneously moving forward, local backwater effects caused by bridges along with unexpected flow paths during structural overflow situations plus wave action upon structural failure must also be considered within flood hazard/risk maps along with planning efforts accordingly. Among other, enlarged or adjusted flow cross-sections, larger spans and enhanced erosion protection are aspects that should find application in future designs (SDG 9 – "Industry, Innovation and Infrastructure").

- **[5] Early warning systems for flooding events need strengthening.**

The flooding event from July 2021 has made it abundantly clear how vital effective early warning systems combined with preparatory behavioral measures are. In this regard effective warning systems require review and further development so they remain functional even amidst local or regional power outages. Recent advancements allow high-resolution so-called impact-based forecasts which enable predictions about inundated areas along with affected buildings. Such forecasts provide crucial information towards disaster management processes. Existing water level predictions at designated monitoring gauges ought to be supplemented through these impact-based forecasts (SDG 13 – "Climate Action").

- **[6] The signaling function of plans and planning processes needs enhancement, heavy rain hazard and risk maps must become publicly accessible.**

Floods and heavy rainfall must be considered together. To date, there has been no regional planning representation of the priority areas for floods for the settlement areas in Rhineland-Palatinate. The simultaneous representation of HQ_{100} areas and affected areas in the event of an extreme flood in regional and land use plans as well as the representation of the water line from the flood in summer 2021 can have an important signaling function for planning. Water legislation (§ 78 Wasserhaushaltsgesetz (2009)) can only be enforced for areas that are being utilized for the first time, but not for brownfield sites or areas that have already been built on. As a result, precautionary measures in already built-up areas must be strengthened through planning and building law as well as informal instruments. When building permits and conversions are granted, homeowners should automatically receive information about flood and heavy rain hazards (SDG 11 – "Sustainable cities and communities", SDG 13 – "Climate action").

- **[7] Flood- and climate-resilient planning and construction needs integration across all levels pertaining spatial development while accounting every aspect influenced by climate change impacts.**

In order to minimize the damage caused by future flood events, it is necessary for all levels – from the federal, state and regional levels to municipal urban land-use planning – to take resilience, climate adaptation and flood prevention into account in settlement and spatial development. This requires concrete objectives and principles as well as planning representations in the plans. These can include specific elements of floodplain management such as the designation of floodplains. Municipal land-use planning and land utilization are essential for strengthening the resilience of municipalities, but they alone are not sufficient for sustainable flood risk management. In many cases, supra-municipal approaches are also necessary. Climate-resilient planning also requires consideration of the future effects of climate change and various resulting phenomena such as flood and drought phases. In this respect, the strengthening of blue-green infrastructures plays an important role, as this can be used to prevent heat, flooding and heavy rainfall (SDG 11 – "Sustainable cities and communities", SDG 13 – "Climate action").

- **[8] Sustainable reconstruction succeeds when stakeholders establish forms of cooperation and work together on an inter-municipal basis. Funding should strengthen this cooperation.**

The establishment or strengthening of suitable coordination structures for the various actors involved in reconstruction must be ensured within a region. Current reconstruction strategies are often conceived and implemented on a very small scale. For climate-resilient settlement structures, reconstruction must strengthen inter-municipal cooperation and take climate change into account. Some cooperation between municipalities exists and pose a positive example (e.g., municipal associations, flood partnerships, water boards). However, the focus is often on a smaller sub-area of a river catchment. In addition to central coordination centers, adaptive approaches should be strengthened making it possible to identify innovative solutions for reconstruction through experimentation and learning. The issue of fair burden-sharing between municipalities should also be considered. Cities that benefit from improved flood protection and water retention in the upper reaches or surrounding areas should also provide certain services for the municipalities upstream in return. The topics of regional cooperation and fair burden sharing should be communicated more strongly and supported within the framework of reconstruction funding and other funding structures (e.g., village renewal or urban development funding), for example through specific funding for inter-municipal cooperation (e.g., merging sports fields to create more space for the river) (SDG 11 – "Sustainable cities and communities", SDG 13 – "Climate action").

- **[9] Intensive preparation of civil protection and water management for rare flood and heavy rainfall events improves the management of these events.**

Catastrophic floods and heavy rainfall events occur rarely, but when they do, they cause devastating damage. In particular, the short warning time of heavy rainfall events and the sometimes laborious procurement of a situation picture make operational hazard defense measures and the possible reduction or prevention of damage more difficult. This is where the technical expertise and responsibilities of water management and disaster control come together. Appropriate and comprehensive joint preparation of disaster prevention and water management can improve the interface work at this point. One effective way of doing this is to develop and jointly analyses model scenarios that depict the specific effects of an event. The water industry's ability to collect and analyze data in real time should be taken into account in disaster prevention work. Equally important is the development of customized, joint training and further education programs for the emergency services (SDG 13 – "Climate action").

- **[10] New protection standards and protection goals for critical and sensitive infrastructures must be defined and awareness of the unavoidable residual risk must be increased.**

In the context of the reconstruction of electricity and communication infrastructures, hospitals, schools, kindergartens, facilities for people with disabilities or retirement homes, which are designated as so-called "critical and sensitive infrastructures" (see Spatial Planning Act [Raumordnungsgesetz (2008) § 2 Para. 2]), higher protection standards and protection goals must be developed. The considerable negative effects of the failure of and damage to such infrastructures – including cascading effects due to the loss of function – mean that these infrastructures and the people in them must be focused on more than before. This also requires higher protection goals that take into account the special

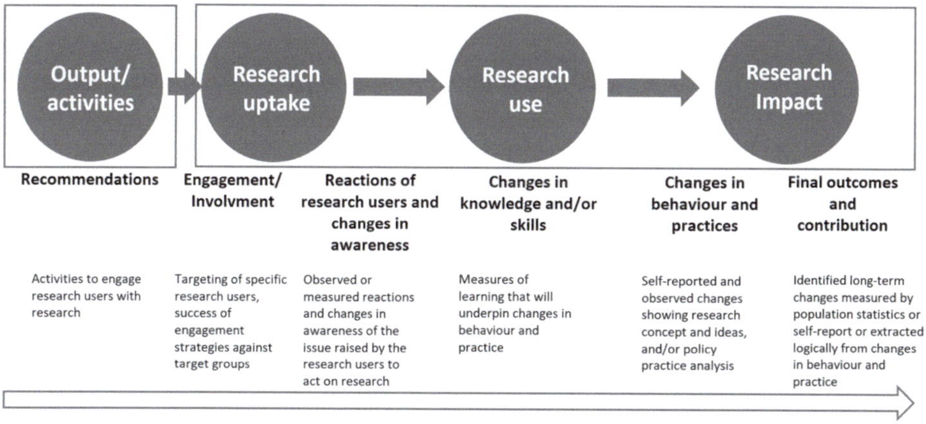

Fig. 1.2 Phases of scientific impact research. (Own figure based on Morton (2015))

importance of these infrastructures and focus on extreme floods. When developing flood prevention concepts, it is essential that local authorities consider differentiated protection targets for existing and new buildings. These should move away from fixed annualization and take into account the sensitivity of the different structures and their users in addition to the hazard intensity (SDG 9 – "Industry, Innovation and Infrastructure", SDG 11 – "Sustainable cities and communities", SDG 13 – "Climate action").

1.2 Effects of the accompanying scientific research

After just three years, the KAHR project can already point to the first documentable effects of the accompanying scientific research. In scientific impact research, a distinction is made between different levels of impact. In addition to the question of which specific activities and results are made available to practitioners or decision-makers, Fig. 1.2 shows that the first phase of scientific impact research involves the uptake of research results, for example in the form of the identification of specific target groups or specific approaches to stakeholders and target groups. In this regard, the science-practice dialogues in the KAHR project were, in retrospect, an important element in informing and sensitizing various stakeholders in the reconstruction process about the current findings and recommendations of the scientific research. A further stage involves the utilization and implementation of scientific findings, data and recommendations by actors, which also lead to learning processes and change processes, whereby the challenges of transferring scientific findings are examined in more detail in ▶ Chap. 21.

The third stage is the actual impact of the research, which is often associated with the long-term effects of the research. Due to its three-year existence, the KAHR project cannot yet point to long-term effects, but the changes made on the basis of KAHR research, such as the decision to relocate the Levana School or new standards for flood-resistant bridge construction, are important indications of the longer-term impact of the research results even beyond the project period.

As the majority of the effects of the 10 scientific KAHR recommendations are not yet to be found in the specialist literature, the findings on the initial effects of the accompanying scientific project are based on the following sources and methods:

- discussions with reconstruction agencies,
- discussions, workshops and map exercises with affected communities and municipalities,
- discussions with planning authorities at local and regional level,
- discussions and workshops with ministries and authorities,
- discussions at scientific and political conferences organized as part of the KAHR project,
- document analyses and
- newspaper analyses.

In numerous cases, more than one source confirmed the use of selected recommendations from the "10 scientific recommendations" published by the KAHR project. Based on the theoretical concept of the "impact chain of research" according to Morton (2015) and other literature (Pfeifer and Helming 2024; Feiertag 2018), we distinguish between "research activities", which are part of the KAHR project's own research process (e.g., consultations, workshops, etc.), and "results", which emerge from the research activities in the form of developed services and products. This also includes the 10 scientific recommendations. In addition, there are results that arise from the activities and "outputs". The impact research also deals with the effects of the results on various (target) groups, such as creating greater awareness of the changing flood and heavy rain risks in the context of climate change. The final long-term effects (impacts) – as a kind of final phase of the research's mode of action – include and imply real changes triggered by the research activities and results. These are often related to long-term changes in terms of social, economic and ecological changes or transformations (Morton 2015; Feiertag 2018; Pfeifer and Helming 2024). Although the KAHR project cannot look back on a long history due to its short duration (three years) and therefore cannot assess its long-term effects, various effects of the accompanying research are already visible and are systematized below on the basis of three questions:

- How were the recommendations communicated and presented?
- What information is available on how these scientific recommendations were taken up and on how they affected reconstruction activities and policies in Rhineland-Palatinate and North Rhine-Westphalia and beyond?
- What factors hinder or promote the consideration of scientific expertise?

Communication and adoption of the 10 scientific recommendations

To ensure that the research results and the 10 recommendations of the KAHR project were perceived and taken up by the stakeholders involved in the reconstruction process, the project presented and documented the recommendations in a science-practice dialogue organized in 2022. The 10 recommendations were also presented in various workshops with the local and state authorities involved in the reconstruction as well as presentations in the Enquete Commission of the Rhineland-Palatinate State Parliament and in presentations in the Federal Building Committee of the Federal Parliament (Bundestag), and selected recommendations were explained in more detail. In addition, interviews in the media were an important instrument for publicizing the scientific recommendations to the wider public and sensitizing private and state actors. Communication was not limited to the publication of the recommendations in a local newspaper, but national media (ARD, ZDF, state parliament hearings, etc.) were also used to sensitize decision-makers at state and federal level. The state and federal levels define important framework conditions for reconstruction, flood protection and spatial planning, so that innovations and changes should also be reflected in these framework conditions.

The visibility of the 10 scientific recommendations of the KAHR project was significantly increased through those numerous discussions with state and local institutions and authorities responsible for reconstruction and through interviews in national and local media (more than 150 interviews). In addition, the recommendations were published on the KAHR project website and are also available on other websites, such as the Federal Parliament's Building Committee (see Birkmann 2022). In addition, individual recommendations were documented in the reports of the Enquete Commissions of the state parliaments in Rhineland-Palatinate and North Rhine-Westphalia (Heidbreder 2023; Schüttrumpf 2022; Thieken 2024).

▪ Changes in knowledge and the basis for decision-making with longer-term consequences

To assess the impact of the 10 KAHR recommendations on changes in knowledge and decision-making, selected recommendations were analyzed in more detail and information was obtained on the extent to which the recommendations had a practical benefit for decision-makers. The following discussion focusses on recommendation no. 10, which relates to critical and sensitive infrastructure, and recommendation no. 4, which deals with the role of bridges. These two recommendations only cover a small part of the total of 10 recommendations, but these two recommendations are very specific. In addition, we have been made aware by local and regional stakeholders that these recommendations have already been incorporated into the decision-making process for selected projects.

To further illustrate how the 10 scientific recommendations were used to change knowledge and decision-making and how this knowledge actually triggered changes in decision-making, we provide a brief overview of the contexts of the issues addressed in recommendations no. 4 and 10. We then discuss specific reconstruction decisions that were influenced by the recommendations and outline associated longer-term impacts.

▪ Critical infrastructure and reconstruction

Numerous critical and sensitive infrastructures, such as water supply and wastewater treatment plants, were severely damaged by the 2021 floods. When rebuilding water supply facilities and sewage treatment plants, it is required that these facilities must be protected against a flood event that statistically occurs once every 100 years (i.e., a HQ_{100}) (current legal standard). The 2021 floods have drastically shown that a level of protection for an HQ_{100} (old) is actually not sufficient, as enormous damage to the corresponding infrastructure has been reported. However, the applicable legal standard is the HQ_{100} standard, whereby in Rhineland-Palatinate, recalculations of the HQ_{100} took place after the 2021 flood event.

Interestingly, the Rhineland-Palatinate Ministry of the Environment and the "Struktur- und Genehmigungsdirektion Nord" (SGD Nord, transl.: structure and approval directorate), referring to KAHR recommendation no. 10, argued that the reconstruction of the water supply plant "Walporzheimer Straße" is a central element of the critical infrastructure and that water supply and disposal is planned with a service life of several decades, so that a higher standard of protection is required. According to the communication between the authorities, this higher standard of protection is also to be financed with funds from the reconstruction fund. An important basis for the argumentation in the corresponding letter between the responsible authorities was the scientific recommendation no. 10 of the KAHR project. SGD Nord and the Ministry of the Environment argued that, on the basis of these scientific findings and the scientific recommendation of the KAHR project, the executing authorities should not only apply the legally prescribed flood protection standard of an HQ_{100} flood event, but also a higher standard of protection, as this is a critical infrastructure with a special need for protection.

This decision-making process on the higher standard of protection to be applied when rebuilding the water supply system shows two important changes in practice that lead to long-term changes. Firstly, the KAHR recommendations are available to decision-makers at local, regional and national level, so that they are recognized at different levels of decision-making. Secondly, the scientific recommendations provide a sound basis of argumentation for raising the level of protection and increasing funding for the reconstruction of the water supply system beyond what is required by law. The long-term effect is a higher level of protection for this infrastructure and for similar infrastructures in the context of reconstruction. Consequently, longer-term effects can be assumed.

▪ Sensitive infrastructures and vulnerable groups: Example Levana School

The 2021 floods damaged numerous sensitive infrastructures that are home to particularly vulnerable user groups that are dependent on the help of third parties in the event of an incident. For these infrastructures, only the HQ_{100} is prescribed as a legally binding protection standard. Furthermore, almost all partially destroyed buildings are protected by the right of continuance, which is why some of the sensitive and critical infrastructures could in principle be rebuilt as they were before the flood. One of the sensitive infrastructures along the Ahr river affected by the 2021 floods is the Levana School, which is analyzed in detail in ▶ Chap. 16. This school for children with disabilities accommodates almost 100 pupils with mental and, in some cases, physical disabilities. An initial risk analysis after the flood event confirmed that it would be possible to rebuild the school in the same location (see Rhein-Zeitung 2024 [newspaper], among others). However, as the district administration was not convinced of this, the KAHR project was asked to carry out a more detailed scientific assessment and investigation (see Truedinger et al. 2023; Birkmann et al. 2023b; Birkmann et al. 2024).

The school is only a few meters away from the course of the Ahr river. The school itself is single-story, i.e., vertical evacuation is not possible, and the main exit of the school faces the Ahr river. In 2021, the school was flooded to a height of approx. 2 m – fortunately, there were no pupils in the school as the event took place at night. In the event of the HQ_{100} recalculated after the flood, the school and the schoolyard, including the evacuation assembly point and paths, are partially flooded up to a height of 1 to 2 m (see Truedinger et al. 2024).

In a statement on the Levana School in 2023 (Birkmann et al. 2023b), which also takes up and specifies KAHR recommendation no. 10, we dealt with these factors in detail and ultimately came out in favor of the recommendation to relocate the school – as the user group is particularly vulnerable, the building offers no vertical evacuation option and, in addition to flooding, the flow velocity in the area of the school on the Ahr river after a bridge is particularly high and poses an additional hazard. A rapid evacuation also does not seem very plausible, as these are children with mental and physical disabilities who cannot carry out this process independently. This opinion was discussed, and in September 2024, the district council finally decided to relocate the district school – also with reference to our scientific statement and the particular importance of protecting sensitive infrastructures (see Bonner General-Anzeiger 2024 [newspaper]).

The scientific statement was also highly relevant for the Ministry in order to finance the relocation with funds from the reconstruction program, as the relocation will involve additional costs of around 10 million euros. From the point of view of resilience research, these funds and additional expenditure are justified, as this is a particularly sensitive infrastructure with a very vulnerable user group and the relocation will mean that the children in the school's care, the school operations and also the building and its facilities will be moved to a more flood-proof location.

■ Flood resilient bridges

The numerous bridges in the Ahr valley contributed massively to an increase in flooding and damage during the 2021 floods. The large quantities of floating debris such as wood, caravans or tanks caused massive blockages on the many old stone bridges, which led to considerable water level rises of up to 2.5 m. Factors such as the height of the bridge, the number of piers, the construction method and the span width have influenced the extent of the blockage. This led to considerable forces and pressures as well as strong currents flowing around the bridges. The bridges were mostly unable to withstand the high forces and flow velocities, with the result that 80 of the 114 bridges on the Ahr river were damaged and some were completely destroyed (Burghardt et al. 2024). A complete analysis of bridge damage on the Ahr, Inde and Vicht rivers, including the causes and extent of the damage, was carried out by Burghardt et al. (2022) and Burghardt et al. (2024). In addition, the effects of bridges on runoff events in 2021 are also analyzed in more detail in ▶ Chap. 10.

Bridges are an important connecting element in the transport sector not only in normal times, but also in the event of a disaster, so their functionality must also be guaranteed in extreme situations. It was therefore necessary to learn from the 2021 flood event and develop recommendations for flood-resilient bridges in low mountain regions. The recommendations for flood-resilient bridges relate to the bridge cross-section, the bridge superstructure, the piers including their foundations, the bridge design, and the construction method. The reconstruction of bridges after a flood is an ideal opportunity not only to adapt the transport infrastructure to a modern transportation system, but also to reduce the risk of damage caused by extreme events. Both aspects need to be considered and implemented in the areas affected by the 2021 floods. However, planners face a number of obstacles in the process, as the old arched stone bridges, for example, characterize the landscape of a tourist region and are often listed as historic monuments.

In order to transfer the topic of "flood-resilient bridges" from research to practice, several workshops were organized with planners, local authorities, districts and the respective federal states. The aim was not only to present the results of the hydro-numerical calculations, but also to discuss the importance of flood-resilient bridges for a climate-adapted future and to incorporate them into renovation planning. This has not only led to an awareness of flood-resilient bridges among the stakeholders, but also to the first plans in Bad Neuenahr-Ahrweiler taking up and implementing these recommendations. Thus, a direct influence of the research on the reconstruction process can be demonstrated. We believe that these decisions and changes, which were initiated and accelerated by the accompanying scientific research, will not only have an impact on the affected infrastructure (schools, water supply, waste disposal facilities, bridges) in the medium and long term, but also on future planning and (re-)construction strategies. As a result, the scientific findings and recommendations have led to changes in reconstruction policy and in the financing methods for specific projects.

1.3 Lessons Learned

Important Lessons Learned from the accompanying scientific research are summarized below. There is no claim to completeness.

- Accompanying scientific research after disasters, such as the flood disaster in Rhineland-Palatinate and North Rhine-Westphalia in the summer of 2021, not only provides scientists with enormous insights into risk drivers, damage potential and discussions on solutions, but also offers the opportunity for scientific data and expertise to flow directly into the reconstruction and rebuilding process.
- Disaster processes that have led to human suffering and considerable damage, such as the flood in the summer of 2021, which devastated the Ahr valley and places along the Inde and Vicht rivers, or the flash floods in the Valencia region of Spain in the fall of 2024, often lead to a loss of trust in institutions and conflict-ridden reconstruction processes. In these processes of reconstruction and rebuilding, scientific accompanying research – especially in larger, interdisciplinary teams – can lead to an objectification of the discussion. As a neutral actor, science can contribute its own recommendations and proposed solutions.
- The accompanying scientific research within the framework of KAHR has shown that changes and innovations are not only important in the regions affected by the disaster, but that the framework regulations – such as laws (BauGB, WHG etc.) or reconstruction funding – are also of considerable importance. The question of whether climate-resilient reconstruction and rebuilding is possible therefore depends not only on the attitude of the actors on the ground, but also on the framework conditions and the question of what the reconstruction funding actually finances.
- A further insight is that scientific accompanying research after such disasters is, as the name already indicates, not only required at the end of the project, e.g., after three years, but that ongoing analyses, user needs from the affected region or country, but also statements for higher levels (e.g., commissions of inquiry) must be incorporated into the reconstruction and rebuilding process. This makes it necessary to adapt research funding not to a rigid end product, but to dynamic, changing requirements and demands.
- Furthermore, it has been shown that scientific accompanying research after disasters must also fulfil a networking function, which on the one hand connects scientific actors in order to avoid duplication of work, and on the other hand also brings together actors from different levels and fields of action. New collaborations have emerged between water management, spatial planning and flood protection (see also ▶ Chap. 13) as well as between different municipalities, whereby new governance structures and processes have also been developed within the framework of inter-municipal collaborations in order to improve flood prevention across administrative boundaries (see also ▶ Chap. 14). These cooperations need to be further strengthened as they are rebuilt or even newly established.
- Another important finding is that reconstruction funding must be significantly changed in order to be able to organize reconstruction and rebuilding more effec-

tively and quickly as well as in a more climate-resilient manner. Particularly in the case of significant damage and areas affected on a large scale, the focus should not be on a collection of individual damaged properties, but rather, as with urban redevelopment areas, entire city districts or areas should be defined that require a new construction and reconstruction concept as part of the reconstruction process. In some cases, the consideration of individual buildings or certain damaged objects does not provide an adequate basis for the consideration of functional interrelationships. In some cases, non-destroyed or undamaged infrastructures must also be included in the reconstruction and new construction in order to promote climate resilience and flood prevention.
- Norms and standards in flood risk management and in the area of risk-based and adaptive spatial planning must also be put to the test. The idea that flood hazard maps or heavy rainfall hazard maps, for example, will remain unchanged in the context of such extreme events and climate change is misguided. More than ever before, stakeholders must expect changes to the relevant specifications in the reconstruction process. Adaptive planning strategies that also allow for readjustment are therefore required. Irrespective of this, there are also limits to adaptation and limits to the adaptability of planning.
- The accompanying scientific research also shows that protection goals and protection standards (for reconstruction and new construction) should be more differentiated and discussed. In our experience, the current orientation towards a design flood HQ_{100} as a protection standard is not sufficient. More differentiated protection standards are necessary, that also take into account the different vulnerability and protection worthiness of land uses, population groups and infrastructures (see also ▶ Chap. 15). A soccer pitch with grass does not need the same level of protection as a school for children with physical and mental disabilities. A hospital or electricity infrastructure should be considered differently than a large parking lot that should not be visited in the event of flooding or heavy rainfall.

Finally, an important conclusion is that the success or failure of accompanying scientific research also depends on the openness and willingness to cooperate of the municipalities and federal states concerned. In addition to the commitment of the researchers involved, it is important that the addressees have a certain openness towards scientific institutions and actors. In the KAHR project, for example, the very good cooperation with the district of Ahrweiler (see also ▶ Chap. 22) and the state of Rhineland-Palatinate offered considerable synergies and potential. Active support from federal ministries can also be very helpful here.

1.4 Further research needs and open questions

Based on the intensive research with 13 partner institutions and in cooperation with numerous practitioners and other research actors, new questions and research needs have also emerged. Without claiming completeness, the topics and questions listed below appear to us to be particularly important for further scientific research and systematic learning from crisis and reconstruction processes following flood disasters.

In the three years following the flood disaster, there have been significant changes to the legal framework for reconstruction and rebuilding, such as the introduction of a reconstruction clause in the Federal Building Code (§ 246c BauGB, see also ► Chap. 15). There is also a new Climate Adaptation Act, which formulates new requirements for risk prevention and adaptation to extreme events (Klimaanpassungsgesetz (KAnG) (2023)). The application and impact of these changes to the legal framework have not yet been adequately recorded in the reconstruction and new development. In many cases, it can be observed that although the factual and mathematical correctness of the use of funds from reconstruction funding – which, at 30 billion euros, is a considerable amount – is carefully checked, it is often not possible to identify examples that can be used as best practice for other reconstruction and rebuilding processes. This means that budget reporting on the use of reconstruction funds predominates. In this respect, the development of a methodology for identifying and evaluating examples of best practice is still an important future research topic. There are also numerous measures and plans that are only now, three years after the disaster, being planned in detail (e.g., locations for flood retention basins) and will only be implemented in practice in the coming years. Selected future research needs are explained in more detail below.

▪ Effects of legal changes for reconstruction

An important future field of research is the analysis of the effects of legal changes for effective, climate-adapted and rapid reconstruction and rebuilding. The first experiences with the use of the reconstruction clause § 246c BauGB in the Ahr valley and selected other regions should be recorded and evaluated. There is also a need for research into the extent to which the Climate Adaptation Act contributes and has so far contributed to strengthening climate resilience in the context of reconstruction processes. Research in this area should examine the extent to which these legal innovations can achieve a better interlinking of climate adaptation and reconstruction and whether the legal changes are also able to accelerate reconstruction and rebuilding processes in a meaningful way (more planning security, bundling of processes). Obstacles to the implementation of the new legal requirements in practice should be identified in order to provide a basis for any necessary tightening up.

▪ Protection and adaptation goals for reconstruction and rebuilding

Another open field of research is the development of innovative approaches for the definition of protection and adaptation targets for reconstruction and rebuilding. Up to now, the HQ_{100} has often been used as the protection standard for future floods in reconstruction. However, the spatial extent of the design floods is changing due to climate change and this approach also fails to take into account the different protection needs of land uses, infrastructures and population groups (see ► Chap. 15). As a result, further research is necessary, including the assessment of the sensitivity and protection worthiness of land uses and infrastructures for planning at regional and municipal level. The Federal Spatial Development Plan for Flood Protection and the planned new Federal Flood Protection Act II show the need for a more differentiated consideration of possible affected uses and areas, but there is still a lack of methods to substantiate this differentiated consideration of vulnerability and risks for different land uses and subject areas such as critical and sensitive infrastructures. In the Ahr valley and more rural regions, for example, the figures for critical infrastructure such as hospitals, which were also severely affected by the flooding, are well below the threshold value of 30,000 full inpatient admissions per year specified in the nationwide BSI Kritis Regulation (Verordnung zur Bestimmung Kritischer Infrastrukturen (2016)). This means that the principles for defining critical and sensitive infrastructure for rural areas must also be put to the test for reconstruction and rebuilding processes following flood disasters.

- **Flood retention basins and areas**

In the Ahr valley, but also in other flood-affected regions in Germany, there are currently considerations to plan and implement technical flood retention areas in the flood areas of 2021 (e.g., on the Ahr, Erft and Inde rivers and their tributaries). In a first step, the KAHR project supported the district of Ahrweiler with studies on land use adaptations and a rough estimate of the theoretical potential of flood retention basins as well as the identification of obstacles to the implementation of such basins (especially land availability and competing uses), whereby it became clear that early land protection is crucial for effective flood prevention (see ▶ Chap. 11). From a scientific point of view, it would now make sense to scientifically monitor the current planning and implementation of flood retention basins in the Ahr valley (19 sites) in particular, as both the further evaluation of the sites as part of the Strategic Environmental Assessment (SEA) and Environmental Impact Assessment (EIA) are important, as well as the question of how sites can be secured in terms of planning (state and regional planning). The structural implementation will take place in the next few years, as issues such as financing and a special-purpose association need to be clarified. It is also necessary to examine which locations should be developed as a priority in terms of effectiveness and acceptance. The systematic investigation and evaluation of larger planning and infrastructure projects, such as technical flood retention basins, in terms of their effectiveness for future flood protection, also in the context of climate change, and the acceptance of such projects has not yet been sufficiently developed.

- **Processing of experiences and exchange of experience between regions with regard to reconstruction, rebuilding, and climate resilience**

As numerous other regions also recorded damage from flooding and heavy rainfall in the summer of 2024, among other things, and reconstruction processes are therefore also necessary there, a stronger exchange of experience between these regions with the findings of the KAHR project would be expedient. It would be particularly useful to pass on important Lessons Learned from the Ahr valley to other regions, for example with regard to the potential for accelerating reconstruction processes or addressing and informing citizens about climate-resilient reconstruction and rebuilding. The formats for this exchange of experience still need to be developed for different actors and levels of government. Country-specific features must be taken into account when it comes to coordination processes and the acceleration of procedures between different levels (federal, state, district, municipal). For this exchange of experience, the effects of scientific advice on practice should also be processed and recommendations for other regions should be specified. Accordingly, a systematic and effective learning and exchange process also requires the preparation of particularly good examples of flood and climate-resilient reconstruction and new construction in various subject areas (bridges, critical and sensitive infrastructures, property protection, etc.). In addition to the best practice examples, experiences could also be documented that should be avoided elsewhere or in other regions and reconstruction processes.

The four selected examples show that the open research questions cover other important fields of action and future topics. There are certainly other topics and research questions that can be added. The contributions and results of the three years of research

1

in the KAHR project compiled in this book offer a good insight into the importance, practical relevance and complexity of scientific accompanying research after such disasters, which are highly relevant not only for the affected regions, but also for numerous stakeholders. This experience should also be made available for future reconstruction programs and efforts after flood disasters. This book represents a contribution to that endeavor.

Acknowledgments On behalf of all project partners, the authors would like to thank the Federal Ministry of Education and Research (BMBF) and the DLR Project Management Agency for their support and funding of the KAHR project, which was financed by the BMBF under the funding code 01LR2102A-M. In this context, special thanks are due to the staff of the BMBF and the DLR Project Management Agency for the collegial exchange and the trusting cooperation during the application process and the entire duration of the project.

References

Birkmann J (2022) Öffentliche Anhörung des Ausschusses für Wohnen, Stadtentwicklung, Bauwesen und Kommunen am 6. Juli 2022 im Deutschen Bundestag. https://www.bundestag.de/dokumente/textarchiv/2022/kw27-pa-bau-flutkatastrophe-901092

Birkmann J, Schüttrumpf H, Trüdinger A, Thieken A, Schüller A, Huck A, Kleber A, Thiebes B, Droste-Franke B, Guse B, Merz B, Brüll C, Kuhlicke C, Stein C, Mudersbach C, Michalski D, Molinari D, Rodriguez-Castro D, Kirsten D, Klopries E, Bachofer F, Steudtner F, Johann G, Lauer H, Luhmann HJ, Arenz HT, Zachert H, Kreibich H, Apel H, Meyer H, Frančik H, Hoppe H, Sauter H, Hasse J, Moßgraber J, Stamm J, Emde K, Laranjeira K, Burghardt L, Kirschbauer L, Madruga de Brito M, Evers M, Disse M, Ravan M, Schaefer M, Garschagen M, Boronowsky M, Paul M, Thiessen N, Koch O, Mahrenholz P, Meier P, Bellanova P, Jüpner R, Reinstädtler S, Vorogushyn S, Schöttmer S, Greiving S, Wolf S, Nietgen T, Hellmund T, Korrmann V, Kron W, Reckhaus Z (2022) 10 Empfehlungen aus Sicht der Wissenschaft zum Thema Wiederaufbau und Zukunftsfähigkeit der flutbetroffenen Regionen. Veröffentlichung im Rahmen des Vorhabens KAHR – Klima-Anpassung, Hochwasser und Resilienz (2021–2025). https://hochwasser-kahr.de/index.php/de/kahr-produkte?view=article&id=83:10-empfehlungen&catid=12

Birkmann J, Schüttrumpf H, Handmer J, Thieken A, Kuhlicke C, Truedinger, Sauter AJ, Klopries EM, Greiving S, Jamshed A, Merz B, Solecki W, Kirschbauer L (2023a) Strengthening resilience in reconstruction after extreme events – Insights from flood affected communities in Germany. Int J Disaster Risk Reduct 96:2023. https://doi.org/10.1016/j.ijdrr.2023.103965

Birkmann J, Trüdinger A, Sauter H, Greiving S (2023b) Stellungnahme zur Levana-Schule als besonders sensible Infrastruktur für die Kreisverwaltung. Stuttgart (unveröffentlicht)

Birkmann J, Truedinger AJ, Schüttrumpf H (2025) The Impact of Science: Uptake of Scientific Recommendations After Extreme Events – Case Study Floods in 2021 in Germany. J Flood Risk Manag 18(3):e70100. https://doi.org/10.1111/jfr3.70100

Bonner General-Anzeiger (2024) Wir kriegen das Wasser nicht weggezaubert. https://ga.de/region/ahr-und-rhein/bad-neuenahr-ahrweiler/wiederaufbau-der-flut-schutz-gegen-hochwasser-v1_aid-121541797

Burghardt L, Schüttrumpf H, Wolf S, Klopries EM (2022) Analyse der Schäden an Brückenbauwerken in Folge des Hochwassers 2021 an der Ahr. Wasser Abfall 24(11):12–17. https://doi.org/10.1007/s35152-022-1346-x

Burghardt L, Klopries E-M, Schüttrumpf H (2024) Structural damage, clogging, collapsing: Analysis of the bridge damage at the rivers Ahr Inde and Vicht caused by the flood of 2021. J Flood Risk Manag. https://doi.org/10.1111/jfr3.13001

Feiertag P (2018) Politikrelevantes Wissen in der Raumplanung: Grenzarbeit zwischen Wissenschaft und Politik in Deutschland, den Niederlanden und der Schweiz. Rohn, Lemgo. ISBN 978-3-946319-20-7.

Heidbreder L (2023) Abschlussbericht der Enquete-Kommission 18/1 „Konsequenzen aus der Flutkatastrophe in Rheinland-Pfalz: Erfolgreichen Katastrophenschutz gewährleisten, Klimawandel ernstnehmen und Vorsorgekonzepte weiterentwickeln" („Zukunftsstrategien zur Katastrophenvorsorge"). Landtag von Rheinland-Pfalz, Drucksache 18/8222, Punkt 9. Online: https://dokumente.landtag.rlp.de/landtag/drucksachen/8222-18.pdf

Klimaanpassungsgesetz (KAnG) der Bundesrepublik Deutschland, BGBl. 2023 I Nr. 393 vom 20.12.2023. https://www.recht.bund.de/bgbl/1/2023/393/VO

Lumbroso D (2007) Review report of operational flood management methods and models, FloodSite Consortium, T17-07-01. http://www.floodsite.net/html/partner_area/project_docs/Task17_report_M17_1review_v1_1.pdf

Mohor GS, Sieg T, Koch O, Buhrmann A (2024) Remote sensing-based mapping of structural building damage in the Ahr valley. J Flood Risk Manag. https://doi.org/10.1111/jfr3.12983

Morton S (2015) Progressing research impact assessment: a "contributions" approach. Res Eval 24(4):405–419. https://doi.org/10.1093/reseval/rvv016

Pfeifer L, Helming K (2024) Effective mission-oriented research: a new framework for systemic research impact assessment. Res Eval. https://doi.org/10.1093/reseval/rvae003

Raumordnungsgesetz der Bundesrepublik Deutschland vom 22. Dezember 2008 (BGBl. I S. 2986), das zuletzt durch Artikel 1 des Gesetzes vom 22. März 2023 (BGBl. 2023 I Nr. 88) geändert worden ist. https://www.gesetze-im-internet.de/rog_2008

Rhein-Zeitung (2024) Weg ist frei: Levana-Schule bekommt neuen Standort im Kreis Ahrweiler. https://www.rhein-zeitung.de/lokales/kreis-ahrweiler/weg-ist-frei-levana-schule-bekommt-neuen-standort-im-kreis-ahweiler_arid-2693591.html

Schüttrumpf H (2022) Stellungnahme im Anhörverfahren der Enquete-Kommission 18/1 „Zukunftsstrategien zur Katastrophenvorsorge" zum Thema „Technischer Hochwasserschutz, Natürliche Hochwasservorsorge, Erosionsschutz, Wasserrückhalt und Retention" am 11. Oktober 2022. Landtag von Rheinland-Pfalz, EK 18/1-78. Online: https://dokumente.landtag.rlp.de/landtag/vorlagen/1-78-18.pdf&ved=2ahUKEwi5xvGEi-mKAxWQ0QIHHRqKGjYQFnoECA0QAQ&usg=AOvVaw0KFFXwFtaZVC_vrBkP2MN4

Thieken A (2024) Stellungnahme im Anhörverfahren der Enquetekommission II: „Krisen- und Notfallmanagement" – durch die Lehren der Vergangenheit die Zukunft sicher gestalten am 03. Juni 2024. Landtag von Nordrhein-Westfalen (https://www.landtag.nrw.de/portal/WWW/dokumentenarchiv/Dokument/MMST18-1487.pdf&ved=2ahUKEwiLof_hi-mKAxXO0QIHHadkIUAQFnoECBIQAQ&usg=AOvVaw0OZky2wspwxRMuZTApdaSn)

Truedinger A, Birkmann J, Fleischhauer M, Ferreira C (2024) Sensitive infrastructures and people with disabilities – Key issues when strenghtening resilience in reconstruction. Egusphere. https://doi.org/10.5194/egusphere-2024-1607

Truedinger AJ, Jamshed A, Sauter H, Birkmann J (2023) Adaptation after extreme flooding events: moving or staying? The case of the Ahr valley in Germany. Sustainability 15(2):1407. https://doi.org/10.3390/su15021407

United Nations Office for Disaster Risk Reduction (2015) Sendai Framework for Disaster Risk Reduction 2015–2030. Veröffentlichung im Rahmen der World Conference on Disaster Risk Reduction, Sendai. Bd 2015 (https://digitallibrary.un.org/record/793460?v=pdf)

Vanneuville W, Kellens W, De Maeyer P, Reniers G, Witlox F (2011) Is 'flood risk management' identical to 'flood disaster management'? IEEE Earthzine 13. See link: https://earthzine.org/is-flood-risk-management-identical-to-flood-disaster-management-3/

Verordnung zur Bestimmung Kritischer Infrastrukturen nach dem BSI-Gesetz (BSI-Kritisverordnung – BSI-KritisV) vom 22. April 2016 (BGBl. I S. 958), die zuletzt durch Artikel 1 der Verordnung vom 29. November 2023 (BGBl. 2023 I Nr. 339) geändert worden ist. https://www.gesetze-im-internet.de/bsi-kritisv/BJNR095800016.html

Wasserhaushaltsgesetz der Bundesrepublik Deutschland vom 31. Juli 2009 (BGBI. I S. 2585), das zuletzt durch Artikel 7 des Gesetzes vom 22. Dezember 2023 (BGBl. 2023 I Nr. 409) geändert worden ist. https://www.gesetze-im-internet.de/whg_2009/BJNR258510009.html

Zenker ML, Bubeck P, Thieken AH (2024) Always on my mind: indications of post-traumatic stress disorder among those affected by the 2021 flood event in the Ahr valley, Germany. Nat Hazards Earth Syst Sci 24:2837–2856. https://doi.org/10.5194/nhess-24-2837-2024

Einführung

Holger Schüttrumpf, Jörn Birkmann, Elena-Maria Klopries, Alessa Trüdinger und Stefanie Stenger-Wolf

J. Birkmann et al. (Hrsg.), *Die Flutkatastrophe 2021*, Sustainable Development Goals (SDG) – Umsetzung in Praxis, Lehre und Entscheidungsprozessen,
https://doi.org/10.1007/978-3-658-48688-4_2

2.1 Handlungsbedarf nach der Flutkatastrophe 2021

Das Hochwasser vom Juli 2021 hat in Nordrhein-Westfalen (NRW) an Erft, Inde, Vicht und Wupper sowie in Rheinland-Pfalz (RLP) an der Ahr zum Verlust von über 180 Menschenleben und ökonomischen Schäden von über 30 Mrd. € geführt (BMI und Bundesfinanzministerium 2021). Damit zählt das Hochwasser 2021 neben der Sturmflutkatastrophe 1962 und dem Elbe-Hochwasser 2002 zu den größten Naturkatastrophen, die Deutschland nach 1945 getroffen haben. Extreme Schadensereignisse waren bereits in der Vergangenheit Auslöser für ein Umdenken im Hochwasser- und Katastrophenschutz sowie eine intensive BMBF-Begleitforschung. Beispiele hierfür sind die KFKI-Forschung nach der Sturmflutkatastrophe 1962 sowie das RIMAX-Programm des BMBF (Bundesministerium für Bildung und Forschung) nach dem Elbe-Hochwasser 2002. Beide Programme, die intensiv durch das BMBF unterstützt wurden bzw. werden, haben zu einem erheblichen Wissensgewinn für die betroffenen Regionen und schlussendlich zu einer Verbesserung des Hochwasser- bzw. Sturmflutschutzes geführt.

Im Juli 2021 waren insbesondere Mittelgebirgsregionen sowie die Übergangsbereiche zum Flachland vom Hochwasser betroffen. Bereits kurz nach dem Hochwasserereignis wurde deutlich, dass neben den umfangreichen Aufräum- und Wiederherstellungsarbeiten auch eine intensive wissenschaftliche Begleitung auf dem Gebiet des Hochwasserrisikomanagements und der vorsorgenden räumlichen Planung erforderlich ist, um die betroffenen Regionen zukunftsfähig und gegenüber zukünftigen Ereignissen resilienter zu gestalten. Wiederaufbauprozesse nach vergangenen Extremereignissen aus anderen Regionen und Ländern zeigen, dass der Wiederaufbau kein linearer und vollständig steuerbarer Prozess ist (z. B. Fu und Gregory 2019; Vahanvati und Rafliana 2019). Demzufolge zielt die den Wiederaufbauprozess begleitende Forschung darauf, die jeweiligen Phasen, Dynamiken und Akteure innerhalb der Wiederaufbauphasen in ausgewählten Regionen und Kommunen zu ermitteln und für ausgewählte Fragen und Prozesse wissenschaftliche Expertise passgenau bereit zu stellen. Das heißt in bestimmten Phasen des Wiederaufbaus sind anhand konkreter Beispiele neue Ansätze und Weiterentwicklungen von Instrumenten, Informationen und Methoden zu testen und in den Prozess einzubeziehen (Birkmann et al. 2023). Dies kann von kleineren Demonstrationsprojekten einer schadloseren Ableitung von Wassermassen durch bestimmte Stadtquartiere auch die Sensibilisierung von Akteuren oder die Weiterentwicklung von Bewertungskriterien für einzelne Maßnahmen und ihre Effektivität, Umsetzbarkeit und Akzeptanz umfassen.

Durch den Wiederaufbau werden zudem vielfach Fakten geschaffen, die sich später nur schwer verändern lassen. Dabei werden die Potenziale, die Hochwassergefährdung und Verwundbarkeit der betroffenen Bevölkerung, Unternehmen oder kritischen Infrastrukturen zu mindern, nur teilweise ausgeschöpft. Ein Blick in die Vergangenheit zeigt, dass ähnliche Hochwasserereignisse bereits mehrfach vorgekommen sind (Roggenkamp und Herget 2015) und aufgrund des Klimawandels zukünftig eher mit häufigeren und noch intensiveren Ereignissen – insbesondere Starkregenereignissen – zu rechnen sein wird (Tradowsky et al. 2023; Koks et al. 2022; Blöschl et al. 2019). Eine wissenschaftliche Begleitforschung des Wiederaufbaus erscheint somit zielführend, um aus dem Hochwasser 2021 die richtigen Lehren zu ziehen und die Fähigkeit zur Bewältigung von Hochwasserereignissen, Resilienz und Anpassungskapazitäten der betroffenen Regionen im Sinne nachhaltiger Raumentwicklung dauerhaft zu verbessern. Eine wichtige Aufgabe des Begleitprojekts bzw. der bearbeitenden Partner liegt u. a. auch darin, in ihrer wissenschaftlichen, beratenden und unterstützenden Rolle den Kommunen und anderen

Akteure vor Ort, den Landkreisen, Bezirksregierungen oder den Ministerien auf Landesebene Chancen und Herausforderungen für einen resilienteren Wiederaufbau in Bezug auf Hochwasserschutz, Starkregenvorsorge- und klimagerechte und adaptive Entwicklung darzustellen und konkrete Strategien und Maßnahmen zu prüfen. Zudem sollen die begleitenden und beratenden Arbeiten Betroffene und Zuständige unterstützen und motivieren, auch neue innovative Konzepte und Lösungen zu erproben mit dem Ziel, in Zukunft besser aufgestellt zu sein als vorher – und das nicht nur in der Hochwasser- bzw. der Starkregenvorsorge, sondern auch in anderen Handlungsfeldern der Zukunftsvorsorge, wie z. B. bezogen auf Klimaschutz.

Der Handlungsbedarf nach dem Hochwasser 2021 ist groß und viele Themenfelder müssen neu bewertet und ggf. neu behandelt werden. So wurden z. B. die ausgewiesenen Überschwemmungsflächen der Hochwassergefahrenkarten vielerorts bei Weitem überschritten (vgl. Thieken et al. 2021) und es kam zu massiven Schäden an Infrastrukturen, insbesondere kritischen und sensiblen Infrastrukturen wie Krankenhäusern, Feuerwehrstationen, Alten-/Pflegeheimen, Polizeiwachen, Schulen, etc. (Koks et al. 2022; Voss et al. 2022; Truedinger et al. 2023). Diese waren nicht ausreichend gegen Hochwasser geschützt und vorbereitet. Warnungen sind teilweise bei der betroffenen Bevölkerung nicht hinreichend oder auch zu spät angekommen und die Vorbereitungskonzepte für solche Notlagen waren teilweise nicht existent (Thieken et al. 2022). Auch die Komplexität des Wiederaufbaus verlangt nach mehr Wissen über die Prozesse, Organisation und Abläufe unterschiedlicher Phasen der Ereignisbewältigung und des Wiederaufbaus sowie der Zielsetzungen und Interessen unterschiedlicher Akteure und entsprechender Synergien und Konflikte.

Daher ist es für die betroffenen Regionen dringend erforderlich, zukünftig Hochwasservorsorge und planerische Vorsorgestrategien zu gewährleisten, die sich nicht starr an Eintrittswahrscheinlichkeiten von Hochwasserereignissen orientieren, sondern auch mögliche zukünftige Veränderungen von Gefahren, Verwundbarkeiten und Risiken in den Blick rücken (Greiving et al. 2023a, b). Ebenso sind Szenarien zu entwickeln, die über den Bemessungsfall hinausgehen und eine Grundlage bieten, um Vorsorge und Katastrophenmanagement besser auf (bisher undenkbare) Ereignisse vorzubereiten. Die existierenden Gefahren- und Risikokarten für ein HQ_{extrem} sind so zu überarbeiten und zu ergänzen, dass sie die relevanten Entscheidungsprozesse möglichst gut unterstützen können. Dabei sind auch mehrere Szenarien mit Versagensfällen bis hin zu einem maximal möglichen Hochwasser denkbar. Auch die Schutzwürdigkeit unterschiedlicher Infrastrukturen und Nutzungen sollte ein höheres Gewicht erhalten.

Dabei geht es auch um die Weiterentwicklung der Schutz- und Planungskonzepte hin zu einem adaptiveren und risikobasierten Ansatz, der durch die Kombination wasserwirtschaftlicher, raumplanerischer, städtebaulicher und sozialwissenschaftlicher Expertise verhindert, dass zukünftige Ereignisse zu solch hohen extremen Auswirkungen in Form von Schäden und Menschenleben führen. Diesbezüglich geht es auch um neue Erkenntnisse über die Möglichkeiten und Hemmnisse, entsprechende Informationen in unterschiedlichen Phasen des Wiederaufbaus für verschiedene Akteure passgenau zur Verfügung zu stellen. Zudem bieten die komplexen und weitreichenden Wiederaufbauprozesse eine einmalige Chance, die Effektivität, Umsetzbarkeit und Akzeptanz verschiedener Strategien und Maßnahmen innerhalb unterschiedlicher Phasen des Wiederaufbaus zu untersuchen. So kann über die klassische Kosten-Nutzen-Betrachtung hinaus wichtiges Methodenwissen generiert und für andere Regionen bereitgestellt werden. In diesem Zusammenhang wird auch auf den am 19.08.2021 als Rechtsverordnung in Kraft getretenen „Bundesraumordnungsplan Hochwasserschutz" verwiesen, der u. a. erstmals in der deutschen Raumplanung das Prinzip der risikobasierten Planung als verbindliches Prinzip eingeführt hat (Greiving et al. 2023a, b).

2.2 Das KAHR-Projekt

Das vom BMBF geförderte Projekt KAHR (**K**lima-**A**npassung, **H**ochwasser und **R**esilienz) unterstützt mit neuesten wissenschaftlichen Erkenntnissen die Aufbaumaßnahmen in den von der Flutkatastrophe im Juli 2021 zerstörten Regionen in Nordrhein-Westfalen und Rheinland-Pfalz. Mit einer Projektlaufzeit von November 2021 bis Februar 2025 wurden in dem Verbundprojekt mit insgesamt 13 Partnern aus Wissenschaft und Praxis Empfehlungen zur Klimaanpassung, der risikobasierten Raumplanung und zum Hochwasserschutz erarbeitet. Ziel ist es, konkrete Strategien und Maßnahmen für einen klimaresilienten und zukunftsorientierten Wieder- und Neuaufbau in den betroffenen Regionen zu schaffen. Als Pilotgebiete wurden in NRW die Gemeinden Stolberg und Eschweiler inkl. des Vicht- und Inde-Einzugsgebiets sowie in RLP die Verbandsgemeinde Altenahr inkl. des Ahr-Einzugsgebiets identifiziert. Die Auswahl weiterer Kommunen mit ergänzenden Untersuchungs- und Beratungsbedarfen erfolgte auf Basis einer vorlaufenden Akteurs- und Regionsanalyse. Bei der Auswahl der Pilotgebiete wurden als zentrales Kriterium die unterschiedlichen Betrachtungsebenen der Verbundpartner berücksichtigt. Während die Wasserwirtschaft üblicherweise einzugsgebietsbezogen ganze Flüsse von der Quelle bis zur Mündung betrachtet, arbeiten Stadt- und Raumplaner auf unterschiedlichen räumlichen Skalen und betrachten dabei ausgewählte Planungsprozesse oder Planungsinstrumente, wie beispielsweise die Bauleitplanung oder die Raumordnung. Sozialwissenschaftliche Studien können ebenfalls auf unterschiedlichen räumlichen Skalen durchgeführt werden, wobei zusammen mit der Raumforschung auch gerade die Betrachtung von Entscheidungs- und Bewertungsprozessen auf der Ebene von Haushalten und Unternehmen sowie kommunalen Akteuren besonders relevant im Kontext des Wiederaufbaus erscheint. Demzufolge sind für die konkrete Zusammenarbeit Akteure und Partner auf der lokalen und regionalen Ebene besonders relevant (z. B. Landkreis, Kommune). Ein zweites Kriterium für die Auswahl war die Betroffenheit der jeweiligen Kommunen bei dem Hochwasser 2021, ein drittes Kriterium ist darüber hinaus die Verfügbarkeit lokaler Ansprechpartner in den ausgewählten Regionen.

Die Struktur des KAHR-Projekts (◘ Abb. 2.1) hat diese Herausforderungen aufgenommen und fachlich abgebildet. Das KAHR-Projekt baut horizontal auf den beiden Themen „Hochwasser-Risiko-Analysen" und „Räumliches Risikomanagement" auf und verzahnt beiden Themen in den jeweiligen Pilotregionen in Rheinland-Pfalz und Nordrhein-Westfalen, die die vertikale Struktur des Projekts bilden. Durch Einbindung ausgewiesener Forschungspartner mit hoher Kompetenz in ihren jeweiligen Forschungsfeldern konnte der Stand der Wissenschaft unmittelbar in den Wiederaufbauprozess implementiert werden. Durch Einbindung lokaler Praxispartner wurde einerseits ein Schwerpunkt auf den Transfer der Ergebnisse in die Pilotregionen gelegt, andererseits konnten dadurch aktuelle Fragestellungen vor Ort direkt in das Forschungskonsortium zurückgespielt werden. Auf diese Weise konnten die Forschungsarbeiten ständig dem Wiederaufbauprozess angepasst und somit optimiert werden. So wurden beispielsweise diverse soziale Aspekte des Hochwasserrisikos und des (Wieder-)Aufbaus beleuchtet (z. B. Truedinger et al. 2023; Thieken et al. 2022; Kuhlicke et al. 2024; Birkmann et al. 2023).

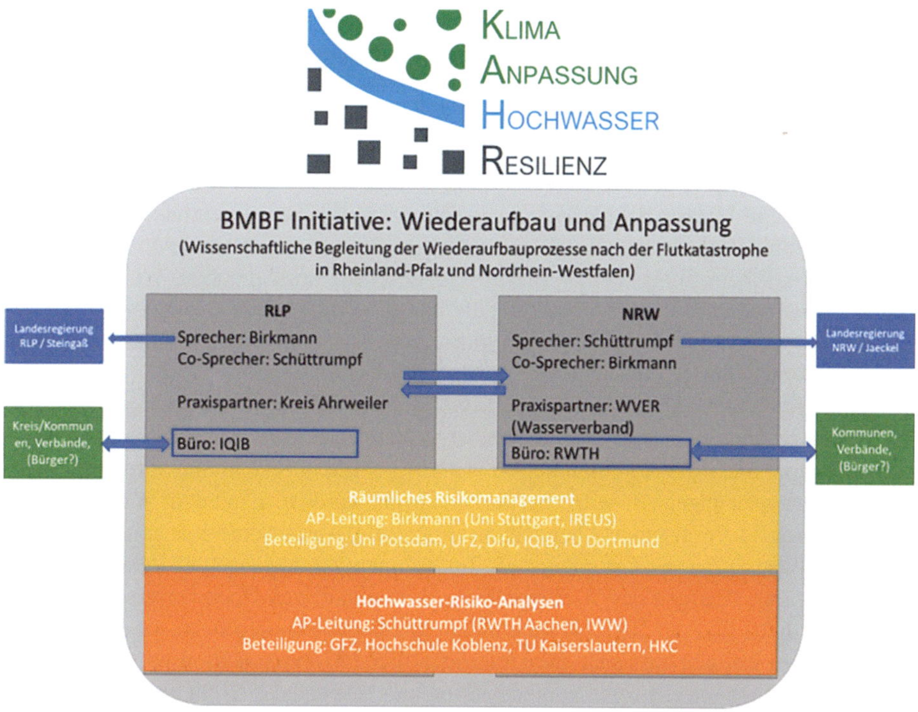

Abb. 2.1 Struktur des Gesamtvorhabens und der beiden Cluster zum Zeitpunkt der Antragstellung

2.3 Kommunikation

Der Wiederaufbau stellt die Kommunen genauso wie die Bürgerinnen und Bürger vor massive Herausforderungen. Im Rahmen des Wieder- und Neuaufbaus sind innovative Ansätze des Hochwasserschutzes und einer klimaresilienten Raum- und Stadtentwicklung zu fördern, da es darum gehen muss, zukünftig besser aufgestellt zu sein. Allerdings ist die Kommunikation zusätzlicher Anforderungen an den Wieder- und Neuaufbau auch ein komplexer Gegenstand, da zahlreiche Akteure (Stadtverwaltungen, Unternehmen) und Bevölkerungsgruppen auch selbst Schäden und konkrete Wiederaufbauprozesse zu bewältigen haben. Dennoch müssen aus wissenschaftlicher Sicht die Lehren aus dem Hochwasserereignis 2021, aber auch die Erfahrungen beim Wiederaufbau und seinen Herausforderungen systematisch dokumentiert und in andere Regionen in Deutschland kommuniziert werden, die sich besser auf zukünftige Extremereignisse einstellen müssen oder auch mit Fragen des Wiederaufbaus nach Hochwassern zu tun haben. Diese Aspekte sind zentrale Herausforderungen für die wissenschaftliche Kommunikation, die in solchen Zeiten schwierig, aber extrem notwendig ist.

Die Wissenschaft beschäftigt sich schon seit Jahrzehnten mit Hochwasserrisiken und Hochwasserschutz sowie in den letzten Jahren auch stärker mit Fragen der Klimaresilienz und Klimaanpassung. Wissenschaft kann neutral und unabhängig Lösungsansätze für die genannten Herausforderungen skizzieren und wissenschaftliche Befunde und Empfehlungen in diesen speziellen Zeiten in die Diskussion einbringen. Erkenntnisse aus der Wissenschaft müssen dabei sowohl direkt an die relevanten Akteure vor Ort, aber auch in die breite Öffentlichkeit kommuniziert werden, um alle Ebenen und verschiedene Akteure gleichzeitig zu erreichen. Ziel der Aktivitäten im Rahmen des vom BMBF geförderten KAHR-Projekts sowieso sonstiger Aktivitäten war es, wissenschaftliche Erkenntnisse in die praktische Umsetzung vor Ort bei Bürgerinnen und Bürgern, Kommunen und Gemeinden sowie auf Landes- und Bundesebene einzubringen und damit der Wissenschaft in Bezug zu Themen des Hochwasserschutzes und der klimaresilienten Entwicklung von Städten und Infrastrukturen eine eigene Stimme zu geben. Dazu haben wir vielfältige Kommunikationsformate aufgebaut und genutzt, um alle betroffenen Ebenen von der individuell betroffenen Person bis zur Bundesebene anzusprechen. Wir konnten so durch die wissenschaftliche Kommunikation einige wichtige Transformations- und Veränderungsprozesse anstoßen und darauf hinweisen, dass der reine Wiederaufbau keinen hinreichenden Schutz vor neuen Extremereignissen bietet.

Der Schutz vor Risiken im Kontext von Hochwasser und Starkregen geht uns alle etwas an und kann nicht auf einen einzelnen Akteur oder eine Institution abgeschoben werden, daher muss die Kommunikation vielfältig, flächendeckend und auf unterschiedliche Zielgruppen angepasst sein. Aus diesem Grund wurde u. a. mit dem Wissenschafts-Praxis-Dialog in Form einer Abendveranstaltung ein Austauschformat mit politischen Entscheidungsträgern vor Ort entwickelt. Im Rahmen von bislang fünf Veranstaltungen wurden zentrale Themen des Wiederaufbaus, des zukünftigen Hochwasserschutzes, der Klima-Resilienz und der Risikokommunikationen gemeinsam erörtert. An dieser Stelle muss erwähnt werden, dass sich die wissenschaftliche Kommunikation vor Ort in einem hoch-emotionalen, da teilweise traumatisierten und belasteten Umfeld abspielt, d. h. oftmals fehlt auch nach solchen Katastrophen das Vertrauen zwischen Institutionen oder Bürgern in bestimmte Teile der Verwaltung und Politik. Zudem verstärkt der langjährige Wiederaufbauprozess die emotionale Belastung der Bevölkerung. Hier ist sowohl eine sachliche wie auch empathische Kommunikation durch die Wissenschaft erforderlich, um gemeinsame Wiederaufbauprozesse zielgerichtet zu aktivieren.

Ein zweites wichtiges Instrument der Kommunikation waren die 2022 und 2023 durchgeführten Wissenschaftskonferenzen, um durch nationale und internationale Vernetzung unter Beteiligung von Universitäten, Verwaltungen und interessierten Gruppen neuste Forschungserkenntnisse von hohem Interesse für die vom Hochwasser betroffenen Regionen auszutauschen. Aus diesen Wissenschaftskonferenzen sind insgesamt drei Special Issues zu unterschiedlichen Themen entstanden, damit auch der weltweite Transfer der Erkenntnisse gesichert ist. Auch weltweit war das Geschehen in der Eifel etwas Besonderes, da ein Industrieland mit einem hohen Hochwasserschutzstandard betroffen war.

Die Medien spielten ebenfalls eine zentrale Rolle in der wissenschaftlichen Kommunikationsstrategie. Unter der Prämisse, dass grundsätzlich jeder von Hochwasser oder Starkregen betroffen sein kann, wurde durch viele Beiträge in Podcasts, im Fernsehen, im Radio und in den Printmedien versucht, die Bevölkerung und die breite Öffentlichkeit für die Themen Hochwasserrisikovorsorge und Klimaresilienz zu sensibilisieren und wichtige Empfehlungen an die Politik und Verwaltung über diese Medien zu kommunizieren. Hohe Zugriffszahlen auf die Beiträge sowie ein vielfältiges Feedback – auch Kritik von einigen Ministerien – zeigen, dass hiermit mehr Personen erreicht werden können, als durch klassische wissenschaftliche Formate. Insgesamt können weit über 150 Medienbeiträge nachgewiesen werden.

Doch auch die Fachcommunity sollte durch die Kommunikation erreicht werden. Hierfür wurden in Zusammenarbeit mit Fachgesellschaften (z. B. DWA, HTG, BWK,

DTK, ARL, SRL) weitere Formate der Wissenschaftskommunikation entwickelt, um die handelnden Personen vor Ort zu erreichen und auch neues Wissen in Fachverbände und Akademien weiterzutragen. Mehr als 100 Vorträge belegen diesen Teil der Kommunikation.

Ein wichtiger Akteur im Rahmen der Kommunikation war auch das Hochwasser-KompetenzCentrum e. V. (HKC) aus Köln mit seinem KAHR-Infomobil. Im Rahmen zahlreicher Vor-Ort-Termine wurden Bürgerinnen und Bürger über die Möglichkeiten der Eigenvorsorge im Hochwasser- und Starkregenfall informiert. Jede interessierte Person hatte die Möglichkeit, am KAHR-Infomobil Hochwasserschutz anzufassen, sich mit Informationsmaterial einzudecken und ggf. auch ganz konkrete Hinweise bzgl. der eigenen Möglichkeiten zu erhalten.

Die Impacts dieser sehr breit angelegten wissenschaftlichen Kommunikation des KAHR-Projekts sind inzwischen fassbar und messbar (siehe ▶ Kap. 23). Hierzu zählen die Umsetzung von Empfehlungen für hochwasserresiliente Brückenbauweisen, Masterpläne für den zukünftigen Hochwasserschutz, Empfehlungen für den besonderen Schutz von kritischen und sensiblen Infrastrukturen im Wiederaufbau, Grundlagen für den technischen und natürlichen Hochwasserrückhalt und vieles mehr. Konkret haben Empfehlungen u. a. zu räumlichen Verlagerungen von Infrastrukturen oder höheren Schutzstandards für kritische Infrastrukturen geführt. Beispielsweise wurde im Herbst 2024 auf Basis von KAHR-Empfehlungen beschlossen, eine Schule für Kinder mit körperlichen und geistigen Behinderungen aus dem Gefahrenbereich der Ahr weg zu verlegen (Au 2024). Auch haben Ministerien wissenschaftliche Empfehlungen von uns genutzt, um höhere Schutzstandards für kritische Infrastrukturen zu genehmigen, die über den gesetzlich vorgeschriebenen Schutzstandard hinausgehen. Damit hat die wissenschaftliche Kommunikation und die interdisziplinäre Zusammenarbeit in dem Forschungsprojekt mit 13 unterschiedlichen Partnern auch zu konkreten Veränderungen am Boden geführt und zu einigen Verbesserungen im Wieder- bzw. Neuaufbau der zerstörten Regionen in Richtung Hochwasserrisikoreduktion und Klimaresilienz. Die verschiedenen Beiträge in diesem Band zeigen dabei die Bedeutung der wissenschaftlichen Begleitforschung auf und verdeutlichen zudem die Notwendigkeit unterschiedliche Handlungsfelder interdisziplinär im Wieder- und Neuaufbauprozess anzugehen, um gemeinsam mit Akteuren aus der Praxis integrierte Strategien und Maßnahmen zu entwickeln, die auch die Synergien zwischen verschiedenen Sektoren und Handlungsfeldern aufzeigen. In dieser Hinsicht beinhaltet der folgende Sammelband, der im Rahmen des KAHR-Projektes entstanden ist, nicht nur für die wissenschaftliche Community interessante Beiträge, sondern bietet auch wichtige Anregungen und Empfehlungen für die Praxis.

2.4 Fazit

Das Hochwasser im Juli 2021 in Nordrhein-Westfalen und Rheinland-Pfalz führte zu über 180 Todesfällen und Schäden von über 30 Mrd. €. In der Folge wurde die Notwendigkeit einer intensiven wissenschaftlichen Begleitung im Wiederaufbau und Hochwasserrisikomanagement erkannt, um betroffene Regionen resilienter zu gestalten. Das KAHR-Projekt, gefördert vom BMBF, unterstützt den Wiederaufbau mit Empfehlungen zur Klimaanpassung und risikobasierten Raumplanung. Es zielt darauf ab, konkrete Maßnahmen für einen klimaresilienten Wiederaufbau in den betroffenen Gebieten zu entwickeln. Die wissenschaftliche Kommunikation spielt hierbei eine zentrale Rolle, um Lehren aus dem Hochwasser zu ziehen und die Öffentlichkeit sowie Fachgemeinschaften für Hochwasservorsorge und Klimaresilienz zu sensibilisieren. Durch vielfältige Kommunikationsformate wie Wissenschafts-Praxis-Dialoge konnten wichtige Transformations- und Veränderungsprozesse angestoßen werden, die u. a. den Schutz kritischer Infrastrukturen und die Anpassung an zukünftige Extremereignisse betreffen.

Zusammenfassung

- Wissenschaftliche Kommunikation ist zentral, um Lehren aus dem Hochwasser zu ziehen und Öffentlichkeit sowie Fachgemeinschaften für Hochwasservorsorge und Klimaresilienz zu sensibilisieren
- Wissenschaftliche Kommunikation dient einer objektiven und fundierten Diskussion und Entscheidungsfindung
- Vielfältige Kommunikationsformate wie Wissenschafts-Praxis-Dialoge und Empfehlungen aus der Wissenschaft führen zu messbaren Impacts
- Wissenschaftliche Begleitung im Wiederaufbau unterstützt Transformations- und Veränderungsprozesse hin zu einer resilienteren Gesellschaft

Danksagung Die Autorinnen und Autoren von ▶ Kap. 2 bedanken sich im Namen aller Projektpartner beim Bundesministerium für Bildung und Forschung (BMBF) und beim DLR Projektträger für die Unterstützung und Förderung des KAHR-Projekts, welches vom BMBF unter dem Förderkennzeichen 01LR2102A-M finanziert wurde. Besonderer Dank gilt in diesem Zusammenhang den Mitarbeiterinnen und Mitarbeitern von BMBF und DLR Projektträger für den kollegialen Austausch sowie die vertrauensvolle Zusammenarbeit während der Antragstellung und der gesamten Projektlaufzeit.

Literatur

Au B (2024) Weg ist frei: Levana-Schule bekommt neuen Standort im Kreis Ahrweiler, Rhein-Zeitung. https://www.rhein-zeitung.de/lokales/kreis-ahrweiler/weg-ist-frei-levana-schule-bekommt-neuen-standort-im-kreis-ahweiler_arid-2693591.html. Zugegriffen: 25. Sept. 2024

Birkmann J, Schüttrumpf H, Handmer J, Thieken A, Kuhlicke C, Truedinger A et al (2023) Strengthening resilience in reconstruction after extreme events – Insights from flood affected communities in Germany. Int J Disaster Risk Reduct 96:103965. https://doi.org/10.1016/j.ijdrr.2023.103965

Blöschl G, Hall J, Viglione A, Perdigão RAP, Parajka J, Merz B et al (2019) Changing climate both increases and decreases European river floods. Nature 573(7772):108–111. https://doi.org/10.1038/s41586-019-1495-6

BMI, Bundesfinanzministerium (2021) Zwischenbericht zur Flutkatastrophe 2021: Katastrophenhilfe, Soforthilfen und Wiederaufbau. Hg. v. Bundesministerium des Inneren, für Bau und Heimat und Bundesministerium der Finanzen. https://www.bundesregierung.de/resource/blob/974430/1963706/613b934d3f359a5118df16755e9e527c/2021-09-27-zwischenbericht-hochwasser-data.pdf?download=1. Zugegriffen: 20. Juli 2022

Fu C, Gregory J (2019) Estimation of an equilibrium model with externalities: post-disaster neighborhood rebuilding. ECTA 87(2):387–421. https://doi.org/10.3982/ECTA14246

Greiving S, Kruse P, Othmer F, Fleischhauer M, Fuchs M (2023a) Implementation of risk-based approaches in Urban land use planning – the example of the city of Erftstadt, Germany. Sustainability 15(21):15340. https://doi.org/10.3390/su152115340

Greiving S, Othmer F, Fleischhauer M (2023b) Die Umsetzung des Prinzips der risikobasierten Planung in der Bauleitplanung am Beispiel der Stadt Erftstadt. Korrespondenz Wasserwirtsch 7(23):417–422

Koks EE, van Ginkel KCH, van Marle MJE, Lemnitzer A (2022) Brief communication: critical infrastructure impacts of the 2021 mid-july western European flood event. Nat Hazards Earth Syst Sci 22(12):3831–3838. https://doi.org/10.5194/nhess-22-3831-2022

Kuhlicke C, de Brito MM, Otto D, Reckhaus Z (2024) Resilienter wiederaufbauen? Erste Thesen zur Rekonfiguration hydrosozialer Territorien nach dem Hochwasser 2021. In: Kabisch S, Rink D, Banzhaf E (Hrsg) Die Resiliente Stadt. Springer, Berlin, Heidelberg, S 267–281

Roggenkamp T, Herget J (2015) Historische Hochwasser der Ahr – Rekonstruktion von Scheitelabflüssen ausgewählter Ahr-Hochwasser. Heimatjahrbuch. Kreis Ahrweiler

Thieken A, Kemter M, Vorogushyn S, Berghäuser L, Sieg T, Natho S, Mohor GS, Petrow T, Merz B, Bronstert A (2021) Extreme Hochwasser bleiben trotz integriertem Risikomanagement eine Herausforderung. Ein Statement zum Hochwasser in Rheinland-Pfalz, 31. Juli 2021. https://www.uni-potsdam.de/en/natriskchange/activities/publications

Thieken AH, Bubeck P, Heidenreich A, von Keyserlingk J, Dillenardt L, Otto A (2022) Performance of the flood warning system in Germany in July 2021 – insights from affected residents

Tradowsky JS, Philip SY, Kreienkamp F, Kew SF, Lorenz P, Arrighi J et al (2023) Attribution of the heavy rainfall events leading to severe flooding in Western Europe during July 2021. Clim Change. https://doi.org/10.1007/s10584-023-03502-7

Truedinger AJ, Jamshed A, Sauter H, Birkmann J (2023) Adaptation after extreme flooding events: moving or staying? The case of the Ahr valley in Germany. Sustainability 15(2):1407. https://doi.org/10.3390/su15021407

Vahanvati M, Rafliana I (2019) Reliability of Build Back Better at enhancing resilience of communities. IJDRBE 10(4):208–221. https://doi.org/10.1108/IJDRBE-05-2019-0025

Voss M, Rüger A, Bock N, Dittmer C, Merkes ST (2022) Die Evakuierung des St.-Antonius-Hospitals Eschweiler während der Flutereignisse im Juli 2021 (Unter Mitarbeit von Universitätsbibliothek der FU Berlin)

Das Hochwasser von Mitte Juli 2021

Jens Reinert, Elena-Maria Klopries und Holger Schüttrumpf

J. Birkmann et al. (Hrsg.), *Die Flutkatastrophe 2021*, Sustainable Development Goals (SDG) – Umsetzung in Praxis, Lehre und Entscheidungsprozessen,
https://doi.org/10.1007/978-3-658-48688-4_3

Im Juli 2021 kam es in Teilen Deutschlands, besonders in Nordrhein-Westfalen und Rheinland-Pfalz, zu einer der schwersten Hochwasserkatastrophen der letzten Jahrzehnte (Junghänel et al. 2021; Bell et al. 2022). Begünstigt durch eine seltene atmosphärische Konstellation setzte sich eine Regenfront über diesen Regionen fest und brachte extreme Niederschlagsmengen. In der Folge wurden in kürzester Zeit die Flüsse zur Gefahr. Besonders betroffen waren in Deutschland die Flussgebiete der Ahr, Erft, Inde und Wupper sowie kleinere Nebenflüsse. Der Deutsche Wetterdienst (DWD) und andere meteorologische Institutionen hatten bereits am 10. Juli erste Warnungen herausgegeben, die sich in den folgenden Tagen intensivierten (DWD 2021). Diese Hinweise deuteten auf ein extremes Wetterereignis hin, das die üblichen Muster von Starkregen und Überflutungen deutlich überschritt. Ein solcher Niederschlag in solch kurzer Zeit übersteigt die natürlichen und baulichen Hochwasserschutzstrukturen. Die Böden waren aufgrund eines nassen Vormonats bereits gesättigt, was den Abfluss während des Hochwasserereignisses beschleunigte. Die Wassermassen konnten kaum noch in den Boden versickern, wodurch die ohnehin großen Wassermengen direkt in die Flüsse und Bäche flossen. Diese Entwicklung führte zu einem schnellen Anstieg der Pegel in vielen Gewässern. Im Rahmen der Aufarbeitung des Ereignisses wurde an verschiedenen Messstationen ein rasantes Überschreiten definierter Warnschwellen identifiziert, was jedoch nicht flächendeckend und zeitnah an die Bevölkerung und Einsatzkräfte weitergeleitet werden konnte. In engen Tälern wie dem Ahrtal wirkte die Landschaft fast wie ein Trichter, der die Wassermassen auf engem Raum kanalisierte und dadurch die Durchflussgeschwindigkeit enorm steigerte. Dieses Zusammenspiel von starkem Oberflächenabfluss und hoher Fließgeschwindigkeit führte zu einem beispiellosen Anstieg der Wasserstände in kürzester Zeit. Die Wellenfront trieb das Wasser mit enormer Geschwindigkeit durch das Tal, was eine Evakuierung oder Vorbereitung auf das Hochwasser an vielen Stellen unmöglich machte. Zusätzlich zur Landschaftsform führte die bestehende Bebauung entlang der Flüsse zu weiteren Engpässen, die die Wassermassen verstärkten und an bestimmten Stellen katastrophale Auswirkungen hatten.

3.1 Chronologie der Wettervorhersage

Das Julihochwasser 2021 war meteorologisch betrachtet bereits Tage vor seinem Höhepunkt am 14. und 15. Juli erkennbar. Der DWD hatte frühzeitig auf die ungewöhnliche Wetterlage hingewiesen, die durch das Tiefdruckgebiet „Bernd“ verursacht wurde. Bereits am 10. Juli 2021 veröffentlichte der DWD in seiner „Wochenvorhersage Wettergefahren“ die ersten Hinweise auf ein „markantes Niederschlagsereignis“. Die meteorologischen Modelle prognostizierten dabei Dauerregen mit Starkregenphasen, die potenziell zu Überflutungen führen könnten (Junghänel et al. 2021). Das Besondere an „Bernd“ war die stationäre Lage des Tiefdruckgebiets, das durch blockierende Hochdruckzellen in seiner Bewegung stark eingeschränkt war. Dies führte dazu, dass dieselben Regionen über Tage hinweg von anhaltenden Regenfällen betroffen waren.

Am 11. Juli konkretisierte der DWD seine Warnungen und sprach von gebietsweise extremen Regenmengen von über 100 l/m^2 innerhalb von 24 h. Die Prognosen fokussierten sich zunehmend auf den westdeutschen Raum, insbesondere auf Nordrhein-Westfalen und Rheinland-Pfalz. Diese Regionen wurden aufgrund der orografischen Gegebenheiten – etwa durch die Mittelgebirgszüge der Eifel und des Sauerlands – als besonders gefährdet identifiziert.

Am 12. Juli 2021 veröffentlichte der DWD eine „Vorabinformation Unwetter", die die potenziell katastrophalen Auswirkungen der bevorstehenden Niederschläge hervorhob (DWD 2021; Reinert et al. 2023). Es wurde explizit darauf hingewiesen, dass die Mengen die Kapazitäten der Bodenaufnahme übersteigen und zu erheblichem Oberflächenabfluss führen könnten. Diese Warnungen richteten sich sowohl an die Bevölkerung als auch an Fachbehörden wie die Wasserwirtschaftsämter und Katastrophenschutzstellen. Zeitgleich erhielten Fachstellen über spezielle Systeme wie das FeWIS (Feuerwehr-Wetter-Informationssystem) und WaWIS (Wasserwirtschafts-Wetter-Informationssystem) detaillierte meteorologische Prognosen. Am Abend des 12. Juli wurden diese Warnungen weiter verschärft, um die möglichen Auswirkungen des bevorstehenden Ereignisses zu verdeutlichen.

Am 13. Juli wurden schließlich „Extreme Unwetterwarnungen" der höchsten Stufe ausgesprochen (DWD 2021). Diese basierten auf aktualisierten Modellläufen, die Dauerregen von bis zu 200 l/m^2 oder mehr in besonders betroffenen Einzugsgebieten von Wupper, Erft, Rur und Ahr vorhersahen. Über den bereits beobachteten Dauerregen hinaus, der die Böden stark gesättigt hatte, wurde vor Starkregen mit Spitzenwerten gewarnt, der besonders in engen Flusstälern zu schnell ansteigenden Wasserständen führen könnte. Diese Vorhersagen wurden in den darauffolgenden Stunden mehrfach an die zuständigen Behörden weitergeleitet.

Am 14. Juli bestätigte sich die Intensität der Vorhersagen, als die Regenmengen tatsächlich in außergewöhnlicher Schwere einsetzten. In der Region fielen binnen 24 h teils über 150 l/m^2. Orte wie Hagen, Köln und das Ahrtal erreichten Werte, die für Wiederkehrintervalle von mehr als 100 Jahren charakteristisch sind (Tradowsky et al. 2023). Besonders problematisch war, dass die hohe Dynamik der Ereignisse die Koordination der Warnkette erheblich beeinträchtigte. Viele Pegelstände stiegen schneller an, als es die hydrologischen Modelle prognostizieren konnten, und lokale Behörden gerieten zunehmend unter Druck, kurzfristig Entscheidungen zu treffen.

Hinweis für Entscheidungsprozesse

Die meteorologischen Warnungen waren aus Sicht der Vorhersagetechnik insgesamt zuverlässig, doch die Interpretation und Weiterleitung dieser Informationen zeigten Schwächen. Eine der größten Herausforderungen lag in der effektiven Kommunikation der Warnungen an die Bevölkerung. Laut einer Untersuchung gaben etwa 30 % der Betroffenen an, keine Warnung erhalten zu haben (Thieken et al. 2023). Dies weist darauf hin, dass der Übergang von meteorologischen Vorhersagen zu konkreten Handlungsanweisungen für die Bevölkerung noch erhebliche Lücken aufweist.

Das Ereignis zeigte, wie entscheidend eine enge Verzahnung zwischen meteorologischen Warnsystemen, hydrologischen Prognosen und dem Katastrophenschutz ist. Während die meteorologischen Modelle die außergewöhnliche Lage früh erkannten, war die Abstimmung zwischen den beteiligten Akteuren nicht ausreichend, um die Dramatik des Ereignisses in vollem Umfang zu antizipieren (DKKV 2024). Die Erkenntnisse aus diesem Ereignis haben grundlegende Diskussionen über Verbesserungen in der Vorhersage- und Warnkette angestoßen. Insbesondere die Verbesserung der Datenübermittlung und die Abstimmung zwischen den beteiligten Fachbehörden stehen dabei im Fokus, um zukünftig besser auf ähnliche Extremsituationen reagieren zu können.

3.2 Einordnung des Niederschlagsverlaufs

3

Vom 12. bis 15. Juli brachte das Tief über Teile Nordrhein-Westfalens und Rheinland-Pfalz Niederschlagsmengen, die sonst in einem gesamten Juli üblich wären – und teils sogar darüber hinausgingen. Dieses Ausmaß war das Ergebnis einer seltenen Kombination von meteorologischen und geografischen Faktoren.

Der Niederschlag intensivierte sich am 14. und 15. Juli und verursachte eine hohe Bodensättigung, wodurch das Wasser nicht mehr in den Boden versickern konnte. Das führte zu einer starken Oberflächenabflussbildung, die besonders steile und stark versiegelte Einzugsgebiete schnell überflutete. Im Ahrtal, einem der am schlimmsten betroffenen Gebieten, schwollen die Flüsse innerhalb weniger Stunden an. Die Hydrologie zeigte in diesen Regionen, dass die vorhandene Wassermenge, bedingt durch die starken Regenfälle, den natürlichen Abfluss deutlich überstieg und das Einzugsgebiet der Ahr, aber auch Nebenflüsse in anderen Regionen, mit einer beispiellosen Geschwindigkeit überflutet wurden.

Bereits am 12. Juli setzte Regen in den westlichen Landesteilen ein, zunächst punktuell und vergleichsweise moderat (Junghänel et al. 2021). Doch mit der langsamen Verlagerung des Tiefdruckkerns verstärkten sich die Niederschläge zunehmend. Am 13. Juli konzentrierte sich die Aktivität auf ein breites Band von der Eifel bis ins Ruhrgebiet. Lokale Regenschauer brachten in wenigen Stunden bis zu $80\,\mathrm{l/m^2}$. Diese ersten Spitzenwerte waren ein Vorbote dessen, was am Folgetag folgte, als das Ereignis seinen Höhepunkt erreichte. Besonders dramatisch war die Lage in der Eifel und im Ahrtal, wo die eng begrenzten Täler die Wassermassen kanalisierten und die Hochwasserdynamik erheblich verstärkten.

Die Intensität der Niederschläge war dabei nur ein Aspekt. Das Besondere an diesem Ereignis war die außergewöhnliche Dauer der Starkregenperiode. Durch die Stationarität des Tiefdruckgebiets, das sich nur langsam bewegte, wurden dieselben Regionen über mehrere Tage hinweg immer wieder von Regenfällen getroffen. Diese Kombination aus Dauerregen und intensiven Schauern führte dazu, dass die Böden bereits nach kurzer Zeit vollständig gesättigt waren und keine weiteren Wassermengen mehr aufnehmen konnten.

Das Extremereignis zeigte, wie stark außergewöhnliche Wetterlagen die hydrologischen Systeme belasten können. Der Übergang von intensiven lokalen Schauern zu flächendeckendem Dauerregen, kombiniert mit den gesättigten Böden und den spezifischen geografischen Gegebenheiten, führte zu einer Situation, die in ihrer Dynamik und ihren Auswirkungen beispiellos war.

3.3 Das System und die Grenzen der Hochwasservorhersage

Die Hochwasservorhersage und -warnung spielten eine zentrale Rolle bei der Einschätzung der Lage vor und während der Katastrophe im Juli 2021. Die Verantwortung für diese Aufgaben lag in Nordrhein-Westfalen beim Landesamt für Natur, Umwelt und Verbraucherschutz (LANUV) und in Rheinland-Pfalz beim Landesamt für Umwelt (LfU). Beide Behörden kombinierten meteorologische Vorhersagen des Deutschen Wetterdienstes (DWD) mit hydrologischen Modellen, um die erwarteten Wasserstände in den Flüssen abzuschätzen.

In Nordrhein-Westfalen begann der Hochwasserinformationsdienst des LANUV bereits am 13. Juli 2021 mit der Veröffentlichung hydrologischer Lageberichte. Diese Berichte kombinierten Wettervorhersagen mit Pegelmessungen und informierten Behörden sowie die Öffentlichkeit über bevorstehende Hochwassergefahren. Der erste Bericht wies bereits auf mögliche Starkregenmengen von bis zu 200 l/m^2 hin und warnte vor stark ansteigenden Wasserständen an Flüssen wie Rur, Erft und Sieg. In den darauffolgenden Tagen wurden insgesamt sechs weitere Lageberichte erstellt, die die Entwicklung der Wasserstände kontinuierlich dokumentierten und die Behörden in die Lage versetzen sollten, präventive Maßnahmen zu ergreifen.

In Rheinland-Pfalz erfolgte eine ähnliche Vorgehensweise. Das Landesamt für Umwelt in Mainz setzte seine Hochwasservorhersagezentrale in den Tagen vor dem Hochwasser in einen 24-Stunden-Betrieb. Von Ende Juni bis Anfang August 2021 wurden Hochwassermeldungen an insgesamt 23 Tagen verbreitet. Am 14. Juli erreichte die Warnlage für die Eifelflüsse, darunter die Ahr, ihre höchste Stufe. Es wurden über verschiedene Kanäle wie KATWARN, NINA und „Meine Pegel" Warnungen verschickt. Die Vorhersagen für Wasserstände, insbesondere an der Ahr, waren jedoch mit erheblichen Unsicherheiten behaftet. Diese resultierten aus der hohen Variabilität der Niederschlagsmengen und -orte sowie den fehlenden Pegeldaten, da viele Messstationen durch das Hochwasser zerstört oder außer Betrieb gesetzt wurden.

Die Hochwasservorhersagen basierten auf komplexen hydrologischen Modellen, die kontinuierlich mit Wetterdaten aktualisiert wurden. In Rheinland-Pfalz wurde beispielsweise das LARSIM-Wasserhaushaltsmodell verwendet, das die Niederschlags- und Abflussdynamik in den Einzugsgebieten abbildet. Für die Pegel Altenahr, Kordel und Prüm zur Lay konnten durch diese Modelle frühzeitig Höchststände prognostiziert werden, jedoch nicht in der Präzision, die für eine gezielte Warnung und Evakuierung erforderlich gewesen wäre. Ein Problem war die Diskrepanz zwischen den simulierten und tatsächlich eingetretenen Wasserständen. Der Höchststand am Pegel Altenahr wurde auf über 10 m geschätzt und lag somit deutlich über den modellierten Werten von maximal 7 m.

Ein weiterer Schwachpunkt war die Kommunikation der Warnungen. Obwohl viele Warnmeldungen rechtzeitig erstellt wurden, erreichten diese nicht immer die Bevölkerung oder wurden nicht korrekt interpretiert. In beiden Bundesländern wird nun an der Weiterentwicklung der Hochwasservorhersagesysteme gearbeitet, um kurzfristig frühzeitig präzisere Warnungen geben zu können. Dazu gehören unter anderem der Ausbau von Pegelmessstationen, die Verbesserung der Datenübertragung sowie die stärkere Verknüpfung von meteorologischen und hydrologischen Informationen.

Die Ereignisse haben verdeutlicht, dass die Hochwasservorhersage ein entscheidender Bestandteil des Risikomanagements ist, jedoch auch erhebliche Herausforderungen birgt, insbesondere bei Extremereignissen wie im Juli 2021 (DKKV 2024).

3.4 Hochwasserentstehung und -verlauf

Wasserstände in den betroffenen Flüssen stiegen rasant an, und vielerorts waren die Pegelmessstationen und die Hochwassermessinfrastruktur entweder unzureichend ausgestattet oder fielen durch die Überlastung aus. Die Pegelstationen, die im Normalfall bei Hochwasservorhersagen helfen, waren teilweise nicht einsatzfähig oder aufgrund der extremen Bedingungen unzugänglich (Reinert et al. 2023; DKKV 2024). Außerdem wurde festgestellt, dass einige der existierenden Pegelstationen an kritischen Flüssen und Nebenflüssen für die Warnungen im Hochwasserfall nicht genutzt wurden. Das erschwerte die Früherkennung der schnell steigenden Wasserstände. Durch die fehlende Möglich-

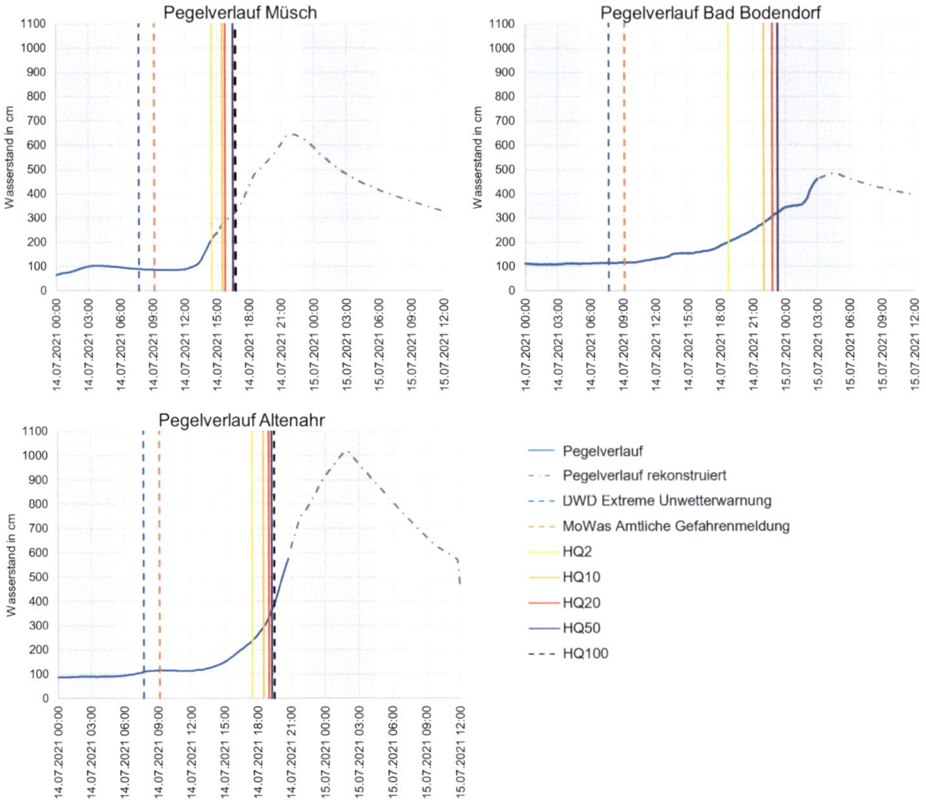

Abb. 3.1 Pegelverläufe Ahr inkl. Überschreitungszeitpunkte HQ2, HQ10, HQ20, HQ50, HQ100; Nachtphase in grau dargestellt. (Eigene Darstellung; Datengrundlage: LfU Rheinland-Pfalz)

keit, die Gefahr in Echtzeit zu überwachen, verzögerte sich auch die Übermittlung der Warnungen an Einsatzkräfte und Bevölkerung erheblich.

Hinweis für Entscheidungsprozesse

Eine große Herausforderung stellte die unzureichende Messinfrastruktur dar. In vielen kleineren Flüssen und Nebenflüssen fehlten Pegelmessstationen, was die Überwachung und Einschätzung der Lage deutlich erschwerte. Pegelmessstationen, die für die Hochwasserinformation genutzt wurden, fielen während des Hochwassers, insbesondere in den Mittelgebirgsregionen, zu einem großen Teil aus oder wurden durch die Wassermassen beschädigt, wodurch die Erfassung der tatsächlichen Wasserstände erheblich beeinträchtigt war.

Hydraulisch gesehen, bewegte sich das Wasser aufgrund der steilen Hänge in bestimmten Gebieten wie der Eifel und den Randgebieten des Sauerlands mit enormer Geschwindigkeit. Die Flüsse und Bäche, die durch natürliche Engstellen oder durch menschliche Bebauung begrenzt waren, konnten die Wassermengen nicht mehr führen. Im Ahrtal etwa verwandelte sich das Flussbett in einen reißenden Strom, der auf seinem Weg Autos,

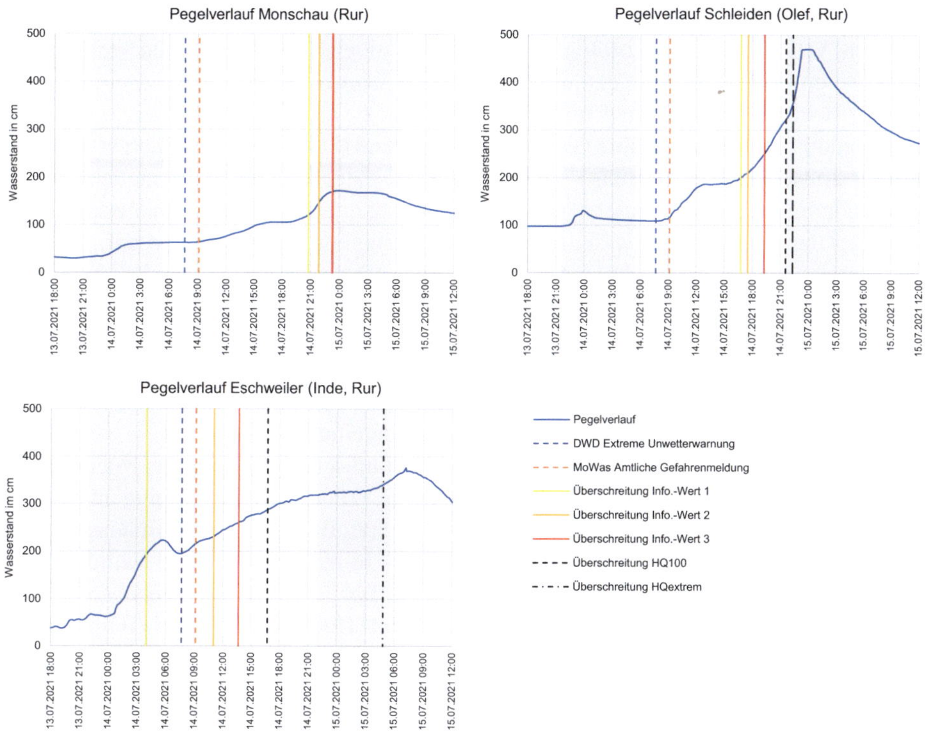

Abb. 3.2 Pegelverläufe Rur (NRW) inkl. Überschreitungszeitpunkte Informationswerte 1–3, HQ100, HQextrem; Nachtphase in grau dargestellt. (Eigene Darstellung; Datengrundlage: LANUV Nordrhein-Westfalen)

Brücken, Gebäude und Straßen mitriss (Ministerium des Innern und für Sport Rheinland-Pfalz 2022). Die Hydraulik des Wassers in diesen engen Tälern führte dazu, dass die Flüsse ihre Ufer überfluteten und sich das Wasser an bestimmten Stellen enorm staute, bevor es sich mit noch größerer Kraft weiter ergoss. Diese schnellen Anstiege (vgl. Abb. 3.1 und 3.2) und Fließgeschwindigkeiten waren nur schwer unter Kontrolle zu bringen und ließen vielen Anwohnern wenig bis gar keine Zeit zur Evakuierung.

3.5 Datengrundlage und Reaktionsfähigkeit

Während der kritischen Phase zwischen dem 12. und 15. Juli stellte der DWD ein breites Portfolio an Warnprodukten bereit. Über Kanäle wie die WarnWetter-App, das Feuerwehr-WetterInformationsSystem (FeWIS) und das WasserwirtschaftsWetterInformationsSystem (WaWIS) erreichten die Warnungen Millionen von Nutzerinnen und Nutzern (DWD 2021). Auch individuelle Beratungen wurden intensiviert: Über 150 individuelle Gespräche mit Katastrophenschutzeinrichtungen fanden allein durch die Außenstelle Essen statt. Diese multilaterale Kommunikationsstrategie zielte darauf ab, sowohl die Bevölkerung als auch Behörden bestmöglich auf die bevorstehende Lage vorzubereiten.

Trotz dieser Bemühungen zeigten die Ereignisse, dass die Übertragung der Warnungen von meteorologischen Gefahren in konkrete hydrologische Warnungen und Handlungsanweisungen an die Bevölkerung an vielen Stellen nicht optimal funktionierte. Die Lehren aus diesen Ereignissen unterstreichen die Notwendigkeit einer noch engeren Verknüpfung meteorologischer und hydrologischer Vorhersagen sowie einer effektiveren Kommunikation entlang der gesamten Warnkette.

Das Hochwasser 2021 hat deutlich gemacht, dass die bisherigen Systeme zur Überwachung und Frühwarnung an ihre Grenzen gestoßen sind (LAWA 2022; Reinert et al. 2023; DKKV 2024). Die Niederschlags- und Wasserstandsdaten konnten die Dringlichkeit und Geschwindigkeit des Ereignisses nicht schnell genug erfassen, was zu einer erheblichen Verzögerung in der Reaktionskette führte. Die Ereignisse zeigten zudem auf, dass die bestehenden meteorologischen und hydrologischen Warnungen besser an die komplexen hydrologischen und hydraulischen Bedingungen der jeweiligen Einzugsgebiete angepasst werden müssen. Die enorme Kraft, mit der das Wasser in den engen Tälern der Eifel floss, und die schnelle Überflutung in städtischen Gebieten haben offengelegt, dass eine umfassende Überprüfung der Hochwasserschutzmaßnahmen und Warnsysteme notwendig ist, um auf zukünftige Ereignisse dieser Art besser vorbereitet zu sein.

Die meteorologischen, hydrologischen und hydraulischen Faktoren des Hochwassers 2021 verdeutlichen, dass das Ereignis über bisherige Erfahrungen hinausging und die Infrastruktur und Warnprozesse an ihre Grenzen brachte. Besonders problematisch waren Verzögerungen in der Übertragung der Warnungen. In manchen Gebieten wurden Warnungen erst in letzter Minute oder gar nicht an die Bevölkerung weitergegeben, und so standen viele Menschen dem plötzlich eintreffenden Hochwasser unvorbereitet gegenüber.

Eine eingehende Analyse der Warn- und Reaktionsprozesse erfolgte im Rahmen des interdisziplinären Forschungsprojekts HoWas2021 (Förderkennzeichen 13N16226 – 13N16231, Projektlaufzeit Dezember 2021 – Dezember 2023), welches vom BMBF gefördert wurde. Das Projekt analysierte zentrale Herausforderungen in den Bereichen Governance und Kommunikation während der Krise. Zu den Schlüsselerkenntnissen gehört die Notwendigkeit, die Warnsysteme robuster zu gestalten und hydrologische Daten in entscheidungsrelevante Informationen zu übersetzen. Diese müssen effektiv kommuniziert werden, sowohl an Entscheidungsträgerinnen und -träger als auch an die betroffene Bevölkerung. Ein weiteres Ergebnis betont die Bedeutung einer besseren Zusammenarbeit zwischen wasserwirtschaftlichen Behörden und Katastrophenschutzstrukturen, um Reaktionszeiten zu verkürzen und die Effektivität der Maßnahmen zu steigern. Das Projekt liefert zudem Handlungsempfehlungen zur Stärkung von Infrastruktur und Kommunikationswegen, um künftige Extremereignisse besser bewältigen zu können. Weitere Informationen und detaillierte Ergebnisse sind in der Schriftenreihe „Governance und Kommunikation im Krisenfall des Hochwasserereignisses im Juli 2021“ (DKKV-Schriftenreihe Nr. 63) verfügbar.

3.6 Fazit

Das Hochwasser im Juli 2021 in Deutschland, insbesondere in Nordrhein-Westfalen und Rheinland-Pfalz, zählt zu den schwersten Flutkatastrophen der letzten Jahrzehnte. Das Tiefdruckgebiet „Bernd“ verursachte extreme Regenmengen von bis zu $200\,l/m^2$ in wenigen Tagen. Besonders betroffen waren das Ahrtal und enge Flusstäler, wo die Wassermassen durch steile Hänge und Bebauung kanalisiert wurden, was zu enormer Geschwindigkeit und Schäden führte. Bereits ab dem 10. Juli warnte der Deutsche Wetterdienst vor der außergewöhnlichen Wetterlage. Dennoch zeigten sich Schwächen in der Verknüpfung meteorologischer und hydrologischer Warnsysteme sowie in der Kommunikation mit Behörden und Bevölkerung. Viele Menschen wurden unzureichend informiert, was schnelle Reaktionen erschwerte. Das Ereignis überstieg die Kapazitäten bestehender Schutzsysteme und unterstreicht die Dringlichkeit einer umfassenden Überarbeitung der Warnstrukturen, um auf zukünftige Extremereignisse besser vorbereitet zu sein.

Zusammenfassung

- Die Übertragung der Warnungen von meteorologischen Gefahren in konkrete hydrologische Warnungen und Handlungsanweisungen für Behörden und die Bevölkerung ist zentral und zeigt derzeit Verbesserungsbedarf.
- Das Ereignis 2021 überstieg die Kapazitäten bestehender Schutz- und Warnsysteme und unterstreicht die Dringlichkeit einer umfassenden Überarbeitung der Warnstrukturen, um auf zukünftige Extremereignisse besser vorbereitet zu sein.
- Eine durchgängige Echtzeitvorhersage gekoppelter meteorologischer, hydrologischer und hydraulischer Vorhersagen mit validierten Schwellwerten sowie einer transparenten Unsicherheitsdarstellung ist erforderlich.
- Klare Zuständigkeiten, eingeübte Entscheidungsprozesse und adressatengerechte Impact-Kommunikation sind zentral und Warnarchitektur muss dringend so überarbeitet werden, dass im Ereignisfall schneller und konsistenter gehandelt werden kann.

Danksagung Die Autorinnen und Autoren von ▶ Kap. 3 bedanken sich beim Bundesministerium für Bildung und Forschung (BMBF) sowie beim DLR Projektträger und dem VDI Technologiezentrum für die Unterstützung und Förderung der Projekte KAHR und HoWas2021. Diese Forschung wurde vom BMBF unter den Förderkennzeichen 01LR2102A-M (KAHR) sowie 13N16226 – 13N16231 (HoWas2021) finanziert.

Literatur

Bell R, Kron W, Thiebes B, Thieken A, Winkhardt-Enz R, Schwarz M, Kleeschulte I (2022) Die Flutkatastrophe im Juli 2021 Ein Jahr danach: Aufarbeitung und erste Lehren für die Zukunft. Deutsches Komitee Katastrophenvorsorge e. V., Bonn

Bund/Länder-Arbeitsgemeinschaft Wasser (LAWA) (2022) Analyse zum Juli-Hochwasser 2021 und Ableitung von Konsequenzen aus Sicht des LAWA-AH. https://www.lawa.de/documents/analyse-zum-juli-hochwasser-2021-barrierefrei_1689857053.pdf. Zugegriffen: 20. Jan. 2025

DKKV (2024) Governance und Kommunikation im Krisenfall des Hochwasserereignisses im Juli 2021. DKKV-Schriftenreihe Nr. 63. DKKV, Bonn (https://dkkv.org/wp-content/uploads/2024/01/HoWas2021_DKKV_Schriftenreihe_63.pdf, zuletzt geprüft am 20.01.2025)

DWD (2021) Bericht zum Ablauf und Umfang der operationellen Warn- und Beratungstätigkeit des Deutschen Wetterdienstes im Vorlauf und während des Unwetters 12.–15. Juli 2021 in NRW und Rheinland-Pfalz, ausgelöst durch das Tief „Bernd". Deutscher Wetterdienst, Offenbach (https://www.dwd.de/DE/presse/hintergrundberichte/berichts_warnchronologie.pdf.pdf?__blob=publicationFile&v=2, zuletzt geprüft am 20.01.2025)

Junghänel T et al (2021) Hydroklimatologische Einordnung der Stark- und Dauerniederschläge in Teilen Deutschlands im Hydroklimatologische Einordnung der Stark- und Dauerniederschläge in Teilen Deutschlands im Zusammenhang mit dem Tiefdruckgebiet „Bernd" vom 12. bis 19. Juli 2021. Deutscher Wetterdienst, Geschäftsbereich Klima und Umwelt

Ministerium des Innern und für Sport Rheinland-Pfalz (2022) Der Wiederaufbau in Rheinland-Pfalz. nach der Naturkatastrophe vom 14./15. Juli 2021. https://wiederaufbau.rlp.de/fileadmin/wiederaufbau/2022/02-Februar/Der_Wiederaufbau_in_Rheinland-Pfalz_2021-2022.pdf. Zugegriffen: 20. Jan. 2025

Reinert J, Wingen M, Klopries E-M, Schüttrumpf H, Dittmer C, Lorenz DF et al (2023) Hochwasserwarnung: Lessons to Learn nach dem Julihochwasser 2021. Korrespondenz Wasserwirtsch 7:428–434. https://doi.org/10.3243/kwe2023.07.003

Thieken AH et al (2023) Performance of the flood warning system in Germany in July 2021 – insights from affected residents. https://nhess.copernicus.org/articles/23/973/2023/. Zugegriffen: 20. Jan. 2025. https://doi.org/10.5194/egusphere-2022-244

Tradowsky JS, Philip SY, Kreienkamp F, Kew SF, Lorenz P, Arrighi J et al (2023) Attribution of the heavy rainfall events leading to severe flooding in Western Europe during July 2021. Clim Change 176(7):90. https://doi.org/10.1007/s10584-023-03502-7

3

Hochwasserbemessung, Extremszenarien und Klimawandel am Beispiel des Ahrtals

Bruno Merz, Heiko Apel, Xiaoxiang Guan, Björn Guse, Li Han, Dung Viet Nguyen und Sergiy Vorogushyn

J. Birkmann et al. (Hrsg.), *Die Flutkatastrophe 2021*, Sustainable Development Goals (SDG) – Umsetzung in Praxis, Lehre und Entscheidungsprozessen,
https://doi.org/10.1007/978-3-658-48688-4_4

Das Ausmaß des Hochwassers und seiner Schäden im Juli 2021 hat die Betroffenen und viele Expertinnen und Experten überrascht. Lassen sich solche Ereignisse im Voraus zuverlässig abschätzen? Hätte es noch schlimmer kommen können? Müssen wir aufgrund des Klimawandels in Zukunft mit noch größeren Hochwasserereignissen rechnen? Welche Maßnahmen sind besonders wirksam, um das Risiko zu verringern? Um diese Fragen zu beantworten, werden Modelle benötigt, die die bei Extremereignissen dominierenden Prozesse realitätsnah und zuverlässig beschreiben. In diesem Kapitel werden am Beispiel des Einzugsgebietes der Ahr innovative Ansätze zur Hochwassermodellierung vorgestellt. Sie ermöglichen es, das Juli-Hochwasser zu verstehen und seine Eintrittswahrscheinlichkeit abzuschätzen. Darüber hinaus zeigt die Anwendung der Modelle, dass es noch schlimmer hätte kommen können. Schon eine leichte Verschiebung des Regenfeldes nach Osten hätte die Abflussspitzen in den Zuflüssen um bis zu 160 % erhöht. Die entwickelten Modelle sind eine hervorragende Grundlage für die Bemessung von Hochwasserschutzmaßnahmen, für Hochwassergefahrenkarten, für Risikoanalysen, für die Untersuchung der Wirkung von Maßnahmen zur Risikominderung und für ein verbessertes Katastrophenmanagement.

4.1 Wie schlimm kann es werden?

Für eine gute Vorbereitung auf Hochwasserereignisse, einen effektiven Schutz von Sachwerten und eine optimale Planung der Infrastruktur sind verlässliche Informationen über Hochwasserabflüsse und deren Eintrittswahrscheinlichkeiten unerlässlich. In diesem Zusammenhang hat das Hochwasser im Juli 2021 erneut eindringlich gezeigt, wie wichtig es ist, im Rahmen des Hochwasserrisikomanagements auch sehr seltene Ereignisse zu berücksichtigen, die größer sind als die technischen Schutzmaßnahmen und/oder größer als die Ereignisse, die die Anwohner erlebt haben oder in Erinnerung haben. Eine weitere wichtige Planungsgrundlage sind Szenarien zukünftiger Hochwasser unter Klimawandel, da Hochwasserschutz einen Zeithorizont von mehreren Jahrzehnten hat.

Diese Aspekte standen im Mittelpunkt des KAHR-Teilprojektes „Hochwassermodellierung Rheinland-Pfalz und NRW", in dem innovative Ansätze (weiter-)entwickelt wurden. Am Beispiel der Ahr wird in diesem Kapitel gezeigt, wie diese Ansätze in der Praxis angewendet werden können. Dieses Kapitel liefert nicht nur spezifische Erkenntnisse für das Ahrtal, sondern beleuchtet auch die heutigen Möglichkeiten und die damit verbundenen Unsicherheiten bei der Abschätzung von Hochwasserrisiken.

4.1.1 Hochwasserwahrscheinlichkeiten – Grundlage für die Bemessung und für Hochwassergefahrenkarten

Hochwasserwahrscheinlichkeiten, d. h. Aussagen darüber, mit welcher Wahrscheinlichkeit ein bestimmter Wasserstand oder Abfluss eintritt, sind zentrale Informationen für das Hochwasserrisikomanagement. Sie bilden die Grundlage für die Risikoanalyse, für die Bemessung von technischen Schutzmaßnahmen und für die in den Hochwassergefahrenkarten dargestellten Überflutungsszenarien. Für die Bemessung technischer Schutzmaßnahmen werden typischerweise Aussagen zu Ereignissen benötigt, die einmal in 100 Jahren erreicht oder überschritten werden. Für bestimmte Zwecke werden kleinere oder größere Ereignisse mit Jährlichkeiten von ca. 10 bis 10.000 Jahren benötigt. Beispielsweise muss für große Talsperren in Deutschland nachgewiesen werden, dass die Tragsicherheit der Absperrbauwerke bei einem 10.000-jährlichen Hochwasser nicht gefährdet ist. Für Hochwassergefahrenkarten werden in der Europäischen Union typischerweise drei Szenarien verwendet, ein häufiges, ein mittleres und ein extremes Szenario. Diese entsprechen in Deutschland Jährlichkeiten von 10, 100 und 200–1000 Jahren. Diese unterschiedlichen Szenarien ermöglichen eine differenzierte Planung, die sowohl alltägliche Hochwasser als auch Extremereignisse berücksichtigt.

Der gebräuchlichste Ansatz zur Ermittlung von Hochwasserwahrscheinlichkeiten basiert auf der Extremwertstatistik. Dabei wird eine statistische Verteilungsfunktion an die beobachteten Extremabflüsse angepasst, um diese in den Bereich großer Jährlichkeiten zu extrapolieren. Ein häufiges Problem dabei ist die relativ kurze Dauer der Beobachtungsreihen an vielen Messstationen. Eine grobe Faustregel besagt, dass die Extrapolation nicht mehr als das 2- bis 3-fache der Beobachtungsperiode betragen sollte. Dies bedeutet, dass mehrere Jahrzehnte an Beobachtungsdaten erforderlich sind, um die typischerweise benötigten Bemessungswerte zuverlässig abzuleiten. Darüber hinaus weisen viele (empirische) Hochwasserhäufigkeitsfunktionen einen schweren Rand (Heavy Tail) auf, d. h. der obere Rand der Verteilung fällt langsamer ab als bei einer exponentiellen Verteilung. Dadurch sind Hochwasserereignisse, die deutlich größer als das größte beobachtete Ereignis sind, relativ leicht möglich (Merz et al. 2022; Vorogushyn et al. 2023). Ein prägnantes Beispiel hierfür ist das Verhalten der Ahr am Pegel Altenahr: Das Hochwasser 2021 übertraf das bisher größte Ereignis der über 70-jährigen Messreihe um das Fünffache (Vorogushyn et al. 2022).

Die abgeleitete Hochwasserhäufigkeitsanalyse basierend auf Langzeitsimulation (Haberlandt und Radtke 2014; Falter et al. 2015; Merz et al. 2024a) ist ein Ansatz, um das Problem der Abschätzung von Hochwasserwahrscheinlichkeiten insbesondere bei kurzen Abflusszeitreihen zu entschärfen. Dabei erzeugt ein sogenannter Wettergenerator sehr lange Klimazeitreihen (z. B. über mehrere tausend Jahre), die als Input für ein hydrologisches Modell dienen. Dessen Simulationsergebnisse werden dann extremwertstatistisch ausgewertet. Vorteile dieser Vorgehensweise sind, dass Klimabeobachtungen oft in höherer räumlicher Auflösung und für längere Zeiträume vorliegen als Abflussbeobachtungen. Durch die Generierung sehr langer Zeitreihen erhöht sich auch die Wahrscheinlichkeit, Ausnahmesituationen, z. B. extremer Niederschlag auf ein Gebiet mit hoher Vorfeuchte, zu modellieren, die in den kürzeren (Abfluss-)Beobachtungen nicht aufgetreten sind. In ▶ Abschn. 4.2 wird das entsprechende Instrumentarium für diesen Ansatz für das Ahr-Einzugsgebiet vorgestellt.

4.1.2 Außergewöhnliche Ereignisse – Planungsgrundlagen für eine resiliente Gesellschaft

Hinweis für Entscheidungsprozesse

Neben den typischen Hochwasserszenarien, z. B. dem 100-jährlichen Ereignis, ist es wichtig, auch außergewöhnliche Ereignisse in das Risikomanagement einzubeziehen (Montanari et al. 2024). Auch wenn sich Schäden durch außergewöhnliche Hochwasserereignisse nicht vollständig vermeiden lassen, kann ein effektives Risikomanagement verheerende Folgen deutlich mindern. Durch gute Vorhersagen sowie Frühwarn- und Evakuierungssysteme können mögliche Todesopfer vermieden werden (Apel et al. 2022; Najafi et al. 2024). Eine durchdachte Raumplanung und ein gezieltes Infrastrukturmanagement sorgen dafür, dass sensible Einrichtungen wie Altenheime und kritische Infrastrukturen wie Kraftwerke entweder außerhalb gefährdeter Gebiete liegen oder entsprechend vor Überflutung geschützt sind. Darüber hinaus können Sicherheits- und Redundanzmaßnahmen entwickelt werden, um den Betrieb kritischer Infrastruktur während eines Hochwassers aufrechtzuerhalten, sowie Pläne zur schnellen Wiederherstellung eines Mindestbetriebsniveaus, falls ein Ausfall unvermeidbar ist (Koks et al. 2022).

Ein Problem bei außergewöhnlichen Ereignissen ist, dass sich Menschen nur schwer in solche Situationen hineinversetzen können. Menschen können die negativen Auswirkungen eines schweren Hochwassers nicht vorhersagen, wenn sie es nicht selbst erlebt haben (Siegrist und Gutscher 2008). Zudem neigen Menschen dazu, die Wahrscheinlichkeit eines Ereignisses danach zu beurteilen, wie leicht sie sich ein solches Ereignis vorstellen können (Availability Bias). So können Menschen Ereignissen, die sie sich nur schwer vorstellen können, eine Wahrscheinlichkeit von Null zuschreiben. Zudem fällt es Menschen und Organisationen schwer, über bedrohliche Aussichten nachzudenken und für Situationen zu planen, die ihnen schaden könnten (Merz et al. 2024b).

Kontrafaktuale sind ein neuer Ansatz in der Naturgefahrenforschung, um plausible außergewöhnliche Ereignisse zu generieren und zu kommunizieren (Woo 2019; Voit und Heistermann 2024). Ein Kontrafaktual beschreibt eine Situation und ihre Konsequenzen, die nicht real war, aber hypothetisch hätte eintreten können. In ▶ Abschn. 4.3 werden räumliche Kontrafaktuale für das Ahr-Einzugsgebiet entwickelt. Es wird untersucht, was passiert wäre, wenn das hochwasserauslösende Tiefdruckgebiet eine etwas andere Zugbahn genommen hätte. Ein solches „Verschieben der Wolke" im Raum ist nicht nur meteorologisch plausibel, sondern auch für Laien sehr leicht nachvollziehbar.

4.1.3 Klimawandelprojektionen – Grundlagen für die Klimaanpassung

Für weite Teile Deutschlands zeigen sich signifikante steigende Trends der Hochwasserabflüsse, die mit dem Klimawandel in Verbindung gebracht werden (Blöschl et al. 2019; Bronstert et al. 2024). Für die Zukunft liegen Projektionen für verschiedene Emissionsszenarien vor, die auf Simulationen mit globalen und/oder regionalen Klimamodellen (GCMs, RCMs) beruhen. Die vom Klimamodell für eine bestimmte Zukunft simulierten Niederschläge und andere Klimavariablen, wie z. B. die Temperatur, werden auf eine höhere räumliche Auflösung skaliert, so dass sie als Input für die hydrologische Modellierung verwendet werden können. Deren Ergebnisse können dann extremwertstatistisch ausgewertet werden, um Abschätzungen der Hochwasserwahrscheinlichkeit unter Klimawandel zu erhalten. Dieses Verfahren hat allerdings zwei Probleme: Zum einen simulieren Klimamodelle, insbesondere globale Klimamodelle, Extremniederschläge nur schlecht (Jong et al. 2023); zum anderen ist die Anzahl der Simulationen begrenzt. Die Abschätzung seltener Hochwasser unter Klimawandel ist daher mit sehr großen Unsicherheiten behaftet.

Ein alternativer Ansatz ist das Herunterskalieren von Klimavariablen aus Klimamodellen mit Hilfe von Wettergeneratoren. Dabei wird ein Wettergenerator so parametrisiert, dass er synthetisches Wetter unter Klimaänderung erzeugt. In ▶ Abschn. 4.2 wird ein solcher Ansatz für das Einzugsgebiet der Ahr vorgestellt, mit dessen Hilfe Hochwasserwahrscheinlichkeiten für die Zukunft unter möglichen Emissionsszenarien abgeschätzt werden können. Der Wettergenerator nutzt dabei Klimavariablen (großräumige atmosphärische Zirkulation und Temperatur), die deutlich besser als Niederschlag von Klimamodellen simuliert werden. Dadurch wird die große Unsicherheit in der Projektion von Starkniederschlägen in den Klimamodellen umgangen.

4.2 Modelle: Konzeption und Implementierung für das Ahrtal

4.2.1 Generierung von langen Klimazeitreihen mit nsRWG

Für ein Gebiet, das die Einzugsgebiete Ahr, Erft und Rur umfasst, wurde der instationäre Wettergenerator des GFZ (nsRWG; non-stationary regional weather generator) (Nguyen et al. 2024) implementiert (◘ Abb. 4.1a). Auf 294 Gitterzellen (räumliche Auflösung: 5 km) liefert der Wettergenerator tägliche Zeitreihen für Niederschlag und Temperatur (Tagesminimum, -maximum, -mittel). nsRWG basiert auf einem multivariaten autoregressiven Modell erster Ordnung und ist in der Lage, die räumliche und zeitliche Korrelation des beobachteten Wetters zu beschreiben.

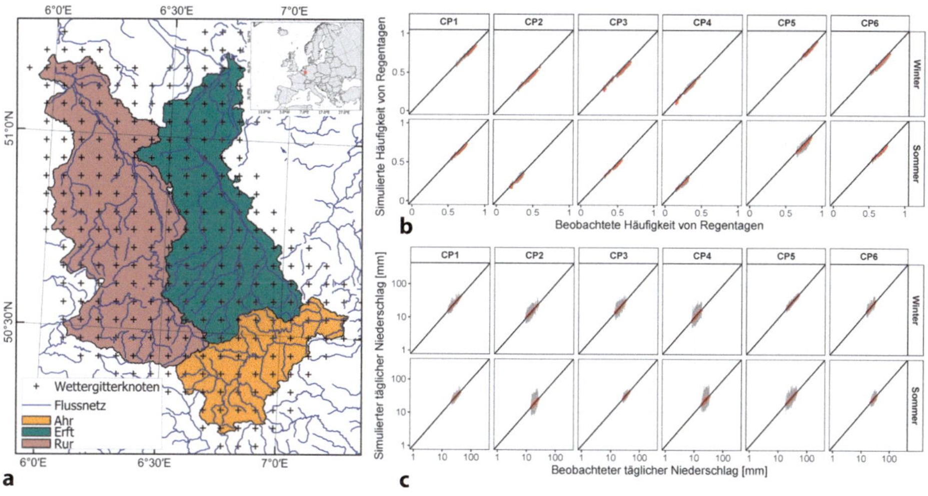

◘ **Abb. 4.1** Domäne des Wettergenerators (**a**) und Vergleich der beobachteten und simulierten Häufigkeit von Regentagen (**b**) und extremen (99,5 Perzentil) täglichen Niederschlagsmengen (**c**). *Rote Punkte* stellen den Median des *grauen Bereichs* dar, der 100 Modellrealisierungen für den Zeitraum 1950–2021 entspricht. Die Niederschlagsdaten werden für 12 Klassen (6 Zirkulationsmuster [circulation patterns, CPs] sowie 2 Jahreszeiten) dargestellt. (**a** eigene Darstellung, Flussnetz Geobasisdaten: © GeoBasis-DE/BKG [2018]. Europakarte erstellt mit Natural Earth Daten; **b**, **c** eigene Darstellung)

nsRWG verwendet die erweiterte Paretoverteilung für Niederschlag, während die Temperaturvariablen durch eine Normalverteilung – nach einer Transformation durch eine Potenzfunktion – beschrieben werden. Um den Klimawandel abbilden zu können, werden die lokalen Niederschlags- und Temperaturverteilungen mit großräumigen atmosphärischen Zirkulationsmustern und der großräumigen mittleren Temperatur konditioniert. (Zirkulationsmuster bzw. „Circulation Pattern CP" beschreiben die großräumige Druckverteilung und Luftbewegung in der Atmosphäre.) Diese großräumigen Informationen werden Beobachtungen, Re-Analysen oder Klimamodellen entnommen. Die Abhängigkeit von Zirkulationsmustern ermöglicht die Berücksichtigung von Änderungen in der atmosphärischen Dynamik. Die Abhängigkeit von der großräumigen Temperatur erlaubt die Berücksichtigung möglicher thermodynamischer Veränderungen, wie z. B. höhere Niederschlagsintensitäten bei gleichem Zirkulationsmuster in einer wärmeren Welt. nsRWG wird mit sechs verschiedenen Zirkulationsmustern, jeweils für Sommer und Winter, betrieben (Details siehe Nguyen et al. 2024).

Die Parametrisierung des Wettergenerators erfolgte mit den HYRAS-Daten (DWD 2024) für den Zeitraum 1. Januar 1950 bis 31. Dezember 2021. Die großräumige Temperatur und atmosphärischen Zirkulationsmuster wurden aus Re-Analysen (ERA5, $dx = 1°$) entnommen. Mittels des Wettergenerators wurden 100 Realisationen synthetischen Wetters für den Zeitraum 1950–2021 erzeugt. Die täglichen Wetterdaten wurden auf Stundenwerte herunterskaliert (Guan et al. 2023). Damit stehen für das Ahr-Einzugsgebiet 7200 Jahre (100 × 72) stündliches (synthetisches) Wetter für die vergangenen Jahrzehnte zur Verfügung.

◘ Abb. 4.1b, c visualisiert die Qualität der mit nsRWG generierten Daten am Beispiel der Häufigkeit von Regentagen und extremen täglichen Niederschlagsmengen (99,5 Perzentil; das 99,5 Perzentil entspricht einer Niederschlagsmenge, die im langjährigen Mittel etwa einmal pro Jahr überschritten wird). Die synthetischen Daten stimmen sehr gut mit den Beobachtungen überein. Die unterschiedliche Häufigkeit von Regentagen (◘ Abb. 4.1b) für verschiedene Zirkulationsmuster (z. B. trockenes Muster CP2 und feuchtes Muster CP5) wird von nsRWG sehr gut wiedergegeben. Auch bei den Extremniederschlägen (◘ Abb. 4.1c) liegen die Punkte sehr nahe an der Diagonalen, so dass der Wettergenerator realistische synthetische Extremniederschläge erzeugt. Die Ergebnisse des Wettergenerators wurden anhand weiterer Indikatoren, wie z. B. der räumlichen Korrelation, überprüft (hier nicht dargestellt). Für alle Indikatoren zeigt nsRWG ein realistisches Verhalten.

4.2.2 Modellierung von Extremabflüssen mit mHM

Die hydrologischen Prozesse (Infiltration, Retention im Einzugsgebiet, verschiedene Abflusskomponenten etc.) werden mit dem mesoskaligen hydrologischen Modell mHM (Samaniego et al. 2010; Kumar et al. 2013) simuliert. mHM ist ein räumlich verteiltes Rastermodell und basiert auf dem Ansatz der „Multiskalen-Parameter-Regionalisierung". Dieser Ansatz erlaubt die Ableitung der Modellparameter für eine räumliche Auflösung der Simulationen von 3 km auf Basis hochaufgelöster, physikalisch beobachtbarer Daten. Dazu gehören Informationen über Topographie, Bodenbeschaffenheit, Geologie und Vegetation. Dieser Ansatz berücksichtigt die kleinräumige räumliche Heterogenität der Eingangsdaten, die mit Hilfe von räumlich konsistenten globalen Parametersätze über

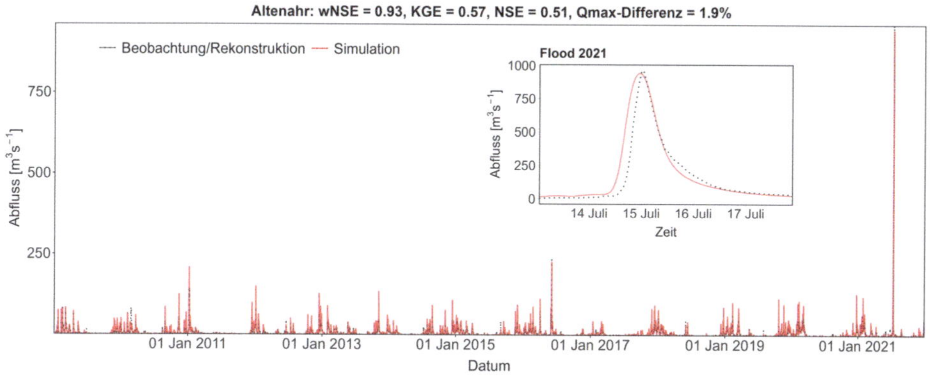

Abb. 4.2 Hydrologische Modellierung mit mHM: Vergleich von Beobachtung und Simulation beispielhaft für den Pegel Ahrtal für die Periode Januar 2010 bis Dezember 2021. Die quantitativen Gütemaße zeigen eine gute bis sehr gute Modellgüte (*wNSE* weighted Nash-Sutcliffe Efficiency mit einem Fokus auf Hochwasser, *KGE* Kling-Gupta Efficiency, *NSE* Nash-Sutcliffe Efficiency). (Eigene Darstellung)

Transferfunktionen verknüpft werden. Die daraus resultierenden Modellparameter werden für die hydrologische Simulation verwendet.

Das Modell mHM wurde für die Einzugsgebiete Ahr, Erft und Rur in stündlicher Auflösung implementiert und stellt damit auch eine der ersten mHM-Anwendungen in stündlicher Auflösung dar. Abb. 4.2 zeigt beispielhaft die Güte des Modells für den Pegel Ahrtal für den Zeitraum 2010–2021. Die Hochwasserereignisse werden sehr gut wiedergegeben; dies gilt sowohl für kleine Ereignisse als auch für das Extremereignis im Juli 2021. Das Modell ist somit in der Lage, sehr unterschiedliche Ereignisse zu simulieren, die vermutlich durch verschiedene Einzugsgebietsprozesse dominiert werden (Vorogushyn et al. 2022).

4.2.3 Modellierung von Überflutungsflächen mit RIM2D

Der Wellenablauf und die Überflutungsflächen werden mit dem zweidimensionalen, rasterbasierten, hydrodynamischen Modell RIM2D simuliert. RIM2D verwendet die vereinfachte Flachwassergleichung ohne konvektive Beschleunigung und erreicht durch die Implementierung auf Graphics Processing Units (GPUs) sehr hohe Rechengeschwindigkeiten (Apel et al. 2022, 2024). RIM2D wird für zwei Abschnitte der Ahr (Abb. 4.3a: Müsch – Altenahr; Altenahr – Sinzig) direkt mit den Abflusszeitreihen von mHM oder mit gemessenen Hydrographen angetrieben.

Abb. 4.3 zeigt die Simulation des Hochwassers vom Juli 2021 mit einer räumlichen Auflösung von 10 m im Bereich Bad Neuenahr – Ahrweiler (Abb. 4.3b) und in der Umgebung von Altenahr (Abb. 4.3c). Als obere Randbedingung wurde die gemessene bzw. rekonstruierte Hochwasserganglinie verwendet. Die Simulation stimmt sehr gut mit der realen Hochwassersituation überein, die anhand von Satellitenbildern und Geländemarken rekonstruiert wurde.

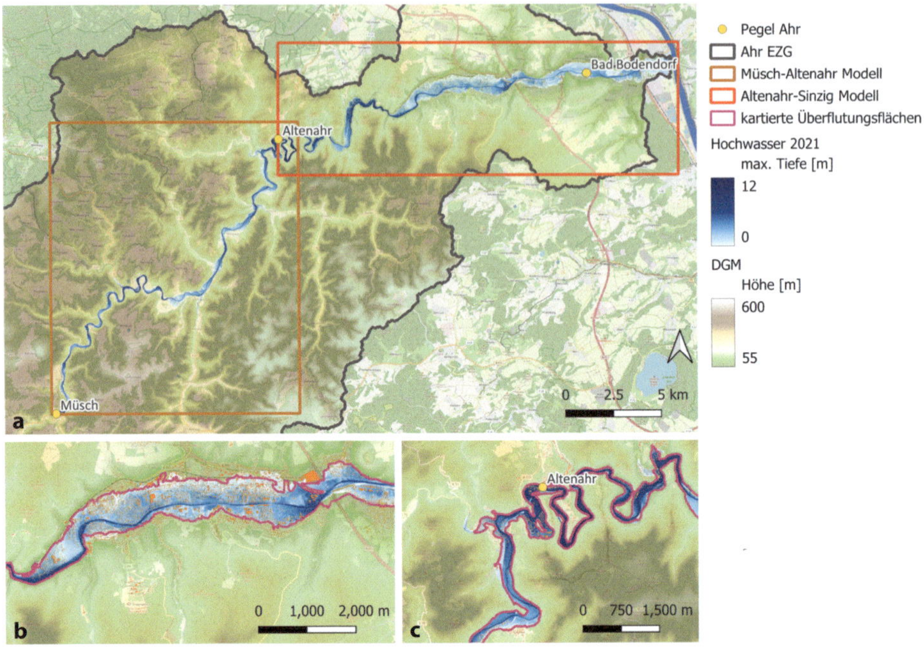

Abb. 4.3 Hydrodynamische Modellierung mit RIM2D. **a** maximale simulierte Wassertiefen in den Modelldomänen Müsch – Altenahr und Altenahr – Sinzig, **b** Vergleich von Beobachtung und Simulation für das Hochwasser 2021 in Bad Neuenahr – Ahrweiler und **c** in der Umgebung von Altenahr. (Hintergrundkarten © OpenStreetMap Mitwirkende, ▶ https://www.openstreetmap.org/copyright. Alle Rechte vorbehalten)

4.3 Ergebnisse und Diskussion

4.3.1 Hochwasserwahrscheinlichkeiten und Bemessungswerte

Mit der Modellkette „Wettergenerator – hydrologisches Modell" wurde eine Langzeitsimulation für das Einzugsgebiet der Ahr durchgeführt. Als Input für das kalibrierte hydrologische Modell mHM wurden 100 Realisationen synthetisches Wetter für den Zeitraum 1950–2021 verwendet. Damit stehen 7200 Jahre stündliche Abflusswerte an beliebigen Stellen im Einzugsgebiet der Ahr zur Verfügung. Aus diesen Daten wurden für den Pegel Altenahr die Jahreshöchstwerte extrahiert (Abb. 4.4). Die Mehrzahl der Jahreshöchstwerte liegt unter 200 m^3/s und entspricht damit den beobachteten Werten. Die Modellkette erzeugt jedoch auch deutlich größere Hochwasser (> 1000 m^3/s). Insgesamt sind sieben Ereignisse größer als das Hochwasser 2021. Dem Hochwasser 2021 kann aufgrund dieser Daten eine empirische Wahrscheinlichkeit von ca. 1000 Jahren zugeordnet werden ($7200/7 \sim 1000$ Jahre). Die Modellkette ist somit in der Lage, das beobachtete Verhalten der Ahr – schwerer Rand der Hochwasserverteilung, d. h. es können Ereignisse auftreten, die deutlich größer sind als der Rest der Hochwasser – plausibel abzubilden.

Abb. 4.5a zeigt die Hochwasserhäufigkeitskurve am Pegel Altenahr, die sich aus den Simulationsdaten der Modellkette ergibt. Hierfür wurden 3000 Jahre Simulationsdaten verwendet, und zwar 100 Realisationen der letzten 30 Jahre (1992–2021). Diese Beschränkung auf 30 Jahre wurde vorgenommen, damit die Ergebnisse einfacher mit

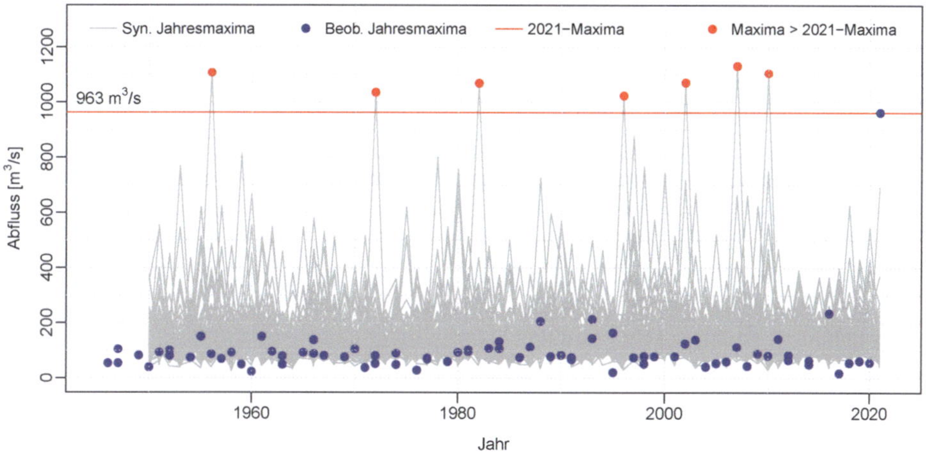

Abb. 4.4 Beobachtete (*blau*) und synthetische (*grau*) Zeitreihen des maximalen jährlichen Abflusses am Pegel Altenahr. *Graue Linien* zeigen 100 Realisationen von jeweils 72 Jahren (1950–2021). *Rote Punkte* markieren größere Ereignisse als der Simulationswert für Juli 2021. (Eigene Darstellung)

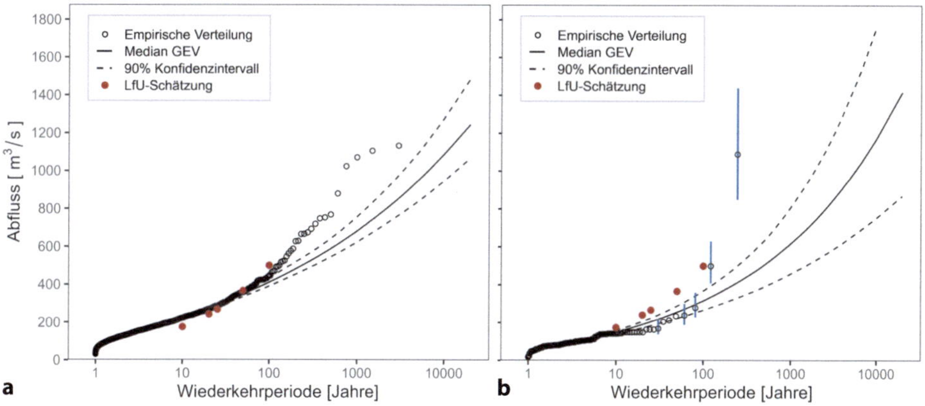

Abb. 4.5 Hochwasserhäufigkeitskurven für den Pegel Altenahr. **a** Empirische Verteilung und angepasste GEV-Verteilung (Generalisierte Extremwertverteilung) abgeleitet aus der Modellkette „Wettergenerator und hydrologisches Modell". *Rote Kreise* zeigen die behördlichen Schätzungen (Wasserportal Rheinland-Pfalz 2024). **b** Empirische Verteilung und angepasste GEV-Verteilung abgeleitet aus den beobachteten systematischen Abflüssen (1947–2019) sowie fünf rekonstruierter historischer Ereignisse. (**a** eigene Darstellung; **b** eigene Darstellung, basierend auf Vorogushyn et al. 2022)

den 30-Jahresperioden verglichen werden können, wie sie bei Klimawandelprojektionen verwendet werden (siehe ▶ Abschn. 4.3.3). Die empirische Verteilung zeigt bei einer Wiederkehrperiode von ca. 100 Jahren eine deutliche Änderung der Steigung; dies deutet auf einen schweren Rand hin. Im oberen Bereich (> 1000 Jahre) ist dagegen der umgekehrte Effekt zu sehen; allerdings liegen dort nur einzelne Werte vor, so dass dieser Effekt nicht zuverlässig erscheint. Die Anpassung der Allgemeinen Extremwertverteilung (GEV) ergibt einen deutlichen schweren Rand. Allerdings weicht die GEV für Jährlichkeiten größer als 100 Jahre deutlich von der empirischen Verteilung ab. Dies ist eine Folge der Inflexibilität der GEV im Vergleich zur empirischen Verteilung. Mit den drei GEV-Parametern ist es nicht möglich, eine S-förmige Verteilung nachzubilden. Die angepasste Verteilung liefert eine Jährlichkeit von ca. 7000 für das Hochwasser 2021.

◘ Abb. 4.5b zeigt ebenfalls die empirische Verteilungsfunktion sowie die angepasste GEV-Verteilung. Diese Funktionen basieren jedoch auf beobachteten Abflussdaten und fünf rekonstruierten historischen Hochwassern. Die beiden GEV-Verteilungen (aus ◘ Abb. 4.5a und 4.5b) stimmen sehr gut überein. Dabei ist zu beachten, dass beide Verteilungen auf unterschiedlichen Daten und Methoden beruhen, einerseits auf beobachteten Klimadaten und einer Modellkette (◘ Abb. 4.5a) und andererseits auf der Statistik von direkt beobachteten und rekonstruierten Abflussdaten (◘ Abb. 4.5b). Die gute Übereinstimmung der beiden Verteilungen erhöht die Glaubwürdigkeit des Ergebnisses. Die Anwendung alternativer Methoden und Daten wird allgemein als Goldstandard zur Erhöhung der Glaubwürdigkeit von Hochwasserwahrscheinlichkeiten und Bemessungswerten angesehen (DWA 2024).

4.3.2 Extremszenarien: räumliche Kontrafaktuale

Ausgangspunkt für die Entwicklung räumlicher Kontrafaktuale für das Einzugsgebiet der Ahr ist die Analyse der Niederschlagsfelder während des Ereignisses vom 14. Juli 2021 in stündlicher Auflösung (◘ Abb. 4.6a). Dazu werden E-OBS Niederschlagsdaten (Cornes et al. 2018) verwendet, die auf eine stündliche Auflösung disaggregiert wurden (Guan et al. 2023). Es wird deutlich, dass einzelne Starkniederschlagszellen zwischen 16 und 19 Uhr das Ahrgebiet nur am westlichen und nördlichen Rand trafen. Eine deutliche Erhöhung der Gebietsniederschläge ist bereits bei einer leichten (und meteorologisch plausiblen) Verschiebung nach Osten zu erwarten. Durch eine schrittweise Verschiebung des Niederschlagsfeldes ausgehend vom beobachteten Referenzniederschlag (S0) wurden insgesamt 24 Kontrafaktuale konstruiert (◘ Abb. 4.6b). In ca. 5 km Schritten wurde das Niederschlagsfeld drei Schritte nach Westen, 13 Schritte nach Osten und nach fünf Schritten Richtung Osten vier Schritte nach Norden und fünf Schritte nach Süden verschoben.

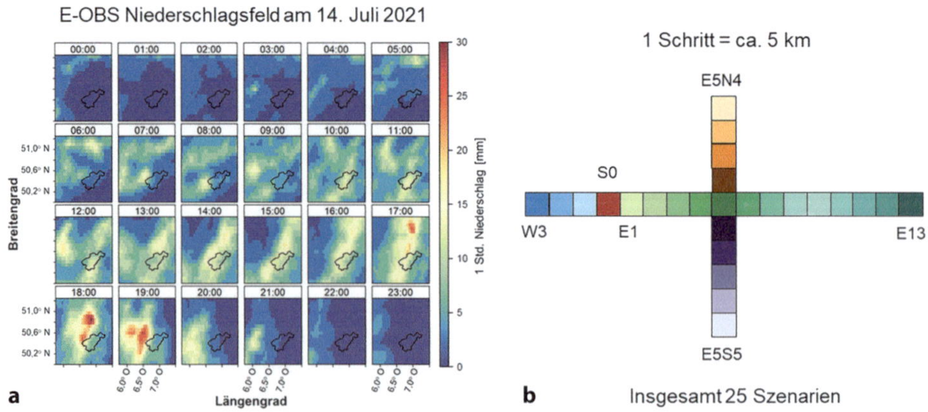

◘ **Abb. 4.6** **a** Stündliche Niederschlagsfelder am 14. Juli 2021 über der Region um das Ahr-Einzugsgebiet (schwarze Kontourlinie) basierend auf disaggregierten E-OBS Daten. **b** Schematische Darstellung der schrittweisen Verschiebung des Niederschlagsfeldes ausgehend vom beobachteten Referenzniederschlag S0. (Verändert nach Vorogushyn et al. 2025)

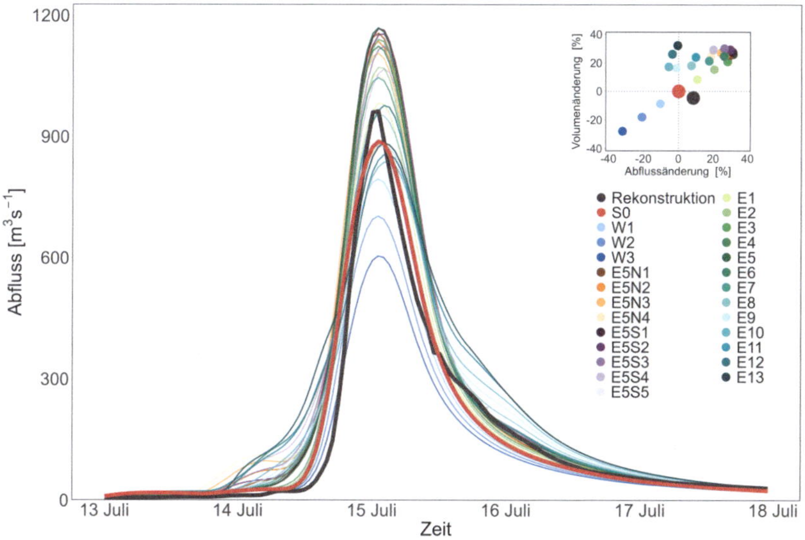

Abb. 4.7 Abflussganglinien von Kontrafaktualen am Pegel Altenahr, Referenzszenario (S0) und Abflussrekonstruktion nach Berkler et al. (2022). Die eingefügte Grafik zeigt die Abfluss- und Volumenänderungen für die 24 Kontrafaktuale im Vergleich zum Referenzszenario S0. (Verändert nach Vorogushyn et al. 2025)

Mit diesen Szenarien wurde das hydrologische Modell mHM für das Einzugsgebiet der Ahr angetrieben.

Die Abflussganglinien am Pegel Müsch sowie an den seitlichen Zuflüssen wurden in Wasserstände umgerechnet und als Antrieb für das Modell RIM2D zur Abschätzung der kontrafaktischen Überflutungen auf der Strecke bis zur Ahrmündung verwendet. Abb. 4.7 zeigt die simulierten kontrafaktischen Abflussganglinien am Pegel Altenahr sowie die prozentualen Änderungen der Abflussscheitel und Abflussvolumina im Vergleich zu S0. Bei einer Verschiebung des Niederschlags um ca. 25 km nach Osten (E5) wird eine maximale Zunahme der Abflussspitze um 32 % bei gleichzeitiger Zunahme des Abflussvolumens um 26 % simuliert (Vorogushyn et al. 2025). Die Verlagerung der Niederschläge nach Westen (W1–W3) verringert die Abflussscheitel und Abflussvolumina. Die Betrachtung weiterer Pegel zeigt, dass einige Zuflüsse noch stärker auf die Niederschlagsverlagerung reagieren als die Pegel am Hauptstrom (Abb. 4.8). Insbesondere der Pegel Denn am Staffelerbach, der im östlichen Teil des Einzugsgebiets der Ahr liegt, zeigt eine Abflusszunahme von über 150 % bei einer Verschiebung des Regenfeldes um ca. 50 km nach Osten. Dieser Teil des Ahreinzugsgebietes war beim Hochwasser 2021 vergleichsweise gering von Niederschlägen betroffen.

Die Ergebnisse der hydrodynamischen Modellierung mit RIM2D zeigen im ungünstigsten Szenario E3 für den Bereich Altenahr eine Zunahme der mittleren maximalen Überflutungstiefen um ca. 1,25 m (Abb. 4.9). Die maximale Zunahme beträgt sogar 1,75 m. Im weiteren Flussverlauf im Bereich Bad Neuenahr-Ahrweiler sind im Szenario E3 maximale Wassertiefenänderungen von immerhin noch 0,19 bzw. 0,44 m zu erwarten (Abb. 4.9). Hier kommt zum einen die Abflachung der Hochwasserwelle und zum anderen das deutlich flachere Gelände zum Tragen. Die Änderungen der Überflutungsflächen betragen für Altenahr 4,2 % und für Bad Neuenahr-Ahrweiler 3,5 % im E3-Szenario. Bei einer „Wolkenverschiebung“ um ca. 15 km nach Westen (W3) werden in beiden Be-

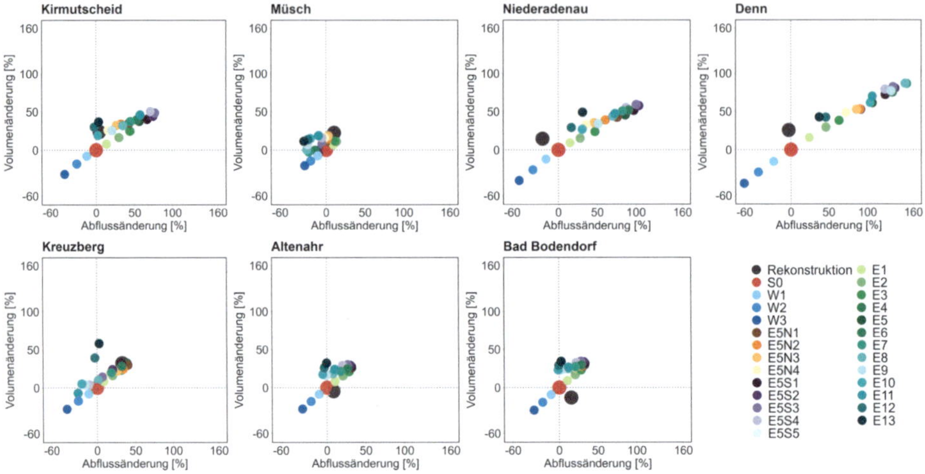

Abb. 4.8 Prozentuale Abfluss- und Volumenänderungen der 24 Kontrafaktuale im Vergleich zum Referenzszenario (S0) an sieben Pegeln im Einzugsgebiet der Ahr. (Verändert nach Vorogushyn et al. 2025)

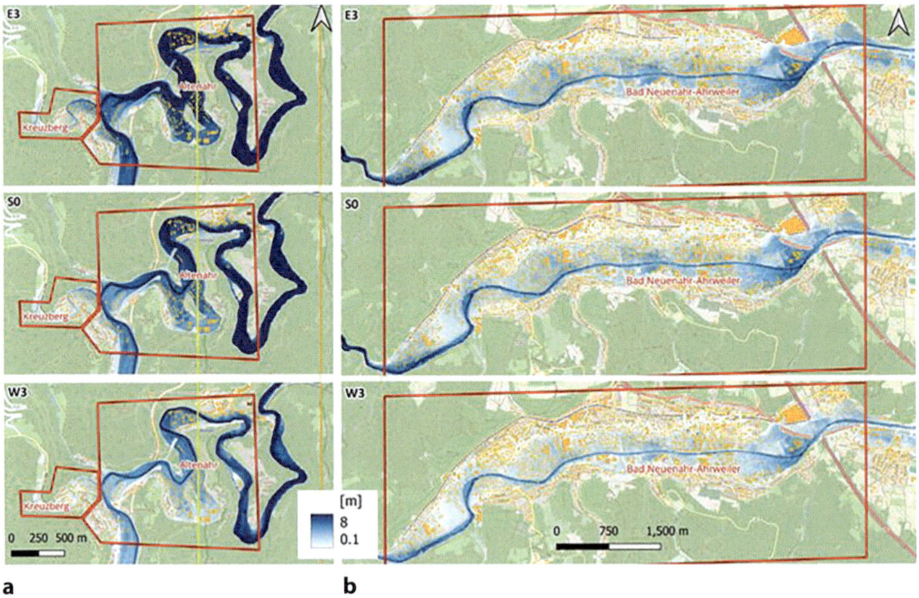

	W3 – S0			E3 – S0		
	Δ mittl. Wst. [m]	Δ max. Wst. [m]	Δ A [%]	Δ mittl. Wst. [m]	Δ max. Wst. [m]	Δ A [%]
Altenahr	-1,50	-2,09	-5,7	+1,25	+1,75	+4,2
Bad Neuenahr-Ahrweiler	-0,48	-1,07	-11,4	+0,19	+0,44	+3,5

Abb. 4.9 Überflutungsflächen und maximale Wassertiefen für drei Szenarien (W3, S0, E3) im Bereich (**a**) Altenahr und (**b**) Bad Neuenahr-Ahrweiler. Die quantitativen Änderungen der Überflutungsflächen (Δ A), der mittleren maximalen Wassertiefe (Δ mittl. Wst.) und der maximalen Wassertiefe (Δ max. Wst.) bei der Verschiebung ca. 15 km nach Osten und Westen sind in der Tabelle zusammengestellt. (Verändert nach Vorogushyn et al. 2025)

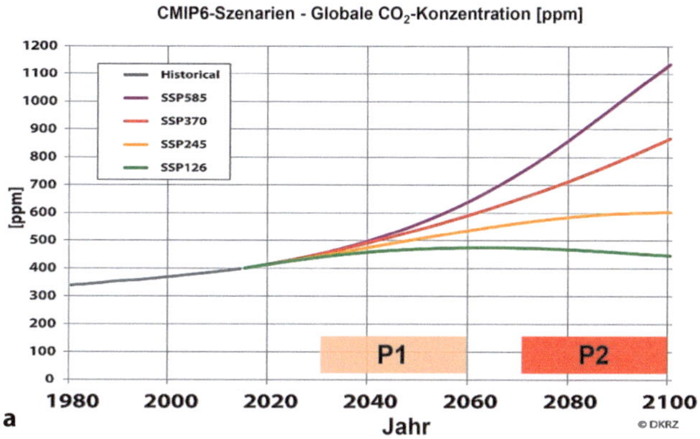

a

	Modellierungszentrum	GCM	Realisierung
1	Canadian Centre for Climate Modelling and Analysis	CanESM5	r1i1p1f1
2	National Center for Atmospheric Research	CESM2	r1i1p1f1
3	Centre National de Recherches Météorologiques and Centre Européen de Recherche et de Formation Avancée	CNRM-CM6-1	r1i1p1f2
4	Geophysical Fluid Dynamics Laboratory	GFDL-ESM4	r1i1p1f1
5	Met Office Hadley Centre and Natural Environment Research Council	UKESM1-0-LL	r1i1p1f1
6	Institute for Numerical Mathematics, Russian Academy of Sciences	INM-CM5-0	r1i1p1f1
7	Institut Pierre-Simon Laplace	IPSL-CM6A-LR	r1i1p1f1
8	Max-Planck-Institut für Meteorologie	MPI-ESM1-2-HR	r1i1p1f1
9	Norwegian Climate Center	MRI-ESM2-0	r1i1p1f1

b

Abb. 4.10 **a** Entwicklung der globalen CO_2-Konzentrationen für verschiedene Emissionsszenarien. **b** Ausgewählte Klimamodelle und ihre entsprechenden Realisationen. (**a** DKRZ 2024; **b** Nguyen et al. 2024)

reichen vergleichbare Abnahmen der maximalen Wassertiefen und Überflutungsflächen simuliert.

4.3.3 Hochwasser unter Klimawandel

Die entwickelten Methoden wurden angewandt, um Hochwasserprojektionen für das Einzugsgebiet der Ahr unter Berücksichtigung des Klimawandels abzuleiten. Dabei wurden vier Szenarien untersucht, die sich aus der Kombination von zwei zukünftigen Zeiträumen und zwei Emissionsszenarien ergeben: die nahe Zukunft (2031–2060) und die ferne Zukunft (2071–2100) sowie die Szenarien SSP245 und SSP585 (DKRZ 2024). Beide Szenarien gehören zu den in der Klimaforschung gängigen Standards für sozioökonomische und klimatische Entwicklungen (Abb. 4.10a). SSP245 repräsentiert ein mittleres Szenario mit moderaten Klimaschutzmaßnahmen, während SSP585 von sehr hohen Treibhausgasemissionen ausgeht und das obere Extrem der Standardszenarien darstellt. Es basiert auf einer fossilen Entwicklung, die von einer zunehmend vernetzten globalen Wirtschaft, einem energieintensiven Lebensstil und starkem Wirtschaftswachstum geprägt ist.

Die Zirkulationsmuster und die großräumige mittlere Temperatur wurden aus neun Klimamodellen für die vier Szenarien entnommen (Abb. 4.10b). Die Auswahl dieser

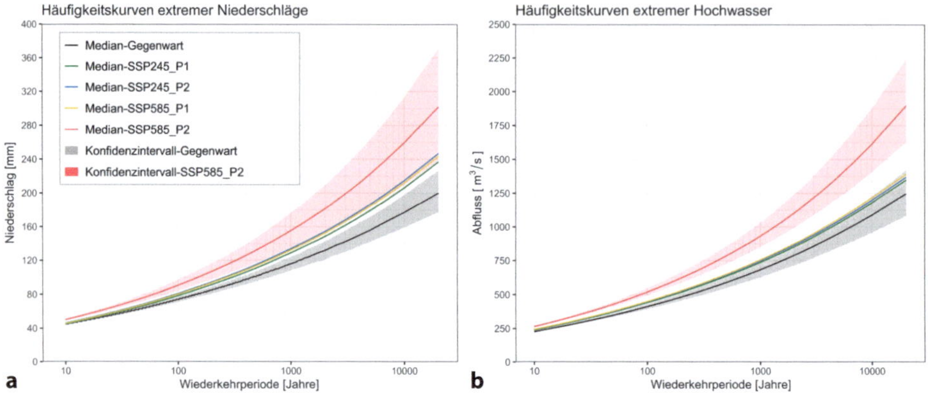

Abb. 4.11 Projektionen unter Klimawandel. **a** Häufigkeitskurven für tägliche Extremniederschläge, gemittelt über das Einzugsgebiet der Ahr. **b** Hochwasserhäufigkeitskurven für den Pegel Altenahr. Der Median ist der mittlere Wert, der sich aus den neun Klimamodellen und der Stichprobenunsicherheit ergibt. P1 steht für 2031–2060 und P2 steht für 2071–2100. Um die Abbildungen nicht zu überfrachten, werden nur die Unsicherheitsbereiche für die Gegenwart und für das Extremszenario SSP585 gezeigt. (**a**, **b** eigene Darstellung)

Modelle erfolgte auf Basis quantitativer und qualitativer Kriterien. Ziel war es, eine Modellauswahl zu erhalten, die das europäische Klima realistisch abbildet und dabei keine übermäßigen Abhängigkeiten zwischen den Modellen aufweist. Detaillierte Informationen zur Modellauswahl sind in Nguyen et al. (2024) zu finden. Mithilfe des Wettergenerators nsRWG wurden jeweils 100 Realisationen täglicher Niederschlags- und Temperaturwerte über Zeiträume von jeweils 30 Jahren erstellt (100 Realisationen × 30 Jahre × 9 Klimamodelle = 27.000 Jahre). Diese Daten wurden anschließend mit dem Verfahren von Guan et al. (2023) auf stündliche Werte herunterskaliert und als Eingangsgröße für das hydrologische Modell mHM verwendet.

Abb. 4.11 zeigt die Projektionen von Extremniederschlägen und Hochwasser unter dem Einfluss des Klimawandels. Der Fokus liegt dabei auf den täglichen Niederschlagsmengen, gemittelt über das Einzugsgebiet der Ahr, sowie auf den Hochwasserscheitelwerten am Pegel Altenahr. Für alle vier betrachteten Szenarien nehmen sowohl die Extremniederschläge als auch die Hochwasserabflüsse im Einzugsgebiet der Ahr unter Klimawandel zu. Allerdings fallen die Zunahmen in drei der Szenarien moderat aus. So steigt beispielsweise der 100-jährliche tägliche Extremniederschlag von derzeit 74 mm auf 81 mm im Szenario SSP245 für die ferne Zukunft. Der 1000-jährliche Extremniederschlag erhöht sich von 116 mm auf 134 mm. Im Worst-Case-Szenario SSP585 hingegen sind starke Zuwächse zu beobachten, mit einem Anstieg von 23 % und 34 % für das 100- bzw. 1000-jährliche Ereignis. Ähnliche Trends zeigen sich bei den Hochwasserabflüssen, die in drei der vier Szenarien nur moderat ansteigen. Im Einklang mit den Veränderungen der Extremniederschläge weist das Szenario SSP585 einen deutlichen Anstieg des 100- und 1000-jährlichen Scheitelabflusses um 43 % und 36 % auf.

4.4 Schlussfolgerungen

Extreme Hochwasserereignisse stellen besondere Herausforderungen an die Modellierung, da Extremereignisse häufig von anderen Prozessen dominiert werden als kleinere Hochwasser und vergleichsweise wenige Beobachtungen vorliegen. Im Rahmen des KAHR-Teilprojektes „Hochwassermodellierung Rheinland-Pfalz und NRW" wurden innovative Ansätze zur Hochwassermodellierung entwickelt und die Ergebnisse den regionalen Akteuren zur Verfügung gestellt.

In diesem Kapitel werden ausgewählte Modellkomponenten und Ergebnisse für das Einzugsgebiet der Ahr vorgestellt:

- Die entwickelte Modellkette erlaubt die Ableitung von Extremhochwassern und deren Eintrittswahrscheinlichkeiten durch Langzeitsimulationen. Die Ergebnisse stimmen mit einer alternativen Methode (Extremwertstatistik unter Verwendung beobachteter und historischer Abflüsse) überein. Diese Übereinstimmung zweier unterschiedlicher Ansätze, die zudem auf unterschiedlichen Daten beruhen, erhöht die Glaubwürdigkeit der Ergebnisse und bildet damit eine solide Grundlage für die Hochwasserbemessung.
- Die räumliche Verschiebung des Niederschlagsfeldes im Juli 2021 führt zu Hochwasserabflüssen, die deutlich über den tatsächlich beobachteten, bereits katastrophalen Werten liegen. Eine plausible Verschiebung des Niederschlagsfeldes um 15–25 km nach Osten erhöht die Abflussscheitel am Pegel Altenahr um 32 % und an den Nebenflüssen um bis zu 160 %. Damit verbunden sind deutlich größere Überflutungsflächen und Wassertiefen, z. B. im Bereich Altenahr eine um 1,75 m höhere maximale Wassertiefe. Das Einzugsgebiet der Ahr hätte also noch stärker betroffen sein können.
- Aufgrund des Klimawandels ist zukünftig mit intensiveren Extremniederschlägen und höheren Hochwasserscheiteln zu rechnen. Für drei der vier betrachteten Szenarien fallen diese Zunahmen moderat aus. Im Worst-Case-Szenario (SSP585, „fossile Entwicklung") sind die Zunahmen jedoch besorgniserregend, z. B. mit einer Zunahme des 100-jährlichen Abflusses um 43 % am Pegel Altenahr.

Die im Rahmen von KAHR entwickelten Modelle sind eine hervorragende Grundlage für die Bemessung von Hochwasserschutzmaßnahmen, die Erstellung von Hochwassergefahrenkarten, die Risikoanalyse und die Untersuchung der Wirkung verschiedener Maßnahmen zur Risikominderung. Sie können aber auch zur Verbesserung des Katastrophenmanagements eingesetzt werden. Die Integration des Überflutungsmodells RIM2D in die operationelle Hochwasservorhersage würde es ermöglichen, nicht nur die Wasserstände an einzelnen Flusspegeln vorherzusagen, sondern auch die zu erwartenden Auswirkungen in der Fläche im Voraus zu simulieren (Apel et al. 2022; Najafi et al. 2024).

Zusammenfassung

- Die Modelle zeigen, dass schon eine leichte Verschiebung des Regenfeldes nach Osten die Abflussspitzen in einigen Zuflüssen der Ahr um bis zu 160 % erhöht hätte.
- Aufgrund des Klimawandels ist künftig mit höheren Extremniederschlägen und Hochwassern zu rechnen. Für drei der vier betrachteten Szenarien fallen diese Zunahmen moderat aus – im Worst-Case-Szenario sind die Zunahmen jedoch besorgniserregend, z. B. mit einer Zunahme des 100-jährlichen Abflusses um 43 % am Pegel Altenahr.
- Die Modelle sind eine wichtige Grundlage für die Bemessung von Schutzmaßnahmen und für ein verbessertes Katastrophenmanagement.

Danksagung Die Autorinnen und Autoren von ▶ Kap. 4 bedanken sich beim Bundesministerium für Bildung und Forschung (BMBF) und beim DLR Projektträger für die Unterstützung und Förderung des KAHR-Projekts. Diese Forschung wurde vom BMBF unter dem Förderkennzeichen 01LR2102A-M finanziert.

Literatur

Apel H et al (2022) Brief communication: impact forecasting could substantially improve the emergency management of deadly floods: case study july 2021 floods in Germany. Nat Hazards Earth Syst Sci 22(9):3005–3014. https://doi.org/10.5194/nhess-22-3005-2022

Apel H, Benisch J, Helm B, Vorogushyn S, Merz B (2024) Fast urban inundation simulation with RIM2D for flood risk assessment and forecasting. Front Water. https://doi.org/10.3389/frwa.2024.1310182

Berkler SB, Bettmann T, Böhm M, Thomas N, Gerlach N, Hengst A, Henrichs Y, Heppelmann T, Iber C, Johst M, Lehmann H, Stickel S, van der Heijden S, Wallisch S (2022) Bericht: Hochwasser im Juli 2021. Landesamt für Umwelt Rheinland-Pfalz, Mainz. https://lfu.rlp.de/fileadmin/lfu/Startseitenbeitraege/2022/Hochwasser_im_Juli2021_1_.pdf

Blöschl G, Hall J, Viglione A, Perdigão RAP, Parajka J, Merz B, Živković N (2019) Changing climate both increases and decreases European river floods. Nature 573(7772):108–111. https://doi.org/10.1038/s41586-019-1495-6

Bronstert A, Apel H, Bormann H, Bürger G, Haberlandt U, Hannappel A, Iber C (2024) Hochwasser und Sturzfluten an Flüssen in Deutschland. In: Klimawandel in Deutschland: Entwicklung, Folgen, Risiken und Perspektiven. Springer, Berlin, Heidelberg, S 109–130

Cornes RC, van der Schrier G, van den Besselaar EJM, Jones PD (2018) An ensemble version of the E-OBS temperature and precipitation data sets. JGR Atmospheres 123:9391–9675. https://doi.org/10.1029/2017JD028200

DKRZ (2024) Die SSP-Szenarien. http://www.dkrz.de/de/kommunikation/klimasimulationen/cmip6-de/die-ssp-szenarien?. Zugegriffen: 12. Sept. 2024

DWA (2024) Entwurf des Merkblatts DWA-M 552 „Stochastische und deterministische Wege zur Ermittlung von Hochwasserwahrscheinlichkeiten". https://de.dwa.de/de/regelwerk-news-volltext/entwurf-merkblatt-dwa-m-552-ermittlung-von-hochwasserwahrscheinlichkeiten.html. Zugegriffen: 8. Sept. 2024

DWD (2024) HYRAS – Hydrometeorologische Rasterdaten. http://www.dwd.de/DE/leistungen/hyras/hyras.html. Zugegriffen: 10. Sept. 2024

Falter D, Schröter K, Dung NV, Vorogushyn S, Kreibich H, Hundecha Y, Apel H, Merz B (2015) Spatially coherent flood risk assessment based on long-term continuous simulation with a coupled model chain. J Hydrol Reg Stud 524:182–193. https://doi.org/10.1016/j.jhydrol.2015.02.021

Guan X, Nissen K, Nguyen VD, Merz B, Winter B, Vorogushyn S (2023) Multisite temporal rainfall disaggregation using methods of fragments conditioned on circulation patterns. J Hydrol Reg Stud 621:129640. https://doi.org/10.1016/j.jhydrol.2023.129640

Haberlandt U, Radtke I (2014) Hydrological model calibration for derived flood frequency analysis using stochastic rainfall and probability distributions of peak flows. Hydrol Earth Syst Sci 18(1):353–365. https://doi.org/10.5194/hess-18-353-2014

Jong B-T, Delworth TL, Cooke WF, Tseng K-C, Murakami H (2023) Increases in extreme precipitation over the Northeast United States using high-resolution climate model simulations. npj Clim Atmos Sci 6:18. https://doi.org/10.1038/s41612-023-00347-w

Koks EE, van Ginkel KCH, van Marle MJE, Lemnitzer A (2022) Brief communication: critical infrastructure impacts of the 2021 mid-July western European flood event. Nat Hazards Earth Syst Sci 22:3831–3838

Kumar R, Samaniego L, Attinger S (2013) Implications of distributed hydrologic model parameterization on water fluxes at multiple scales and locations. Water Resour Res 49(1):360–379. https://doi.org/10.1029/2012WR012195

Merz B, Basso S, Fischer S, Lun D, Blöschl G, Merz R, Schumann A (2022) Understanding heavy tails of flood peak distributions. Water Resour Res 58(6):e2021WR030506. https://doi.org/10.1029/2021WR030506

Merz B, Nguyen D, Guse B, Kreibich H, Sairam N, Apel H, Farrag M, Han L, Vorogushyn S (2024a) Kontinuierliche, räumlich verteilte Langzeitsimulation zur Abschätzung des Hochwasserrisikos für Deutschland. Hydrol Wasserbewirtsch 68(4):220–233. https://doi.org/10.5675/HyWa_2024.4_2

Merz B, Nguyen VD, Guse B, Han L, Guan X, Rakovec O, Vorogushyn S (2024b) Spatial counterfactuals to explore disastrous flooding. Environ Res Lett 19(4):44022. https://doi.org/10.1088/1748-9326/ad22b9

Montanari A, Merz B, Blöschl G (2024) HESS Opinions: the sword of Damocles of the impossible flood. Hydrol Earth Syst Sci 28(12):2603–2615. https://doi.org/10.5194/hess-28-2603-2024

Najafi H, Shrestha PK, Rakovec O, Apel H, Vorogushyn S, Kumar R, Samaniego L (2024) High-resolution impact-based early warning system for riverine flooding. Nat Commun 15(1):3726. https://doi.org/10.1038/s41467-024-48065-y

Nguyen VD, Vorogushyn S, Nissen K, Brunner L, Merz B (2024) A non-stationary climate-informed weather generator for assessing of future flood risks. ESS Open Archive. https://doi.org/10.22541/essoar.171742646.65272251/v1 (in review)

Samaniego L, Kumar R, Attinger S (2010) Multiscale parameter regionalization of a grid-based hydrologic model at the mesoscale. Water Resour Res 46:W5523. https://doi.org/10.1029/2008WR007327

Siegrist M, Gutscher H (2008) Natural hazards and motivation for mitigation behavior: people cannot predict the affect evoked by a severe flood. Risk Anal 28:771–778

Voit P, Heistermann M (2024) A downward-counterfactual analysis of flash floods in Germany. Nat Hazards Earth Syst Sci 24(6):2147–2164. https://doi.org/10.5194/nhess-24-2147-2024

Vorogushyn S, Apel H, Kemter M, Thieken AH (2022) Analyse der Hochwassergefahrdung im Ahrtal unter Berücksichtigung historischer Hochwasser. Hydrol Wasserbewirtsch 66(5):244–254. https://doi.org/10.5675/HyWa_2021.5_2

Vorogushyn S, Guse B, Macdonald E, Wietzke L, Merz B (2023) Wie entstehen überraschende Extremhochwasser? Einflussfaktoren der tail heaviness bei Hochwasserverteilungen. Hydrol Wasserbewirtsch 47(5):264–274

Vorogushyn S, Han L, Apel H, Nguyen VD, Guse B, Guan X, Rakovec O, Najafi H, Samaniego L, Merz B (2025) It could have been much worse: spatial counterfactuals of the July 2021 flood in the Ahr valley, Germany. Nat Hazards Earth Syst Sci 25(6):2007–2029. https://doi.org/10.5194/nhess-25-2007-2025

Wasserportal Rheinland-Pfalz (2024) Jährlichkeiten Abfluss für den Pegel Altenahr. https://geodaten-wasser.rlp-umwelt.de/wasserstand/2718040300/jaehrlichkeiten. Zugegriffen: 14. Jan. 2025

Woo G (2019) Downward counterfactual search for extreme events. Front Earth Sci 7:340

Bevölkerungsbefragungen und Experteninterviews in stark betroffenen Gebieten koordinieren

Philip Bubeck, Zora Reckhaus, Christian Kuhlicke, Lisa Dillenardt und Annegret Thieken

J. Birkmann et al. (Hrsg.), *Die Flutkatastrophe 2021*, Sustainable Development Goals (SDG) – Umsetzung in Praxis, Lehre und Entscheidungsprozessen,
https://doi.org/10.1007/978-3-658-48688-4_5

5.1 Einleitung: Warum sind Befragungen notwendig?

Der Umgang mit (potenziellen) Hochwasserereignissen in Deutschland hat in den letzten Jahrzehnten einen grundlegenden Wandel erfahren. Bis in die 1980er Jahre hinein vertraute man darauf, Überflutungen mit Hilfe baulicher Schutzmaßnahmen ingenieurtechnisch bewältigen zu können (Becker et al. 2007). Die schweren Hochwasserereignisse entlang von Rhein und Oder in den 1990er Jahren und das Elbehochwasser im August 2002 zeigten jedoch Grenzen dieser Strategie auf. Angesichts von zahlreichen Deichbrüchen und hohen Schäden wurde der traditionelle Ansatz des *Hochwasserschutzes* zunehmend durch integrierte Risikomanagement- und Vorsorgekonzepte abgelöst. Letztere berücksichtigen, dass Deiche versagen können und ergänzen daher technische Hochwasserschutzmaßnahmen mit zusätzlichen Maßnahmen, die darauf abzielen, das Restrisiko von Überflutungen zu mindern (Kreibich et al. 2015; Kuhlicke et al. 2020; Hudson und Thieken 2022). So verabschiedete 1995 die Bund/Länder-Arbeitsgemeinschaft Wasser (LAWA) neue Leitprinzipien für das Hochwasserrisikomanagement. Die LAWA ist das Arbeitsgremium der Umweltministerkonferenz (UMK) zur Diskussion und Koordinierung von Fragen der Wasserwirtschaft und des Wasserrechts in den Bundesländern, dem seit 2005 auch der Bund angehört. In diesen Leitlinien wurde die Bedeutung von komplementären Risikomanagementstrategien einschließlich Raumplanung, schadensminderndem Verhalten der Bevölkerung und Versicherungslösungen hervorgehoben (LAWA 1995). Viele dieser Aspekte wurden in die Wasser-, Raumordnungs- und Baugesetze der Länder und des Bundes aufgenommen und finden sich auch in der europäischen Hochwasserrisikomanagementrichtlinie (2007/60/EG) wieder.

Dieser Wandel seit den 1990er Jahren stellt sowohl wissenschaftliche Ansätze zur Analyse und Bewertung der vielschichtigen Aspekte des Hochwasserrisikos als auch dessen integriertes Management vor erweiterte und neue Herausforderungen (Bubeck et al. 2016). Lag der Fokus lange allein auf Seiten der Hochwassergefährdung, gilt es nun, auch die Exposition, Vulnerabilitäten und individuelle Vorsorgestrategien (potenziell) Betroffener und weiterer Akteure (z. B. Kommunen, Unternehmen, Betreiber von [sozialen oder kritischen] Infrastrukturen, Versicherungsgesellschaften) in den Blick zu nehmen (Merz et al. 2014). So sollen die Bevölkerung und Unternehmen in gefährdeten Gebieten einerseits durch geeignete und zumutbare Vorsorgemaßnahmen, wie Rückstauklappen, die Sicherung oder den Austausch von Ölheizungen und hochwasserangepasste Bauweisen und Gebäudenutzungen, einen aktiven Beitrag zur Schadenreduzierung leisten. Seit 2009 ist diese Pflicht zur Eigenvorsorge im Wasserhaushaltsgesetz (WHG § 5) verankert. Für die Verbesserung und Bewertung des Managements bedarf es auf staatlicher und wissenschaftlicher Seite andererseits Erkenntnisse darüber, ob und wie gut bestehende Frühwarnsysteme funktionieren, wie Menschen und Unternehmen Hochwasserereignisse bewältigen, welche Vorsorgemaßnahmen akzeptiert bzw. bevorzugt werden und vor welchen Herausforderungen betroffene Kommunen oder Infrastrukturbetreiber stehen.

Derartige Daten, die für die Umsetzung integrierter Risikomanagementstrategien entscheidend sind, werden nicht mit etablierten Messmethoden und Dokumentationen erfasst, sondern in Deutschland anlassbezogen mit Hilfe von Bevölkerungsbefragungen und Experteninterviews erhoben. Der Mehrwert solcher Daten zeigt sich in der wissenschaftlichen Literatur, die die oben genannte Vielzahl der Themen widerspiegelt und unter anderem das Vorsorgeverhalten der Bevölkerung und die Wirksamkeit schadensreduzierender Maßnahmen (z. B. Kreibich et al. 2005; Hudson et al. 2014; Grothmann und Reusswig 2006), die Risikowahrnehmungen der Bevölkerung (Osberghaus und Hinrichs

2021), Konflikte im Hochwasserrisikomanagement (Kuhlicke et al. 2016), die Rahmung von Unsicherheiten und Nichtwissen und Anpassungsprozessen (Kuklicke und Demeritt 2016), das Funktionieren der Frühwarnung (Thieken et al. 2023), die physische und mentale Erholung nach Schadensereignissen (Kienzler et al. 2015; Bubeck und Thieken 2018; Köhler et al. 2023) und die Akzeptanz von Maßnahmen und des Wiederaufbaus (Truedinger et al. 2023) umfasst.

5.2 Wie wurde in KAHR befragt?

Um die Situation der Bevölkerung, von Unternehmen und sozialen Infrastrukturen nach dem Hochwasser im Juli 2021 und deren Perspektiven auf den Wiederaufbau in das KAHR-Projekt einfließen zu lassen, wurden mehrere Befragungen zu unterschiedlichen Themen durchgeführt. Hierzu zählen (siehe auch ▶ Kap. 9, ◘ Tab. 9.2, und ▶ Kap. 19, ◘ Tab. 19.1):

1. Eine postalisch beworbene Online-Befragung[1] im Landkreis Ahrweiler im Juni und Juli 2022, d. h. ein Jahr nach dem Ereignis, zu den Themen Auswirkungen des Hochwassers, Risikowahrnehmung, mentale Gesundheit und potenzielle Einflussfaktoren, soziale Vulnerabilitäten, Stand des Wiederaufbaus und Meinungen zum Hochwasserrisikomanagement (vgl. Truedinger et al. 2023; Zenker et al. 2024).
2. Eine postalisch beworbene Online-Befragung[1] im Landkreis Ahrweiler im November und Dezember 2022, d. h. anderthalb Jahre nach dem Ereignis, zu den Themen mentale Gesundheit und potenzielle Einflussfaktoren, strukturelle und finanzielle Schäden an Gebäuden und Inventar sowie potenzielle Einflussfaktoren, Stand des Wiederaufbaus, Risikowahrnehmung und Eigenvorsorge.
3. Zu denselben Themen wurde parallel eine über Meta beworbene Befragung in Nordrhein-Westfalen (NRW) durchgeführt (vgl. Dillenardt und Thieken 2024; Sairam et al. 2024).
4. Befragung von Besuchern des Infomobils des HochwasserKompetenzCentrums e. V. (HKC) zur Risikowahrnehmung sowie zu Informationsbedarfen und Vorsorgeverhalten (vgl. ▶ Kap. 19 und 20).
5. Zwischen November 2022 bis Januar 2023 wurden hochwassergeschädigten Unternehmen in NRW und Rheinland-Pfalz (RP) telefonisch zu Sachschäden, Betriebsunterbrechungen und -einschränkungen sowie zur Wiederherstellung und Vorsorge befragt (vgl. ▶ Kap. 8).
6. Personen, die an den Bevölkerungsbefragungen im Jahr 2022 teilgenommen haben, wurden abschließend gefragt, ob sie an einer Wiederholungsbefragung teilnehmen würden. Diese Befragung wurde im Mai und Juni 2024 online durchgeführt, um Entwicklungen des Wiederaufbaus und der mentalen/gesundheitlichen Situation der Betroffenen zu erfassen.
7. Experteninterviews mit Akteuren der sozialen Infrastruktur in Bad Neuenahr-Ahrweiler, Stolberg und Eschweiler. Im Rahmen einer Panelstudie wurden 21 Interviews im Jahr 2022 und 15 Interviews mit den gleichen Interviewpartnern im Jahr 2023 geführt. Zusätzlich wurden im Jahr 2023 noch drei weitere Interviews durchgeführt, um die Perspektiven aus dem Jahr 2022 zu ergänzen (vgl. ▶ Kap. 17).

---

1 Es bestand zudem die Möglichkeit, sich den Fragebogen zusenden zu lassen. Auf diese Weise wurde gewährleistet, dass auch Menschen ohne Online-Zugang an der Befragung teilnehmen konnten.

5

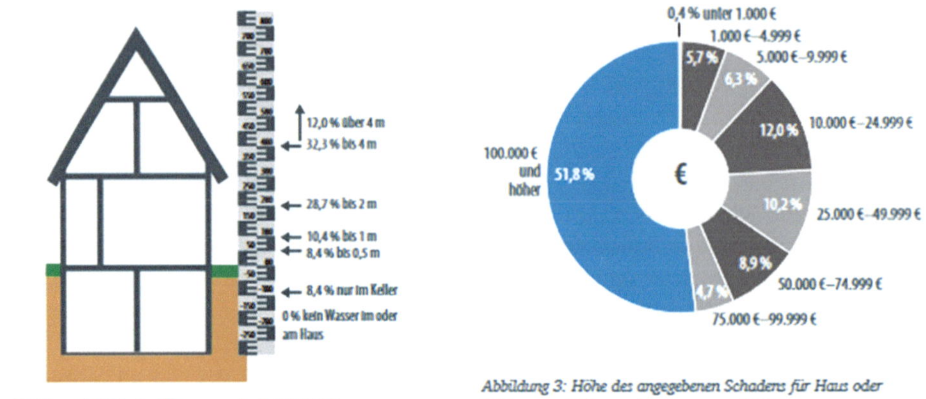

Die Folgen belasten Haushalte auch ein Jahr nach der Flut

Die Flut traf viele der befragten Haushalte sehr schwer. Die Hälfte der Haushalte erlitt finanzielle Schäden von mehr als 100.000 Euro. Bei nahezu jedem Haushalt stand das Wasser zumindest im Keller, bei 47,5 % stand das Wasser außen an der Hauswand bis zu 2 Meter hoch oder gar darüber (44,4 %). Etwas mehr als 40 % der Befragten mussten aufgrund der Zerstörungen ihre Häuser verlassen. Von ihnen hatten 15,2 % ein Jahr nach dem Ereignis immer noch nicht die Möglichkeit, in ihr Zuhause zurückzukehren. 90 % der Befragten gaben an, dass das geschädigte Gebäude bzw. der geschädigte Hausrat noch nicht vollständig repariert oder ersetzt werden konnte. Dies lag häufig am Fehlen von Handwerker:innen und/oder Material.

Abbildung 2: Höhe des Wasserstandes bei 498 Befragten.

Abbildung 3: Höhe des angegebenen Schadens für Haus oder Wohnung und Hausrat bei insgesamt 492 Befragten.

Abb. 5.1 Beispiel eines Flyers des KAHR-Projekts mit Ergebnissen einer Haushaltsbefragung. (Eigene Darstellung, ▶ https://hochwasser-kahr.de/images/pdf/FlyerIGUuIREUSBroschuere_KAHR_final.pdf)

Diese Befragungen lieferten wichtige Erkenntnisse zur Bewältigung des Ereignisses und zu Prioritäten des Wiederaufbaus (Truedinger et al. 2023; Birkmann et al. 2023), Schäden (vgl. ▶ Kap. 7 und 8), Anzeichen posttraumatischer Belastungsstörungen Betroffener (siehe ▶ Kap. 9; Zenker et al. 2024; Sairam et al. 2024) und privater Eigenvorsorge (siehe ▶ Kap. 19; Dillenardt und Thieken 2024). Die Ergebnisse flossen zudem in Informationsangebote für Kommunen und private Haushalte ein (siehe Abb. 5.1).[2]

5.3 Vorschläge für die Koordinierung von Befragungen in schwer betroffenen Regionen

Wie ▶ Kap. 3 aufgezeigt hat, handelt es sich beim Ereignis vom Juli 2021 um das Binnenhochwasser mit den schwersten Auswirkungen in Deutschland seit Jahrzehnten. Entsprechend groß war und ist die Aufmerksamkeit von Öffentlichkeit, Politik und Wissenschaft. Während Befragungen wichtige Erkenntnisse liefern (siehe ▶ Abschn. 5.2), können diese aber auch mit einem erheblichen zeitlichen Aufwand und emotionalen Belastungen für die Betroffenen einhergehen – und dies in Zeiten des Wiederaufbaus, die ohnehin mit großen Belastungen und Unsicherheiten verbunden sind. Bei Befragungen im Rah-

2 Weitere Flyer unter anderem zu den Haushaltsbefragungen finden sich auf der Webseite des KAHR-Projekts: ▶ https://www.hochwasser-kahr.de/index.php/de/informationen#info_06, zuletzt aufgerufen am 26.07.2024.

men des KAHR-Projekts wurden die Befragten daher explizit auf die Gefahr einer Re-Traumatisierung hingewiesen und an mehreren Stellen der Befragung auf mögliche Anlaufstellen der psycho-sozialen Hilfe, wie z. B. das Trauma Hilfe Zentrum, hingewiesen (siehe ▶ Kap. 9, Box 9.1). Eine bessere Versorgung könnte im Rahmen von Vor-Ort-Befragungen mit professioneller psychologischer Begleitung, z. B. in Zusammenarbeit mit Hilfsorganisationen (vgl. Apel und Coenen 2021), erreicht werden. Gleichzeitig begrüßen es viele Betroffene, dass sie nach ihren Erfahrungen und Einschätzungen befragt werden. Gespräche über das Erlebte können auch entlasten. Dies haben gerade auch die Gespräche vor Ort am Infomobil des HKC gezeigt (vgl. ▶ Kap. 20).

Im Laufe des KAHR-Projektes deutete sich aber auch eine Ermüdung der Bevölkerung an, d. h. eine sinkende Beteiligung an Befragungen. Neben dem KAHR-Projekt zog das Ereignis andere Wissenschaftler:innen und Studierende an, die für ihre Forschungsfragen und Abschlussarbeiten eigene Befragungen durchführten. Insbesondere Menschen in stark geschädigten Regionen wurden mehrfach kontaktiert. Diese zusätzliche Belastung von Betroffenen berührt auch einen ethischen Aspekt: Hochwassergeschädigte Haushalte und Unternehmen sind im Vergleich zu Personen in anderen Gebieten durch die Hochwasserauswirkungen benachteiligt; durch die Teilnahme an hochwasserbezogenen Befragungen, Diskussionsgruppen, Veranstaltungen etc. kann eine weitere zeitliche Belastung und Benachteiligung entstehen, die andere Bevölkerungsgruppen nicht im selben Maße haben. Um dem entgegenzuwirken, könnten zukünftig Formen der Aufwandsentschädigung entwickelt werden.

Hinweis für Entscheidungsprozesse

Die Erfahrungen des KAHR-Projektes zeigen, dass es daher wichtig ist, Befragungen der Bevölkerung und Experteninterviews inhaltlich und zeitlich sowie zwischen verschiedenen Akteuren und Projekten zu koordinieren, um die Beanspruchung Betroffener – insbesondere in sehr stark betroffenen Gebieten – möglichst gering zu halten. So wurden Fragen und Informationsbedarfe unterschiedlicher Projektpartner in gemeinsame Fragebögen und Interviewleitfäden integriert, um Doppelbefragungen zu vermeiden. Zudem sollte darauf geachtet werden, dass die durchgeführten Befragungen den geltenden wissenschaftlichen, ethischen und datenschutzrechtlichen Standards entsprechen, um so das Erkenntnisinteresse einerseits und die Belastung der Befragten andererseits auszugleichen.

Um Befragungen inhaltlich und zeitlich zu koordinieren, haben sich die im KAHR-Projekt angelegten Projektbüros in NRW und RP als sehr hilfreich herausgestellt. Die Aufgabe der Projektbüros bestand darin, als Schnittstelle zwischen Wissenschaft, Politik und regionalen Akteuren zu fungieren. Die regionale Verankerung der Projektbüros ermöglichte es, die Bedarfe lokaler Stakeholder direkt zu erfassen und passgenaue Beteiligungsformate und Experten-Veranstaltungen zu konzipieren. Zu den Aufgaben zählte zudem der Aufbau und die Pflege eines projektinternen Wissens- und Informationsmanagements und der Transfer von aktuellen Forschungsergebnissen der Verbundpartner in Wiederaufbauprozesse. Im Rahmen dieser Funktion konnten die Projektbüros dazu beitragen, Informationen zu Befragungen zu bündeln und Dopplungen zu vermeiden.

Um wissenschaftliche Mindeststandards zu gewährleisten, könnten in Zukunft die Ethikkommissionen und Datenschutzbeauftragten der Universitäten eine wichtigere Rolle einnehmen. Diese Kommissionen sind inzwischen standardmäßig an Universitäten etabliert, sodass in betroffenen Regionen keine zusätzlichen Stellen oder Gremien geschaffen werden müssten, sondern auf bestehenden Strukturen aufgebaut werden könnte. Von diesen Stellen geprüfte und freigegebene Befragungen weisen ein wichtiges Gütesiegel auf, das den Betroffenen auch kommuniziert werden sollte, damit sie die Qualität einer Befragung einschätzen können.

Basierend auf den Erfahrungen des KAHR-Projektes empfehlen wir, regionale Projektbüros auch zukünftig in größeren Verbundvorhaben zu schweren Ereignissen zu verankern, die eine solche Schnittstellenfunktion wahrnehmen können. Die Projektbüros könnten z. B. verstärkt darauf achten, dass für Befragungen, die im Namen von durch öffentliche Forschungsgelder finanzierten wissenschaftlichen Vorhaben durchgeführt werden, die Zustimmung einer Ethikkommission vorliegt. Unter einem solchen „Siegel" verbreitete Befragungen würden Mindestanforderungen entsprechen und könnten somit auf ein erhöhtes Vertrauen seitens der Bevölkerung und eine gute Beteiligung hoffen. Zu berücksichtigen ist allerdings, dass dies mit einem erheblichen zeitlichen Aufwand verbunden sein kann, der eine sehr schnelle und zeitkritische Befragung entgegensteht. Dies gilt es, im Einzelfall abzuwägen. Der Rat für Sozial- und Wirtschaftsdaten (RatSWD) empfiehlt generell die Einrichtung einer wissenschaftlichen Koordinierungsstelle, die im Fall von Krisen als zentrale Anlaufstelle für gemeinsame Aktivitäten bereitsteht (RatSWD 2024).

Um die Nach- und Doppelnutzung von Befragungsdaten zu ermöglichen, sollten Datensätze nach Projektende im Sinne der Leitlinien für die Beschreibung, Speicherung und Veröffentlichung wissenschaftlicher Daten (FAIR-Prinzipien) in Repositorien veröffentlicht werden (Wilkinson et al. 2016). Das Akronym FAIR wird wie folgt aufgelöst: **F**indable – auffindbar, **A**ccessible – zugänglich, **I**nteroperable – interoperabel, **R**eusable – wiederverwendbar. Belange des Daten- und Persönlichkeitsschutzes sind hierbei zu beachten und unterstreichen nochmals die Vorteile einer Vorab-Prüfung von Befragungen durch eine Ethikkommission.

5.4 Weitere Empfehlungen für ein evidenzbasiertes Hochwasserrisikomanagement

Eine grundsätzliche Schwierigkeit bei Befragungen in Deutschland, und insbesondere im Themenkomplex Hochwasser, ist die Erstellung einer repräsentativen Stichprobe der zu untersuchenden Grundgesamtheit. Hier wurden im Rahmen des KAHR-Projekts unterschiedliche Wege gegangen, auch in Abhängigkeit der Unterstützung durch öffentliche Stellen. Im Landkreis Ahrweiler (RP) konnten zwei Befragungen mit Unterstützung des Landkreises durchgeführt werden. Dabei wurde jeweils ein Drittel aller Betroffenen randomisiert ausgewählt, die eine Soforthilfe direkt nach Ereignis beantragt hatten (Truedinger et al. 2023; Zenker et al. 2024; Dillenardt und Thieken 2024; Sairam et al. 2024). Ein Anschreiben der Landrätin verdeutlichte die Unterstützung der Befragung durch die lokale Politik, während gleichzeitig gewährleistet wurde, dass die personenbezogenen Daten den Forschungspartnern nicht zugänglich waren. Dieses Vorgehen hatte den großen Vorteil einer Zufallsauswahl und der Unterstützung durch öffentlich Stellen. Auch die Bürgermeister und Bürgermeisterinnen wurden über die Befragungen informiert. Da dieses Vorgehen nicht überall möglich war, wurde Befragungen in NRW vor allem mit Hilfe von Meta-Anzeigen oder anderen Projektpartnern aus NRW beworben. Die Vergleichbarkeit der Ergebnisse weist darauf hin, dass beide Befragungen zu robusten Ergebnissen führten (siehe ▶ Kap. 9). Wissenschaftlich sollte aber eine zufallsgenerierte Stichprobe angestrebt werden.

Nach schweren Überflutungen sind in den letzten Jahren wiederholt hohe Summen für staatliche Sofort- und Wiederaufbauhilfen aus Steuermitteln aufgebracht worden. Nach den Fluten vom Juli 2021 etwa bewilligten Bund und Länder das Sondervermögen „Aufbauhilfe 2021" in Höhe von 30 Mrd. € (siehe § 4 Abs. 1 Aufbauhilfefonds-Errichtungsgesetz 2021[3]). Um zukünftig die Datenerhebung nach Schadensereignissen zu verbessern, könnte in Verbindung mit staatlich gewährten Zahlungen an Privatpersonen, Unternehmen und Infrastrukturbetreiber auch auf das Erkenntnisinteresse der Öffentlichkeit und der Wissenschaft hingewiesen werden, verbunden mit der Bitte, sich nach Möglichkeit an Befragungen und/oder Experteninterviews zu beteiligen.

Ein deutlicher Fortschritt für ein evidenzbasiertes Hochwasserrisikomanagement wäre ein verbesserter Zugang zu Registerdaten für die Forschung in Deutschland, wie er auch vom RatSWD gefordert wird (RatSWD 2023). Der Begriff Register bezieht sich dabei auf Verzeichnisse oder Datenbanken, in denen gesetzlich vorgesehene Daten zu Personen erfasst werden. Bislang sind vorhandene Registerdaten grundsätzlich nicht für eine Nutzung durch die Wissenschaft vorgesehen. Auch die Verknüpfung verschiedener Registerdaten untereinander und mit anderen Datenquellen ist nur sehr beschränkt möglich (RatSWD 2023), obwohl dies eine evidenzbasierte, räumlich hoch aufgelöste Analyse und damit ein evidenzbasiertes und effizienteres Hochwasserrisikomanagement ermöglichen würde. Für die bessere und statistisch valide Erhebung von Befragungsdaten ist daher abschießend zu empfehlen, Registerdaten für Forschungszwecke – zumindest in geförderten Projekten und bestimmten Krisensituationen – zu öffnen.

Wie eingangs erwähnt hat das Hochwasserrisikomanagement in den vergangenen Jahrzehnten einen Wandel durchlaufen, der neue wissenschaftliche Fragen und Methoden in diesem Feld eröffnet hat. Allerdings sind verschiedene Befragungen oft nicht vergleichbar, auch wenn sie ähnliche Themen adressieren, was die Zusammenführung und gemeinsame Auswertung über zeitliche und räumliche Grenzen hinweg erschwert. Mittlerweile liegen aber so viele Erfahrungen vor, dass standardisierte Formulierungen von Fragen und Antwortskalen entwickelt werden sollten. Eine solche Systematisierung und Standardisierung der Datenerhebung könnte die Qualität, Datenverknüpfung und langfristige Datennutzung nochmals steigern (RatSWD 2024).

Zusammenfassung

- Bevölkerungsbefragungen und Experteninterviews liefern wichtige Informationen für das integrierte Hochwasserrisikomanagement.
- Derartige Daten werden nicht mit etablierten Messmethoden und Dokumentationen erfasst, sondern in Deutschland anlassbezogen mit Hilfe von Bevölkerungsbefragungen und Experteninterviews erhoben.
- Erhebungen liefern wichtige Erkenntnisse, z. B. zur Bewältigung des Ereignisses, zu Prioritäten des Wiederaufbaus, Schäden, Anzeichen posttraumatischer Belastungsstörungen Betroffener und privater Eigenvorsorge.
- Gleichzeitig können solche Erhebungen eine zusätzliche zeitliche und emotionale Belastung für betroffene Bürger:innen und Einrichtungen darstellen. Daher sollte in stark von Extremereignissen betroffenen Regionen unter anderem eine verstärkte Koordina-

3 Aufbauhilfefonds-Errichtungsgesetz 2021 vom 10. September 2021 (BGBl. I S. 4147), siehe: ▶ https://www.gesetze-im-internet.de/aufbheg_2021/BJNR414710021.html.

tion und Integration verschiedener Erhebungen durch regionale Projektbüros sowie ein verbesserter Zugang zu amtlichen Registerdaten erfolgen.

Danksagung Die Autorinnen und Autoren von ▶ Kap. 5 bedanken sich beim Bundesministerium für Bildung und Forschung (BMBF) und beim DLR Projektträger für die Unterstützung und Förderung des KAHR-Projekts. Diese Forschung wurde vom BMBF unter dem Förderkennzeichen 01LR2102A-M finanziert.

Literatur

Apel D, Coenen M (2021) Physical symptoms and health-care utilization in victims of the 2013 flood disaster in Germany – A longitudinal study of health-related flood consequences and evaluation of psycho-social support. Int J Disaster Risk Reduct 52:101949

Becker G, Aerts JCJH, Huitema D (2007) Transboundary flood management in the Rhine basin: challenges for improved cooperation. Water Sci Technol 56(4):125–135

Birkmann J, Schüttrumpf H, Handmer J, Thieken A, Kuhlicke C, Truedinger A, Kirschbauer L (2023) Strengthening resilience in reconstruction after extreme events – Insights from flood affected communities in Germany. Int J Disaster Risk Reduct 96:103965

Bubeck P, Thieken AH (2018) What helps people recover from floods? Insights from a survey among flood-affected residents in Germany. Reg Environ Change 18:287–296

Bubeck P, Aerts JCH, de Moel H, Kreibich H (2016) Preface: Flood-risk analysis and integrated management. Nat Hazards Earth Syst Sci 16(4):1005–1010

Dillenardt L, Thieken AH (2024) Individual Flood Risk Adaptation in Germany: Exploring the Role of Different Types of Flooding. EGUsphere, S 1–28

Grothmann T, Reusswig F (2006) People at risk of flooding: why some residents take precautionary action while others do not. Nat Hazards 38:101–120

Hudson P, Thieken AH (2022) The presence of moral hazard regarding flood insurance and German private businesses. Nat Hazards 112(2):1295–1319

Hudson PGMB, Botzen WJW, Kreibich H, Bubeck P, Aerts JCJH (2014) Evaluating the effectiveness of flood damage mitigation measures by the application of propensity score matching. Nat Hazards Earth Syst Sci 14(7):1731–1747

Kienzler S, Pech I, Kreibich H, Müller M, Thieken AH (2015) After the extreme flood in 2002: changes in preparedness, response and recovery of flood-affected residents in Germany between 2005 and 2011. Nat Hazards Earth Syst Sci 15(3):505–526

Köhler L, Masson T, Köhler S, Kuhlicke C (2023) Better prepared but less resilient: the paradoxical impact of frequent flood experience on adaptive behavior and resilience. Nat Hazards Earth Syst Sci 23(8):2787–2806

Kreibich H, Thieken AH, Petrow T, Müller M, Merz B (2005) Flood loss reduction of private households due to building precautionary measures – lessons learned from the Elbe flood in August 2002. Nat Hazards Earth Syst Sci 5(1):117–126

Kreibich H, Bubeck P, Van Vliet M, De Moel H (2015) A review of damage-reducing measures to manage fluvial flood risks in a changing climate. Mitig Adapt Strateg Glob Change 20:967–989

Kuhlicke C, Callsen I, Begg C (2016) Reputational risks and participation in flood risk management and the public debate about the 2013 flood in Germany. Environ Sci Policy 55:318–325

Kuhlicke C, Seebauer S, Hudson P, Begg C, Bubeck P, Dittmer C, Bamberg S (2020) The behavioural turn in flood disaster risk management and its implication for future research and policy. Wires Water 7(3):e1418

Kuklicke C, Demeritt D (2016) Adaptive and risk-based approaches to climate change and the management of uncertainty and institutional risk: The case of future flooding in England. Glob Environ Chang 37:56–68

LAWA (1995) Leitlinien für einen zukunftsweisenden Hochwasserschutz. Hochwasser – Ursachen und Konsequenzen. Länderarbeitsgemeinschaft Wasser, Stuttgart

Merz B, Aerts J, Arnbjerg-Nielsen K, Baldi M, Becker A, Bichet A, Blöschl G, Bouwer LM, Brauer A, Cioffi F, Delgado JM, Gocht M, Guzzetti F, Harrigan S, Hirschboeck K, Kilsby C, Kron W, Kwon H-H, Lall U, Merz R, Nissen K, Salvatti P, Swierczynski T, Ulbrich U, Viglione A, Ward PJ, Weiler M, Wilhelm B, Nied M (2014) Floods and climate: emerging perspectives for flood risk assessment and management. Nat Hazards Earth Syst Sci 14:1921–1942

Osberghaus D, Hinrichs H (2021) The effectiveness of a large-scale flood risk awareness campaign: evidence from two panel data sets. Risk Analysis 41(6):944–957

RatSWD (2023) Positionspapier des RatSWD: Nutzung von Registerdaten für Zwecke der Forschung sichern. Berlin (https://www.konsortswd.de/wp-content/uploads/RatSWD-Positionspapier-Nutzung-von-Registerdaten.pdf)

RatSWD (2024) RatSWD-Positionspapier: Forschung in Krisensituationen: Forderungen an Wissenschaft und Politik. Berlin (https://www.konsortswd.de/wp-content/uploads/RatSWD_Positionspapier_Forschung_in_Krisensituationen.pdf)

Sairam N, Buch A, Zenker ML, Dillenardt L, Coenen M, Thieken AH, Jung-Sievers C (2024) Health-related Quality of Life and everyday functioning in the flood-affected population in Germany – a case study of the 2021 Floods in West Germany

Thieken AH, Bubeck P, Heidenreich A, Von Keyserlingk J, Dillenardt L, Otto A (2023) Performance of the flood warning system in Germany in July 2021 – insights from affected residents. Nat Hazards Earth Syst Sci 23(2):973–990

Truedinger AJ, Jamshed A, Sauter H, Birkmann J (2023) Adaptation after extreme flooding events: moving or staying? The case of the Ahr Valley in Germany. Sustainability 15(2):1407

Wilkinson MD, Dumontier M, Aalbersberg IJ, Appleton G, Axton M, Baak A, Mons B (2016) The FAIR Guiding Principles for scientific data management and stewardship. Sci Data 3(1):1–9

Zenker ML, Bubeck P, Thieken AH (2024) Always on my mind: indications of post-traumatic stress disorder among those affected by the 2021 flood event in the Ahr valley, Germany. EGUsphere, S 1–30

Aufnahme nur kurzfristig erfassbarer Geländebefunde nach Hochwasserereignissen

Stefanie Stenger-Wolf, Holger Schüttrumpf, Elena-Maria Klopries, Catrina Brüll, Harish Selvam, Frank Lehmkuhl, Alexandra Weber, Philipp Schulte, Piero Bellanova, Jan Schwarzbauer, Henner Hollert, Sabrina Schiwy, Sarah Johann, Susanne Kozerke und Eva Vonden

J. Birkmann et al. (Hrsg.), *Die Flutkatastrophe 2021*, Sustainable Development Goals (SDG) – Umsetzung in Praxis, Lehre und Entscheidungsprozessen,
https://doi.org/10.1007/978-3-658-48688-4_6

6.1 Anlass der Aufnahme kurzlebiger Daten

Nach einem Hochwasser werden Daten zur Analyse des Hergangs des Flutgeschehens benötigt. Die Daten sollen insbesondere Hinweise auf Schadensursachen geben, mit dem langfristigen Ziel der Entwicklung von Strategien zur Minderung der Auswirkungen. Seltene Hochwasser ($> HQ_{100}$) führen in der Regel zu Schäden in anthropogen genutzten Überschwemmungsgebieten und zur Verlagerung großer Mengen Sediment. Mit einer Hochwasserkatastrophe geht einher, dass die Lage zunächst unübersichtlich und das Schadensausmaß nicht direkt bekannt ist. Unter Umständen erreichte das Hochwasserereignis auch neue Pegelspitzenwerte und ein zuvor unbekanntes Ausmaß an Zerstörungen, wie es im Juli 2021 der Fall war (siehe ▶ Kap. 3).

6.1.1 Rekonstruktion des hydraulischen Abflussgeschehens

Die Aufnahme von sichtbaren Spuren der Flut helfen bei der späteren Rekonstruktion des Abflussgeschehens und können eine erste Einschätzung des Hochwasserereignisses bezüglich seiner Seltenheit ermöglichen. Von besonderem Interesse sind außerdem Abweichungen von einem erwarteten Hergang des Hochwassers, also eines zuvor bereits aufgezeichneten oder simulierten Events. Selbst wenn das Hochwasserereignis von bestehenden Pegeln erfasst werden konnte, kann durch Verklausungen und weitere hydraulische Prozesse die Situation von dem vorher Bekannten bzw. Berechneten abweichen.

Vor der Vor-Ort-Begehung sind Informationen zu Überschwemmungsgebieten und -tiefen in Hochwassergefahren- und Risikokarten zu finden. Diese wurden über hydro-numerische Simulationen ermittelt, was immer mit Vereinfachungen und Ungenauigkeiten einhergeht. Daher ist die Rekonstruktion eines eingetretenen vorabsimulierten Szenarios vor Ort ebenfalls von Interesse. Sowohl die tatsächlich eingetretene Ausbreitung der Überflutung und Wasserstände, als auch Fließgeschwindigkeiten sind hier wichtige Informationen. Denn hohe hydraulische Kräfte aufgrund hoher Fließgeschwindigkeiten und Wasserstände führen schnell zu großen Schäden in betroffenen Gebieten.

Hinweis für die Praxis

Die eingetretenen Maximalwasserstände lassen sich im Nachgang eines Hochwasserereignisses gut durch Spuren der Hochwasserwelle selbst, sogenannte Flutmarken, bestimmen (◘ Abb. 6.1). Diese kommen dadurch zustande, dass Hochwasserwellen nicht aus klarem Wasser bestehen, sondern Sedimente, Schadstoffe und Treibgut mittransportiert werden, die beim Absinken des Wasserstandes Ablagerungen an Gebäuden und anderen Strukturen hinterlassen. Durch die hohen Fließgeschwindigkeiten finden sich ebenfalls Spuren an der Vegetation. So werden durch das Hochwasser beispielsweise an einer Hecke Blätter bis zu einer gewissen Höhe gleichmäßig nach innen gedrückt. Die Richtung der weggedrückten Blätter ist dann ein Hinweis auf die Fließrichtung, aber nicht auf die Geschwindigkeit. Außerdem finden sich nach einem Hochwasser häufig Feinsedimentablagerungen auf Blättern, die überflutet waren, womit sich die Überflutungshöhe zusätzlich validieren lässt (◘ Abb. 6.1). An Zäunen und Bäumen kann zudem Treibgut zurückbleiben. Fließgeschwindigkeiten hingegen lassen sich im Nachhinein nur schwer bestimmen.

Um diese abzuschätzen, können Videoaufnahmen herangezogen werden, die während des Hochwassergeschehens angefertigt wurden. Hier ist man in der Regel auf Anwohner und Anwohnerinnen sowie Einsatzkräfte angewiesen, die das Hochwassergeschehen unmittelbar miterlebt und dokumentiert haben. Häufig finden sich weitere hilfreiche Videos und Bildmaterial auf Social-Media-Plattformen.

Abb. 6.1 Spuren der Hochwasserwelle an der Vegetation. (Foto: Holger Schüttrumpf, 31.07.2021)

Hinweis für die Praxis

Nicht zuletzt sollte eine umfangreiche Fotodokumentation von Gebäude- und Infrastrukturschäden sowie von geomorphologischen Veränderungen am Gewässerbett und Akkumulationskörpern auf den Auen angefertigt werden. Erosion und Akkumulation von Sedimenten ermöglichen Rückschlüsse auf die Lage des Hauptstromes, und Schadensmuster geben Hinweise auf hydraulische Kräfte, die auf das jeweilige betroffene Gebäude eingewirkt haben (Abb. 6.2).

Abb. 6.2 Abgelöste Verklinkerung an einer Hauswand gibt Hinweise auf hohe hydraulische Einwirkungen während des Hochwassers. (Foto: Stefanie Stenger-Wolf, 18.08.2021)

6.1.2 Rekonstruktion der Schadstoffausbreitung

Viele Schadstoffe lagern sich an Sedimenten an und werden mit diesen mittransportiert. Über das Sammeln und anschließende Auswerten von Sedimentproben lassen sich Rückschlüsse auf die Schadstoffbelastung nach dem Hochwasser sowie auf Stoffströme bilden. Bei der Überschwemmung anthropogen genutzter Auen werden nicht nur Gebäude und Infrastruktur zerstört, es werden auch verschiedenste Schadstoffe aus unterschiedlichsten Quellen von den Wassermassen erfasst und weiter nach Unterstrom (stromabwärts) transportiert (Crawford et al. 2021). In urbanen Gebieten kann es zum Aufschwimmen und Verdriften von Heizöltanks, die sich im Keller, aber auch neben Gebäuden befinden können, kommen. Kommt es zur Beschädigung der Tanks, resultieren Leckagen in einer Ölkontamination in überfluteten Bereichen (Abb. 6.3). Außerdem werden in Kellern, Garagen, Lagerhallen sowie in an Industriestandorten gelagerte Farben, Lacke, Treibstoffe und andere Gefahrenstoffe von eintretenden Wassermassen erfasst. Schadstoffeintrag aus von den Wassermassen erfassten Kraftfahrzeugen werden über Auen in Gewässer transportiert. Nicht zuletzt kommt es zum Eintrag von Abwasser aus überlasteten Kanalsystemen in die aufnehmenden Vorfluter und teilweise sogar zur Überflutung von Kläranlagen. In nahe am Gewässer gelegenen Gewerbe- und Industriegebieten ist das Eintragspotenzial von Schadstoffen im flutbedingten Havariefall besonders hoch. Neben dem Eintrag einer diversen Mischung aus verschiedenen Schadstoffen, zu denen polyzy-

Abb. 6.3 Kennzeichnung von potentieller Ölkontamination an einem betroffenen Wohngebäude. (Foto: Harish Selvam, 18.08.2021)

klische aromatische Kohlenwasserstoffe (PAHs) und Pharmazeutika gehören, werden bei hohen Abflüssen zudem Altlasten auf Auenbereichen erodiert (Esser et al. 2022; Lehmkuhl et al. 2022a). Somit wird bei jedem Flutereignis eine individuelle und komplexe Mischung verschiedenster remobiliserter und neu eingetragener Schadstoffe freigesetzt, die mehr oder weniger starke toxische Wirkungen haben kann. Jedes Flutereignis ist daher ein einmaliges Ereignis hinsichtlich potenzieller Gefahren und muss dementsprechend individuell betrachtet werden. Aus Gründen des Bevölkerungs- und Umweltschutzes ist eine Identifikation und Bewertung der Schadstoffablagerungen von Interesse.

Auch eine Beprobung und anschließende Untersuchung der Wasserphase kann erste Aufschlüsse über die Kontamination der betroffenen Region liefern. Dies ist jedoch in Anbetracht der üblichen Beprobungsverfahren (Stichproben, maximal 24 h Mischproben über Großvolumen Festphasen Extraktion) und der anschließend unbedingt erforderlichen zeitnahen Probenaufbereitung nur eine kurze Momentaufnahme der Belastung der Wassersäule mit limitierter Aussagekraft im Vergleich zur Untersuchung der Schwebstoffe und Sedimente. Sofern sowohl Wasser- als auch Schwebstoff- oder Sedimentphase beprobt werden, sollten die unterschiedlichen Anforderungen hinsichtlich Probenahmegefäße sowie Auf- und Weiterbereitung des jeweiligen Probentyps beachtet werden.

6.1.3 Kurzlebigkeit der Felddaten

Nach einer Hochwasserkatastrophe steht die Arbeit der Katastrophenschutz und Spontanhelfende an erster Stelle. Gleichzeitig gibt es Daten, die nur innerhalb weniger Tage oder Wochen aufgenommen werden können, weil sie witterungsbedingt oder durch Aufräumarbeiten verloren gehen. Hierzu zählen Sedimentablagerungen, die durch Aufräumarbeiten beseitigt oder zerfahren sowie zertreten werden. Durch erneute Niederschläge oder eine hohe Sonneneinstrahlung kommt es ebenfalls zur Veränderung der Ablagerungen. Flutmarken können durch die Sonne getrocknet, durch erneute Niederschläge weggewaschen und bei Aufräumarbeiten entfernt werden. Außerdem lassen sie sich häufig nur im situativen Kontext zuordnen. Ob es sich um eine anderweitige horizontale Verschmutzung einer Hauswand oder eine Flutmarke handelt, lässt sich manchmal nur an weiteren Spuren auf ähnlicher Höhe erkennen (◘ Abb. 6.4). Auch Spuren an der Vegetation lassen sich ohne andere Hinweise oft nur schwer einordnen.

6

Die Beprobung von Flutsedimenten und das Einmessen von Flutmarken müssen folglich zeitnah nach dem Hochwasser erfolgen. Da insbesondere zentrale Infrastrukturen wie Brücken bereits wenige Tage nach dem Hochwasser zumindest provisorisch repariert werden, ist auch eine Fotodokumentation nur in einem kurzen Zeitfenster möglich.

◘ **Abb. 6.4** An den Gebäuden sind Spuren des Hochwassers auf gleicher Höhe zu erkennen. (Bildquelle: Seckler und Wolf 2024)

6.2 Herausforderungen und Empfehlungen für die Aufnahme kurzlebiger Daten nach einem Katastrophenhochwasser

Nach einem extremen Hochwasser ist der optimale Einsatz der kurzfristig verfügbaren Datenaufnehmenden essentiell. Insbesondere wenn mehrere Einzugsgebiete betroffen sind, so wie es Mitte Juli 2021 der Fall war, können umfangreiche Ortsbegehungen und Lageaufnahmen unter Umständen zu lange dauern und Daten können in der Zwischenzeit verloren gehen. Es müssen folglich möglichst direkt Bereiche angesteuert werden, in denen eine Datenaufnahme sehr wahrscheinlich erfolgreich sein wird. Dies sind meist die Bereiche, in denen eine hohe Betroffenheit vorliegt. Als wichtigste Datengrundlage für die Vorbereitung dienen bestehende Hochwassergefahren und -risikokarten. Hier können potentielle Schadens-Hot-Spots anhand von simulierten Wassertiefen und Fließgeschwindigkeiten ausgemacht werden.

Beim Aufsuchen dieser „Schadens-Hot-Spots" darf bei aller Notwendigkeit wissenschaftlicher Aufarbeitung der Ereignisse die dramatische Situation für die betroffene Bevölkerung nicht in Vergessenheit geraten. Extreme Hochwasserereignisse können für große Teile der Bevölkerung traumatisch sein, wie für das Hochwasser 2021 an der Ahr nachgewiesen wurde (Zenker et al. 2024). Ist die Befragung von direkt Betroffenen zur Bewertung der Geländebefunde zwingend nötig, so sind solche Gespräche dringend mit einem größtmöglichen Maß an Vorsicht zu führen (vgl. ► Kap. 9). Der Arbeit von Einsatzkräften und Helfenden ist selbstverständlich zu jedem Zeitpunkt Vorrang zu gewähren.

6.2.1 Aufnahme von Flutmarken

Um in der Datenauswertung anhand von Flutmarken einen Überflutungsbereich aufspannen zu können, sollten ausreichend Flutmarken mit einer hohen Genauigkeit aufgenommen werden. Dabei werden interessante Bereiche möglichst lückenlos aufgenommen, denn das Ausspiegeln von Flutmarken ist mit zunehmender Entfernung von der Flutmarke mit Unsicherheiten verbunden. Büttner (2013) empfiehlt insbesondere für die Kalibrierung von 2D-hydraulischen Modellen eine Flutmarkendichte von mindestens 20 Messpunkten pro km^2 sowie eine möglichst gleichmäßige Verteilung der Flutmarken. Es ist empfehlenswert, die Aufnahme von Flutmarken mit bestehenden Hochwassergefahrenkarten, in die ein grobes Raster eingetragen wird, vorzubereiten. Vor Ort sollte eine sorgfältige Auswahl an Messpunkten getroffen werden, da Schwankungen in den anhand von Flutmarken abgeschätzten Wasserstandshöhen die größten Ungenauigkeiten bei einer späteren Modellkalibrierung ausmachen können (Neal et al. 2009). Wenn die Kapazitäten dies zulassen, empfiehlt es sich, möglichst viele Datenpunkte aufzunehmen, damit Ungenauigkeiten im Mittel weniger ins Gewicht fallen.

Im urbanen Raum ist es zielführend, im Vorhinein Straßenzüge, die begangen werden sollen, grob auf Kartenmaterial zu markieren. Besonders interessant ist die Aufnahme von Flutmarken an Gebäuden, an denen bereits historische Hochwassermarken angebracht sind (◘ Abb. 6.5). Außerhalb von Ortschaften können ebenfalls Bauwerke wie Pegelhäuser, Mühlen oder landwirtschaftlich genutzte Gebäude identifiziert werden. An Brücken aufgenommene Flutmarken sind mit Vorsicht zu betrachten, da Brückenpfeiler teilweise bereits im abschüssigen Gelände des Uferbereichs stehen. Dadurch können

Abb. 6.5 Flutmarke in Walporzheim an der Ahr. Am Gebäude befindet sich eine ungefähre Kennzeichnung des Wasserstandes von 1804 und Harish Selvam zeigt den anhand von Flutmarken abgeschätzten Wasserstand des Hochwassers von Mitte Juli 2021 an. (Foto: Jan Oetjen, 2021)

dort aufgenommene Flutmarken tiefer liegend erscheinen und sich in der Auswertung als Ausreißer bemerkbar machen. Sind keine geeigneten baulichen Strukturen vorhanden, an denen nach Flutmarken gesucht werden kann, können ebenfalls Treibgut an Bäumen und Zäunen sowie weitere Spuren an der Vegetation (siehe ▶ Abschn. 6.1.1) aufgenommen werden. Flutmarken können in der Regel gut mit einem Gliedermaßstab oder Maßband eingemessen werden. Sollte die Flutmarke den Messbereich überschreiten, kann eine Person als Referenzhöhe dienen (Abb. 6.5). Für die Aufnahme der Flutmarken empfiehlt sich das Abfotografieren mit einem Handy oder einer Kamera, die die GPS-Position des Bildes als sogenannten Geo-Tag aufnehmen kann. Ideal ist das Aufnehmen von drei Bildern, eines aus einer Entfernung, die das Wiedererkennen des Standorts zulässt, eins aus einer mittleren Nähe, durch die man die Position der Flutmarke verorten kann, und eins aus unmittelbarer Nähe zwecks Ablesens der Skala.

6.2.2 Probenahme von Flutsedimenten

Die Probenahme nach einem Katastrophenhochwasser kann in Abstimmung mit mehreren Institutionen der Wasserwirtschaft und Forschungsinstitutionen stattfinden, was die Verfügbarkeit an Vergleichsdaten und Ressourcen verbessert. Außerdem kann unter Umständen auf bestehende Protokolle aus früheren Studien zurückgegriffen werden.

Abb. 6.6 Probenahme im Uferbereich an der Inde. (Foto: Eva Vonden, 04.08.2021)

Für die Auswahl von Bereichen, in denen Sediment- und Wasserproben genommen werden sollen, sind insbesondere betroffene Schadstoffemittenten von Interesse. Flussabwärts von Industrien und historischen Industriestandorten sind ebenfalls erhöhte Schadstoffkonzentrationen zu erwarten. Zur genaueren Eingrenzung von Probenahmestandorten können Fließgeschwindigkeiten in Hochwassergefahrenkarten wichtige Informationen liefern. Typischerweise lagern sich vermehrt Sedimente in Bereichen mit reduzierter Fließgeschwindigkeit ab. In urbanen Bereichen, die aufgrund von Gebäuden eine sehr hohe hydraulische Rauheit haben, lagern sich ebenfalls große Mengen an Sedimenten ab. Zudem sind urbane Bereiche Schadstoffemittenten, beispielsweise für Schadstoffe aus dem Kanalsystem, Haushaltschemikalien, Heizöl oder Treibstoffe. Proben aus Gebäuden selbst oder der unmittelbaren Umgebung von Gebäuden liefern aufschlussreiche Daten. In betroffenen Kellern sammeln sich große Mengen Sediment, die oftmals ein umfassendes Bild über das lokale Schadstoffinventar ermöglichen. Außerdem lässt sich eine Belastung von Sedimentproben anhand eines Ölgeruchs oder der Kennzeichnung von Ölschäden an betroffenen Gebäuden ausmachen. Auf Auen mit Grünland lässt sich die Aufwuchshöhe (Nettohöhe der Ablagerung) der Sedimente gut an dem Bewuchs feststellen (Abb. 6.6).

Basierend auf der analytischen Auswertungsmethode der Wasser- oder Sedimentprobe muss ein geeignetes Probenahmegefäß gewählt werden, da beispielsweise keine Weichmacher oder Metalle aus den Gefäßen in die Probe übergehen dürfen, die das

Abb. 6.7 Lagerung von Flutsedimentproben im Kühlraum des IWW. Die Proben befinden sich in provisorischen Behältnissen, wie handelsüblichen Gefrierbeuteln und Einmachgläsern. (Foto: Stefanie Stenger-Wolf, 26.08.2021)

anschließende Ergebnis verfälschen könnten. Es wird daher empfohlen, ein Notfallsortiment geeigneter Probenahmegefäße, wie z. B. gereinigte Glas- oder Aluflaschen vorzuhalten. Idealerweise werden diese Probenahmegefäße direkt in einer Kiste gelagert, die lediglich in den Kofferraum verladen werden muss. Probenahmeequipment, wie etwa eine Handschaufel oder Einmalhandschuhe, Desinfektionsmittel sowie vorbereitete Probenahmeprotokolle, kann hier direkt bereitgehalten werden. Die Realität sieht häufig anders aus, und helfenden Institutionen und Personen, die mit der Probenahme nicht vertraut sind, sind pragmatische Instruktionen zu geben, die auch Kompromisse enthalten. So wurden für die Sammlung von Sedimentproben Mitte Juli 2021 gespülte Einmachgläser sowie handelsübliche Gefrierbeutel verwendet. Die Proben wurden am Lehrstuhl und Institut für Wasserbau und Wasserwirtschaft bis zur weiteren Untersuchung kühl und dunkel gelagert. Zudem fand eine frühzeitige Verständigung bezüglich einer möglichst einheitlichen Protokollierung und Kennzeichnung der verschiedenen Proben statt. Proben, die in nicht optimalen Probenahmegefäßen genommen wurden, oder zwischenzeitlich nicht sachgemäß gelagert werden konnten, können häufig dennoch untersucht und ausgewertet werden. Mögliche Verfälschungen der Ergebnisse durch das Probenahmegefäß oder eine unsachgemäße Lagerung können beachtet werden, wenn beides lückenlos dokumentiert ist (Abb. 6.7). Es ist zwingend notwendig die Probenahmesituation fotografisch festzuhalten. Insbesondere wenn, wie oben beschrieben, „helfende Institutionen" miteinbezogen werden, muss es im Nachhinein eine Möglichkeit geben, um zu überprüfen, ob überhaupt geeignetes Material entnommen wurde.

Gerade bei der Schadstoffbelastung von Sedimenten mit Schwermetallen sind Vergleichswerte notwendig, da es Einzugsgebiete mit hohen geogenen Hintergrundkonzentrationen gibt. Somit ist auch die Aufnahme von unbelasteten Sedimenten von Interesse sowie die Beprobung von Bereichen, zu denen es bereits Informationen aus früheren Studien gibt. Das Vorliegen von Vergleichsdaten aus früheren Studien ermöglicht die Einordnung der Analyseergebnisse der Flutsedimente.

Über die frühzeitige Abstimmung und Zusammenarbeit verschiedener Interessensgruppen (z. B. unterschiedliche Forschungsfelder wie Geologie, Hydrologie, Ökotoxikologie) kann ebenfalls die fachübergreifende Verwendbarkeit der Proben und die Vergleichbarkeit der Ergebnisse, beispielsweise über das Anwenden von vergleichbaren Analyseverfahren, sichergestellt werden (Bellanova et al. 2024; Johann et al. 2024; Schwanen et al. 2022; Weber und Lehmkuhl 2024; Weber et al. 2023).

6.2.3 Allgemeine Vorbereitung des Feldmess- und Probenahmeeinsatzes

Nach der Auswahl der zu begehenden Bereiche für die Aufnahme von Flutmarken und die Probensammlung, ist die Erreichbarkeit sicherzustellen. Orientiert werden sollte sich an öffentlichen Parkplätzen, die sich im Optimalfall nicht in unmittelbarer Gewässernähe befinden und wahrscheinlich nicht durch das Hochwasser betroffen sind. Hier können die offiziellen Hochwassergefahren- und Risikokarten herangezogen werden. Idealerweise liegt der Parkplatz nicht nur außerhalb des Überschwemmungsbereiches eines HQ_{100}, sondern auch eines $HQ_{Extrems}$. Gegebenenfalls sollten Alternativen eingeplant und Wegstrecken in Kauf genommen werden. Da Straßen und Brücken in stark betroffenen Überschwemmungsgebieten möglicherweise nicht mehr passierbar sind, sollten alternative Zufahrtwege geplant sein.

Generell wird empfohlen, betroffene Bereiche erst zu besuchen, wenn dies wieder gefahrlos möglich ist, die Pegelstände also rückläufig sind. Jedoch kann eine erste Aufnahme der Lage, die erste Informationen liefert, mittels Drohne bereits während des Hochwasserereignisses von sicheren Standorten außerhalb des Überschwemmungsgebietes aus möglich sein. Hier empfiehlt sich eine Drohne, die über „Real-time kinematic positioning" (RTK), verfügt, da so auf Kontrollpunkte am Boden verzichtet werden kann (▪ Abb. 6.8). Drohnenflüge bedürfen in der Regel Genehmigungen und sind mit sicherheitsbedingten Einschränkungen verbunden. Die generelle Fluggenehmigung sollten vorab vorliegen, und Flugrouten sollten so geplant werden, dass die sicherheitsbedingten Einschränkungen eingehalten werden. Hier empfiehlt sich die frühzeitige Abstimmung und Zusammenarbeit mit offiziellen Stellen. Daten können häufig zwischen verschiedenen Institutionen ausgetauscht werden, sodass durch eine frühzeitige Abstimmung Doppelarbeit vermieden werden kann.

Für die persönliche Sicherheit wird das Tragen von Einmalhandschuhen und eines Mundschutzes sowie das Mitführen von Desinfektionsmittel empfohlen, da vorab nicht bekannt ist, welche Schadstoffe oder auch Keime aus Kläranlagen während des Hochwassers transportiert wurden. Ebenfalls wird das Tragen von langer Kleidung und Sicherheitsschuhen empfohlen, um im Bereich von Trümmern und in unwegsamem Gelände geschützt zu sein. Aus Sicherheits- und Praktikabilitätsgründen wird das Arbeiten in Teams von mindestens zwei Personen empfohlen. Ein korrektes Verhalten vor Ort ist ebenfalls essentiell. Die Privatsphäre und Rechte betroffener Personen sind jederzeit zu wahren und eine Gefährdung von Betroffenen durch die Behinderung von Hilfskräften ist absolut zu vermeiden. Für die Betroffenen ist ein absoluter Ausnahmezustand eingetreten, viele Betroffene wurden traumatisiert oder befinden sich bei einer Datenaufnahme unmittelbar nach dem Ereignis immer noch in einem Schock-Zustand. Alle an der Datenaufnahme beteiligten Personen sollten dies während der gesamten Datenaufnahme

Abb. 6.8 Letzte Vorbereitungen vor dem Start der Drohnenaufnahmen. (Foto: Stefanie Stenger-Wolf, 25.10.2022)

präsent haben und sich entsprechend respektvoll verhalten. Die Bedürfnisse Betroffener haben immer Vorrang und auf die Erfassung von Daten ist in bestimmten Situationen zu verzichten.

Personen sollten nie ohne vorherige Zustimmung fotografiert werden. Auch bei Wohngebäuden ist es ratsam, vorher zu fragen, insbesondere wenn diese von innen fotografiert werden. Zudem sollten an Betroffene keine Versprechungen bezüglich einer Datenauswertung gemacht werden, da diese gegebenenfalls nicht eingehalten werden können. Alternativ können offizielles Informationsmaterial oder der Verweis an offizielle Stellen vorbereitet und verteilt werden. Gespräche mit Betroffenen vor Ort können wertvolle Informationen aus erster Hand liefern, und häufig stehen Betroffene der Datenaufnahme positiv gegenüber, da sie die Verbesserung der zukünftigen Hochwasservorsorge unterstützen möchten. So können in Interviews mit Betroffenen weitere wichtige Daten und Hinweise gesammelt werden. Betroffene können häufig ungefähre Wasserstände rekonstruieren (Abb. 6.9).

Jeder Messeinsatz führt zu neuen Erfahrungen und Lehren für den nächsten Messeinsatz. Die hier beschriebenen Empfehlungen zur Aufnahme kurzlebiger Daten (Tab. 6.1) sind also als Empfehlungen zu sehen.

Abb. 6.9 Anwohner zeigt den Höchstwasserstand während des Hochwassers an. (Foto: Elena-Maria Klopries, 03.08.2021)

Hinweis für die Praxis

Tab. 6.1 Übersicht der Empfehlungen für die Aufnahme kurzlebiger Daten basierend auf den Erfahrungen der Autorinnen und Autoren

Ziel	Empfehlung	Ausrüstung	Datengrundlagen
Rekonstruktion des Überschwemmungsgebiets und der maximalen Wasserstände anhand von Flutmarken	– Erstellen eines Rasters anhand des bei einem HQ_{Extrem} überfluteten Gebiets, in welchem zu begehende Straßenzüge, einzelne Gebäude und Bereiche markiert sind – Raster mit einer Flutmarkendichte von 20 Flutmarken pro km^2	Zollstock, Maßband, Kamera oder Mobiltelefon mit einer Geotag-Funktion bei Bildaufnahme	Hochwassergefahren- und Risikokarten

6

Tab. 6.1 (Fortsetzung)

Ziel	Empfehlung	Ausrüstung	Datengrundlagen
Luftbildaufnahme des Überschwemmungsgebietes sowie der Betroffenheit zu einem bestimmten Zeitpunkt	– Befliegung des Gebiets mit einer geeigneten Drohne in möglichst hoher Höhe – Festlegen von Flugrouten, über die betroffene Bereiche aufgenommen werden können, das Gewässer und Gebäude aber nicht direkt überfolgen werden	Drohne, idealerweise mit RTK, Genehmigung	Hochwassergefahren- und Risikokarten
Bestimmung der Schadstoffverteilung	– Auswahl von Gebieten flussabwärts von potentiellen Schadstoffquellen – Auswahl von Gebieten, für die es bereits Informationen gibt – Vorhalten von geeigneten Probenahmegefäßen – frühzeitige Abstimmung mit anderen Institutionen zwecks Vorgehens und Protokollierung – Lagerung und Prozessierung der Proben zwischen unterschiedlichen Disziplinen abzustimmen	Kiste mit Probenahmegefäßen und Probenahmeausrüstung	Hochwassergefahren- und Risikokarten
Erreichen der betroffenen Gebiete	– Anstreben von öffentlichen Parkplätzen und Vorhalten von Alternativvorschlägen – Vorbereitung von alternativen Routen	Fahrzeug, je nach Situation ein einfacher PKW oder ein geländegängiges Fahrzeug	Hochwassergefahren- und Risikokarten
Persönliche Sicherheit	– Erst nach Abklingen der Pegel stark betroffene Bereiche besuchen – Tragen von Einmalhandschuhen, Mundschutz, Sicherheitsschuhen und langer Kleidung	Mundschutz, Einmalhandschuhe, Sicherheitsschuhe, lange Kleidung, Desinfektionsmittel	Pegelinformationen

6.3 Auswertung der nach dem Hochwasser von Mitte Juli 2021 aufgenommenen Daten

Beim Hochwasser von Mitte Juli 2021 wurden ab dem 14.07.2021 Daten in flutbetroffenen Regionen aufgenommen. Das Team aus Forschenden verschiedener Institutionen der RWTH Aachen University und Goethe Universität Frankfurt begann zunächst in Einzugsgebieten, die schnell erreichbar waren, und in denen bereits Daten aus früheren Forschungsarbeiten vorlagen. Der Schwerpunkt der Datenerhebung lag auf dem Einzugsgebiet der Inde, es wurden aber auch Daten entlang der Wurm, Rur und Urft aufgenommen (Abb. 6.10). In den ersten Wochen nach der Flut fanden ebenfalls Datenaufnahmen im kleineren Maßstab im Einzugsgebiet der Erft und Wupper statt. In Zusammenarbeit mit weiteren internationalen Forschungseinrichtungen, beispielsweise aus den USA und

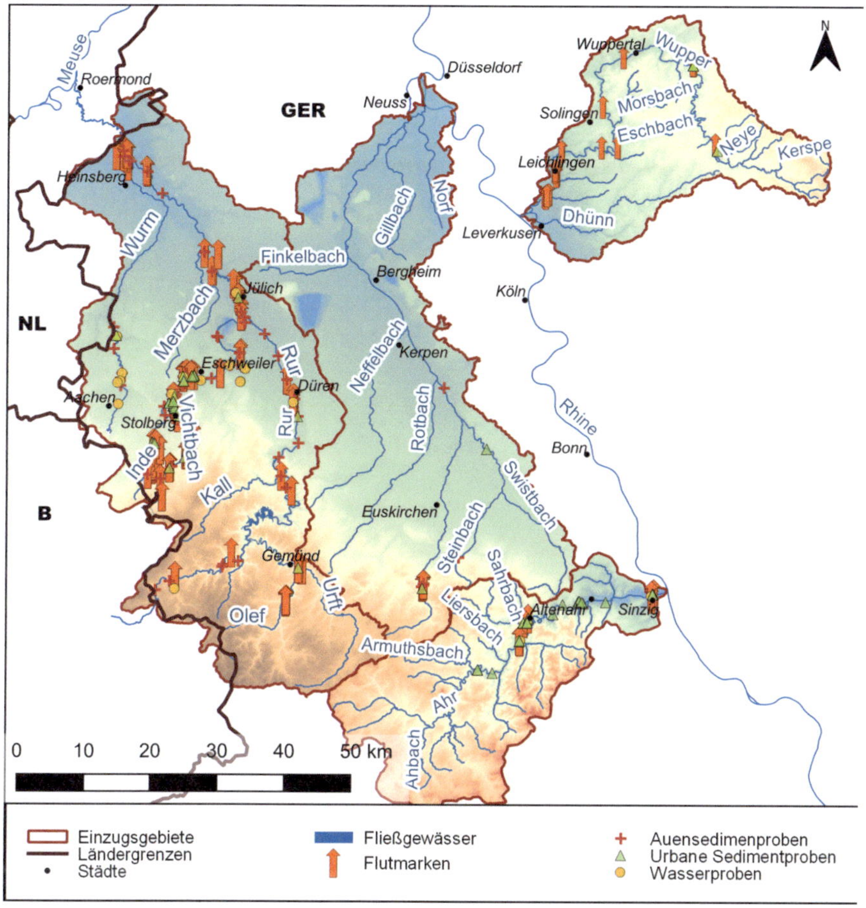

Abb. 6.10 Übersichtskarte der im Rahmen unserer Studien untersuchten vom Hochwasser betroffenen Flusssysteme in Nordrhein-Westfalen und Rheinland-Pfalz (Deutschland), mit allen Probenahmestellen und Hochwassermarkenbeobachtungen. (Eigene Darstellung)

den Niederlanden, wurden ab August 2021 ebenfalls Daten im Einzugsgebiet der Ahr aufgenommen.

Nach dem Hochwasserereignis wurden zunächst Grundlagendaten, wie die Flutmarken ausgewertet, und Erfahrungsberichte verfasst (Lehmkuhl et al. 2022b, 2022c; Schüttrumpf 2021; Schüttrumpf et al. 2022a, 2022b, 2022c). Die Flutmarken wurden geoinformationsgestützt flächendeckend im Rahmen einer Masterarbeit ausgewertet. Für die Rekonstruktion von Überflutungsflächen wurden aus dem vorhandenen Datensatz Gebiete ausgewählt, in denen eine Anzahl von mindestens zehn Flutmarken und eine Messdichte von mindestens 20 Flutmarken je km^2 vorliegen (Abb. 6.11; Kisseler et al. 2023).

Nach der Erstellung einer Datenbasis mit Grundlagendaten konnten diese für weitere Analysen verwendet werden, beispielsweise für die Kalibrierung eines hydronumerischen Modells zur Ermittlung der Sedimentverteilung in Eschweiler (Weber et al. 2023). Des Weiteren wurden die entnommenen Sediment- und Wasserproben in einer breiten

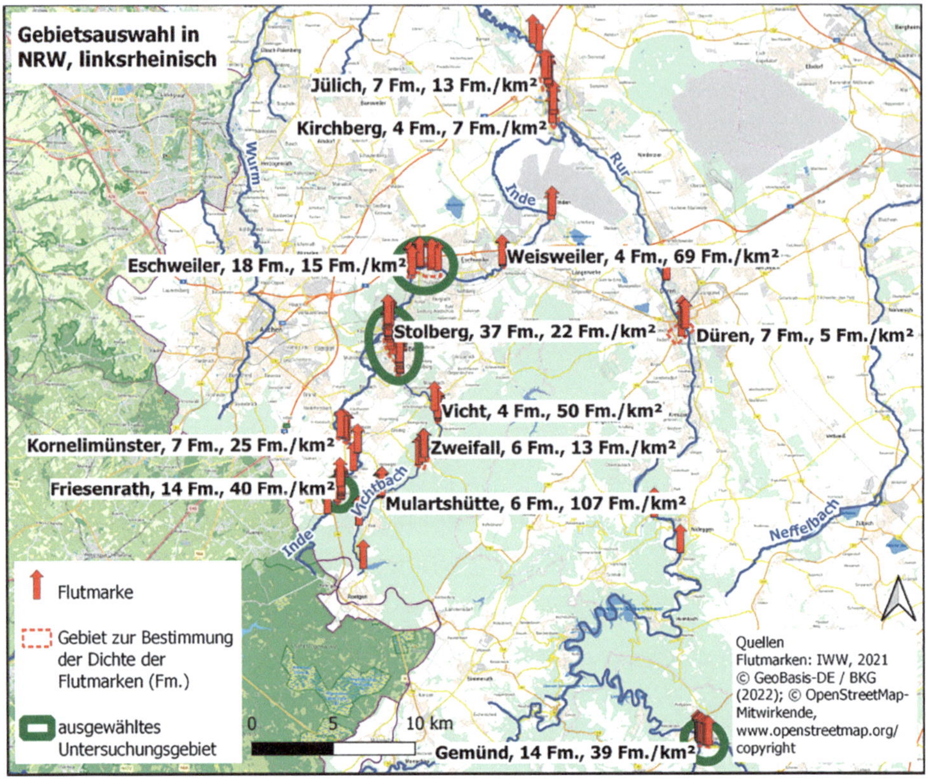

Abb. 6.11 Flutmarkendichte und Gebietsauswahl in Nordrhein-Westfalen. (Bildquelle: Masterarbeit Kisseler, 2022)

Biotestbatterie hinsichtlich einer potentiellen akuten und mechanismusspezifischen Toxizität untersucht. Ein Fokus lag dabei auf der Untersuchung der embryotoxischen und teratogenen Effekte in Zebrabärblingsembryonen und -larven. Zudem wurde in Zelltests untersucht, ob eine potentielle endokrine, mutagene und dioxinähnliche Toxizität in den Proben vorliegt, die durch die komplexen Mischungen hervorgerufen werden können. Die vorherige Zusammenarbeit zwischen verschiedenen Forschungsinstitutionen bei der Datenaufnahme bildete die Grundlage für weitere größtenteils interdisziplinäre Auswertungen (Bellanova et al. 2022, 2024; Johann et al. 2024; Keßels et al. 2024; Schwanen et al. 2022; Weber und Lehmkuhl 2024; Weber et al. 2023).

6.4 Fazit

Nach einem Hochwasserereignis ist die Erfassung von Daten entscheidend, um den Hergang der Flut zu analysieren und langfristige Strategien zur Schadensminderung zu entwickeln. Erhobene Daten helfen, die Ursachen von Schäden in anthropogen genutzten Überschwemmungsgebieten zu verstehen sowie Schadstoffeinträge und -ausbreitungen einzuschätzen. Insbesondere bei seltenen Hochwasserereignissen kann es schwierig sein, das Ausmaß des Schadens sofort zu erfassen.

Die Rekonstruktion des hydraulischen Abflussgeschehens erfolgt durch die Analyse von Flutmarken und anderen Spuren der Hochwasserwelle. Diese Daten liefern wichtige Informationen über Wasserstände und Fließgeschwindigkeiten, welche für die Beurteilung der Schwere des Ereignisses entscheidend sind. Die Rekonstruktion des Überschwemmungsgebiets ist ebenfalls für die Validierung von hydronumerischen Modellierungen interessant, denn es können lokale Abweichungen durch Verklausungen auftreten.

Ein weiterer wichtiger Aspekt ist die Rekonstruktion der Schadstoffausbreitung während einer Überschwemmung. In urbanen Gebieten können durch das Hochwassergeschehen verschiedene Schadstoffe (re)mobilisiert werden, was eine Identifikation ihrer Ablagerungen notwendig macht. Dabei liefern vor allem Sedimentproben Rückschlüsse auf die Schadstoffbelastung nach dem Hochwasser. Eine chemische und ökotoxikologische Untersuchung kann dabei wichtige Informationen zur Schadstoffherkunft liefern und zu einer Einordnung der möglichen Effekte beitragen.

Die Kurzlebigkeit dieser Felddaten ist eine Herausforderung: Witterungsbedingungen und Aufräumarbeiten können dazu führen, dass wichtige Informationen schnell verloren gehen. Daher müssen Daten zeitnah erfasst werden, um ihre Gültigkeit zu gewährleisten.

Empfehlungen für die Datenerhebung umfassen eine sorgfältige Planung und Vorbereitung der Feldmessungen sowie den Einsatz geeigneter Technologien wie Drohnen zur Dokumentation der betroffenen Gebiete. Die interdisziplinäre Zusammenarbeit verschiedener Institutionen kann zudem wertvolle Vergleichsdaten liefern und die Effizienz erhöhen. Bei der Probenahme sollten potenziell gefährdete Bereiche priorisiert werden, insbesondere in Bezug auf Schadstoffemittenten. Eine korrekte Lagerung und Dokumentation der Proben sind essentiell für spätere Analysen.

Die gesammelten Daten aus dem Hochwasser im Juli 2021 dienen als Grundlage für weitere interdisziplinäre Forschungsarbeiten und Modellkalibrierungen zur besseren Vorhersage zukünftiger Hochwasserereignisse.

Zusammenfassung

- Im Anschluss an ein Hochwasser ist die zeitige Erfassung von Geländebefunden entscheidend, denn diese interdisziplinären Daten helfen, Ursachen von Schäden in anthropogen genutzten Überschwemmungsgebieten zu verstehen und Schadstoffausbreitungen zu identifizieren.
- Da Schadstoffe oft sedimentgebunden transportiert werden, liefern Flutsedimente Informationen über eine Kontamination überschwemmter Flächen.
- Die Rekonstruktion des hydraulischen Abflussgeschehens, insbesondere nach Pegelausfällen, erfolgt durch die Analyse von Flutmarken, welche Informationen über Ausbreitung und Tiefe der Überschwemmung bieten.
- Videomaterial von der Flut kann zudem wichtige Informationen über Fließgeschwindigkeiten liefern.
- Die Kurzlebigkeit der Felddaten aufgrund von Witterung und Aufräumarbeiten stellt eine Herausforderung dar. Daher müssen Daten zeitnah erfasst werden.
- Die gesammelten Daten zum Hochwasser im Juli 2021 bilden die Grundlage für Modellkalibrierungen und weitere Forschungsarbeiten zur Verbesserung der Hochwasservorsorge in Deutschland und Europa.

Danksagung Wir bedanken uns für die folgenden Förderungen, welche die kurzfristige Datenaufnahme sowie die Auswertungen und weitere Verwertung der Daten ermöglicht haben: Machbarkeitsstudie zur Beurteilung der Schadstoffbelastung von Sedimenten infolge des Juli-Hochwassers 2021 im Übergang vom Mittelgebirge zum Tiefland, Deutsche Forschungsgemeinschaft (DFG) – Projektnummer 496274914, Machbarkeitsstudie zur Beurteilung der ökotoxikologischen und toxikologischen Belastung von Sedimenten infolge des Juli-Hochwassers 2021 im Übergang vom Mittelgebirge zum Tiefland, Deutsche Forschungsgemeinschaft (DFG) – Projektnummer 497800446, Wissenschaftliche Begleitung der Wiederaufbauprozesse nach der Flutkatastrophe in Rheinland-Pfalz und Nordrhein-Westfalen – Klimaanpassung, Hochwasser und Resilienz (KAHR), BMBF Förderinitiative 01LR2102H.

Literatur

Bellanova P, Schwarzbauer J, Reicherter K (2022) Toxic European Summer Flood; Dispersion of organic pollutants along the Vicht and Inde rivers, Germany https://doi.org/10.5194/egusphere-egu22-8550

Bellanova P, Schwarzbauer J, Reicherter K (2024) Inventory of aqueous and sediment-associated organic pollutants released by the 2021 flood in the Vicht–Inde catchment, Germany. Environ Sci Eur. https://doi.org/10.1186/s12302-024-00925-6

Büttner O (2013) 2D-Modellierung der Hydraulik und des Feinsedimenttransportes von Extremhochwässern in urbanen Gebieten. TU Braunschweig, Braunschweig (Dissertation)

Crawford SE, Brinkmann M, Ouellet JD, Lehmkuhl F, Reicherter K, Schwarzbauer J, Bellanova P, Letmathe P, Blank LM, Weber R, Brack W, van Dongen JT, Menzel L, Hecker M, Schüttrumpf H, Hollert H (2021) Remobilization of pollutants during extreme flood events poses severe risks to human and environmental health. J Hazard Mater 421:126691. https://doi.org/10.1016/j.jhazmat.2021.126691

Esser V, Wolf S, Schwanen C, Schulte P, Brüll C, Lehmkuhl F, Schüttrumpf H (2022) Schadstoffverlagerungen bei Hochwasserereignissen: Ergebnisse aus Langzeitstudien an der neuen Inde im Kontext des Juli-Hochwassers. In: RWTH (Hrsg) RWTH Themen: Hochwasser – Beiträge zu Risiken, Folgen und Vorsorge. Forschungsmagazin

Johann S, Düster M, Bellanova P, Schwarzbauer J, Weber A, Wolf S, Schüttrumpf H, Lehmkuhl F, Hollert H (2024) Dioxin-like and estrogenic activity screening in fractionated sediments from a German catchment after the 2021 extreme flood. Environ Sci Eur. https://doi.org/10.1186/s12302-024-00989-4

Keßels J, Wolf S, Römer W, Dörwald L, Schulte P, Lehmkuhl F (2024) Enormous headward and gully erosion in a floodplain area reclaimed for open-cast lignite mining during the July 2021 flood in the Inde River valley (Western Germany). Environ Sci Eur. https://doi.org/10.1186/s12302-024-00997-4

Kisseler E, Wolf S, Sauter H, Schüttrumpf H (2023) Reduktion der Vulnerabilität gegenüber Hochwasser – Empfehlungen zur Flächennutzung auf Basis von Flutmarken des Juli-Hochwassers 2021. Korrespondenz Wasserwirtsch 2023(8):498–506. https://doi.org/10.3243/kwe2023.08.003

Lehmkuhl F, Esser V, Schulte P, Weber A, Wolf S, Schüttrumpf H (2022a) Sediment pollution and morphodynamics of an extreme event: Examples from the July 2021 flood event from the Inde River catchment in North Rhine-Westphalia https://doi.org/10.5194/egusphere-egu22-5114

Lehmkuhl F, Schüttrumpf H, Schwarzbauer J, Brüll C, Dietze M, Letmathe P, Völker C, Hollert H (2022b) Assessment of the 2021 summer flood in Central Europe. Environ Sci Eur. https://doi.org/10.1186/s12302-022-00685-1

Lehmkuhl F, Weber A, Esser V, Schulte P, Wolf S, Schüttrumpf H (2022c) Fluviale Morphodynamik und Sedimentkontamination bei Extremereignissen: Das Juli-Hochwasser 2021 im Inde-Einzugsgebiet (Nordrhein-Westfalen). Korrespondenz Wasserwirtsch 2(15):422–427

Neal JC, Bates PD, Fewtrell TJ, Hunter NM, Wilson MD, Horritt MS (2009) Distributed whole city water level measurements from the Carlisle 2005 urban flood event and comparison with hydraulic model simulations. J Hydrol Reg Stud 368(1–4):42–55. https://doi.org/10.1016/j.jhydrol.2009.01.026

Schüttrumpf H (2021) Das Juli-Hochwasser 2021 in NRW – Ein erster Erfahrungsbericht. Wasser Abfall 23:14–17

Schüttrumpf H, Birkmann J, Brüll C, Burghardt L, Johann G, Klopries E-M, Lehmkuhl F, Schüttrumpf A, Wolf SS (2022a) Die Flutkatastrophe 2021 im Ahrtal und ihre Folgen für den zukünftigen Hochwasserschutz. Wasser Brunnen 14:43–49 (https://publications.rwth-aachen.de/record/850877)

Schüttrumpf H, Birkmann J, Brüll C, Burghardt L, Johann G, Klopries E, Lehmkuhl F, Schüttrumpf A, Wolf S (2022b) Herausforderungen an den Wiederaufbau nach dem Katastrophenhochwasser 2021 in der Eifel. In: Technische Universität Dresden, Institut für Wasserbau und technische Hydromechanik (Hrsg) Nachhaltigkeit im Wasserbau – Umwelt, Transport, Energie. Dresdner Wasserbauliche Mitteilungen 68. Technische Universität Dresden, Institut für Wasserbau und technische Hydromechanik, S 5–16

Schüttrumpf H, Brüll C, Klopries E-M, Lehmkuhl F, Wolf SS (2022c) Das Hochwasser 2021 in Nordrhein-Westfalen und Rheinland-Pfalz: Beobachtungen und Erfahrungen. In: Institut für Siedlungswasserwirtschaft der RWTH Aachen University (Hrsg) Gewässerschutz, Wasser, Abwasser, Wasserwirtschaft im Klimawandel: digitale 55. Essener Tagung für Wasserwirtschaft. Gesellschaft zur Förderung des Instituts für Siedlungswasserwirtschaft an der RWTH Aachen e. V., (https://publications.rwth-aachen.de/record/847668)

Schwanen C, Bellanova P, Schwarzbauer J (2022) The 2021 Flood Disaster in Germany; Distribution, remobilization and accumulation of organic pollutants along the natural floodplains of the Rur river https://doi.org/10.5194/egusphere-egu22-12558

Seckler C, Wolf S (2024) Zur Machbarkeit eines Tunnels zur Hochwasserentlastung von der Ahr in den Rhein. Wasser Abfall 26(5):51–57

Weber A, Lehmkuhl F (2024) Mixed response of trace element concentrations in fluvial sediments to a flash flood in a former mining area. Environ Sci Eur. https://doi.org/10.1186/s12302-024-00926-5

Weber A, Wolf S, Becker N, Märker-Neuhaus L, Bellanova P, Brüll C, Hollert H, Klopries E-M, Schüttrumpf H, Lehmkuhl F (2023) The risk may not be limited to the flooding – Polluted high flood sediments pose a health threat to the unaware public. Environ Sci Eur. https://doi.org/10.1186/s12302-023-00765-w

Zenker M-L, Bubeck P, Thieken AH (2024) Always on my mind: indications of post-traumatic stress disorder among those affected by the 2021 flood event in the Ahr valley, Germany. Nat Hazards Earth Syst Sci 24(8):2837–2856. https://doi.org/10.5194/nhess-24-2837-2024

Gebäudeschäden

Guilherme Samprogna Mohor, Michael Schäfer,
Aaron Buhrmann und Lothar Kirschbauer

J. Birkmann et al. (Hrsg.), *Die Flutkatastrophe 2021*, Sustainable Development Goals (SDG) – Umsetzung in Praxis, Lehre und Entscheidungsprozessen,
https://doi.org/10.1007/978-3-658-48688-4_7

7.1 Einleitung

Nach Flutereignissen wie Hochwasser oder Überschwemmungen durch Starkregen stehen viele Menschen vor erheblichen finanziellen Herausforderungen, insbesondere aufgrund der Schäden an ihrem Eigentum. Gebäude sind in besonderem Maße anfällig für gravierende Schäden, die sowohl die Struktur als auch die Funktionalität beeinträchtigen können, wenn sie Flutereignissen ausgesetzt sind. Das Zusammenwirken intensiver Regenfälle, steigender Wasserstände und unzureichender Drainage führt in der Konsequenz häufig zu Überflutungen, welche unmittelbare und langfristige Auswirkungen auf Bauwerke haben. Die Auswirkungen auf Gebäude sind vielfältig und reichen von beschädigten Fundamenten und Wänden bis hin zu beeinträchtigten elektrischen, energetischen und sanitären Anlagen. Um effektive Schutzmaßnahmen entwickeln und Sanierungsstrategien planen zu können, ist ein Verständnis dieser Zusammenhänge von entscheidender Bedeutung. Daher ist die Schadensdokumentation essenziell, z. B. für die Dimensionierung von Kompensationsfonds, die Kosten-Nutzen-Analyse der Vorsorgemaßnahmen und das Verständnis der Schadensprozesse, um flutbedingte Risiken zukünftig zu verringern. Jedoch steht eine vollständige Dokumentation der Schäden an individuellen Gebäuden nach dem Hochwasser 2021 auch drei Jahre später noch aus. Denn aufgrund der großen räumlichen Ausdehnung des Hochwasserereignisses, das sich über 12 Landkreise in zwei Bundesländern erstreckte, konnte keine Institution eine detaillierte Schadenserfassung an einzelnen Häusern und Objekten vor Ort durchgeführt werden. Die gewählte Methodik bedient sich daher einer großräumigen Kartierung eines Großteils des Ahrtals auf Basis von Luftbildern, welche manuell und (teil-)automatisiert analysiert wurden.

In diesem Kapitel werden Schadensbilder an Gebäuden erläutert, die in ausgewählten betroffenen Gemeinden durch Luftbilder und Befragungen erhoben wurden. Anhand dieser Daten wurden zwei Analysen durchgeführt: (1) der Zusammenhang zwischen strukturellen Schadensklassen und der Gebäudeposition relativ zum Flussverlauf und (2) der Zusammenhang zwischen strukturellen Schadensklassen und finanziellen Schäden.

7.2 Beschreibung typischer Schadensbilder

Zu den am häufigsten auftretenden Schadensarten im Ahrtal gehörten Überflutungen durch eindringendes Wasser, Kontaminationen, Anprallschäden und Unterspülungen, die jeweils spezifische Sanierungs- und Präventionsmaßnahmen erfordern. Beobachtungen in den betroffenen Gemeinden legen den Schluss nahe, dass der Standort eines Gebäudes maßgeblich für Art und Schwere des Schadens ist. So zeigen sich in vielen Orten die gravierendsten Schadensbilder (vollständige Zerstörung) in unmittelbarer Nähe zum Gewässer. Ob die Schadensgrade mit zunehmender Entfernung des Gebäudes vom Gewässer abnehmen, soll hier quantitativ untersucht werden (s. ► Abschn. 7.4.2). Die beschriebenen Beobachtungen zum Schadensbild im Ahrtal decken sich dabei mit den Ergebnissen von Wüthrich et al. (2024) und Maiwald et al. (2022).

Die meisten der im Folgenden beschriebenen, durch Hochwasser verursachten Schadensbilder können mit hohen Kosten verbunden sein, beispielsweise wenn eine durch Überflutung durchfeuchtete Dämmung einer vollständigen Erneuerung bedarf. Die Beseitigung daraus entstandener Schäden kann insbesondere kostspielig und zeitintensiv sein, da die Gebäude-/Wärmedämmung in vielen Fällen großflächig mit der Gebäudestruktur verbunden ist. Sollte das eindringende Wasser zusätzlich kontaminiert sein, wird der Sanierungsaufwand noch deutlich größer.

Der Begriff der Kontamination bezeichnet die Verschmutzung von Wasser, Boden und Luft durch Schadstoffe, welche mit dem Flutwasser transportiert wurden und beispielsweise in den Putz der Hausfassaden eindrangen. Zu den häufigsten Schadstoffen zählen beispielsweise Haushaltschemikalien, Öl, Treibstoffe sowie schwermetallhaltige industrielle Abfälle. Diese können ein Risiko für die Gesundheit von Menschen darstellen und langfristig eine Schädigung der Umwelt zur Folge haben. Die Beseitigung solcher Kontaminationen kann mitunter aufwendig sein und erfordert häufig spezialisierte Reinigungs- und Sanierungsmaßnahmen.

Anprallschäden resultieren aus der kinetischen Energie, welche durch mitgerissene Gegenstände (z. B. Gastanks), Trümmer und Geröll auf Bauwerke und Infrastrukturen übertragen wird. Entsprechende Schäden stellen ein besonderes Problem dar, da sie die strukturelle Integrität der Gebäude gefährden und das Risiko von Einstürzen erhöhen. Die Beseitigung solcher Schäden ist oft komplex und erfordert eine genaue Untersuchung der betroffenen Strukturen, um sicherzustellen, dass alle Schäden identifiziert und behoben werden.

Eine Unterspülung liegt vor, wenn der Boden unter Bauwerken und Straßen durch strömendes Wasser erodiert. Dies resultiert in einer Beeinträchtigung der Stabilität der betroffenen Strukturen, welche im Extremfall zum vollständigen Zusammenbruch des Gebäudes führen kann. Erosionsschäden sind in der Regel nur schwer prognostizierbar und gehen mit einem beträchtlichen finanziellen Aufwand für die Instandsetzung und Sicherung der betroffenen Bauwerke einher. Unterspülung als Schadensart ist im Nachhinein durch Luftbilder alleine nur bedingt abzuleiten, falls die betroffenen Gebäude in Folge der Unterspülung einstürzen oder abgerissen werden. Dieser Umstand unterstreicht die in ▶ Kap. 6 beschriebenen Notwendigkeit zur Ergebung von kurzlebigen Daten im Krisenfall.

7.3 Methodisches Vorgehen

Fernerkundungsdaten erlauben neue automatisierte oder semi-automatisierte Ansätze für eine schnelle und räumlich breite Erfassung der Schäden. Die nach dem Hochwasserereignis von 2021 im Ahrtal durchgeführte Studie von Samprogna Mohor et al. (2024) untersuchte, ob eine schnelle, teilautomatisierte Schadensdokumentation mittels Luftbilder eine verlässliche Methode darstellt. Mit Hilfe einer standardisierten Schadensklassifikation und Luftbildern der Region wurden strukturelle Schäden an 612 Gebäuden durch drei unabhängige Bewerter kartiert. Mittels einer Analyse der Interrater-Reliabilität nach O'Connor und Joffe (2020) konnte gezeigt werden, dass die Datenerhebung anhand von Luftbildern eine gute Übereinstimmung mit einer unabhängigen Vor-Ort-Kartierung aufweist. Darüber hinaus belegte die Fallstudie, dass die standardisierten Schadensklassen für Dritte verständlich und anwendbar sind und dass die Kartierung von strukturellen Gebäudeschäden anhand von Luftbildern eine geeignete Methode darstellt, um strukturelle Gebäudeschäden in größeren Gebieten schnell zu erfassen. Mit der genannten Methode wurden über 7000 betroffene Gebäude im Ahrtal kartiert. In etlichen Gemeinden wurde darüber hinaus die Lage der Gebäude in Bezug auf das nächste Gewässer erfasst, da diese einen Einfluss auf Gebäudeschäden haben kann. Diese Informationen können wiederum beim Wiederaufbau hilfreich sein, um geeignete Vorsorgemaßnahmen gezielt auszuwählen. Geographische Informationssysteme (GIS) ermöglichen die Verknüpfung von Basiskatasterdaten und neu erfassten Daten sowie die Erstellung räumlicher Daten. Die OpenStreetMap-Gebäudepolygone (OpenStreetMap contributors 2021) dienten als Basisdaten, um die Schäden sowie weitere Attribute zu kartieren.

Neben der Auswertung der Luftbilder dienten zwei Befragungen, die im Rahmen des KAHR-Projekts in der betroffenen Region durchgeführt wurden, als weitere Datengrundlage. Eine Umfrage wurde zwischen 2022 und 2023 durch das HochwasserKompetenz-Centrum e. V. (HKC 2025; Forschung für Nachhaltigkeit I FONA 2025) durchgeführt. Über die Antworten konnten sowohl der Gebäudeschaden als auch die Position relativ zum Gewässer bestimmt werden. Die zweite Befragung wurde zwischen November und Dezember 2022 durch die Universität Potsdam durchgeführt und umfasste sowohl strukturelle als auch finanzielle Gebäudeschäden (Dillenardt und Thieken 2024).

Die Erfassung von Gebäudeschäden erfordert eine einheitliche Beschreibung, die für verschiedene Erfasser:innen verständlich und anwendbar sein sollte. Im Folgenden werden die Schadensskala nach Maiwald und Schwarz (2019) sowie die verwendeten Luftbilder- und Umfragedaten beschrieben. Die Skala wurde von Maiwald und Schwarz (2022) erweitert, um auch Schäden an Fachwerkgebäuden zu klassifizieren. Diese Erweiterung wurde jedoch in dieser Studie nicht berücksichtigt.

7.3.1 Vereinheitlichte Schadensbeschreibung für strukturelle Gebäudeschäden

Die Klassifizierung der im Flutgebiet entstandenen Gebäudeschäden erfolgte anhand der Schadensklassifikation nach Maiwald und Schwarz (2019). Diese stellt ein differenziertes System zur Bewertung von Gebäudeschäden, insbesondere in Folge von Naturkatastrophen, dar. Die Unterteilung der Schäden erfolgt in sechs verschiedene Klassen, wobei der Umfang und die Schwere der Beschädigungen als Kriterien herangezogen werden (◘ Abb. 7.1). Obgleich die Schadensbeschreibung auch Schäden nach Kontamination sowie nicht-strukturelle Schäden umfasst, orientiert sich die externe visuelle Bewertung an den erforderlichen Maßnahmen, beispielsweise dem Austausch von Ausbauteilen, nicht- oder tragenden Wänden oder einem erforderlichen Abriss. Die Klassifikation gewährleistet eine standardisierte Erfassung und Bewertung von Gebäudeschäden, was für die Schadensregulierung, die Planung von Instandsetzungsmaßnahmen sowie die Risikobewertung von essenzieller Bedeutung ist (Maiwald und Schwarz 2019).

Die Einteilung der Schadensgrade erfolgt in die Klassen D1 bis D6. Im Kern basiert das System auf einer Differenzierung zwischen leichten, mittleren, schweren und sehr schweren Schäden. Leichte Schäden (D1) manifestieren sich in der Regel in Form von kosmetischen Beeinträchtigungen oder geringfügigen strukturellen Beschädigungen

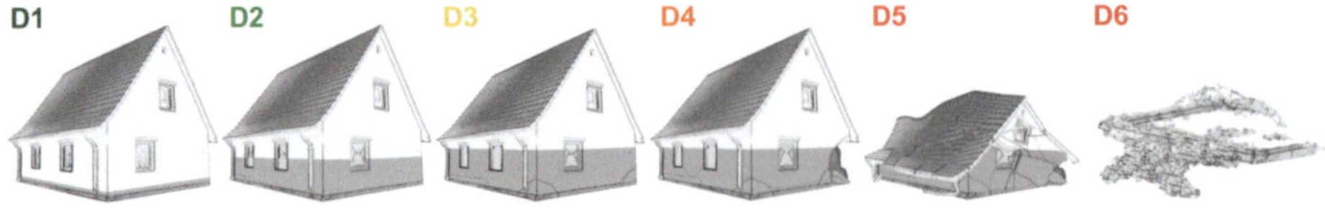

◘ **Abb. 7.1** Abgebildete Schadensklassen nach Maiwald und Schwarz (2019)

Kriterium	Beobachtung/Maßnahme	Schadensgrad					
		D1	D2	D3	D4	D5	D6
Bauphysikalischer Schaden	Durchfeuchtung tragender und nicht tragender Wände und der Geschossdecken	x	x	x	x	x	x
Chemischer Schaden	Verschmutzungen (Schlamm, Ablagerungen)	x	x	x	x	x	x
	Kontaminationen (Öl, Chemikalien)		x	x	x	x	x
Mechanischer Schaden	Eingedrückte Fenster und Türen *oder Geländer/Brüstungen*		x	x	x	x	x
	Leichte Risse in tragenden Wänden		x	x	x	x	x
	Unterspülte Fundamente		x	x	x	x	x
	Teilabgedeckte Dächer		x	x	x	x	x
	Größere Risse und/oder Verformungen in tragenden Wänden und Decken			x	x	x	x
	Setzungen			x	x	x	x
	Einsturz von nicht tragenden Wänden			x	x	x	x
	Einsturz von Bauteilen (tragende Wände, Decken, *Dächer*)				x	x	x
	Kollaps oder Einsturz von größeren Gebäudeteilen					x	x
	Bauwerk vollständig weggeschwemmt, umgestürzt oder vom Fundament verschoben						x

Abb. 7.2 Beschreibung der Schadensklassen und gekennzeichnete Ergänzungen bzw. Abweichungen in Anlehnung an Maiwald und Schwarz (2019)

an Gebäuden, welche ohne größeren Aufwand repariert werden können. Die Kategorie der mittelschweren Schäden (D2 und D3) umfasst gravierendere strukturelle Beeinträchtigungen, welche eine umfangreichere Instandsetzung erforderlich machen, jedoch die Statik des Gebäudes nicht substanziell gefährden. Schwere Schäden (D4) betreffen wesentliche strukturelle Elemente des Gebäudes und bedeuten in der Regel eine Gefährdung der Standsicherheit, sodass umfassende Sanierungsarbeiten erforderlich sind. Sehr schwere Schäden (D5 und D6) implizieren eine derartige Beeinträchtigung der Bausubstanz, dass eine Wiederherstellung oft nicht mehr wirtschaftlich sinnvoll oder technisch umsetzbar ist. Dies hat in der Praxis häufig einen Abriss oder eine Räumung des Grundstücks sowie einen Neubau zur Folge. Allerdings können auch leichte Schäden gemäß dieser Klassifikation, wie beispielsweise die Durchfeuchtung, Verschmutzung oder Kontamination von nicht tragenden Wänden, in der Praxis mit hohen Kosten für die Betroffenen und Eigentümer:innen einhergehen (Maiwald und Schwarz 2019).

Die vorliegende Klassifikation erlaubt es Fachleuten aus den Bereichen Bauwesen, Versicherungswirtschaft, Katastrophenmanagement und anderen relevanten Bereichen eine gemeinsame Sprache bei der Beurteilung und Bewältigung von Bauschäden zu verwenden. Dies fördert die Effizienz und Effektivität im Umgang mit solchen Schäden. Im Rahmen des BMBF-Projekts KAHR wurde diese Schadensklassifikation eingesetzt, um den durch die Flut beschädigten Gebäuden Schadensgrade zuzuordnen. Aufgrund der Tatsache, dass bei der Erhebungsmethode nicht alle Kriterien aus der Systematik nach Maiwald und Schwarz (2019) eindeutig bestimmbar waren, wurden diese um Schadensbilder wie „Eingedrückte Geländer/Brüstungen" oder „Teilabgedeckte Dächer" ergänzt. Diese Ergänzungen sind in Abb. 7.2 blau-kursiv dargestellt; in Rot sind Merkmale gekennzeichnet, die in Luftbildern nicht oder nur schwer erkennbar sind.

7.3.2 Verwendete Daten

7.3.2.1 Luftbilddaten

Im Rahmen der Schadenskartierung wurden zwei Befliegungsdatensätze verwendet: Zum einen wurden Orthophotos des Ahrtals verwendet, die das Landesamt für Vermessung und Geobasisinformation (LVermGeo) Rheinland-Pfalz regelmäßig (i. d. R. alle zwei Jahre) durch Befliegungen erhebt. Des Weiteren führte das LVermGeo kurz nach dem Ereignis eine Sonderbefliegung durch. Die freien Daten bieten eine Vorher-Nachher-Bildergalerie, die auf der Website verfügbar ist (GeoBasis-DE / LVermGeoRP 2021a).

Zum anderen entwickelte das Unternehmen GeoFly GmbH mit Hilfe von Schrägbildern, die aus einer Nadir-Position plus vier Schrägbildern pro Aufnahme abgeleitet wurden, ein texturiertes 3D-Modell des Ahrtals. Die Auflösung des Modells beträgt 6,5 cm pro Pixel und umfasst den Bereich von Müsch bis Sinzig. Die Daten können über den browserbasierten Dienst „ObViewSly" im Schrägbild-Viewer eingesehen werden (GeoFly 2023). Für die im Weiteren beschriebene Auswertung der GeoFly Daten wurde die erweiterte Navigation von virtualcitysystems GmbH benutzt.

7.3.2.2 Umfragedaten des Hochwasser Kompetenz Centrums

Das HKC führte im Rahmen seiner vor Ort Beratungstätigkeit mit dem KAHR-Infomobil zwischen Mai 2022 und Mai 2023 eine Befragung der vom Hochwasser 2021 Betroffenen in NRW und RLP durch (HKC 2025; Forschung für Nachhaltigkeit I FONA 2025). Der Fragebogen konnte manuell in Papierform oder digital über einen QR-Code als Online-Version ausgefüllt werden. Die Befragung wurde in Zusammenarbeit mit der Hochschule Koblenz erstellt, die Fragen zu Schadensbildern und zur Lage von Gebäuden in Relation zum Hochwasser (◘ Abb. 7.3) erhoben hat. In der aus Datenschutzgründen anonymisierten Umfrage wurde neben dem Schaden der Objekte auch deren Position im Verhältnis zum Gewässer abgefragt. Den Teilnehmenden wurde dazu eine schematische Karte an die Hand gegeben, mit der diese Charakteristik angegeben werden konnte. Ein Rückschluss auf den tatsächlichen Standort war damit jedoch nicht möglich. In ausgewählten Gemeinden wurde dieser relativ zum Gewässer gelegenen Standort auch in der Luftbildauswertung bestimmt. Insgesamt standen so für die Analyse des Zusammenhangs zwischen Gebäudestandort und zugehörigen Schäden 1602 Gebäudedaten aus der Luftbildauswertung und 357 Datensätze aus der Umfrage des HKC zur Verfügung.

7.3.2.3 Haushaltsbefragung der Universität Potsdam

Die Universität Potsdam hat eine Befragung durchgeführt, die sich unter anderem mit den Merkmalen des Hochwassers, der Frühwarnung und den entstandenen Auswirkungen befasste (siehe ► Kap. 5 und 6 für eine weitere Beschreibung und Diskussion der Befragung, sowie Dillenardt und Thieken 2024). Der Fragebogen basiert auf früheren Befragungen, die in Deutschland nach Hochwasserereignissen durchgeführt wurden (Thieken et al. 2017). Im stark von der Flut betroffenen Landkreis Ahrweiler in Rheinland-Pfalz (RLP) wurden Briefe mit einem Online-Zugang zum Fragebogen von der Kreisverwaltung an eine Zufallsstichprobe von ca. 5250 Haushalten verschickt, die Soforthilfe beantragt hatten. In Nordrhein-Westfalen (NRW) wurde über Facebook für die Umfrage geworben. Die Umfrage lief zwischen November und Dezember 2022, währenddessen wurden insgesamt 609 Fragebögen ausgefüllt (277 in RLP, 332 in NRW). Für die vorgenommenen Analysen wurden aus den Befragungsdaten nur die Antworten zu den strukturellen und finanziellen Schäden herangezogen, sodass insgesamt 402 Datensätze zur Verfügung

- Hintergrundkarte: ©GeoBasis-DE / LVermGeoRP (2021), Digitale Orthophotos (historischer Stand 2021), https://geoshop.rlp.de/historische_luftbilder/orthophotos_historisch_2021.html, dl-de/by-2-0 (www.govdata.de/dl-de/by-2-0)

Abb. 7.3 Klassifikation des Gebäudestandorts. (Eigene Darstellung. Orthophotos: © GeoBasis-DE / LVermGeoRP (2021b), dl-de/by-2-0, ▶ http://www.lvermgeo.rlp.de)

standen. Aus diesen Daten kann der Zusammenhang zwischen strukturellen und finanziellen Gebäudeschäden abgeleitet werden, damit eine schnelle Schätzung der finanziellen Schäden entsprechend der kartierten strukturellen Schäden berechnet werden kann.

7.4 Statistische Analyse der Gebäudeschäden

7.4.1 Schadensanalyse nach Orts-/Gemeinde

Nach der Fallstudie von Samprogna Mohor et al. (2024), in welcher Schäden für 612 Gebäude sowohl vor Ort als auch durch Luftbilder unabhängig voneinander bewertet wurden, wurden weitere 7197 der betroffenen Gebäude in sechs Ortsgemeinden bzw. Gemeinden im Ahrtal kartiert. Dabei wurden alle betroffenen Gebäude in jenen direkt an der Ahr gelegenen Gemeinden zwischen Altenahr und Sinzig kartiert. Die Tab. 7.1 zeigt die Anzahl der kartierten Schadensklassen nach Gemeinde. Die meisten Gebäude

Tab. 7.1 Anzahl kartierter Schadensklassen nach Gemeinde

Orts-/Gemeinde	D1	D2	D3	D4	D5	D6	Summe pro Gemeinde
Altenahr	42	57	258	60	29	42	488
Mayschoß	45	86	114	17	10	11	283
Rech	59	52	25	14	2	19	171
Dernau	88	277	324	38	13	18	758
Bad Neuenahr-Ahrweiler	1839	1884	851	97	14	18	4703
Sinzig	408	310	49	13	3	11	794
Summe pro Schadensklasse	2481	2666	1621	239	71	119	7197

7

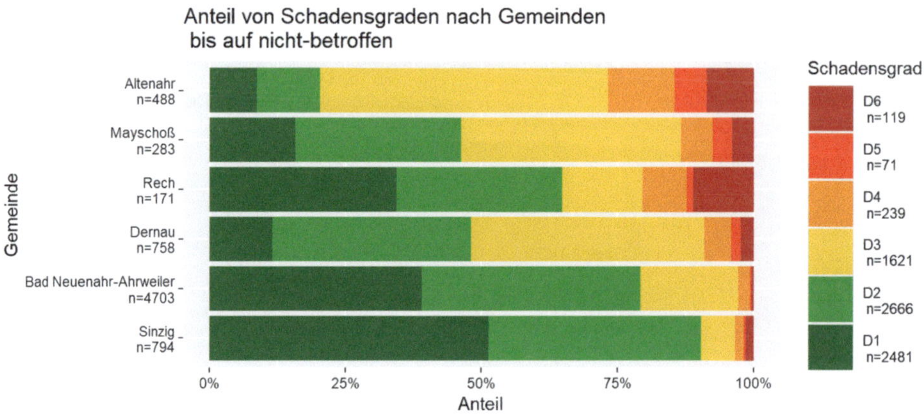

Abb. 7.4 Anteil von kartierten Schadensklassen nach Gemeinden. (Eigene Darstellung)

sind Wohngebäude. Gebäude mit gemischter oder rein kommerzieller Nutzung wurden nicht separat gekennzeichnet.

Abb. 7.4 zeigt zusätzlich die relative Verteilung der Schadensgrade pro untersuchter Gemeinde. Die Gemeinden sind dabei in Fließrichtung entlang des Flussverlaufs angeordnet. Während in den Gemeinden Altenahr, Mayschoß und Rech die Ahr stark mäandriert und das Tal vergleichsweise schmal ist, weitet sich dieses ab dem Stadtgebiet Bad Neuenahr-Ahrweiler. Dies macht sich in einem geringeren Anteil an schweren Schäden für die Richtung Ahrmündung gelegenen Ortsgemeinden Bad Neuenahr-Ahrweiler und Sinzig bemerkbar. Das breitere Tal führte dabei zu einer geringeren Fließgeschwindigkeit, flacheren Wasserständen aber gleichzeitig auch zu größeren Überflutungsflächen. Die insgesamt höhere Anzahl an betroffenen Gebäuden lässt sich mit der Größe der Stadt Bad Neuenahr-Ahrweiler im Vergleich zu den übrigen Ortschaften erklären.

Das untersuchte Gebiet ist in Abb. 7.5 unten dargestellt. Beispielhaft zeigt die Abb. 7.5 oben die räumliche Verteilung der kartierten Schadensklassen in der ausgewählten Ortsgemeinde Dernau. Bereits in dieser grafischen Darstellung lässt sich der Zusammenhang zwischen Gebäudeposition relativ zum Gewässer und dem Schadensgrad erkennen. In Dernau traten die leichten Schäden (D1) entfernt von der Ahr am Rand des Überflutungsgebiets auf (Abb. 7.5).

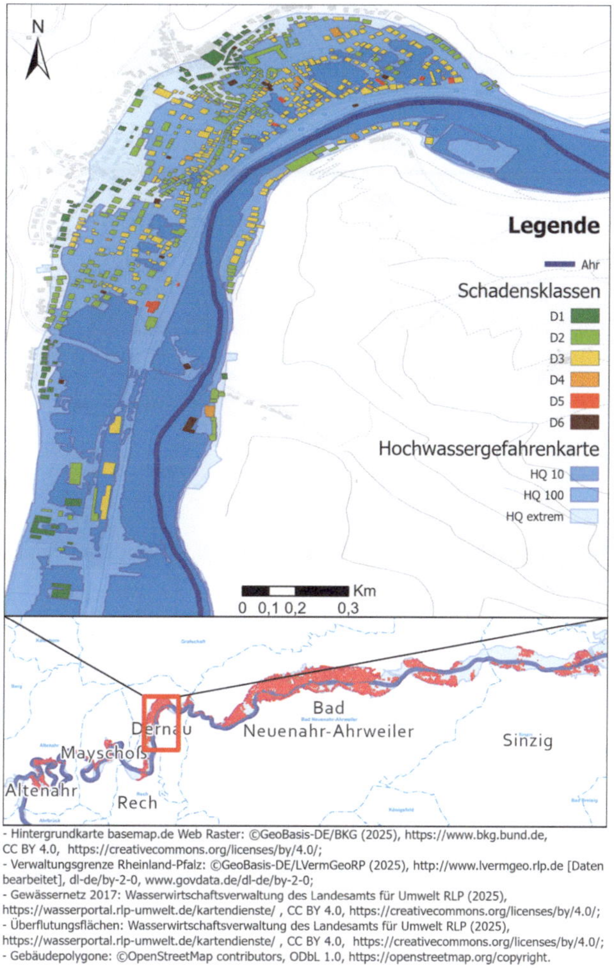

Abb. 7.5 Karte des Untersuchungsgebietes von Altenahr bis Sinzig (*unten*) und beispielhafte Darstellung der klassifizierten Gebäude nach strukturellen Schadensklassen in der Ortsgemeinde Dernau (*oben*). (Eigene Darstellung. Hintergrundkarte: © GeoBasis-DE / BKG (2025); Verwaltungsgrenze Rheinland-Pfalz: © GeoBasis-DE / LVermGeoRP (2025); Gewässernetz 2017: Wasserwirtschaftsverwaltung des Landesamts für Umwelt RLP (2025a); Überflutungsflächen: Wasserwirtschaftsverwaltung des Landesamts für Umwelt RLP (2025b); Gebäudepolygone: © OpenStreetMap contributors (2022), ODbL 1.0, ► https://openstreetmap.org/copyright)

7.4.2 Zusammenhang zwischen Gebäudeposition und Schadensgrad

In diesem Kapitel werden die Schadensgrade mit der Gebäudeposition in Verbindung gebracht. Dabei konnte ein signifikanter Zusammenhang zwischen der Positionierung der Gebäude relativ zum Gewässer und dem Schadensausmaß im Ahrtal gezeigt werden. Dieser Zusammenhang manifestiert sich sowohl in der Auswertung der Luftbilder als auch in den Umfragen des HKC. Es lässt sich konstatieren, dass der Schadensgrad D6, der eine vollständige Zerstörung des Gebäudes impliziert, signifikant häufig an der Position auftritt, die sich in nächster Nähe zum Fluss befindet und direkt von der Strömung getroffen wird. Andererseits lässt sich feststellen, dass der Schadensgrad D6 umso seltener auftritt, je weiter das Gebäude vom Fluss entfernt liegt. Diese Beobachtung lässt sich durch Anprallschäden erklären, welche am gravierendsten sind, wenn die Flutwelle, gegebenenfalls mit Treibgut, ungebremst mit hoher Geschwindigkeit auf das Gebäude trifft. Dies trifft zu, wenn das Gebäude in der vordersten Reihe steht und direkt von der Flut-

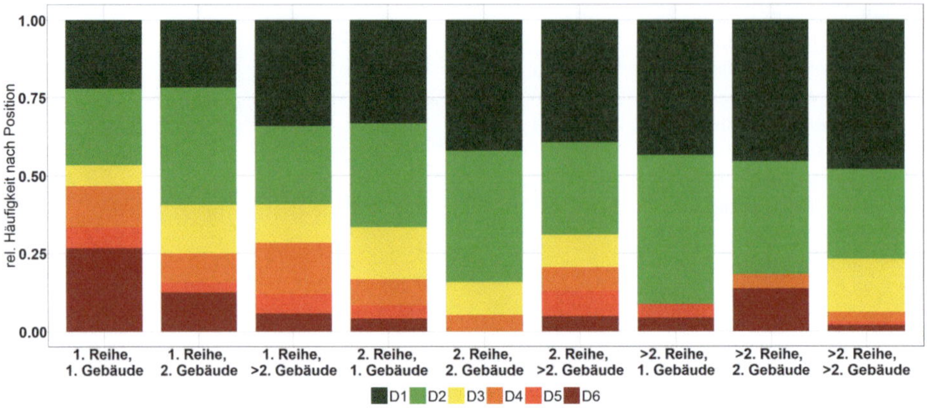

Abb. 7.6 Schadensanalyse anhand der Luftbilder. (Eigene Darstellung)

welle angeströmt wird (vgl. Abb. 7.6, 1. Reihe, 1. Gebäude; Kirschbauer und Schäfer 2024).

In ähnlicher Weise nimmt die Häufigkeit schwerwiegender Schadensgrade (D4 bis D6) in den weiter zurückliegenden Reihen ab (Abb. 7.6, > 2. Reihe, X. Gebäude). Dort sind typischerweise die Schadensgrade D1 bis D3 vorherrschend. Dies lässt den Schluss zu, dass die Gebäude in den hinteren Reihen durch die davorstehenden Gebäude vor Anprall geschützt werden bzw. dort die Fließgeschwindigkeit und die Wassertiefe geringer ist. Des Weiteren zeigt die Auswertung, dass die Wahrscheinlichkeit einer Unterspülung abnimmt, je weiter das Gebäude vom Gewässer entfernt steht. Es sei an dieser Stelle erneut darauf verwiesen, dass auch die Schadensgrade D1 bis D3 mit signifikanten Kosten für die Sanierung verbunden sein können, beispielsweise im Falle einer Durchfeuchtung der Fassadendämmung.

Dieses Muster der Schadensverteilung ist sowohl in der Luftbildauswertung als auch in der HKC-Befragung zu finden. Es sei jedoch darauf hingewiesen, dass die Befragten der Umfrage des HKC in der Regel Flutbetroffene waren, deren Häuser nicht vollständig zerstört wurden (Abb. 7.7). Daher lässt sich annehmen, dass der Schadensgrad D6 (vollständige Zerstörung des Bauwerks) in den Daten des HKC unterrepräsentiert ist.

7.4.3 Zusammenhang zwischen strukturellen und finanziellen Schäden

Aus den Daten der Haushaltsbefragung der Universität Potsdam (Dillenardt und Thieken 2024) wurde der Zusammenhang zwischen strukturellen und finanziellen Gebäudeschäden ermittelt. Aus den Antworten der Befragten konnten unter anderem der Schadensgrad des Gebäudes sowie die finanziellen Aufwendungen zur Behebung dieses Schadens abgeleitet werden. Aus allen Antworten wurden der Median und der Mittelwert (Durchschnitt) der absoluten finanziellen Schäden pro Schadensklasse (D1–D6) berechnet. Durch die geringere Anzahl von Gebäuden in höheren Schadensklassen werden die Werte allerdings unsicherer. Daher wurden zwei weitere Modelle abgeleitet, mit dem Ziel, die Gesamt-Gebäudeschäden in den kartierten Gemeinden abzuschätzen: ein einfaches lineares Quadrat-

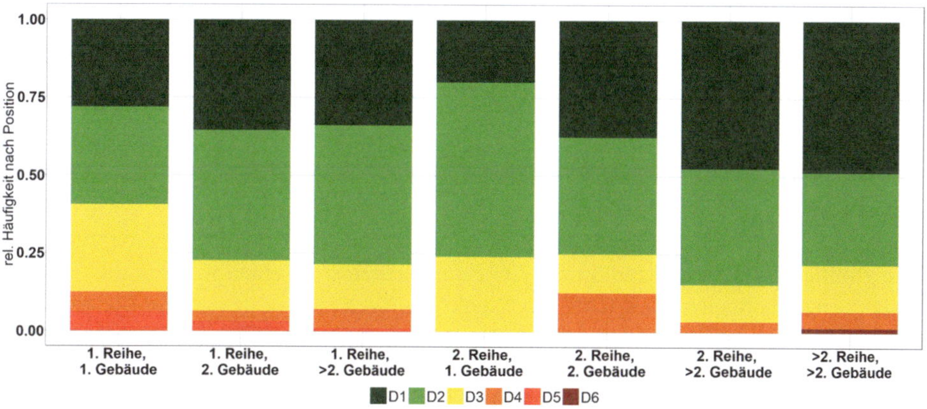

Abb. 7.7 Schadensanalyse anhand der Umfrage des HKC. (Eigene Darstellung)

Tab. 7.2 Geschätzte finanzielle Gebäudeschäden je Gemeinde in Euro nach vier Modellansätzen

Gemeinde	Modell 1: Median	Modell 2: Mittelwert	Modell 3: Monotones Modell	Modell 4: Quadratwurzelmodell
Altenahr	141.778.000	220.180.038	181.115.889	176.465.249
Mayschoß	66.872.000	111.454.015	87.493.372	85.332.040
Rech	31.933.000	53.087.802	46.780.525	46.270.271
Dernau	176.528.000	298.479.242	231.413.977	225.751.723
Bad Neuenahr-Ahrweiler	707.281.000	1.438.690.648	1.032.123.454	1.011.859.893
Sinzig	95.873.000	212.958.834	151.314.151	149.049.501
Gesamt (im untersuchten Gebiet)	1.220.265.000	2.334.850.579	1.730.241.368	1.694.728.677

wurzelmodell und ein monotones Modell, bei dem jede höhere Stufe einen nichtlinearen additiven Effekt hat. Trainiert wurde das monotone Bayes'sche Modell mit der probabilistischen Programmiersprache „Stan“ (Stan Development Team 2024), die mit dem Paket „brms“ (Bürkner 2017), einer „Stan“-Schnittstelle für „R“, implementiert wurde. Für die Berechnung der finanziellen Gebäudeschäden wurden aus den vier Modellen jeweils nur die Mittelwerte der Parameter pro Schadensklasse verwendet. Dabei wurde die Anzahl der Gebäude pro Schadensklasse mit dem durchschnittlichen monetären Wert pro Schadensklasse multipliziert. Tab. 7.2 zeigt die geschätzten Gesamtschäden je Gemeinde.

Die geschätzten Gebäudeschäden aus diesen Modellen berücksichtigen keine Schäden an gewerblichen Objekten, die in der Regel deutlich höhere Schäden durch eine Flut erleiden (siehe ▶ Kap. 8). Darüber hinaus sind in den Modellen die Schäden an Hausrat und Inventar nicht berücksichtigt, sodass die Modelle die aufgetretenen Schäden eher unterschätzen. Erwartungsgemäß ergeben die Schätzungen mit dem Mittelwert die höchsten Gesamtschäden. Die beiden abgeleiteten Modelle liefern sehr ähnliche Ergebnisse, während der Ansatz mit dem Median immer eine konservative Schätzung ergibt, da der

Median die höchsten Werte nicht berücksichtigt. Ein Abgleich mit den tatsächlich gemeldeten Gebäudeschäden ist, unter anderen aus Datenschutzgründen, leider nicht möglich. Dies wäre aber wichtig, um zu entscheiden, welches Modell in Zukunft, z. B. für eine schnelle Schadensabschätzung im Ereignisfall, am besten verwendet werden könnte.

7.5 Fazit

Die großflächige Dokumentation von Gebäudeschäden stellt einen wesentlichen, jedoch bislang unzureichend berücksichtigten Aspekt des Risikomanagements dar. Die beschriebenen Studien aus dem BMBF-geförderten Projekt KAHR zeigen, dass eine systematische Erfassung von Schadensgraden mittels Fernerkundungsdaten möglich ist. Des Weiteren besteht die Möglichkeit mittels Befragungsdaten die Zusammenhänge zwischen strukturellen Schäden und finanziellen Schäden sowie der Lage des Gebäudes zu ermitteln. Auf Basis von Luftbildern wurden etwa 7200 betroffene Gebäude zwischen den Gemeinden Altenahr und Sinzig im Ahrtal kartiert.

Diese Studie zeigte, dass die relativ zeitnah erfassten strukturellen Gebäudeschäden allein im Untersuchungsgebiet einem finanziellen Schaden zwischen 1,2 und 2,3 Mrd. EUR entsprechen. Obwohl die Methode eine Unterschätzung der realen Schäden wahrscheinlich macht, da Kosten für z. B. Hausrat und Grundstücksräumung/-reinigung durch die Modelle nicht berücksichtigt werden, sind die ermittelten Werte von Nutzen. Ein Abgleich mit den tatsächlich gemeldeten Gebäudeschäden ist bedauerlicherweise nicht möglich, wäre jedoch von entscheidender Bedeutung, um zu eruieren, welches Modell in Zukunft, beispielsweise für eine zeitnahe Schadensabschätzung im Ereignisfall, die besten Ergebnisse liefert.

Hinweis für Entscheidungsprozesse

Die öffentliche Verwaltung könnte damit zukünftig aus Luftbildaufnahmen eine schnelle Übersicht über die Schadensbilder erlangen und darüber hinaus diese auch für weitere langfristige Analysen nutzen.

Die Schadensbilder innerhalb der Gemeinden variieren stark. Die Lage der Gebäude zum Fluss ist ein Faktor, der das Risiko ihrer Beschädigung beeinflusst. Diese Information ist sowohl bei der Planung des Wiederaufbaus als auch bei der Konzeption von Vorsorgemaßnahmen zu berücksichtigen. Eine hohe Vulnerabilität ist für Gebäude in unmittelbarer Nähe zum Gewässer zu beobachten, sodass dort die Umsetzung besonderer baulicher Maßnahmen erforderlich ist, um zukünftigen Ereignissen standhalten zu können.

Hinweis für Entscheidungsprozesse

Die durchgeführten Analysen belegen den Wert von Befragungs- und Fernerkundungsdaten für das Risikomanagement, sowohl im Nachgang zu einem Ereignis als auch in der Vorbereitungsphase, da beispielsweise besonders gefährdete Gebäudepositionen identifiziert wurden. Daher wird empfohlen, die Datenerfassung mit Luftbildern von gefährdeten

Gebieten kontinuierlich zu betreiben und Gefährdungsanalysen auch auf Basis der Gebäudeposition relativ zum Fluss durchzuführen.

Zusammenfassung

- Zur systematischen Erfassung und Analyse von im Ahrtal im Jahr 2021 entstandenen Schäden an ca. 7200 privaten Wohngebäuden wurden sowohl Luftbild- als auch Umfragedaten manuell und teilautomatisiert analysiert.
- Zur Gewährleistung einer einheitlichen Schadensklassifikation wurde die Systematik nach Maiwald und Schwarz (2019) mit sechs Schadensklassen (D1 bis D6) herangezogen.
- Erhebungen vor Ort identifizierten Durchfeuchtung, Kontamination, Anprall und Unterspülung als die häufigsten Schadensbilder im Ahrtal.
- Die Auswertung ergab, dass die geringsten Schadensgrade D1 und D2 am häufigsten auftraten. Es konnte festgestellt werden, dass mit zunehmendem Schadensgrad in allen sechs untersuchten Gemeinden eine Abnahme der Häufigkeit zu verzeichnen ist.
- Darüber hinaus konnte nachgewiesen werden, dass die gravierendsten Schäden D5 und D6 am häufigsten an Gebäuden auftreten, die in direkter Nähe zu den Gewässern stehen.
- Des Weiteren wurden die finanziellen Schäden mit vier verschiedenen Ansätzen pro Gemeinde abgeschätzt. In Bezug auf die Simulation der finanziellen Schäden wiesen die vier verwendeten Modelle lediglich geringe Abweichungen auf.
- Die Untersuchungen zeigen die Bedeutung dieser Schadensdaten sowohl für die Aufarbeitung von Flutereignissen als auch für die Vorbereitung für einen besseren Schutz vor diesen.

Danksagung Die Autoren von ▶ Kap. 7 bedanken sich bei der virtualcitysystems GmbH und der Geofly GmbH für das 3D-Stadtmodell sowie beim HKC und Oliver Koch für die Erhebung der Daten. Außerdem bedanken sich die Autoren beim Bundesministerium für Bildung und Forschung (BMBF) und beim DLR Projektträger für die Unterstützung und Förderung des KAHR-Projekts. Diese Forschung wurde vom BMBF unter dem Förderkennzeichen 01LR2102A-M finanziert.

Literatur

Bürkner P-C (2017) brms: An R Package for Bayesian Multilevel Models Using Stan. J Stat. https://doi.org/10.18637/jss.v080.i01

Dillenardt L, Thieken AH (2025) Individual flood risk adaptation in Germany: exploring the role of different types of flooding. Nat Hazards Earth Syst Sci 25:3257–3278. https://doi.org/10.5194/nhess-25-3257-2025

Forschung für Nachhaltigkeit | FONA (2025) Informieren, beraten, sensibilisieren – unterwegs mit dem KAHR-Infomobil für mehr privaten Hochwasserschutz und Eigenvorsorge. Forsch Nachhalt | Fona (https://www.fona.de/de/aktuelles/nachrichten/2022/221124_KAHR-Infomobil_Hochwasserschutz.php, zuletzt geprüft am 06.01.2025)

GeoBasis-DE/BKG (2025) basemap.de Web Raster. https://www.bkg.bund.de. Creative Commons 4.0, https://creativecommons.org/licenses/by/4.0/. Online verfügbar unter https://basemap.de

GeoBasis-DE/LVermGeoRP (2021a) Ahr Hochwasser Side-by-Side. https://mapcache.de/ahr/. Zugegriffen: 21. Apr. 2023

GeoBasis-DE/LVermGeoRP (2021b) *Digitale Orthophotos 2021*: http://www.lvermgeo.rlp.de. Data licence Germany – attribution – Version 2.0: http://www.govdata.de/dl-de/by-2-0. Online verfügbar unter https://geoshop.rlp.de/historische_luftbilder/orthophotos_historisch_2021.html

GeoBasis-DE/LVermGeoRP (2025) Verwaltungsgrenze Rheinland-Pfalz. Gemeinden. http://www.lvermgeo.rlp.de. Data licence Germany – attribution – Version 2.0, http://www.govdata.de/dl-de/by-2-0. Online verfügbar unter https://lvermgeo.rlp.de/geodaten-geoshop/open-data

GeoFly (2023) OBVIEWSLY – Der Schrägbildviewer. Demoversion. https://obviewsly.de/. Zugegriffen: 21. Apr. 2023

HKC (2025) KAHR-Infomobil. HKC. https://hkc-online.de/de/Infomobil/KAHR-Infomobil. Zugegriffen: 6. Jan. 2025

Kirschbauer L, Schäfer M (2024) Analyse von Gebäudeschäden und Empfehlungen für hochwasserresiliente Gebäude. Wasser Abfall 26(5):46–50 (https://www.springerprofessional.de/analyse-von-gebaeudeschaeden-und-empfehlungen-fuer-hochwasserres/27057234)

Maiwald H, Schwarz J (2019) Unified damage description and risk assessment of buildings under extreme natural hazards. Mauerwerk 23(2):95–111. https://doi.org/10.1002/dama.201910014

Maiwald H, Schwarz J, Kaufmann Ch, Abrahamczyk L (2022) Das Hochwasser 2021: Ingenieuranalyse der Bauwerksschäden. Bautechnik 99:12. https://doi.org/10.1002/bate.202200062

O'Connor C, Joffe H (2020) Intercoder reliability in qualitative research: debates and practical guidelines. Int J Qual Methods 19:1989922. https://doi.org/10.1177/1609406919899220

OpenStreetMap contributors (2021) OpenStreetMap Database. https://download.geofabrik.de/ (https://www.openstreetmap.org/copyright)

OpenStreetMap contributors (2022) OpenStreetMap, available under the Open Database License (ODbL). https://www.openstreetmap.org/copyright

Samprogna Mohor G, Sieg T, Koch O, Buhrmann A, Maiwald H, Schwarz J, Thieken AH (2024) Remote sensing-based mapping of structural building damage in the Ahr valley. J Flood Risk Manag. https://doi.org/10.1111/jfr3.12983

Stan Development Team (2024) Stan Reference Manual. https://mc-stan.org/. Zugegriffen: 20. Mai 2024

Thieken AH, Kreibich H, Müller M, Lamond J (2017) Data collection for a better understanding of what causes flood damage – experiences with telephone surveys. In: Molinari D, Ballio F, Menoni S (Hrsg) Flood damage survey and assessment. Wiley American Geophysical Union, S 95–106

Wasserwirtschaftsverwaltung des Landesamts für Umwelt RLP (2025a) Onlinedienst Gewässernetz 2017. Creative Commons 4.0 „CC by". https://creativecommons.org/licenses/by/4.0/ (https://wasserportal.rlp-umwelt.de/kartendienste)

Wasserwirtschaftsverwaltung des Landesamts für Umwelt RLP (2025b) Überflutungsflächen, Onlinedienst Hochwassergefahrenkarte. Creative Commons 4.0 „CC by". https://creativecommons.org/licenses/by/4.0/ (https://wasserportal.rlp-umwelt.de/kartendienste)

Wüthrich D et al (2024) Field survey assessment of flood loads and related building damage from the July 2021 event in the Ahr valley (Germany). J Flood Risk Manag. https://doi.org/10.1111/jfr3.13024

Wirtschaftliche Auswirkungen: eine Unternehmensbefragung gewährt detaillierte Einblicke

Annegret Thieken, Guilherme Samprogna Mohor, Ravi Kumar Guntu, Meike Müller und Heidi Kreibich

J. Birkmann et al. (Hrsg.), *Die Flutkatastrophe 2021*, Sustainable Development Goals (SDG) – Umsetzung in Praxis, Lehre und Entscheidungsprozessen,
https://doi.org/10.1007/978-3-658-48688-4_8

8.1 Einleitung

Hinweis für die Praxis

Überschwemmungen und Hochwasser können vielfältige Auswirkungen auf die Wirtschaft haben: Zunächst einmal sind direkte Schäden an Gebäuden, Betriebseinrichtungen, Fahrzeugen oder an Waren, Produkten und Lagerbeständen von Unternehmen zu nennen. In Schadensbilanzen werden oft nur diese direkten Sachschäden berichtet. Vergangene Hochwasser in Deutschland zeigen, dass Schadensanteile der gewerblichen Wirtschaft am direkten Gesamtschaden pro Bundesland und Hochwasserereignis variieren: Beim Juni-Hochwasser 2013 berichtete Bayern einen Gesamtschaden von 1,3 Mrd. €, davon entfielen 32,4 % auf die gewerbliche Wirtschaft. Sachsen hingegen meldete 2013 einen höheren Gesamtschaden von 1,9 Mrd. €, davon entfielen aber nur 13,9 % auf die gewerbliche Wirtschaft (vgl. Thieken et al. 2016). Daher sind für jedes Ereignis eigene Erhebungen nötig.

Weiterhin spielen für Unternehmen Unterbrechungen und Einschränkungen ihres Betriebes eine erhebliche Rolle. In Befragungen nach dem Hochwasser im Juni 2013 gaben 88 % der geschädigten Unternehmen an, von Betriebsunterbrechungen betroffen gewesen zu sein (vgl. Thieken et al. 2016). Weitere Einschränkungen und indirekte Schäden entstehen beispielsweise durch Lieferschwierigkeiten von Zulieferern. Diese indirekten Schäden können meistens nur grob abgeschätzt werden, aber eine ähnliche Größenordnung wie direkte Schäden annehmen, wobei sich große Unterschiede zwischen den Wirtschaftssektoren zeigen (Sieg et al. 2019).

Für das Hochwasser im Juli 2021 wurde im März 2022 insgesamt von direkten Schäden in Höhe von 33,1 Mrd. € ausgegangen (BMI und BMF 2022); dazu kommen 0,3 Mrd. € Einsatzkosten sowie geschätzte 7,1 Mrd. € indirekte Schäden (Trenczek et al. 2022). Die Industrie- und Handelskammer Aachen gab ca. 15.000 betroffene Unternehmen in Nordrhein-Westfalen (NRW) an (zitiert in Trenczek et al. 2022). BMI und BMF (2022) gehen von 7000 betroffenen Unternehmen in NRW und 3000 in Rheinland-Pfalz (RLP) aus. Der Gesamtverband der Versicherer (GDV) meldete 27.000 versicherte Schadensfälle von Firmen (Gewerbe, Industrie) und Technischen Versicherungen, davon 19.000 mit einem Schadensaufwand von 2,4 Mrd. € aus NRW und 4000 mit einem Schadensaufwand von 0,9 Mrd. € aus RLP (GDV 2023).

Von den direkten Schäden entfallen je nach Bundesland nur 6 bis 12 % auf die gewerbliche Wirtschaft (vgl. BMI und BMF 2022). Neben Schäden in Privathaushalten tragen insbesondere hohe Schäden an Infrastrukturen zum hohen Gesamtschaden bei. Allein die Deutsche Bahn berichtete von direkten Schäden in Höhe von 1,3 Mrd. €.[1]

Was diese hohen Schäden für einzelne Unternehmen oder Wirtschaftssektoren bedeuten, bleibt unklar. Daher werden in diesem Kapitel auf Basis einer Telefonbefragung von geschädigten Unternehmen in NRW und RLP detailliertere Einblicke in die Typen und das Ausmaß von Schäden in Unternehmen im Juli 2021 gegeben. Zudem werden Einflussfaktoren diskutiert. Abschließend werden Aspekte des Wiederaufbaus vorgestellt.

1 ▶ https://flut-aufbau.deutschebahn.com/; zuletzt abgerufen am 13.09.2024.

8.2 Unternehmensbefragung zum Hochwasser 2021

Zwischen dem 8. November 2022 und dem 31. Januar 2023 wurden vom Hochwasser betroffene Unternehmen in NRW und RLP im Auftrag des Deutschen GeoForschungsZentrums und der Deutschen Rückversicherung AG von einem privaten Befragungsinstitut telefonisch zu den Hochwasserauswirkungen im Juli 2021, potenziellen Einflussfaktoren, dem Wiederaufbau und zur betrieblichen Vorsorge befragt. Vorgehen und Fragebogen orientierten sich an früheren Erhebungen nach Hochwasserereignissen in Deutschland (vgl. Thieken et al. 2017). Insgesamt konnten 434 Unternehmen mit einer durchschnittlichen Dauer von 42 Minuten befragt werden. Darüber hinaus gab es 608 Verweigerungen, 76 Befragungen, die entweder abgebrochen wurden oder während des Erhebungszeitraums nicht vereinbart werden konnten, sowie 1886 Unternehmen, die telefonisch nicht erreicht werden konnten. Damit wurde insgesamt ein guter Rücklauf von 14 % erreicht. In drei Fällen stellte sich später heraus, dass sich die Angaben in der Befragung nicht auf eine Betriebsstätte, sondern auf mehrere Standorte (u. a. Bildungseinrichtungen, Verwaltungsgebäude) bezogen; sie wurden deswegen bei den vorliegenden Auswertungen ignoriert. Letztlich wurden 431 Fälle verarbeitet.

Von den befragten Unternehmen kamen 258 (60 %) aus NRW und 173 (40 %) aus RLP. Dies spiegelt die in BMI und BMF (2022) erwähnte Verteilung von hochwasserbetroffenen Unternehmen in beiden Bundesländern wider.

In Anlehnung an die Wirtschaftsklassifikation WZ2008 wurden alle befragten Unternehmen einem von fünf übergeordneten Wirtschaftssektoren zugeordnet: (1) Landwirtschaft, Forstwirtschaft, Fischerei (n = 14); (2) produzierendes Gewerbe, Ver-/Entsorgung, Baugewerbe (n = 81); (3) Handel, Gastgewerbe und Verkehr, inkl. Instandhaltung und Reparatur von Kraftfahrzeugen (n = 126); (4) Finanzen, Vermietung und Unternehmensdienstleistungen (n = 81) sowie (5) öffentliche und private Dienstleistungen einschließlich Einrichtungen des Bildungs-, Gesundheits- und Sozialwesens (n = 129). ◘ Abb. 8.1 zeigt die Verteilung im Datensatz für beide Bundesländer.

Weiterhin wurden die Unternehmen nach Größe eingeteilt. Da Umsatzzahlen zwar abgefragt, aber oft nicht berichtet wurden, wurde die Anzahl der Beschäftigten als Maß

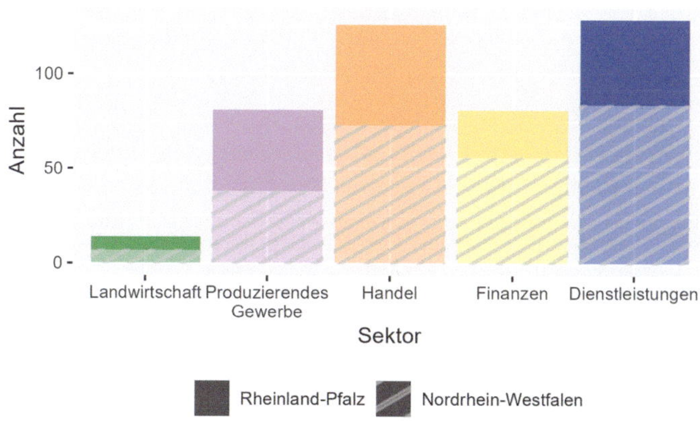

◘ **Abb. 8.1** Anzahl der befragten Unternehmen nach Sektor, unterteilt nach Bundesland. (Eigene Darstellung basierend auf dem Datensatz „Überflutungen in Deutschland im Juli 2021: Schäden von Unternehmen", Deutsches GeoForschungsZentrum, Deutsche Rückversicherung AG, 2022, unveröffentlicht)

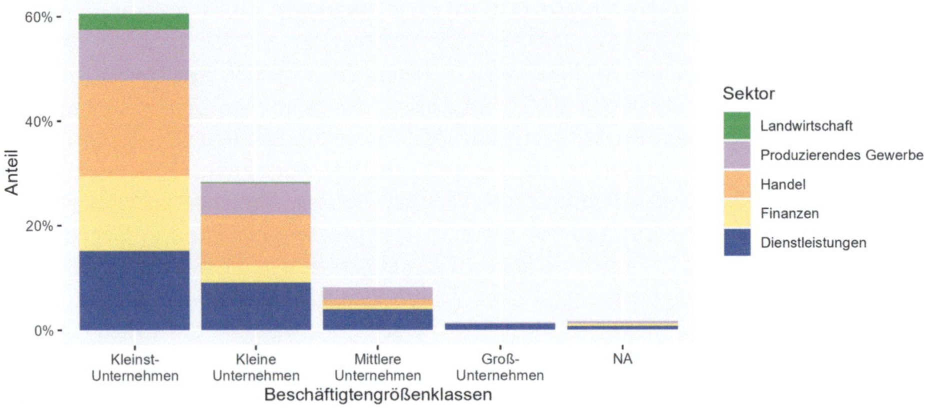

Abb. 8.2 Anteil der befragten Unternehmen nach Anzahl der Beschäftigten als Maß für die Betriebsgröße, unterteilt nach Sektoren. (Eigene Darstellung basierend auf dem Datensatz „Überflutungen in Deutschland im Juli 2021: Schäden von Unternehmen", Deutsches GeoForschungsZentrum, Deutsche Rückversicherung AG, 2022, unveröffentlicht)

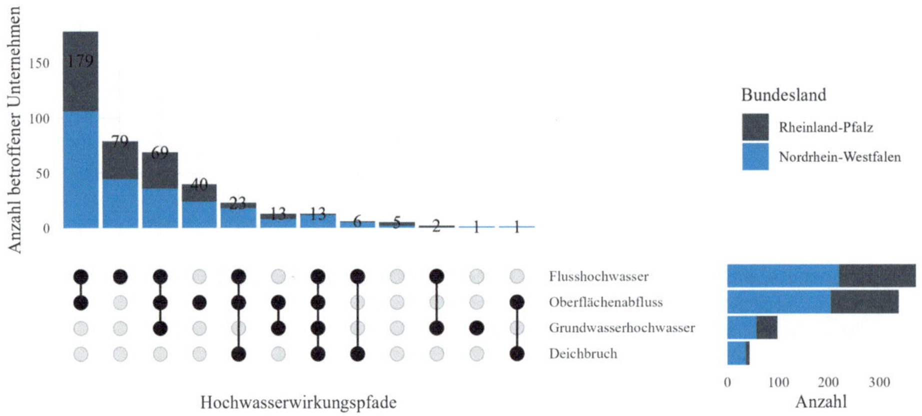

Abb. 8.3 Verteilung der berichteten Hochwasserwirkungspfade. (Eigene Darstellung basierend auf dem Datensatz „Überflutungen in Deutschland im Juli 2021: Schäden von Unternehmen", Deutsches GeoForschungsZentrum, Deutsche Rückversicherung AG, 2022, unveröffentlicht)

für die Betriebsgröße verwendet. Die Mehrheit der befragten Unternehmen wird demnach als Kleinstunternehmen mit bis zu neun Beschäftigten klassifiziert (n = 261; vgl. Abb. 8.2). Kleine Unternehmen mit 10 bis zu 49 Beschäftigten und mittlere Unternehmen mit 50 bis zu 249 Beschäftigten sind mit 122 bzw. 35 Fällen im Datensatz vertreten. Große Unternehmen mit mehr als 249 Beschäftigten haben hingegen nur selten teilgenommen (n = 6). In sieben Fällen lag keine Information zur Zahl der Beschäftigten vor. Die Verteilung der Betriebe nach Beschäftigten ist in beiden Bundesländern ähnlich.

Die Hochwassereinwirkung auf die Betriebsstandorte war nicht überall gleich. Das Ereignis im Juli 2021 war in hohem Maße durch schnell abfließendes Wasser (Sturzfluten) und Ausuferungen von Gewässern (Flusshochwasser) gekennzeichnet. Die Befragten konnten die erlebte Dynamik standardisierten Beschreibungen verschiedener Wirkungspfade zuordnen, wobei das Ereignis meistens als Kombination aus Flusshochwasser und

Überflutungen durch oberflächlich abfließendes Wasser eingestuft wurde (◘ Abb. 8.3). Eine eindeutige Unterscheidung zwischen Sturzfluten und Überflutungen nach Starkregen, z. B. durch Versagen der Kanalisation, war durch die textliche Pfadbeschreibung schwierig; daher wurden diese beiden Wirkungspfade in ◘ Abb. 8.3 als „Oberflächenabfluss" zusammengefasst. Einige Befragte gaben zudem an, von Grundwasserhochwasser oder von Deichbrüchen betroffen gewesen zu sein.

8.3 Wie wurden Unternehmen im Juli 2021 durch Hochwasser geschädigt?

8.3.1 Allgemeine Auswirkungen

Überschwemmungen verursachen nicht nur direkte Schäden an Strukturen und Produkten durch hydrodynamische und hydraulische Kräfte sowie chemische Verunreinigungen. Sie führen auch zu Versorgungs- und Verkehrsstörungen und damit zu Beeinträchtigungen von Lieferketten sowie teilweisen oder vollständigen Betriebsunterbrechungen, die von Umsatzeinbußen bis zur Insolvenz führen können (z. B. Thieken et al. 2016). ◘ Abb. 8.4 zeigt die Häufigkeit solcher Auswirkungen in den befragten Unternehmen, eingeteilt nach Sektoren.

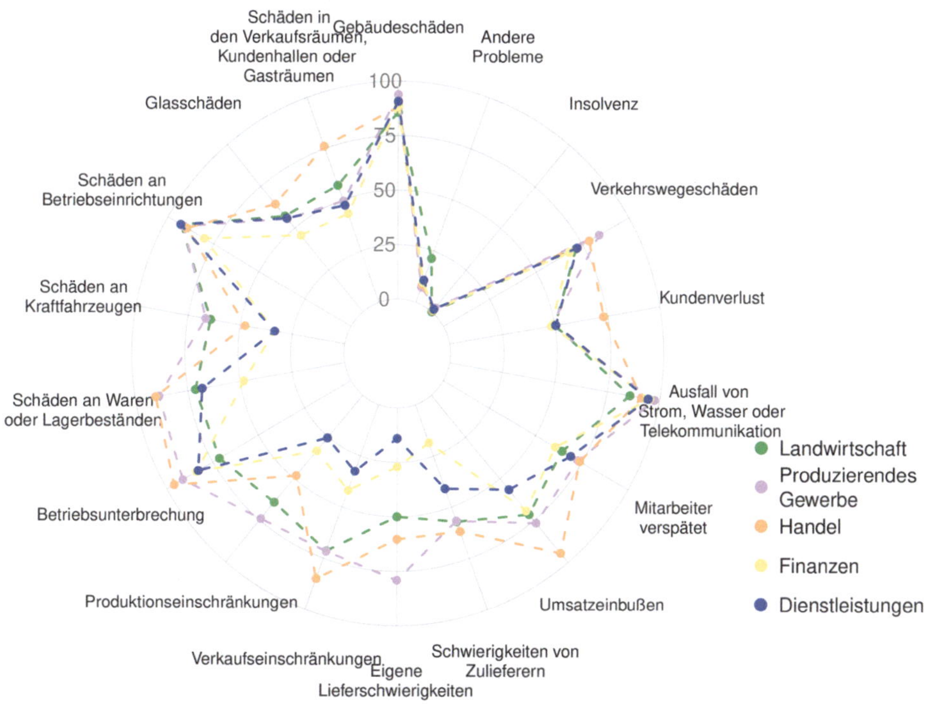

◘ **Abb. 8.4** Häufigkeit der Auswirkungen des Ereignisses auf die Befragten nach Sektor. (Eigene Darstellung basierend auf dem Datensatz „Überflutungen in Deutschland im Juli 2021: Schäden von Unternehmen", Deutsches GeoForschungsZentrum, Deutsche Rückversicherung AG, 2022, unveröffentlicht)

Ausfälle von Strom, Wasser oder Telekommunikation und Gebäudeschäden wurden am häufigsten, d. h. von 94 % aller befragten Unternehmen genannt. Mit Ausnahme der Insolvenz, die in weniger als 2 % der Fälle eingetreten ist, traten alle anderen genannten Probleme bei mindestens 43 % der befragten Unternehmen auf.

Im Sektor „Handel, Gastgewerbe und Verkehr" haben Unternehmen am häufigsten unter Betriebsunterbrechungen und Verkaufseinschränkungen gelitten, was Umsatzeinbußen und Kundenverluste nach sich zog. Kundenverluste hatte vor allem dieser Sektor. Zudem berichteten 96 % der befragten Unternehmen im Sektor „Handel, Gastgewerbe und Verkehr" von Betriebsunterbrechungen; zum Zeitpunkt der Befragung, d. h. ca. 18 Monate nach dem Hochwasser, dauerten diese bei 15 % immer noch an. Der Finanzsektor (Unternehmensdienstleistungen) erlitt im Vergleich weniger Sachschäden und hatte weniger Probleme mit externen Lieferketten, aber die Unternehmen berichteten häufig von Ausfällen von Strom, Wasser oder Telekommunikation sowie von Betriebsunterbrechungen. 97 % der Betriebe im produzierenden Gewerbe hatten ebenfalls unter Ausfällen von Strom, Wasser oder Telekommunikation zu leiden, waren am häufigsten von Produktionseinschränkungen betroffen und hatten Probleme mit den eigenen Lieferketten. Der größte Unterschied zwischen den Sektoren besteht beim Auftreten von eigenen Lieferschwierigkeiten, wobei das produzierende Gewerbe davon fünfmal häufiger betroffen war als der Dienstleistungssektor (◘ Abb. 8.4).

8

Die Covid-19-Pandemie, die zum Zeitpunkt des Hochwassers und danach noch akut war, hat nach Meinung vieler Unternehmen die Auswirkungen des Hochwassers leicht (20,2 %) oder stark (23,2 %) erhöht. 46,2 % der befragten Unternehmen sahen hingegen keinen Einfluss durch Covid-19, während 7,9 % meinten, die Pandemie habe die Hochwasserauswirkungen (stark) verringert. Die Gründe für diese Einschätzungen wurden nicht abgefragt, aber es wurde analog zu den Hochwasserauswirkungen erhoben, von welchen corona-bedingten Auswirkungen die Unternehmen betroffen waren. Die fünf häufigsten Nennungen waren Ausfall von Mitarbeiter:innen aufgrund von Krankheit und/oder Quarantäne (62,4 %), Umsatzeinbußen (52,9 %), Verlust von Kunden (37,1 %), Verkaufseinschränkungen (37,1 %) und Betriebsunterbrechungen (35,3 %), wobei Mehrfachnennungen möglich waren. 11,4 % der befragten Unternehmen gaben an, Covid-19 habe keine Probleme in ihrem Unternehmen verursacht.

Im Mittel gaben die 431 befragten Unternehmen drei corona-bedingte und elf hochwasserbedingte Auswirkungen (von 10 bzw. 18 Optionen) an, wobei beide Angaben korrelieren (Spearman Rangkorrelationskoeffizient: 0,399, $p < 0{,}001$). Eine einfaktorielle Varianzanalyse (ANOVA) mit einem Post-hoc-Test nach Scheffé zeigt, dass sich die Anzahl der corona-bedingten Auswirkungen – im Gegensatz zur Anzahl der Hochwasserauswirkungen – in den Klassen zur Einschätzung der Verstärkung oder Abschwächung der Hochwasserauswirkungen durch Covid-19 signifikant unterscheidet: Unternehmen, die keinen Einfluss von Covid-19 auf die Hochwasserauswirkungen wahrgenommen haben, gaben im Mittel 2,3 corona-bedingte Auswirkungen an, wohingegen Unternehmen, die von einer Verstärkung der Hochwasserauswirkungen durch Covid-19 ausgingen, im Mittel von 3,9 corona-bedingten Auswirkungen berichteten ($p < 0{,}001$). Eine erhöhte Anzahl von corona-bedingten Auswirkungen, nämlich im Mittel 3,6, gaben auch die Unternehmen an, die eine (leichte) Verringerung der Hochwasserauswirkungen durch Covid-19 annahmen ($p = 0{,}05$). Eine Mehrfachgefahrenbelastung kann sich somit sehr unterschiedlich auswirken. So könnten reduzierte Warenbestände aufgrund von Covid-19-Eindämmungsmaßnahmen zu einer Verringerung der Hochwasserschäden, z. B. im Handel, geführt haben, erhöhte Lagerbestände aufgrund von Umsatzeinbußen im produzierenden Gewerbe jedoch zu einer Erhöhung.

Unter den Wirtschaftssektoren war der Sektor „Handel, Gastgewerbe und Verkehr" signifikant stärker sowohl von corona-bedingten Auswirkungen als auch von Hochwasserauswirkungen betroffen als die anderen Sektoren mit Ausnahme der Landwirtschaft, für die aber nur wenige Fälle vorliegen. Darüber hinaus berichtete das produzierende Gewerbe signifikant mehr Hochwasserauswirkungen als die beiden Dienstleistungssektoren. Getestet wurden diese Unterschiede mit einer einfaktoriellen Varianzanalyse mit einem Post-hoc-Test nach Scheffé.

In Bezug auf die Betriebsgröße zeigen sich insgesamt nur wenige Unterschiede im Hinblick auf die erlittenen Hochwasserauswirkungen. Die (wenigen) befragten Großunternehmen hatten am häufigsten Sachschäden (außer an Kraftfahrzeugen), Unterbrechungen, Lieferschwierigkeiten und Verkehrsprobleme, während Kleinstunternehmen am häufigsten Kunden und Umsätze verloren haben.

8.3.2 Sachschäden

In der Befragung wurden für drei Kategorien von direkten Sachschäden konkrete Schadenssummen erfasst: (1) Schäden an Gebäuden, (2) Schäden an der Betriebseinrichtung und (3) Schäden an Waren, Produkten und Lagerbeständen. In den meisten Fällen (ca. 62 %) erreichten die direkten Schäden einen sechsstelligen Wert, bei ca. 24 % der Unternehmen kam es sogar zu Schäden in Millionenhöhe. ◘ Tab. 8.1 und 8.2 zeigen die Schäden je Sektor oder Betriebsgröße als Mittelwerte und Mediane. Die Mittelwerte sind

◘ **Tab. 8.1** Durchschnittliche absolute Schäden an Unternehmensvermögen nach Sektor (Angaben in Euro; Werte in Klammern sind Mediane, n bezeichnet die Anzahl der Unternehmen, die in die Berechnung der Mittelwerte und Mediane eingingen)

Sektor	Gebäude	Betriebseinrichtungen	Waren, Produkte und Lagerbestände
Landwirtschaft	635.636 (500.000) n = 11	284.667 (160.000) n = 12	111.944 (25.000) n = 9
Produzierendes und verarbeitendes Gewerbe & Baugewerbe	1.466.985 (450.000) n = 66	1.573.458 (200.000) n = 66	574.277 (50.000) n = 64
Handel, Gastgewerbe & Verkehr	768.041 (325.000) n = 88	396.596 (150.000) n = 101	182.299 (50.000) n = 101
Finanz-, Vermietungs- und Unternehmensdienstleistungen	608.865 (295.000) n = 52	133.822 (30.000) n = 59	58.521 (15.000) n = 23
Öffentliche und private Dienstleistungen	1.574.380 (400.000) n = 71	523.286 (80.000) n = 90	134.732 (10.000) n = 63
Insgesamt	1.093.204 (350.000) n = 289	616.804 (100.000) n = 328	253.875 (31.500) n = 260

Tab. 8.2 Durchschnittliche absolute Schäden an Unternehmensvermögen nach Betriebsgröße (Angaben in Euro; Werte in Klammern sind Mediane, n bezeichnet die Anzahl der Unternehmen, die in die Berechnung der Mittelwerte und Mediane eingingen)

Unternehmensgröße	Gebäude	Betriebseinrichtungen	Waren, Produkte und Lagerbestände
Kleinstunternehmen (1–9 Beschäftigte)	711.459 (250.000) n = 167	297.854 (50.000) n = 203	159.422 (30.000) n = 154
Kleine Unternehmen (10–49 Beschäftigte)	908.482 (500.000) n = 83	541.898 (150.000) n = 96	134.470 (31.500) n = 82
Mittlere Unternehmen (50–249 Beschäftigte)	2.838.103 (1.350.000) n = 29	3.630.652 (600.000) n = 23	1.503.250 (150.000) n = 20
Großunternehmen (> 249 Beschäftigte)	7.350.000 (1.700.000) n = 4	160.000 (200.000) n = 3	55.000 (10.000) n = 3

8

stark von einzelnen hohen Werten beeinflusst; daher sind die Mediane durchweg geringer. Der Sektor „produzierendes Gewerbe und Baugewerbe“ hat die höchsten Werte für alle drei Schadensarten; allerdings weist der Dienstleistungssektor ebenfalls sehr hohe Gebäudeschäden auf (Tab. 8.1).

Im Hinblick auf unterschiedliche Betriebsgrößen zeigen Großunternehmen die höchsten Gebäudeschäden; dies gilt nicht für die anderen beiden Arten von Sachschäden (Tab. 8.2). Da nur wenige Großunternehmen an der Befragung teilgenommen haben, sind diese Werte aber nicht repräsentativ, zumal die befragten Großunternehmen fast ausschließlich dem Dienstleistungssektor zuzuordnen sind (vgl. Abb. 8.2).

8.3.3 Ausmaß von Betriebsunterbrechungen und -einschränkungen

Verheerende Ereignisse wie das Hochwasser im Juli 2021 können in Unternehmen jahrelang spürbar sein und die Betriebsabläufe empfindlich stören. 77 % (oder 332) der befragten Unternehmen hatten eine Betriebsunterbrechung (BU), die zum Zeitpunkt der Befragung bereits beendet war (Abb. 8.5 links). Weitere 11 % der Unternehmen (dies entspricht 47 Betrieben) litten zum Zeitpunkt der Befragung, d. h. 18 Monate nach den Überflutungen, hingegen immer noch unter einer BU. Von diesen konnten 33 Unternehmen abschätzen, wie lange diese vermutlich noch andauern werde (Abb. 8.5 rechts). Viele dieser Unternehmen gehen von Unterbrechungen von bis zu drei Jahren nach dem Hochwasser aus.

Von Betriebseinschränkungen bis zum Befragungszeitpunkt berichteten 60 % (oder 257) der Unternehmen (Abb. 8.5 links). 38 % (oder 164 Betriebe) hatten 18 Monate nach dem Hochwasser immer noch Einschränkungen, von denen 132 Unternehmen die voraussichtliche Dauer abschätzen konnten (Abb. 8.5 rechts). Hier sind Werte von bis

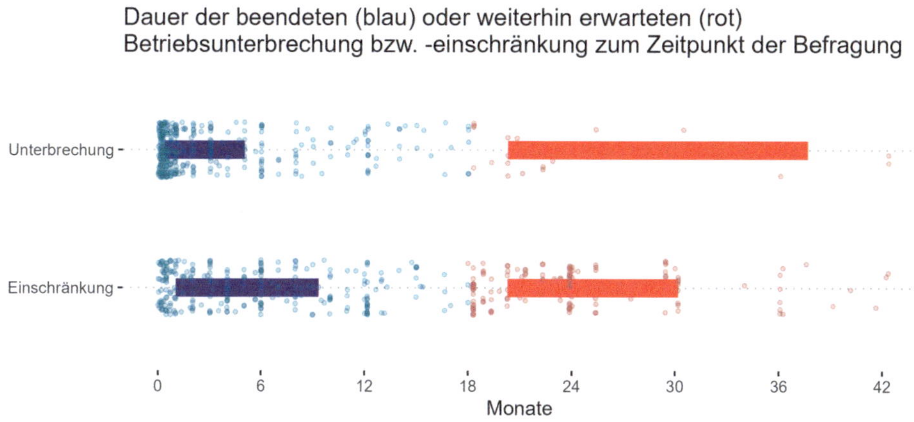

Abb. 8.5 Dauer der beendeten (*dunkelblau*; *links*) oder weiterhin erwarteten (*rot*; *rechts*) Betriebsunterbrechung bzw. -einschränkung zum Zeitpunkt der Befragung. Die Balken repräsentieren die Werte zwischen dem 25- und 75-Perzentil, die Punkte Einzelwerte; maximale Werte von bis zu 10 bis 11 Jahren sind nicht dargestellt. (Eigene Darstellung basierend auf dem Datensatz „Überflutungen in Deutschland im Juli 2021: Schäden von Unternehmen", Deutsches GeoForschungsZentrum, Deutsche Rückversicherung AG, 2022, unveröffentlicht)

Tab. 8.3 Durchschnittliche finanzielle Schäden durch Betriebsunterbrechungen nach Sektor und Unternehmensgröße (Angaben in Euro; Werte in Klammern sind Mediane, n bezeichnet die Anzahl der Unternehmen, die in die Berechnung der Mittelwerte und Mediane eingingen)

Sektor	Betriebsunterbrechung	Unternehmensgröße	Betriebsunterbrechung
Landwirtschaft	22.400 (15.000) n = 5	Kleinstunternehmen (1–9 Beschäftigte)	139.931 (30.000) n = 143
Produzierendes und verarbeitendes Gewerbe & Baugewerbe	384.136 (100.000) n = 44	Kleine Unternehmen (10–49 Beschäftigte)	311.173 (100.000) n = 74
Handel, Gastgewerbe & Verkehr	299.370 (100.000) n = 84	Mittlere Unternehmen (50–249 Beschäftigte)	703.250 (200.000) n = 16
Finanz-, Vermietungs- und Unternehmensdienstleistungen	81.678 (30.000) n = 45	Großunternehmen (> 249 Beschäftigte)	400.000 (500.000) n = 3
Öffentliche und private Dienstleistungen	170.038 (32.500) n = 60	Insgesamt	235.458 (60.000) n = 238

zu insgesamt 2,5 Jahren keine Seltenheit. Einzelne Unternehmen gehen von Einschränkungen von bis zu zehn oder elf Jahren aus.

Mit den Betriebsunterbrechungen gehen finanzielle Auswirkungen einher, die in Tab. 8.3 pro Sektor und Betriebsgröße zusammengefasst sind. Es zeigt sich, dass Betriebsunterbrechungen bei der Hälfte der Unternehmen im produzierenden Gewerbe sowie im Sektor „Handel, Gastgewerbe und Verkehr" Schäden in sechsstelliger Höhe verursacht haben (der Median beträgt 100.000 €). Davon waren vor allem mittlere Unternehmen betroffen, aber auch kleine Betriebe und Kleinstunternehmen berichten im Durchschnitt sechsstellige Werte.

8.3.4 Einflussfaktoren auf das Schadensausmaß

Um Hochwasserschäden zukünftig mindern zu können, ist es wichtig, die Schadensprozesse besser zu verstehen. Bisher beziehen sich die meisten Erkenntnisse zu Schadensprozessen auf Schäden von Privathaushalten durch Flusshochwasser (Gerl et al. 2016; Thieken et al. 2022). Die Analyse der Schäden auf Basis der Daten aus der Unternehmensbefragung zum Hochwasser 2021 soll dazu beitragen, Lücken im Verständnis der Einflussfaktoren für Schäden in Unternehmen zu schließen. Für die Analyse wurden 19 potenziell wichtige Einflussfaktoren aus den Bereichen Gefährdung, Notmaßnahmen, Vorsorge und Unternehmenscharakteristik ausgewählt (◘ Tab. 8.4).

◘ Abb. 8.6 zeigt die paarweisen Korrelationen zwischen den Einflussfaktoren und den verschiedenen Schadenstypen als Spearman-Rangkorrelationskoeffizienten. Es werden nur signifikante Korrelationen (p-Wert < 0,05) angezeigt. Die untersten fünf Zeilen enthalten die Korrelationen mit den folgenden Schadenstypen: relative Schäden an Gebäuden (bloss), relative Schäden an der Betriebseinrichtung (eloss), relative Schäden an Waren, Produkten und Lagerbeständen (gsloss), Dauer der Betriebsunterbrechung (bit) und Dauer der Betriebseinschränkung (brt).

8

Hohe Korrelationen bestehen insbesondere zwischen jeweils den Sachschadenstypen (bloss, eloss, gsloss) und den Gefährdungsparametern Wassertiefe (wd) und Fließgeschwindigkeit (v), dem Erfolg der Notmaßnahmen (ms) und den Unternehmenscharakteristika Größe des Betriebsgeländes (sp) und Anzahl der Mitarbeiter (emp) (◘ Abb. 8.6).

Mit der Betriebsunterbrechungszeit zeigen neben dem Wasserstand (wd) und der Fließgeschwindigkeit (v) auch die Kontamination (con) und Frühwarnzeit (wt) relativ hohe Korrelationen. Interessanterweise zeigen zusätzlich zu diesen Faktoren auch Faktoren, die die Notmaßnahmen und Vorsorge charakterisieren, hohe Korrelationen mit der Dauer der Betriebseinschränkung. Dies sind insbesondere das Vorhandensein eines Notfallplans (ep), Erfolg der Notmaßnahmen (ms), Hochwassererfahrung (fe) und der Indikator für Vorsorge (pr) (◘ Abb. 8.6).

Offensichtlich bestehen auch relativ viele Korrelationen zwischen den verschiedenen Einflussfaktoren (◘ Abb. 8.6). Um diese Wechselwirkungen zwischen den Einflussfaktoren zu berücksichtigen und möglichst robust die wichtigsten Einflussfaktoren auf die Schäden zu ermitteln, wurde ein datengesteuerter Ansatz verwendet, bei dem verschiedene Modelle (Random Forest, Elastic Net, Extreme Gradient Boosting [XGBoost]) mit den empirischen Befragungsdaten trainiert wurden. Es wurden zehn Wiederholungen einer zehnfachen Kreuzvalidierung auf der Grundlage einer zufälligen Partitionierung durchgeführt. Es wurde jeweils das Modell mit dem geringsten mittleren absoluten Fehler ausgewählt, um die Wichtigkeit des Einflussfaktors abzuleiten. Die Wichtigkeit für jeden Einflussfaktor auf Basis der drei Modelle ist in ◘ Abb. 8.7 dargestellt.

Die Wassertiefe (wd) ist der wichtigste Einflussfaktor für alle Schadenstypen (◘ Abb. 8.7), was mit früheren Analysen übereinstimmt (z. B. Schoppa et al. 2020; Sieg et al. 2017). Außerdem hatte beim Hochwasser 2021 die Fließgeschwindigkeit (v) einen sehr wichtigen Einfluss auf die Schäden; für die Betriebseinrichtungsschäden ist sie der zweitwichtigste Faktor (◘ Abb. 8.7b). Dies ist bei großräumigen, langsam ansteigenden Flussüberschwemmungen nicht der Fall; dort sind z. B. Kontaminationen wichtiger für die

Tab. 8.4 **Liste** der analysierten Einflussfaktoren für Sturzflutschäden in Unternehmen. Die Skalierung der Einflussfaktoren, d. h. der Variablentyp ist wie folgt charakterisiert: „k" steht für kontinuierliche rational skalierte, „o" für ordinal skalierte und „n" für nominal skalierte Variablen

	Abkürzung	**Einflussfaktor**	**Typ**	**Ausprägung**
Gefährdung	wd	Wassertiefe	*k*:	1 bis 963 cm über Grund
	d	Überschwemmungsdauer	*k*:	1 bis 1200 h
	v	Fließgeschwindigkeit	*o*:	1 = niedrige bis 3 = reißende Fließgeschwindigkeit
	con	Kontamination	*o*:	0 = keine bis 4 = schwere Kontamination
Notmaßnahmen	wt	Frühwarnzeit	*k*:	0 bis 168 h
	ws	Quelle der Warnung	*o*:	0 = keine Warnung bis 4 = offizielle behördliche Warnung
	ew	Frühwarnung erhalten	*n*:	0 = nein, 1 = ja
	me	Notmaßnahmen durchgeführt	*n*:	0 = nein, 1 = ja
	ep	Notfallplan vorhanden	*n*:	0 = nein, 1 = ja
	kh	Wissen über die Gefährdung	*n*:	0 = nein, 1 = ja
	ms	Erfolg der Notmaßnahmen	*o*:	0 = keine, 1 = komplett ineffektive bis 4 = sehr effektive Maßnahmen umgesetzt
Vorsorge	fe	Hochwassererfahrung	*o*:	0 = keine bis 5 = rezente Hochwassererfahrung
	pr	Vorsorgeindikator	*o*:	0 = keine/geringe, 1 = mittlere, 2 = sehr gute Vorsorge
	in	Versicherung	*n*:	0 = nein, 1 = ja
Unternehmenscharakteristik	sp	Größe des Betriebsgeländes	*k*:	100 bis 50.000 m^2
	sec	Sektor	*n*:	1 = Landwirtschaft, 2 = Verarbeitendes Gewerbe, 3 = Handel, 4 = Finanzwesen, 5 = Dienstleistungen
	ss	Räumliche Situation	*o*:	1 = mehrere Gebäude, 2 = ganzes Gebäude, 3 = ein oder mehrere Stockwerke, 4 = weniger als ein Stockwerk
	own	Eigentumsverhältnisse	*n*:	1 = Eigentümer, 2 = Mieter, 3 = teilweise Eigentum/gemietet
	emp	Anzahl Mitarbeiter	*k*:	1 bis 920
Schäden	bloss	Relativer Gebäudeschaden (= Schaden [€]/Gebäudewert [€])	*k*:	Schadensgrad zwischen 0 (kein Schaden) und 1 (Totalschaden)
	eloss	Relativer Betriebseinrichtungsschaden	*k*:	Schadensgrad zwischen 0 und 1
	gsloss	Relativer Schaden an Waren, Produkten und Lagerbeständen	*k*:	Schadensgrad zwischen 0 und 1
	bit	Dauer Betriebsunterbrechung	*k*:	0 bis 901 Tage
	brt	Dauer Betriebseinschränkung	*k*:	0 bis 540 Tage

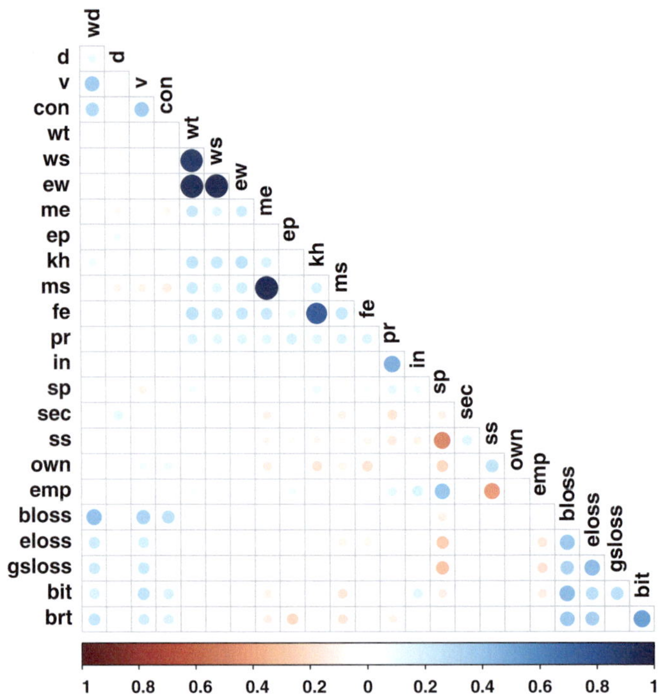

Abb. 8.6 Signifikante (p-Wert < 0,05) Spearman-Rangkorrelationskoeffizienten der 24 Variablen, d. h. zwischen den 19 potenziellen Einflussfaktoren und den fünf Schadenstypen (vgl. Tab. 8.4). (Eigene Darstellung basierend auf dem Datensatz „Überflutungen in Deutschland im Juli 2021: Schäden von Unternehmen", Deutsches GeoForschungsZentrum, Deutsche Rückversicherung AG, 2022, unveröffentlicht)

Höhe des Schadens (Kreibich et al. 2009; Sieg et al. 2017). Außerdem spielt die Größe des Unternehmens, charakterisiert durch die Größe des Betriebsgeländes (sp) und die Anzahl der Mitarbeitenden (emp), eine entscheidende Rolle (Abb. 8.7). Auch wichtig ist, wie erfolgreich die Notmaßnahmen umgesetzt werden konnten (ms) und die damit im Zusammenhang stehende Frühwarnzeit (wt), die für Gebäude- und Betriebseinrichtungsschäden in der Wichtigkeit direkt nach der Anzahl der Mitarbeitenden kommen (Abb. 8.7a, b). Die Dauer der Betriebsunterbrechung wird vor allem durch Gefährdungsparameter bestimmt (d. h., Wassertiefe [wd], Fließgeschwindigkeit [v] und Dauer der Überschwemmung [d], vgl. Abb. 8.7d).

Hinweise für die Praxis

Interessant ist, dass vier der sechs wichtigsten Einflussfaktoren für die Dauer der Betriebseinschränkung sich auf Notmaßnahmen und Vorsorge beziehen: Vorhandensein eines Notfallplans (ep), Vorsorgeindikator, der als gewichtete Summe der vor dem Ereignis ergriffenen Vorsorgemaßnahmen berechnet wird (pr), Wissen über die Gefährdung (kh) und Erfolg der Notmaßnahmen (ms) (Abb. 8.7e). Dies deutet darauf hin, dass gute Vorsorge und Notmaßnahmen selbst bei Extremereignissen wie 2021 Unternehmenseinschränkungen verkürzen können.

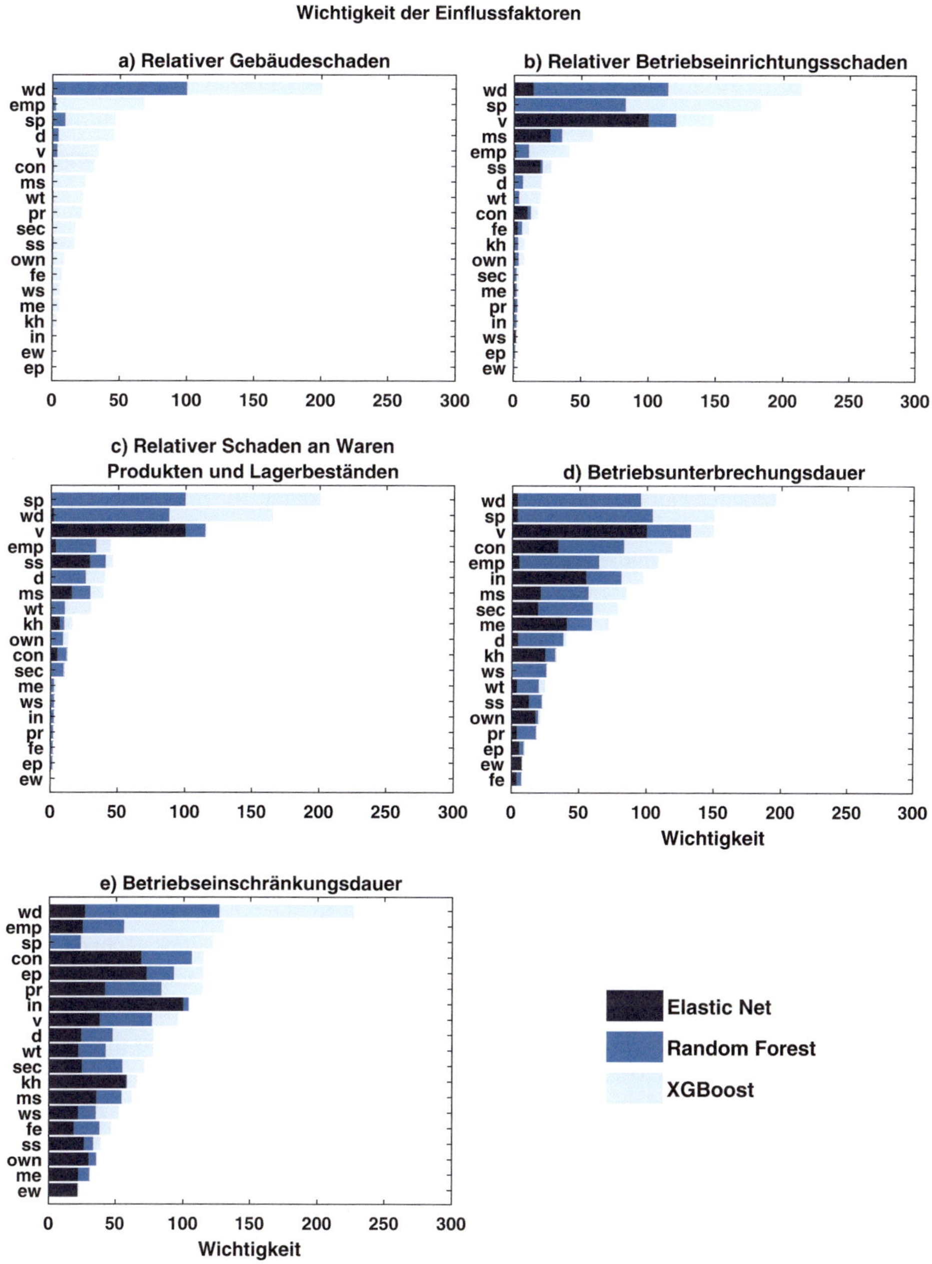

Abb. 8.7 Wichtigkeit der Einflussfaktoren für die relativen Sachschäden von Unternehmen an (**a**) Gebäuden, (**b**) Betriebseinrichtungen und (**c**) Waren, Produkten und Lagerbeständen sowie für die Dauern der (**d**) Betriebsunterbrechung und (**e**) Betriebseinschränkung. Die X-Achse zeigt die Wichtigkeit der Einflussfaktoren auf Basis der drei Modelle (Random Forest, Elastic Net, Extreme Gradient Boosting [XGBoost]). (Eigene Darstellung basierend auf dem Datensatz „Überflutungen in Deutschland im Juli 2021: Schäden von Unternehmen", Deutsches GeoForschungsZentrum, Deutsche Rückversicherung AG, 2022, unveröffentlicht)

8.4 Wiederaufbau

Der vorherige Abschnitt hat gezeigt, dass viele betroffene Unternehmen mit enormen direkten Hochwasserschäden und weiteren Auswirkungen umgehen mussten. In diesem Abschnitt wird daher thematisiert, wie weit der Wiederaufbau zum Zeitpunkt der Befragung, d. h. ca. 18 Monate nach dem Hochwasser vorangeschritten war, wie Schäden kompensiert wurden und welche Hochwasservorsorgestrategie die Unternehmen nach dem Hochwasser verfolgen, die – wie oben gezeigt wurde – den Schaden minimieren können.

8.4.1 Stand des Wiederaufbaus zum Zeitpunkt der Befragung

Zum Zeitpunkt der Befragung berichtete über die Hälfte der befragten Unternehmen (51,3 %, n = 221), dass der direkte Schaden (an Gebäuden, Betriebseinrichtungen etc.) noch nicht gänzlich behoben sei. Demgegenüber war weniger als ein Viertel der Unternehmen (21,3 %) in der Lage, die Schäden innerhalb von sechs Monaten nach dem Ereignis zu beseitigen, weitere 14,4 % benötigten bis zu zwölf Monate, 12,5 % länger als ein Jahr. Unternehmen in den Sektoren „Finanz- und Unternehmensdienstleistungen" sowie „Handel, Gastgewerbe und Verkehr" tendierten zu kürzeren Zeiten als Unternehmen anderer Sektoren. Auch mittlere und große Unternehmen benötigten tendenziell länger zur Schadensbeseitigung als kleinere Betriebe. Die unterschiedlichen Zeiten des Wiederaufbaus spiegeln somit im Wesentlichen die Unterschiede in den berichteten Schäden wider. Allerdings konnten hier mit einer Varianzanalyse (d. h. einer einfaktoriellen Varianzanalyse mit einem Post-hoc-Test nach Scheffé) keine signifikanten Unterschiede zwischen den Sektoren oder Betriebsgrößen festgestellt werden.

Auch die Einschätzungen auf die Frage, wie lange es noch dauere, bis die Hochwasserauswirkungen nicht mehr im Unternehmen spürbar seien, zeigen ein breites Spektrum: Während etwa ein Viertel der Unternehmen (24,4 %) angab, dass die Auswirkungen zum Zeitpunkt der Befragung oder bereits davor gar nicht mehr spürbar waren, rechnen etwa 5 % der Betriebe (4,9 %) damit, dass ein Zustand wie vor dem Ereignis nie mehr eintreten werde. Weitere 5,1 % gaben einen Zeitraum von mindestens zehn Jahren an. Insgesamt scheinen Unternehmen aus dem produzierenden Gewerbe stärker betroffen gewesen zu sein und benötigten tendenziell auch länger, um Schäden zu beseitigen und zum normalen Betrieb zurückzukehren. Allerdings konnten mit einer Varianzanalyse auch hier keine signifikanten Unterschiede zwischen den Sektoren oder Betriebsgrößen festgestellt werden.

8.4.2 Finanzierung des Wiederaufbaus durch Versicherungen und staatliche Aufbauhilfen

Die anfallenden Kosten für die Beseitigung und Reparatur der direkten Schäden sowie für eine Kompensation der Schäden infolge von Betriebsunterbrechungen müssen Unternehmen selbst tragen, falls sie zum Zeitpunkt des Ereignisses nicht versichert sind oder es keine staatlichen Aufbauhilfen gibt. Nach dem Hochwasser vom Juli 2021 wurden großzügige Hilfen mit einer Gesamtsumme von 30 Mrd. € von der Bundesregierung und den

Tab. 8.5 Angaben zum Versicherungsstatus der befragten Unternehmen und zu Änderungen nach dem Hochwasser im Juli 2021

Versicherungsstatus	Unternehmen (n = 431)	Änderung der Versicherungskonditionen nach dem Hochwasser; Mehrfachantworten möglich	Versicherte Unternehmen (n = 231)
VOR dem Hochwasser versichert	53,6 % (n = 231)	Keine Änderung	48,5 % (n = 112)
NACH dem Hochwasser versichert	14,6 % (n = 63)	Erhöhte Prämie	27,7 % (n = 64)
Geplant für die kommenden 6 Monate	7,2 % (n = 31)	Erhöhter Selbstbehalt	6,9 % (n = 16)
Nicht beabsichtigt	11,8 % (n = 51)	Erteilung von Auflagen	2,6 % (n = 6)
Nicht versicherbar	6,1 % (n = 26)	Vertragskündigung	2,6 % (n = 6)
Keine Angabe/weiß nicht	6,7 % (n = 29)	Keine Angabe/weiß nicht	18,6 % (n = 43)

Bundesländern zur Verfügung gestellt. Auch geschädigte Unternehmen waren antragsberechtigt. Viele der 431 befragten Unternehmen haben Soforthilfe (45,7 %) oder Spenden (28,7 %) erhalten. Deutlich weniger Unternehmen (14,2 %) haben Wiederaufbauhilfen beantragt.

Von den 431 befragten Unternehmen waren vor dem Hochwasser 231 (53,6 %) versichert. Weitere 63 (14,6 %) haben sich nach dem Hochwasser versichert, 31 Betriebe (7,2 %) planten dies für die kommenden sechs Monate, während 51 Unternehmen (11,8 %) dies nicht beabsichtigten. Zudem berichteten 26 Unternehmen (6,1 %), dass sie nicht versicherbar seien. Dies kann der Fall sein, wenn die Betriebsstätten in Gebieten mit einer sehr hohen Überflutungswahrscheinlichkeit liegen. Nach dem Hochwasser wurde bei 64 versicherten Unternehmen die Versicherungsprämie erhöht, bei 16 der Selbstbehalt, sechs Unternehmen wurden Auflagen erteilt und weiteren sechs Unternehmen wurde der Vertrag seitens der Versicherung gekündigt (vgl. Tab. 8.5).

Ob sich die berichteten Schäden, die Kompensationszahlungen sowie die Zufriedenheit mit dem Kompensationsprozess zwischen versicherten und nicht-versicherten Unternehmen unterscheiden, wurde mit einer einfaktoriellen Varianzanalyse untersucht. Es ergeben sich signifikante Unterschiede ($p < 0{,}05$) für die Gebäudeschäden (aber nur leicht signifikante [$p < 0{,}1$] beim Schadensgrad), für die Schäden an der Betriebseinrichtung (nicht aber beim Schadensgrad), für die Höhe der Schadenskompensation und vor allem in der Zufriedenheit mit der Abwicklung der Schadenskompensation (Tab. 8.6). In der Dauer der Betriebsunterbrechung zeigt sich nur eine leichte Signifikanz, für die BU-Schadenshöhe keine (vgl. Tab. 8.6). Die Zahlen legen nahe, dass Unternehmen mit höheren Sachwerten eher versichert waren und daher auch höhere Sachschäden und Entschädigungszahlungen berichteten. Allerdings konnten im Hinblick auf die Betriebsgröße und die Angaben zu den Sachwerten keine signifikanten Unterschiede festgestellt werden. Dennoch verschwindet der Unterschied zwischen versicherten und nicht-versicherten Unternehmen in den (relativen) Schadensgraden, sodass man davon ausgehen kann, dass kein sogenanntes moralisches Risiko („moral hazard"), das die Hochwasserschäden verstärkt, vorliegt. Damit wird der Befund von Hudson und Thieken (2022) bestätigt.

Tab. 8.6 Vergleich von versicherten und nicht-versicherten Unternehmen

Variable	Mittelwert der versicherten Unternehmen (n = 231)	Mittelwert der nicht-versicherten Unternehmen (n = 171)	P-Wert	Eta^2
Gebäudeschäden	1.438.726 €	571.096 €	0,013	0,022
Schadensgrad Gebäude	0,50	0,41	0,094	0,015
Schäden an der Betriebseinrichtung	836.497 €	358.450 €	0,034	0,014
Schadensgrad Einrichtung	0,73	0,70	0,436	0,002
Dauer der Betriebsunterbrechung bis Nov/Dez 22	109 Tage	81 Tage	0,065	0,011
Schadenshöhe der Betriebsunterbrechung	272.173	177.099	0,141	0,009
Höhe der Schadenskompensation	735.737 €	141.133 €	0,037	0,016
Anteil der Unternehmen ohne Schadenskompensation	6,9 % (= 16 Betriebe)	30,4 % (= 52 Betriebe)	–	–
Zufriedenheit mit der Abwicklung der Kompensation von 1 (sehr zufrieden) bis 6 (sehr unzufrieden)	2,56 (n = 195)	3,29 (n = 105)	< 0,001	0,043
Proaktive Unternehmen (nach dem Hochwasser)	81 % (= 187 Betriebe)	80,1 % (= 137 Betriebe)	–	–
Wegzug von Unternehmen	8,2 % (= 19 Betriebe)	6,4 % (= 11 Betriebe)	–	–
Gänzlich passive Unternehmen	0,4 % (= 1 Betrieb)	8,2 % (= 14 Betriebe)	–	–

8

Auffällig ist, dass nur wenigen versicherten Unternehmen (n = 16) noch keine Schadenskompensation ausgezahlt wurde, während dies bei ca. 30 % der nicht-versicherten Betriebe der Fall ist (Tab. 8.6). Bei einigen dieser Betriebe ist die Schadensregulierung noch nicht abgeschlossen.

Hinweise für die Praxis

Besonders auffällig ist weiterhin der stark signifikante Unterschied der versicherten und nicht-versicherten Unternehmen mit der Zufriedenheit der Abwicklung der Kompensation. Obwohl in beiden Teilgruppen die volle Bandbreite der Antwortskala ausgeschöpft wird, sind die versicherten Unternehmen im Mittel zufriedener als die nicht-versicherten Betriebe (Tab. 8.6). Gründe für die Unzufriedenheit (in beiden Gruppen) sind unklare Zuständigkeiten und Abläufe, komplexe und lange Formulare, Abhängigkeit von Gutachtern und die lange Dauer des Gesamtprozesses. Die Befunde stellen die Abwicklung und Effektivität der hohen staatlichen Hilfen durchaus in Frage. Ähnliche Ergebnisse zeigten sich bei den Hochwassern 2002 und 2013 für geschädigte Privathaushalte (DKKV 2015).

Im Hinblick auf schadensmindernde Maßnahmen und Vorsorge konnten keine signifikanten Unterschiede festgestellt werden. Es wurden lediglich leicht signifikante Unterschiede ($p = 0{,}090$) bei der Zustimmung zur Aussage „Betriebliche Vorsorge gegen Überflutungsschäden ist wichtig, aber andere Belange haben momentan Vorrang" identifiziert: Auf einer Skala von 1 (ich stimme voll zu) bis 6 (ich stimme gar nicht zu) antworteten versicherte Unternehmen mit 2,71, nicht-versicherte mit 2,46, d. h., hier ist eine leichte Tendenz zum Aufschub der betrieblichen Vorsorge zu sehen.

Allerdings ergeben sich bei den realisierten Vorsorgestrategien nur leichte Unterschiede zwischen versicherten und nicht-versicherten Unternehmen (◘ Tab. 8.6). Ein sehr hoher Anteil von über 80 % der versicherten und nicht-versicherten Unternehmen hat proaktive Maßnahmen zur Schadensminderung vor oder nach dem Hochwasser umgesetzt. Bei nicht-versicherten Unternehmen gibt es einen höheren Anteil von passiven Unternehmen, die gar keine Vorsorge oder Notfallplanungen ergreifen, während versicherte Unternehmen etwas häufiger wegziehen (vgl. ◘ Tab. 8.6). Zusammenfassend ist festzustellen, dass Versicherung sich eher nicht auf das Vorsorgeverhalten von Unternehmen auswirkt.

8.5 Zusammenfassung

Hochwasserereignisse können die gewerbliche Wirtschaft direkt aufgrund von Schäden an Gebäuden, Betriebseinrichtungen oder Waren, Produkten und Lagerbeständen treffen, aber auch indirekt durch Unterbrechungen oder Einschränkungen der wirtschaftlichen Aktivitäten. Auf Basis einer Telefonbefragung von 431 geschädigten Unternehmen aus Nordrhein-Westfalen und Rheinland-Pfalz konnten Art und Höhe der Schäden in Betrieben aus verschiedenen Wirtschaftssektoren sowie Einflussfaktoren auf die Höhe von Schäden und Unterbrechungsdauern für das Hochwasser im Juli 2021 ermittelt werden. Die Auswertungen zeigen, dass insbesondere Betriebe aus dem produzierenden Gewerbe mit hohen Schäden konfrontiert waren. Unternehmen im Sektor „Handel, Gastgewerbe und Verkehr" hatten mit den meisten Auswirkungen zu kämpfen. Während der Wasserstand, Fließgeschwindigkeit und die Größe des Betriebes die wichtigsten Einflussfaktoren für die Höhe der direkten Sachschäden darstellen, sind für die Dauer der Betriebsunterbrechungen oder -einschränkungen auch die Vorwarnzeit bzw. Vorbereitungen auf den Ereignisfall (z. B. durch einen Notfallplan) relevant. Weiterhin zeigt sich, dass ca. ein Drittel der nicht-versicherten Unternehmen trotz der hohen staatlichen Wiederaufbauhilfen keine Kompensationszahlungen erhalten hat. Nicht-versicherte Unternehmen sind zudem deutlich unzufriedener mit der Schadensabwicklung als versicherte Betriebe. Dies bedeutet, dass sich Eigenvorsorge – durch Notfallpläne oder Versicherung – auch bei einem extremen Hochwasser lohnt. Erfreulicherweise ist unter den befragten Unternehmen nur ein sehr kleiner Anteil von weniger als 5 % gänzlich passiv beim Umgang mit Hochwasserrisiken.

Zusammenfassung

- Hochwasserereignisse können die gewerbliche Wirtschaft direkt aufgrund von Schäden an Gebäuden, Betriebseinrichtungen oder Waren, Produkten und Lagerbeständen treffen, aber auch indirekt durch Unterbrechungen oder Einschränkungen der wirtschaftlichen Aktivitäten.

- Die Auswertung einer Telefonbefragung von 431 geschädigten Unternehmen aus Nordrhein-Westfalen und Rheinland-Pfalz zeigt, dass insbesondere Betriebe aus dem produzierenden Gewerbe mit hohen Schäden konfrontiert waren. Unternehmen im Sektor „Handel, Gastgewerbe und Verkehr" hatten mit den meisten Auswirkungen zu kämpfen.
- Während der Wasserstand, Fließgeschwindigkeit und die Größe des Betriebes die wichtigsten Einflussfaktoren für die Höhe der direkten Sachschäden darstellen, sind für die Dauer der Betriebsunterbrechungen oder -einschränkungen auch die Vorwarnzeit bzw. Vorbereitungen auf den Ereignisfall (z. B. durch einen Notfallplan) relevant.
- Weiterhin zeigt sich, dass ca. ein Drittel der nicht-versicherten Unternehmen trotz der hohen staatlichen Wiederaufbauhilfen keine Kompensationszahlungen erhalten hat.
- Nicht-versicherte Unternehmen sind zudem deutlich unzufriedener mit der Schadensabwicklung als versicherte Betriebe. Dies bedeutet, dass sich Eigenvorsorge – durch Notfallpläne oder Versicherung – auch bei einem extremen Hochwasser lohnt.
- Unter den befragten Unternehmen ist nur ein sehr kleiner Anteil von weniger als 5 % gänzlich passiv beim Umgang mit Hochwasserrisiken.

Danksagung Die Autorinnen und Autoren von ▶ Kap. 8 bedanken sich beim Bundesministerium für Bildung und Forschung (BMBF) und beim DLR Projektträger für die Unterstützung und Förderung des KAHR-Projekts. Diese Forschung wurde vom BMBF unter dem Förderkennzeichen 01LR2102A-M finanziert.

Literatur

BMI, BMF (2022) Bericht zur Hochwasserkatastrophe 2021: Katastrophenhilfe, Wiederaufbau und Evaluierungsprozesse. https://www.bmi.bund.de/SharedDocs/downloads/DE/veroeffentlichungen/2022/abschlussbericht-hochwasserkatastrophe.html. Zugegriffen: 16. Sept. 2024

DKKV (2015) Das Hochwasser im Juni 2013. DKKV-Schriftenreihe Band, Bd 53. Deutsches Komitee Katastrophenvorsorge, Bonn (https://dkkv.org/publikation/dkkv-schriftenreihe-nr-53/ (letzter Zugriff: 16.09.2024))

GDV (2023) Zahlen und Fakten: Flutkatastrophe „Bernd" Juli 2021. Gesamtverband der Versicherer, Berlin (https://www.gdv.de/resource/blob/137346/c81a749e02503cb41a3e5a61966f3d31/pdf-schadenbilanz-data.pdf (letzter Zugriff: 16.09.2024))

Gerl T, Kreibich H, Franco G, Marechal D, Schröter K (2016) A review of flood loss models as basis for harmonization and benchmarking. PLoS ONE 11(7):e159791. https://doi.org/10.1371/journal.pone.0159791

Hudson P, Thieken AH (2022) The presence of moral hazard regarding flood insurance and German private businesses. Nat Hazards 112:1295–1319. https://doi.org/10.1007/s11069-022-05227-9

Kreibich H, Piroth K, Seifert I, Maiwald H, Kunert U, Schwarz J, Merz B, Thieken AH (2009) Is flow velocity a significant parameter in flood damage modelling? Nat Hazards Earth Syst Sci 9(5):1679–1692. https://doi.org/10.5194/nhess-9-1679-2009

Schoppa L, Sieg T, Vogel K, Zöller G, Kreibich H (2020) Probabilistic flood loss models for companies. Water Resour Res 56(9):e2020W–27649. https://doi.org/10.1029/2020WR027649

Sieg T, Vogel K, Merz B, Kreibich H (2017) Tree-based flood damage modeling of companies: damage processes and model performance. Water Resour Res 53(7):6050–6068. https://doi.org/10.1002/2017WR020784

Sieg T, Schinko T, Vogel K, Mechler R, Merz B, Kreibich H (2019) Integrated assessment of short-term direct and indirect economic flood impacts including uncertainty quantification. PLoS ONE 14(4):e212932. https://doi.org/10.1371/journal.pone.0212932

Thieken A, Bessel T, Kienzler S, Kreibich H, Müller M, Pisi S, Schröter K (2016) The flood of June 2013 in Germany: how much do we know about its impacts? Nat Hazards Earth Syst Sci 16:1519–1540. https://doi.org/10.5194/nhess-16-1519-2016

Thieken A, Kreibich H, Müller M, Lamond J (2017) Data collection for a better understanding of what causes flood damage – experiences with telephone surveys. In: Molinari D, Menoni S, Ballio F (Hrsg) Flood damage survey and assessment: new insights from research and practice. AGU, Wiley, S 95–106 https://doi.org/10.1002/9781119217930.ch7. ISBN 978-1-119-21792-3 (Chapter 7)

Thieken AH, Samprogna Mohor G, Kreibich H, Müller M (2022) Compound inland flood events: different pathways, different impacts and different coping options. Nat Hazards Earth Syst Sci 22(1):165–185. https://doi.org/10.5194/nhess-22-165-2022

Trenczek J, Lühr O, Eiserbeck L, Leuschner V (2022) Schäden der Sturzfluten und Überschwemmungen im Juli 2021 in Deutschland. Eine ex-post-Analyse. Projektbericht. Prognos, Düsseldorf (https://www.prognos.com/sites/default/files/2022-07/Prognos_KlimawandelfolgenDeutschland_Detailuntersuchung%20Flut_AP2_3b_.pdf (letzter Zugriff: 16.09.2024))

Die Auswirkungen des Hochwassers 2021 auf die mentale Gesundheit der Betroffenen

Marie-Luise Zenker, Philip Bubeck und Annegret Thieken

Die Abbildung 9.4 wurde ersetzt, da die Farbkennzeichnung des Texts „Verletzung/Erkrankung“ fehlerhaft war. Der Text wird jetzt korrekt zweifarbig – rosa und grün – dargestellt.

J. Birkmann et al. (Hrsg.), *Die Flutkatastrophe 2021*, Sustainable Development Goals (SDG) – Umsetzung in Praxis, Lehre und Entscheidungsprozessen,
https://doi.org/10.1007/978-3-658-48688-4_9

9.1 Einleitung

Hochwasserereignisse haben in Deutschland in den letzten Jahren zehntausende Menschen direkt betroffen und vermehrt zu schwerwiegenden Auswirkungen geführt. Besonders die Bilder von Schäden an Gebäuden, Besitztümern und kritischer Infrastruktur, die mit enormen Kosten verbunden sein können, bleiben vielen Menschen im Gedächtnis. Allein seit 2002 traten fünf Ereignisse auf, die mehr als 2 Mrd. € an Sachschäden verursacht haben (Kron et al. 2022). Hierzu zählen die Hochwasser 2002 und 2013 an Elbe und Donau (Thieken et al. 2016), Starkregenereignisse in Baden-Württemberg und Bayern 2016 (Laudan et al. 2017) und Überschwemmungen in Bayern und Baden-Württemberg im Frühsommer 2024.[1] Mit mehr als 30 Mrd. € an Schäden sticht das Ereignis im Juli 2021, das in Deutschland vornehmlich Rheinland-Pfalz (RLP) und Nordrhein-Westfalen (NRW) betraf, deutlich heraus (Truedinger et al. 2023). Um die Auswirkungen von Hochwasserereignissen auf die menschliche Gesundheit zu beziffern, wird vorwiegend die Anzahl an Todesopfern und Verletzten herangezogen (Petrucci et al. 2019; DKKV 2015). Mit 190 Todesopfern in Deutschland und mehr als 750 Verletzten allein in RLP stellt das Ereignis vom Juli 2021 auch diesbezüglich einen traurigen Rekord auf (Thieken et al. 2023; DKKV 2022).

9

Während die Anzahl an Todesopfern in der Regel gut dokumentiert ist, ist über die Auswirkungen von Hochwasserereignissen auf die mentale Gesundheit weit weniger bekannt, obwohl sich in den letzten Jahren vermehrt die Erkenntnis durchgesetzt hat, dass derartige Extremereignisse eine schwerwiegende und langfristige Belastung für die Betroffenen darstellen können (Zhong et al. 2018). So gaben 45 % in einer Betroffenenbefragung nach dem Hochwasser 2013 in Deutschland an, dass sie das Ereignis auch nach ca. 45 Monaten, also nach fast vier Jahren, immer noch (sehr) stark belaste.[2] Andere Studien zeigen, dass die Zahl der Menschen, die nach Hochwasserkatastrophen an psychischen Störungen leiden, im Gegensatz zur Gesamtbevölkerung erheblich ansteigt (Zenker et al. 2024; Paranjothy et al. 2011). Dies können unter anderem posttraumatische Belastungsstörungen (PTBS), Depressionen oder Angstzustände sein, wobei die PTBS die am häufigsten auftretende psychische Erkrankung nach einem traumatischen Erlebnis darstellt (Dreßing und Foerster 2021).

Auslöser für eine PTBS können potenziell traumatische Erlebnisse sein, welche durch die folgenden Charakteristika gekennzeichnet sind (DSM-IV: American Psychiatric Association 1994): (1) Die Person hat ein Ereignis erlebt oder miterlebt, das für sie selbst oder für eine andere Person eine ernsthafte Bedrohung durch Tod oder schwere Verletzungen darstellte. (2) Die Reaktion auf dieses Ereignis umfasste intensive Angst, Hilflosigkeit oder Entsetzen.

Eine PTBS kann das Leben der Erkrankten stark beeinträchtigen. Zu den typischen Symptomen gehören das Wiedererleben der Situation, das Vermeiden von Gefühlen, Schlaflosigkeit, Angstzustände und Aggressionen. Darüber hinaus geht eine PTBS häufig mit anderen Symptomen und psychischen Erkrankungen wie Sucht, Depressionen oder Angststörungen einher. Auch früher diagnostizierte Krankheiten können wieder auftreten oder sich verschlimmern. Die Schwere des Hochwassers vom Juli 2021 und dessen Auswirkungen lassen vermuten, dass es entsprechende psychische Reaktionen ausgelöst

1 ▶ https://www.gdv.de/gdv/medien/medieninformationen/hochwasser-suedeutschland-schadenschaetzung-versicherer-178816; zuletzt abgerufen am 10.01.2025.

2 ▶ https://hochwasser-kahr.de/images/pdf/Flyer_KAHR_online_01_neu.pdf.

haben könnte. Dies zeigt sich anhand der hohen Anzahl an Todesopfern und Verletzten, durch die hohen Wasserstände sowie durch die außerordentlich hohen Schäden an Hausrat und Gebäuden. Bei 47,5 % der ein Jahr nach dem Ereignis Befragten im Landkreis Ahrweiler hatte das Wasser an der Außenseite der Hauswand eine Höhe von 2 Metern oder mehr erreicht. Etwa die Hälfte der Haushalte gab an, finanzielle Schäden von über 100.000 € erlitten zu haben.

Genauere Erkenntnisse dazu, in welchem Umfang das Erleben potenziell traumatischer Überflutungen zu psychischen Erkrankungen der Betroffenen führen könnte, gibt es in Deutschland bisher nur wenige. Eine davon ist die Studie von Apel und Coenen (2021) nach dem Hochwasser 2013, welche eine PTBS-Prävalenz von 20,4 % ein Jahr nach dem Ereignis, 10,8 % nach anderthalb Jahren und 8,7 % nach zwei Jahren feststellte. Zudem zeigten Augustin et al. (2024) nach dem Hochwasserereignis 2021 in Ahrweiler eine zunehmende Anzahl psychischer Verhaltensstörungen und Verletzungen auf Basis von Krankenkassendaten. Informationen zu möglichen Prävalenzen im Nachgang von extremen Überflutungen können wichtige Hinweise für die Bereitstellung einer ausreichenden psychosozialen Versorgung betroffener Regionen liefern und stellen, vor allem vor dem Hintergrund zunehmend häufigerer und stärkerer Hochwasserereignisse aufgrund des Klimawandels (Tradowsky et al. 2023), eine wichtige Entscheidungsgrundlage für ein effektives Risikomanagement dar. Anders als beispielweise Australien oder England verfügt Deutschland bislang über kein umfassendes Konzept zur Erfassung und Bewältigung psychischer Auswirkungen von Katastrophen (DH Emergency Preparedness Division 2009; National Mental Health Commission 2023). Dies ist jedoch für die Erreichung des Entwicklungsziels 3 „Gesundheit und Wohlergehen" der Vereinten Nationen von zentraler Bedeutung, insbesondere des SDG 3.4, das die Förderung der mentalen Gesundheit als wichtige Aufgabe beinhaltet.

Im Rahmen des KAHR-Projektes wurde daher untersucht, welche Auswirkungen ein derart schweres Hochwasser auf die mentale Gesundheit von Bürger:innen hat, die direkt von dem Hochwasser beeinträchtigt wurden. Mit Hilfe mehrerer zeitlich versetzter Befragungen, u. a. im besonders schwer betroffenen Landkreis Ahrweiler, wird in diesem Kapitel folgenden Fragen nachgegangen:

- Wie viele der Betroffenen zeigen Anzeichen einer PTBS (Prävalenz)?
- Wie verändert sich die Prävalenz im Laufe der Zeit?
- Welche Schutz- bzw. Risikofaktoren beeinflussen das mögliche Auftreten einer PTBS?
- Hat die geographische Region einen Einfluss auf PTBS?

9.2 Mentale Gesundheit erfassen

Sich unmittelbar nach einem potenziell traumatischen Ereignis unwohl zu fühlen und negative psychische Symptome zu erleben, ist häufig und weist nicht zwangsläufig auf eine psychische Störung hin. In der Regel lassen diese Probleme im Laufe der Zeit nach und hängen von der Bewältigungsfähigkeit jeder und jedes Einzelnen ab (UHZ 2023). Der Gesundheits- bzw. Krankheitszustand einer Person wird dabei sowohl durch Risikofaktoren, denen sie ausgesetzt ist, als auch durch ihre Resilienz (Schutzfaktoren), also ihre Widerstandsfähigkeit, beeinflusst (Bonanno 2004; Mason et al. 2010).

Wie stark Menschen nach einem Hochwasserereignis mental belastet sind und ob Anzeichen für eine psychische Erkrankung vorliegen, kann beispielsweise mit Hilfe epidemiologischer Screening-Skalen in standardisierten Befragungen ermittelt werden. In der Literatur liegt eine Vielzahl klinisch validierter Skalen unterschiedlicher Länge zur Erfassung allgemeinen psychischen Stresses und spezifischen psychischen Erkrankungen vor (Galatzer-Levy et al. 2018). Ein einfaches Instrument zur Erfassung unspezifischen psychologischen Stresses ist z. B. die „Kessler Distress Scale (K6)" (Kessler et al. 2002). Depressionssymptome können unter anderem mit der „Hospital Anxiety and Depression Scale (HADS)" und der „Centre for Epidemiologic Studies Depression Scale (CES-D)" erfasst werden (Zigmond und Snaith 1983; Radloff 1977). Anzeichen für das Vorliegen einer PTBS können z. B. mit der „Impact of Event Scale – Revised (IES-R)" erhoben werden (Weiss 2007).

Für die in diesem Kapitel dargestellten Ergebnisse wurde die deutsche Fassung einer kurzen epidemiologischen Screening-Skala für PTBS verwendet (Siegrist und Maercker 2010). Dabei wird sich auf die vierte Version des Diagnostischen und Statistischen Manuals psychischer Störungen (DSM-IV) bezogen, welche eine standardisierte Klassifikation und Diagnosekriterien für psychische Störungen beinhaltet, die von Fachleuten im Bereich der psychischen Gesundheit verwendet werden (American Psychiatric Association 1994). Die Skala wies in ihrer englischen Version eine sehr gute und mehrfach überprüfte Testgüte auf (Breslau et al. 1999; Siegrist und Maercker 2010). Mit dem Messinstrument wird erfasst, wie oft sieben Items (Symptome), unter anderem Gedanken- und Gefühlsvermeidung, Interessensverlust und Schlafstörungen, in den letzten vier Wochen aufgetreten sind (◘ Tab. 9.1). Die vierstufige Antwortskala reicht von „überhaupt nicht" bis „fünfmal pro Woche/fast immer". Ein Symptom gilt als vorhanden, wenn Befragte die höchste oder zweithöchste Antwortmöglichkeit angeben. Indem die Ergebnisse aller sieben Symptome addiert werden, wird ein PTBS-Score errechnet, der von 0 bis 7

9

◘ Tab. 9.1 Die sieben Fragen (Items) der kurzen epidemiologischen Screening-Skala für PTBS nach Siegrist und Maercker (2010) auf einer Antwortskala von „überhaupt nicht"; „einmal pro Woche oder seltener/manchmal"; „2–4-mal pro Woche/die Hälfte der Zeit"; „5-mal pro Woche/fast immer". Die Fragen zu den Items beziehen sich dabei auf die letzten 4 Wochen

Nummer	Frage [Item/Symptom]
1	Haben Sie sich bemüht, Aktivitäten, Menschen oder Orte zu meiden, die Sie an das Erlebnis erinnern? [Gedanken- und Gefühlsvermeidung]
2	Haben Sie deutlich weniger Interesse an Aktivitäten, die vor dem Erlebnis wichtig für Sie waren, oder haben Sie sie deutlich seltener unternommen? [Interessensverlust]
3	Fühlten Sie sich den Menschen Ihrer Umgebung deswegen entfremdet und anders? [Entfremdungsgefühl]
4	Fühlten Sie sich abgestumpft oder taub (z. B. nicht weinen können oder sich unfähig fühlen, liebevolle Gefühle zu erleben)? [Abgestumpftheit, Taubheit]
5	Hatten Sie das Gefühl durch das Erlebnis, dass sich Ihre Zukunftspläne und Hoffnungen nicht erfüllen werden (z. B., dass Sie keine Kinder haben oder dass Sie keinen Erfolg im Beruf haben würden)? [Nichterfüllte Zukunftspläne]
6	Hatten Sie Ein- oder Durchschlafstörungen seit dem Erlebnis (d. h., vorher noch nicht)? [Schlafstörungen]
7	Waren Sie nervös oder schreckhaft (z. B., wenn jemand hinter Ihnen Geräusche macht)? [Nervosität]

reicht. Ein Hinweis auf eine PTBS liegt gemäß der Screening-Skala vor, wenn bei Befragten insgesamt vier oder mehr Symptome vorliegen. Der Schwellenwert von vier ist aus empirischen Studien abgeleitet und wird angewendet, da er nachweislich am besten zwischen Patienten und Patientinnen mit und ohne PTBS unterscheidet (Breslau et al. 1999; Kimerling et al. 2006; Siegrist und Maercker 2010). Es ist darauf hinzuweisen, dass die Ergebnisse der Screening-Skala dennoch keine medizinische Diagnose darstellen.

9.3 Die Befragungen von betroffenen Privathaushalten nach dem Hochwasser 2021

Im Sommer und Herbst 2022 wurden Privathaushalte im Landkreis Ahrweiler sowie in den betroffenen Regionen NRWs zu ihren Erfahrungen während und nach dem Hochwasserereignis vom Juli 2021 befragt (▫ Tab. 9.2 und ▫ Abb. 9.1). Für beide Befragungen, d. h. ein und anderthalb Jahre nach dem Ereignis, konnten im Landkreis Ahrweiler jeweils 5250 Betroffene angeschrieben werden, die zuvor Soforthilfe beantragt hatten (im Folgenden LKA1 und LKA1.5). Darüber hinaus wurde die Befragung anderthalb Jahre nach dem Ereignis auch in NRW durchgeführt, wobei eine alternative Methode zur Stichprobengenerierung angewendet werden musste, da ein Vorgehen wie im Landkreis Ahrweiler nicht möglich war. So wurden stattdessen Anzeigen auf Meta/Facebook geschaltet. Beide Befragungen wurden dabei online durchgeführt (siehe ► Kap. 5), wobei auch die Möglichkeit bestand, sich bei den Befragungen im Landkreis Ahrweiler den Fragebogen in Papierform zusenden zu lassen. Der Datensatz aus NRW wurde anhand der dreistelligen Postleitzahl in zwei Teilregionen aufgeteilt: die stark betroffene (Vor-)Eifel-Region (im Folgenden: Eifel1.5) sowie alle weiteren Gebiete aus NRW (im Folgenden: Weitere1.5).

▫ **Tab. 9.2** Charakterisierung der Haushaltsbefragungen nach dem Hochwasserereignis 2021

	1 Jahr nach dem Ereignis Juni – August 2022	**1 ½ Jahre nach dem Ereignis November – Dezember 2022**		
Region	Landkreis Ahrweiler (RLP) **(LKA1)**	Landkreis Ahrweiler (RLP) **(LKA1.5)**	NRW	
			Stark betroffene (Vor-)Eifelregion **(Eifel1.5)**	Weitere Regionen **(Weitere1.5)**
Stichprobe	1/3 der Haushalte, die Soforthilfe beantragt haben → jeweils ca. 5250 Haushalte		Meta/Facebook Werbung	
Rücklauf	516 (10 %)	277 (7 %)	332*	
Befragungsthemen	Persönliche Betroffenheit vom Hochwasser 2021 Erholung mentale Gesundheit und potenzielle Einflussfaktoren soziale Vulnerabilität Meinungen zum Hochwasserrisikomanagement	Mentale Gesundheit und potenzielle Einflussfaktoren strukturelle Schäden und potenzielle Einflussfaktoren Stand des Wiederaufbaus Eigenvorsorge Risikowahrnehmung		

* aufgrund der Stichprobengenerierung ist keine Rücklaufquote berechenbar

9

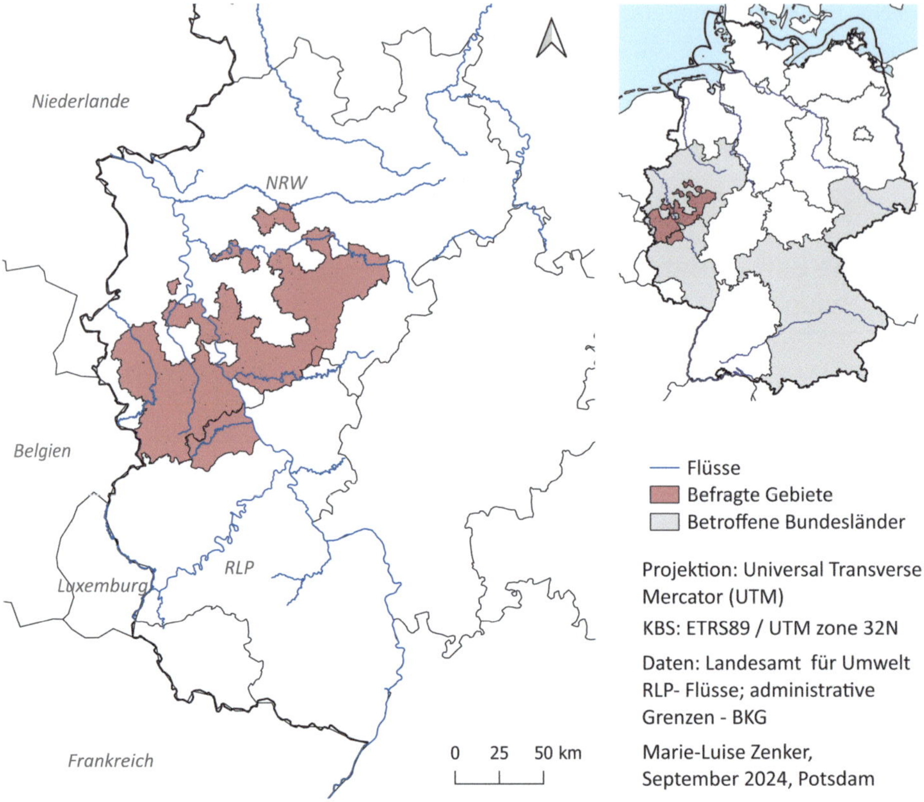

Abb. 9.1 Übersicht über die befragten Gebiete – Landkreis Ahrweiler, Eifel- und Voreifelregion sowie weitere betroffene Gebiete in NRW. (Genutzte Datenquellen: © GeoBasis-DE / BKG (2024); Wasserwirtschaftsverwaltung des Landesamts für Umwelt RLP (2024))

Um die mentale Belastung der Betroffenen zu erfassen, wurde die oben beschriebene kurze epidemiologische Screening-Skala von Siegrist und Maercker (2010) implementiert (siehe ▶ Abschn. 9.2).

9.4 Erkenntnisse zu Auswirkungen des Hochwasserereignisses von 2021 auf die mentale Gesundheit

9.4.1 Erlebte Symptome einer PTBS bei den Befragten

Die erfragten Symptome aus der kurzen PTBS-Screening-Skala geben erste Hinweise darauf, wie stark die Betroffenen zu verschiedenen Befragungszeitpunkten nach dem Ereignis belastet waren. Die Antworten der Befragten sind in der Abb. 9.2 dargestellt. Diese Abbildung zeigt die sieben PTBS-Symptome der Befragten ein (LKA1) sowie anderthalb Jahre nach dem Hochwasserereignis (LKA1.5; Eifel1.5; Weitere1.5) auf Grundlage der vier Antwortmöglichkeiten von „5-mal pro Woche/fast immer“ bis „überhaupt nicht“. Das oberste Item stellt dabei das im LKA1 am häufigsten genannte

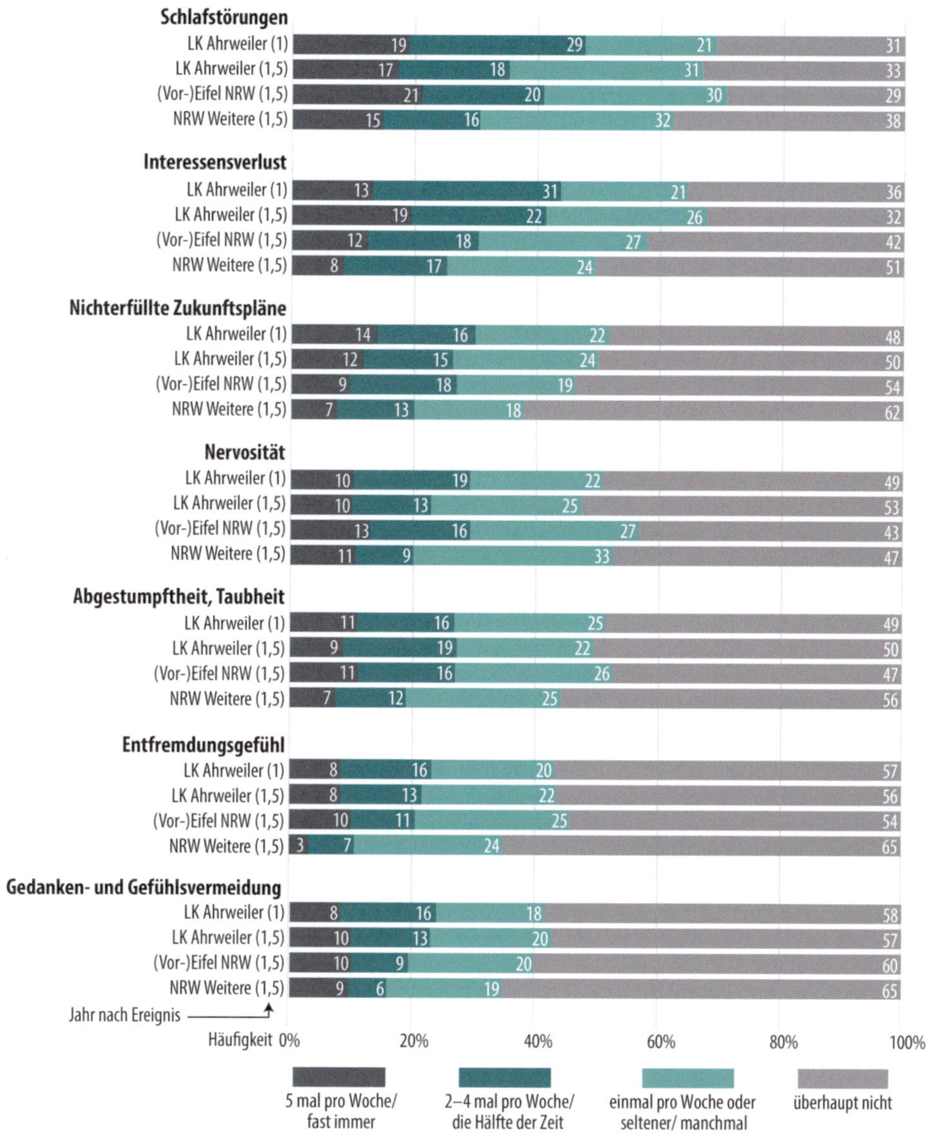

Abb. 9.2 PTBS-Symptome bei Befragten im Landkreis Ahrweiler ein Jahr nach dem Hochwasserereignis (n = 411) sowie anderthalb Jahre nach dem Hochwasserereignis im Landkreis Ahrweiler (n = 232), der Eifel- und Voreifelregion NRWs (n = 171) und weiteren betroffenen Regionen NRWs (n = 96) nach der PTBS-Screeningskala von Siegrist und Maercker (2010), sortiert nach dem am häufigsten aufgetretenen Symptom im LKA1 von oben nach unten. (Eigene Darstellung)

Symptom dar, während das unterste Item das am seltensten angegebene Item im LKA1 ist. Die Abbildung lässt sowohl Rückschlüsse auf zeitliche (ein und anderthalb Jahre nach dem Hochwasser) als auch räumliche Aspekte (LKA1.5, Eifel1.5, Weitere1.5) zu.

In der ersten Befragung ca. einem Jahr nach dem Ereignis (LKA1) wiesen im besonders stark betroffenen Landkreis Ahrweiler knapp 68 % der Befragten mindestens eines der sieben Symptome auf. Bei 5 % der Befragten traten sogar alle sieben Symptome auf.

Tab. 9.3 Mittelwerte und in Klammern die zugehörigen Standardabweichungen der Itemausprägungen von 1–4 („überhaupt nicht" bis „5-mal pro Woche/fast immer") inklusive einer Deutschen Vergleichsstudie von Siegrist und Maercker (2010), sortiert nach dem am häufigsten aufgetretenen Symptom im LKA1 von oben nach unten

Item/Symptom	LKA1 ($n = 411$)	LKA1.5 ($n = 232$)	Eifel1.5 ($n = 171$)	Weitere1.5 ($n = 96$)	Vergleich Deutsche Gesamtbevölkerung ($n = 475$)
Schlafstörungen	2,36 (1,11)	2,19 (1,08)	2,33 (1,11)	2,07 (1,06)	1,91 (1,05)
Interessensverlust	2,21 (1,07)	2,28 (1,12)	2,01 (1,05)	1,83 (0,99)	1,77 (1,03)
Nichterfüllte Zukunftspläne	1,96 (1,09)	1,88 (1,05)	1,82 (1,03)	1,65 (0,96)	1,62 (0,98)
Nervosität	1,91 (1,04)	1,80 (1,85)	1,99 (1,06)	1,83 (0,99)	1,77 (0,97)
Abgestumpftheit, Taubheit	1,89 (1,04)	1,85 (1,01)	1,91 (1,04)	1,71 (0,94)	1,53 (0,89)
Entfremdungsgefühl	1,74 (0,98)	1,73 (0,98)	1,76 (1,00)	1,48 (0,77)	1,48 (0,85)
Gedanken- und Gefühlsvermeidung	1,73 (1,00)	1,75 (1,02)	1,69 (1,00)	1,60 (0,97)	1,99 (1,09)

9

Etwa 70 % der Befragten berichteten mindestens einmal pro Woche über Schlafprobleme, wobei diese bei knapp 20 % fast täglich auftraten (siehe Abb. 9.2). Viele Befragte (65 %) gaben außerdem an, dass sie Aktivitäten, die ihnen vor der Überschwemmung wichtig waren, nicht mehr genießen konnten.

Wie in Tab. 9.3 ersichtlich, liegt das Auftreten von PTBS-Symptomen in allen vier Stichproben deutlich über den Werten einer repräsentativen Studie in Deutschland, die den Ist-Zustand der erwachsenen Bevölkerung erfasst hat (Maercker et al. 2008; Siegrist und Maercker 2010). Eine Ausnahme stellt jedoch das Vermeiden von Aktivitäten, Menschen und Orten dar, die die Betroffenen an das Ereignis erinnern. Dies könnte damit zusammenhängen, dass der laufende Wiederaufbau, finanzielle Probleme und Versicherungsangelegenheiten sowie sichtbare Schäden an Gebäuden und Infrastruktur es äußerst schwierig machen, die enormen Auswirkungen des Hochwasserereignisses zu ignorieren. Darüber hinaus erhielt das Hochwasserereignis weiterhin ein hohes Maß an Aufmerksamkeit in nationalen und lokalen Medien. Auch Projekte zum Wiederaufbau und Forschungsvorhaben hielten das Ereignis bei den Betroffenen im Gedächtnis. Es lässt sich außerdem feststellen, dass die Mittelwerte der PTBS-Symptome in der Stichprobe Weitere1.5 am niedrigsten sind und das Gefühl der Entfremdung sogar dem Niveau der Repräsentativstudie entspricht.

9.4.2 Anzeichen einer PTBS bei den Befragten

Die einzelnen Symptome einer PTBS zeigen bereits die erheblichen psychischen Belastungen der Hochwasserbetroffenen. Allerdings stellen, wie eingangs gezeigt, einzelne Symptome allein noch kein Anzeichen für eine PTBS dar. Damit eine Indikation für eine PTBS vorliegt, müssen für mindestens vier der Symptome die beiden höchsten Antwort-

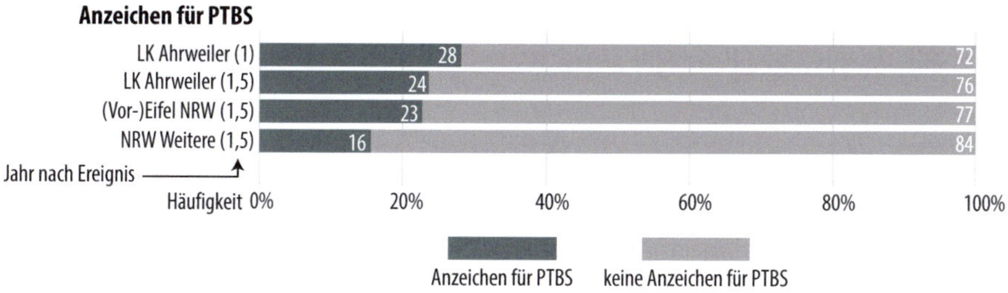

Abb. 9.3 Prävalenzwerte für PTBS unter den Befragten ein Jahr nach dem Hochwasserereignis im Landkreis Ahrweiler und anderthalb Jahre nach dem Hochwasserereignis im Landkreis Ahrweiler, in der (Vor-)Eifelregion NRWs und weiteren betroffenen Gebieten in NRW. (Eigene Darstellung)

kategorien durch die Befragten berichtet werden (siehe ▶ Abschn. 9.2). Die so ermittelte Prävalenz gibt Aufschluss darüber, bei wie viel Prozent der Befragten Anzeichen für eine PTBS vorliegen.

Wie in Abb. 9.3 zu sehen ist, lag die Prävalenz in der Stichprobe LKA1 ein Jahr nach dem Hochwasserereignis bei 28,2 %. Ein halbes Jahr später zeigten noch 23,7 % der Befragten Anzeichen für eine PTBS (LKA1.5). Der Abfall der Prävalenz im Landkreis Ahrweiler könnte dabei auf einen leichten Rückgang im Laufe der Zeit hindeuten. Im Hinblick auf die Datenerhebung könnte dies allerdings auch auf eine andere Stichprobenzusammensetzung zurückzuführen sein. In der NRW-Eifel-Stichprobe (Eifel1.5) lag die Prävalenz anderthalb Jahre nach dem Ereignis bei sehr ähnlichen 22,8 %. Dies zeigt, dass die Eifelregion in beiden Bundesländern sehr stark betroffen war. Eine ähnliche Topografie führte zu ähnlichen Hochwassercharakteristika wie z. B. hohen Fließgeschwindigkeiten. Zudem wurden in beiden Bundesländern die in den Gefahrenkarten ausgewiesenen Gefährdungszonen für extreme Hochwasser (HQextrem) vielerorts deutlich überschritten. Dies verdeutlicht die Schwere und Seltenheit des Ereignisses. Mit einer Prävalenz von 16 % weisen die weiteren Regionen in NRW (Weitere1.5), in denen die Auswirkungen des Hochwassers insgesamt deutlich geringer waren, nochmals niedrigere Werte auf (Abb. 9.3).

Die berechneten Prävalenzwerte für PTBS sind deutlich höher als in der deutschen Gesamtbevölkerung. Laut einer repräsentativen Befragung aus dem Jahr 2016 wiesen lediglich 1,5 % der Gesamtbevölkerung Anzeichen einer PTBS auf (Maercker et al. 2018). Allerdings könnte neben dem Hochwasserereignis auch die COVID-19-Pandemie zur erhöhten PTBS-Prävalenz beigetragen haben. Ein Blick in die bestehende Literatur zur PTBS nach Hochwasserereignissen zeigt eine hohe Heterogenität – mit Ergebnissen, die von 2,6 bis 52 % reichen (Keya et al. 2023; Golitaleb et al. 2022). Diese Variation kann unter anderem durch unterschiedliche Untersuchungszeitpunkte nach dem Ereignis, die Schwere des Ereignisses, Unterschiede in geographischen Räumen und Kulturen sowie durch die Methode zur Messung von PTBS und die zugrunde liegenden Definitionen bedingt sein. Dennoch lassen sich die berechneten Prävalenzen mit dem Durchschnittswert der Metaanalyse von Golitaleb et al. (2022) vergleichen, der bei 29 % liegt.

9.4.3 Persönliche Erfahrungen und Charakteristika als potenzielle Einflussfaktoren für eine PTBS

Es gibt eine Vielzahl potenzieller Faktoren, die einen Einfluss auf die Entwicklung einer PTBS haben können. Zudem sind nicht alle Menschen gleichermaßen gefährdet, negative mentale gesundheitliche Auswirkungen durch ein Hochwasserereignis zu erleiden. In der Literatur wird bereits eine Vielzahl von Einflussfaktoren genannt, die sich entweder allgemein auf mentale Gesundheit nach Hochwasser (Keya et al. 2023; Fernández et al. 2015; Stanke et al. 2012) oder speziell auf Anzeichen einer PTBS nach Hochwasser beziehen (Golitaleb et al. 2022). Diese Studien umfassen verschiedene Aspekte, die sowohl vor als auch während und nach dem potenziell traumatischen Ereignis eine Rolle spielen könnten. Zu den Vorbedingungen gehören insbesondere soziodemografische Faktoren. Zu jenen, die das Risiko für mentale Probleme erhöhen, gehören beispielsweise das weibliche Geschlecht, eine schlecht gestellte wirtschaftliche Situation eines Haushalts und Vorerkrankungen (Graham et al. 2019; Paranjothy et al. 2011; Mason et al. 2010). Im Gegensatz dazu sind die Befunde hinsichtlich des Alters einer Person als Risikofaktor heterogen (Stanke et al. 2012; Galea et al. 2007).

Neben diesen persönlichen Vorbedingungen können auch die Charakteristika des Hochwasserereignisses selbst, wie der Wasserstand und die Fließgeschwindigkeit, entscheidend sein (Bubeck und Thieken 2018; Lamond et al. 2015). Zudem können persönliche Erfahrungen des Hochwassers, sog. Stressoren, einen erheblichen Einfluss auf die psychische Gesundheit der Betroffenen haben. Beispiele sind die Notwendigkeit zur Rettung, Ungewissheit über den Verbleib von Angehörigen oder Freunden, der Verlust einer geliebten Person oder zugezogene Verletzungen durch das Hochwasser (Galea et al. 2007).

Auch nach dem Hochwasser wirken weiterhin Faktoren auf die psychische Gesundheit ein, insbesondere im Zusammenhang mit dem Wiederaufbauprozess. Das Erleiden erheblichen finanziellen Verlustes ist dabei der am häufigsten berichtete Faktor. Weitere Faktoren sind die Unterbrechung von kritischen Infrastrukturen und Dienstleistungen, fehlende oder unzureichende Versicherungsdeckung und der Mangel an Fachkräften (Bubeck und Thieken 2018; Galea et al. 2007; Jermacane et al. 2018; Paranjothy et al. 2011). Des Weiteren beeinflussen unterschiedliche Bewältigungsmechanismen sowie persönliche Gefühle und Wahrnehmungen ebenfalls die psychische Gesundheit nach einem solchen Ereignis (Bubeck und Thieken 2018; Bonanno et al. 2011). Obwohl bereits eine beträchtliche Menge relevanter Literatur existiert, sind die Zusammenhänge zwischen diesen Faktoren komplex. Studien stimmen nicht immer über die Bedeutung der Einflussfaktoren oder die Richtung des Einflusses überein.

Um Einflussfaktoren für PTBS nach dem Hochwasserereignis 2021 in den betroffenen Regionen von RLP und NRW zu identifizieren, wurden binäre logistische Regressionsanalysen für die Befragungswellen ein Jahr (LKA1) sowie anderthalb Jahre (LKA1.5, Eifel1.5, Weitere1.5 zusammen in einem Datensatz) nach dem Ereignis durchgeführt. Auf Grundlage der Erkenntnisse aus der Literatur und mit einem explorativen Ansatz wurden potenzielle Einflussfaktoren herangezogen und in vier Kategorien unterteilt:

1. Vorbedingungen,
2. Hochwassercharakteristika,
3. Stressoren und
4. Auswirkungen und Bewältigung.

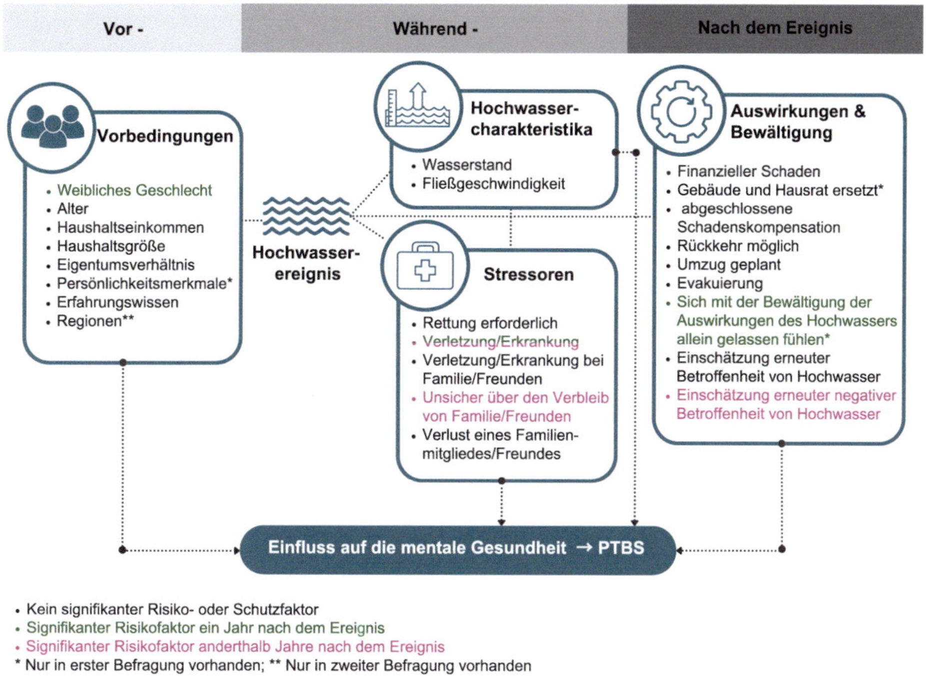

Abb. 9.4 Zusammenfassung der untersuchten potenziellen Einflussfaktoren der Befragten vom Hochwasser 2021 im Hinblick auf das Auftreten von PTBS. Signifikanzniveau $p < 0{,}05$. (Eigene Darstellung)

In der Abb. 9.4 sind alle untersuchten potenziellen Einflussfaktoren im Hinblick auf das Auftreten von PTBS dargestellt. Die vier Kategorien werden den zeitlichen Dimensionen vor, während und nach dem Hochwasserereignis zugeordnet, in denen sie jeweils wirken.

Die Analysen zeigen, dass bei den Vorbedingungen das Geschlecht ein signifikanter Einflussfaktor ein Jahr nach dem Hochwasserereignis ist. Frauen zeigten in der Stichprobe LKA1 ein höheres Risiko, Anzeichen für eine PTBS zu entwickeln. Dies stimmt mit anderen Studien überein, die das Geschlecht als wichtigen Faktor für psychische Belastungen identifiziert haben (Mason et al. 2010; Paranjothy et al. 2011). Dies kann verschiedene Gründe haben und wird oft durch soziale Normen begründet. Einerseits würden Frauen dazu neigen, empfindsamer zu sein und ihre Gefühle leichter ausdrücken zu können. Sie übernähmen typischerweise die Hauptverantwortung für ihre Kinder und sorgten sich im Allgemeinen mehr um ihre Gemeinschaft (Tunstall et al. 2006). Andererseits werden Männer oft als stark beschrieben, die ihre Verletzlichkeiten nicht offenbaren oder gar darüber sprechen würden (Seidler et al. 2019). Diese Stereotype könnten im Kontext der Geschlechterdebatte weiter diskutiert werden, da sich diese mit der Zeit verändern könnten. So zeigte das Geschlecht anderthalb Jahre nach dem Ereignis keinen signifikanten Einfluss mehr auf. Weitere Faktoren, die den Haushalt beschreiben, wie Alter oder Eigentumsverhältnisse, konnten Anzeichen für eine PTBS in dieser Studie nicht erklären.

Zusätzlich wurde anhand der Befragung anderthalb Jahre nach dem Ereignis untersucht, ob die verschiedenen Regionen – der Landkreis Ahrweiler, die Eifel- und Voreifelregion in NRW sowie weitere betroffene Regionen in NRW – einen Einfluss auf die

Entwicklung einer PTBS haben. Die durchgeführten Regressionsanalysen ergaben jedoch keinen signifikanten Einfluss der Regionen auf das Auftreten einer PTBS. Jedoch deuten ähnliche Ergebnisse auf Gemeinsamkeiten in den Regionen Landkreis Ahrweiler und Eifel1.5, die insbesondere durch eine ähnliche Exposition und Schwere des Ereignisses begründet werden können. Im Gegensatz dazu wurde bei der Stichprobe Weitere1.5 eine geringere Prävalenz festgestellt. Dies lässt zum einen darauf schließen, dass die Auswirkungen auf die mentale Gesundheit mit der Schwere des Ereignisses und der Exposition zunehmen. Zum anderen aber könnten die Regionen bereits durch andere Faktoren, wie sozio-demographische Merkmale oder Bewältigungsmechanismen beschrieben worden sein, die in allen Regionen gleichermaßen gelten. Darüber hinaus könnten die drei definierten Regionen eine zu starke Generalisierung der Befragungsdaten darstellen, wodurch bestimmte Details bzw. Nuancen verloren gehen könnten.

Das Ausmaß des Hochwasserereignisses und zum großen Teil auch dessen Auswirkungen werden durch die Hochwassercharakteristika bestimmt. Die in den Befragungen erfassten Eigenschaften Wasserstand und Fließgeschwindigkeit erwiesen sich allerdings nicht als signifikant für die Beschreibung einer PTBS. Jedoch zeigen die Ergebnisse, dass insbesondere Stressoren, die während des Ereignisses auftreten, einen Einfluss auf die Entwicklung einer PTBS haben. Beide Befragungen belegen, dass vor allem Personen gefährdet sind, Anzeichen von PTBS zu entwickeln, die während des Hochwasserereignisses Verletzungen erlitten oder erkrankt sind. Dies entspricht der PTBS-Definition (DSM-4: American Psychiatric Association 1994), in der das (Mit-)Erleben eines Ereignisses im Vordergrund steht. Eine Verletzung kann als greifbare Evidenz gelten (Galea et al. 2007; Wiseman et al. 2013). Dieser Risikofaktor impliziert die Wichtigkeit, die mentale Gesundheit jener Personen zu überwachen, die während eines Hochwasserereignisses verletzt wurden bzw. erkrankt sind, und ihnen die notwendige Behandlung anzubieten, um nicht nur ihr körperliches, sondern auch ihr psychisches Wohlbefinden zu stabilisieren. Diese Erkenntnisse betonen die Bedeutung von SDG 3.8, das den universellen Zugang zu qualitativ hochwertigen Gesundheitsdiensten, einschließlich psychischer Versorgung, für alle Menschen fordert.

Darüber hinaus kann die Unsicherheit über den Verbleib von Familie oder engen Freunden eine erhebliche Belastung für die Betroffenen darstellen. In der LKA1-Stichprobe war jedoch der Anteil der Personen, die über den Verbleib ihrer Liebsten im Unklaren waren, bei denjenigen mit und ohne Anzeichen von PTBS ähnlich hoch ($n_{\text{PTBS-Anzeichen}} = 75{,}8\,\%$; $n_{\text{keine PTBS-Anzeichen}} = 66{,}7\,\%$). Dies deutet auf generell hohe Fallzahlen hin, stellte somit aber ein Jahr nach dem Ereignis keinen signifikanten Einflussfaktor für die Entwicklung einer PTBS dar. In der Befragung anderthalb Jahre nach dem Ereignis hingegen waren Befragte, die beim Ereignis über den Verbleib ihrer Familie oder engen Freunde unsicher waren, einem erhöhten Risiko ausgesetzt, Anzeichen von PTBS zu zeigen.

In der Zeit des Wiederaufbaus und der Erholungsphase zeigten die Betroffenen der Stichprobe LKA1, die sich mit der Bewältigung der Auswirkungen des Hochwasserereignisses allein gelassen fühlten, eine 50 % höhere Wahrscheinlichkeit, Anzeichen für eine PTBS zu entwickeln. Etwas mehr als die Hälfte der Befragten (54,4 %) berichteten über Schwierigkeiten bei der Suche nach verfügbaren Baugutachtern, Handwerkern oder Materialien. Darüber hinaus scheint die Finanzierung des Wiederaufbaus für viele eine Belastung darzustellen, denn 18 % berichteten, dass die Auszahlung unzureichend war oder dass die Versicherung den Schaden nicht anerkannte und somit nicht auszahlte (10,1 %). Außerdem mussten 14,8 % darauf warten, dass ihre Vermieter oder die Eigentümergemeinschaft tätig wurden, bevor Reparaturen getätigt werden konnten. Diese Angaben deuten auf Schwierigkeiten während der Wiederaufbau- und Erholungsphase hin und sind ein wichtiger Anknüpfungspunkt für das Management des Wiederaufbaus nach dem Ereignis.

Des Weiteren zeigte die Sorge, erneut negativ von einer Überflutung betroffen zu sein, die Tendenz zur Erhöhung des Risikos, Anzeichen von PTBS zu entwickeln. Während dieser Faktor ein Jahr nach dem Ereignis nur leicht signifikant war, erwies dieser sich nach anderthalb Jahren als stark signifikanter Einflussfaktor. Diese Ergebnisse machen deutlich, dass das Erleben eines potenziellen Traumas die negative Wahrnehmung des zukünftigen Hochwasserrisikos schüren und mit der Angst vor Wiederholung verbunden sein kann (Mason et al. 2010; Puechlong et al. 2020).

Während die zuvor genannten Faktoren das Risiko für psychische Probleme erhöhen können, können soziale Netzwerke als Schutzfaktor wirken. Studien belegen, dass ein Unterstützungssystem aus Freunden und Nachbarn vor psychischen Problemen während des Bewältigungsprozesses schützen kann (Stanke et al. 2012; Tunstall et al. 2006).

9.5 Fazit & Handlungsempfehlungen

Schwere Hochwasserereignisse können langfristige Auswirkungen auf die mentale Gesundheit der Betroffenen haben. Dies zeigen auch Befragungen Betroffener nach dem Hochwasser im Juli 2021 in RLP und NRW, in denen ein kurzes epidemiologisches Screening-Tool für PTBS in Deutschland im Kontext von Überflutungen eingesetzt wurde. Die Anzeichen für das Vorliegen einer PTBS waren mit 16 bis 28 % ein Jahr bis anderthalb Jahre nach dem Ereignis deutlich höher, als dies laut repräsentativen Befragungen der Gesamtbevölkerung zu erwarten wäre. Im zeitlichen Verlauf scheinen Symptome und Prävalenz von PTBS leicht zurückzugehen. Auch lassen sich regionale Unterschiede zwischen stark und weniger stark getroffenen Gebieten ausmachen. Wichtige Risikofaktoren sind das weibliche Geschlecht, sich während des Hochwasserereignisses zu verletzen oder schwer zu erkranken und die Unsicherheit über den Verbleib von Familie oder engen Freunden. Nach der Flut ist es das Gefühl, bei der Bewältigung der Hochwasserauswirkungen auf sich allein gestellt zu sein, sowie die Angst vor erneuten negativen Folgen eines Hochwasserereignisses.

Die hier vorgestellten Ergebnisse zeigen, dass solch schwere Hochwasser zu zusätzlichen gesundheitsbezogenen Versorgungsbedarfen führen, die gesonderte Aufmerksamkeit erfordern. Psychosoziale und therapeutische Angebote sollten angepasst werden, um der hohen Prävalenz und den Risikofaktoren im Zusammenhang mit PTBS in der betroffenen Bevölkerung Rechnung zu tragen. Hierfür sollten niedrigschwellige Angebote geschaffen werden, wie z. B. eine Bereitstellung von Beratungs- und Betreuungsangeboten sowie Krisenintervention durch die psychosozialen Dienste der Kommunen oder – im Falle sehr schwerer Ereignisse – extra eingerichtete Anlaufstellen und aufsuchende Hilfen.

Ein gelungenes Beispiel hierfür ist das Trauma Hilfe Zentrum, das im besonders schwer getroffenen Landkreis Ahrweiler für die Betroffenen eingerichtet wurde.

Beispiel aus der Praxis

Um die langfristige Verbesserung der psychosozialen Betreuung in der stark betroffenen Region nach dem schweren Hochwasserereignis im Juli 2021 zu unterstützen, wurde das **Trauma Hilfe Zentrum im Ahrtal** etabliert. Ein multiprofessionelles Team aus Sozialarbeiter:innen, Pflegekräften, ärztlichem und psychologischem Fachpersonal bietet anonyme und kostenlose Beratungen an und unterstützt bei der Vermittlung weiterführender Behandlungen. Für diejenigen, die das Zentrum aus persönlichen Gründen nicht besuchen können, wird auch eine aufsuchende Hilfe angeboten. Neben individuellen Sprechstunden bietet das Trauma Hilfe Zentrum zudem Informationsveranstaltungen rund um das Thema Trauma, Gruppenangebote zum Thema Stressbewältigung sowie Trauma Info Gruppen an. Bis September 2022 wurden insgesamt 1475 Einzelberatungen durchgeführt und 70 Gruppenangebote wahrgenommen. Auch mehr als zwei Jahre nach dem Ereignis wird weiterhin Unterstützung in Anspruch genommen (Scharping, persönliche Kommunikation, 4. März 2024). Gefördert wird das Trauma Hilfe Zentrum durch das Ministerium für Wissenschaft und Gesundheit des Landes RLP.

Hinweis für die Praxis

Zudem sollten besonders diejenigen nach dem Hochwasserereignis berücksichtigt werden, die währenddessen verletzt wurden oder dadurch schwer erkrankt sind. Es wäre ratsam, bei der Untersuchung körperlicher Folgen auch eine langfristige psychische Unterstützung sowie ein Monitoring des allgemeinen Gesundheitszustandes und des Wohlbefindens anzubieten. Der nur langsame Rückgang der Belastung, der sich mit anderen Studien deckt, weist zudem darauf hin, dass entsprechende Beratungs- und Betreuungsangebote mittel- bis langfristig in schwer betroffenen Gebieten angeboten und aufrechterhalten werden müssen. Ebenfalls zu empfehlen sind Maßnahmen, die darauf abzielen, dass sich Betroffene nicht alleingelassen fühlen, wie z. B. die Förderung des Austauschs untereinander oder mit verantwortlichen Akteuren und die gezielte Einbindung vulnerabler Gruppen mit schwachen sozialen Netzwerken (▶ Kap. 18 und 19). Darüber hinaus wären gezielte Unterstützungsangebote zu Themen des Wiederaufbaus sinnvoll, um den Betroffenen den Umgang mit den hohen bürokratischen Hürden zu erleichtern. Ungelöste Finanzierungsfragen oder die erfolglose Suche nach Sachverständigen und Handwerker:innen können den Genesungsprozess erheblich verzögern.

Da Erkenntnisse zu diesem Thema in Deutschland bislang rar sind, leistet unsere Studie einen wichtigen Beitrag zum besseren Verständnis von mentaler Belastung nach Hochwasserereignissen. Insgesamt besteht weiterer Forschungsbedarf, unter anderem zu Bewältigungsstrategien oder zu Kausalitäten verschiedener Einflussfaktoren. Darüber hinaus sind Langzeitstudien erforderlich, um den Prozess der negativen mentalen Hochwasserauswirkungen auf die Betroffenen besser zu verstehen.

Schließlich schlagen wir vor, die psychischen Gesundheitsauswirkungen von Überschwemmungen besser in das Hochwasserrisikomanagement zu integrieren. So sollte ein Rahmen für die Unterstützung der psychischen Gesundheit von Betroffenen nach Überschwemmungen geschaffen werden, um in der Wiederaufbauphase Betroffene gezielt und effektiv unterstützen zu können. Die vorliegenden Forschungsergebnisse liefern hierfür wertvolle Anknüpfungspunkte.

Zusammenfassung

- Befragungen betroffener Haushalte im Anschluss an Hochwasserereignisse können wertvolle Informationen über das Ausmaß der mentalen Belastung und möglichen Einflussfaktoren liefern.
- Die Befragung nach dem Hochwasser 2021 zeigt bei Betroffenen deutlich stärkere PTBS-Anzeichen als in der deutschen Gesamtbevölkerung.
- Die mentale Belastung scheint mit der Zeit leicht abzunehmen, bleibt jedoch auf hohem Niveau.
- Die identifizierten Risikofaktoren treten in allen Phasen des Ereignisses auf: vor, während und nach dem Hochwasser.
- Das anhaltende Auftreten von PTBS-Anzeichen und die vorhandenen Risikofaktoren verdeutlichen den Bedarf an langfristiger psychologischer Versorgung, der im integrierten Hochwasserrisikomanagement berücksichtigt werden sollte.

Danksagung Die Autorinnen und Autoren von (▶ Kap. 9) bedanken sich beim Bundesministerium für Bildung und Forschung (BMBF) und beim DLR Projektträger für die Unterstützung und Förderung des KAHR-Projekts. Diese Forschung wurde vom BMBF unter dem Förderkennzeichen 01LR2102A-M finanziert. Zudem danken wir allen Teilnehmenden der Befragung für ihre Mitwirkung.

Literatur

American Psychiatric Association (Hrsg) (1994) Diagnostic and statistical manual of mental disorders. DSM-IV, 4. Aufl. American Psychiatric Publishing

Apel D, Coenen M (2021) Mental health and health-related quality of life in victims of the 2013 flood disaster in Germany – A longitudinal study of health-related flood consequences and evaluation of institutionalized low-threshold psycho-social support. Int J Disaster Risk Reduct. https://doi.org/10.1016/j.ijdrr.2021.102179

Augustin J, Andrees V, Czerniejewski A, Dallner R, Schulz CM, Mezger NCS (2024) Auswirkungen des Ahrtal-Hochwassers auf die Gesundheit der lokalen Bevölkerung – eine Analyse auf Grundlage von GKV-Routinedaten. Bundesgesundheitsblatt Gesundheitsforschung Gesundheitsschutz 67(1):5–13. https://doi.org/10.1007/s00103-023-03809-x

Bonanno GA (2004) Loss, trauma, and human resilience: have we underestimated the human capacity to thrive after extremely aversive events? Am Psychol 59(1):20–28. https://doi.org/10.1037/0003-066X.59.1.20

Bonanno GA, Westphal M, Mancini AD (2011) Resilience to loss and potential trauma. Annu Rev Clin Psychol 7:511–535. https://doi.org/10.1146/annurev-clinpsy-032210-104526

Breslau N, Peterson EL, Kessler RC, Schultz LR (1999) Short screening scale for DSM-IV posttraumatic stress disorder. Am J Psychiatry 156(6):908–911. https://doi.org/10.1176/ajp.156.6.908

Bubeck P, Thieken AH (2018) What helps people recover from floods? Insights from a survey among flood-affected residents in Germany. Reg Environ Change 18(1):287–296. https://doi.org/10.1007/s10113-017-1200-y

Bundesamt für Kartographie und Geodäsie (BKG): GeoBasis-DE: Verwaltungsgebiete 1 : 5 000 000, Stand 01.01. (VG5000 01.01.): https://gdz.bkg.bund.de/index.php/default/verwaltungsgebiete-1-5-000-000-stand-01-01-vg5000-01-01.html [Daten], dl-de/by-2-0, https://www.govdata.de/dl-de/by-2-0, https://sgx.geodatenzentrum.de/web_public/gdz/datenquellen/Datenquellen_vg_nuts.pdf. Zugegriffen: 5. Jan. 2024

DH Emergency Preparedness Division (2009) Emergency Planning Guidance: Planning for the Psychosocial and Mental Health Care of People Affected by Major Incidents and Disasters, S 94

DKKV (2015) Das Hochwasser im Juni 2013: Bewährungsprobe für das Hochwasserrisikomanagement in Deutschland. DKKV-Schriftenreihe Nr. 53. Bonn

DKKV (2022) Die Flutkatastrophe im Juli 2021 in Deutschland. Ein Jahr danach: Aufarbeitung und erste Lehren für die Zukunft. DKKV-Schriftenreihe Nr. 62. Bonn

Dreßing H, Foerster K (2021) Traumafolgestörungen in ICD-10, ICD-11 und DSM-5: Diagnosekriterien und ihre Bedeutung für die gutachtliche Praxis. Forens Psychiatr Psychol Kriminol 15:47–53. https://doi.org/10.1007/s11757-020-00645-6

Fernández A, Black J, Jones MK, Wilson L, Salvador-Carulla L, Astell-Burt T, Black D (2015) Flooding and mental health: a systematic mapping review. PLoS ONE. https://doi.org/10.1371/journal.pone.0119929

Galatzer-Levy IR, Huang SH, Bonanno GA (2018) Trajectories of resilience and dysfunction following potential trauma: a review and statistical evaluation. Clin Psychol Rev 63:41–55. https://doi.org/10.1016/j.cpr.2018.05.008

Galea S, Brewin CR, Gruber M, Jones RT, King DW, King LA, McNally RJ, Ursano RJ, Petukhova M, Kessler RC (2007) Exposure to hurricane-related stressors and mental illness after Hurricane Katrina. Arch Gen Psychiatry 64(12):1427–1434. https://doi.org/10.1001/archpsyc.64.12.1427

Golitaleb M, Mazaheri E, Bonyadi M, Sahebi A (2022) Prevalence of Post-traumatic Stress Disorder After Flood: A Systematic Review and Meta-Analysis. Front Psychiatry 13:890671. https://doi.org/10.3389/fpsyt.2022.890671

Graham H, White P, Cotton J, McManus S (2019) Flood- and Weather-Damaged Homes and Mental Health: An Analysis Using England's Mental Health Survey. IJERPH 16(18):3256. https://doi.org/10.3390/ijerph16183256

Jermacane D, Waite TD, Beck CR, Bone A, Amlôt R, Reacher M, Kovats S, Armstrong B, Leonardi G, Rubin JG, Oliver I (2018) The English National Cohort Study of Flooding and Health: The change in the prevalence of psychological morbidity at year two. BMC Public Health 18(1):330. https://doi.org/10.1186/s12889-018-5236-9

9

Kessler RC, Andrews G, Colpe LJ, Hiripi E, Mroczek DK, Normand SL, Zaslavsky AM (2002) Short screening scales to monitor population prevalences and trends in non-specific psychological distress. Psychol Med 32(6):959–976

Keya TA, Leela A, Habib N, Rashid M, Bakthavatchalam P (2023) Mental Health Disorders Due to Disaster Exposure: A Systematic Review and Meta-Analysis. Cureus 15(4):e37031. https://doi.org/10.7759/cureus.37031

Kimerling R, Ouimette P, Prins A, Nisco P, Lawler C, Cronkite R, Moos RH (2006) Brief report: Utility of a short screening scale for DSM-IV PTSD in primary care. J Gen Intern Med 21(1):65–67. https://doi.org/10.1111/j.1525-1497.2005.00292.x

Kron W, Bell R, Thiebes B, Thieken A (2022) The July 2021 flood disaster in Germany. High-level Experts and Leaders Panel on Water and Disasters, S 12–44

Lamond JE, Joseph RD, Proverbs DG (2015) An exploration of factors affecting the long term psychological impact and deterioration of mental health in flooded households. Environ Res 140:325–334. https://doi.org/10.1016/j.envres.2015.04.008

Laudan J, Rözer V, Sieg T, Vogel K, Thieken AH (2017) Damage assessment in Braunsbach 2016: Data collection and analysis for an improved understanding of damaging processes during flash floods. Nat Hazards Earth Syst Sci 17(12):2163–2179. https://doi.org/10.5194/nhess-17-2163-2017

Maercker A, Forstmeier S, Wagner B, Glaesmer H, Brähler E (2008) Posttraumatische Belastungsstörungen in Deutschland. Ergebnisse einer gesamtdeutschen epidemiologischen Untersuchung. Nervenarzt 79(5):577–586. https://doi.org/10.1007/s00115-008-2467-5

Maercker A, Hecker T, Augsburger M, Kliem S (2018) ICD-11 Prevalence Rates of Posttraumatic Stress Disorder and Complex Posttraumatic Stress Disorder in a German Nationwide Sample. J Nerv Ment Dis 206(4):270–276. https://doi.org/10.1097/NMD.0000000000000790

Mason V, Andrews H, Upton D (2010) The psychological impact of exposure to floods. Psychol Health Med 15(1):61–73. https://doi.org/10.1080/13548500903483478

National Mental Health Commission (2023) NATIONAL DISASTER MENTAL HEALTH AND WELLBEING FRAMEWORK. Supporting Australians' mental health through disaster. National Mental Health Commission, S 25 (https://nema.gov.au/sites/default/files/inline-files/28108%20NEMA%20National%20Disaster%20Mental%20Health%20and%20Wellbeing%20Framework%20FA%20v4.pdf)

Paranjothy S, Gallacher J, Amlôt R, Rubin GJ, Page L, Baxter T, Wight J, Kirrage D, McNaught R (2011) Psychosocial impact of the summer 2007 floods in England. BMC Public Health 11(1):145. https://doi.org/10.1186/1471-2458-11-145

Petrucci O, Papagiannaki K, Aceto L, Boissier L, Kotroni V, Grimalt M, Vinet F (2019) MEFF: The database of Mediterranean flood fatalities (1980 to 2015). J Flood Risk Manag 12(2):e12461. https://doi.org/10.1111/jfr3.12461

Puechlong C, Weiss K, Le Vigouroux S, Charbonnier E (2020) Role of personality traits and cognitive emotion regulation strategies in symptoms of post-traumatic stress disorder among flood victims. Int J Disaster Risk Reduct 50:101688. https://doi.org/10.1016/j.ijdrr.2020.101688

Radloff LS (1977) The CES-D scale: A self-report depression scale for research in the general population. Appl Psychol Meas 1(3):385–401

Seidler ZE, Rice SM, Ogrodniczuk JS, Oliffe JL, Shaw JM, Dhillon HM (2019) Men, masculinities, depression: Implications for mental health services from a Delphi expert consensus study. Prof Psychol Res Pract 50(1):51–61. https://doi.org/10.1037/pro0000220

Siegrist P, Maercker A (2010) Deutsche Fassung der Short Screening Scale for DSM-IV Posttraumatic Stress Disorder. Aktueller Stand der Validierung. Trauma Gewalt 4:208–213

Stanke C, Murray V, Amlôt R, Nurse J, Williams R (2012) The effects of flooding on mental health: outcomes and recommendations from a review of the literature. PLoS Curr 4:e4f9f1fa9c3cae. https://doi.org/10.1371/4f9f1fa9c3cae

Thieken A, Bubeck P, Zenker ML (2023) Fatal incidents during the flood of July 2021 in North Rhine-Westphalia, Germany: what can be learnt for future flood risk management? J Coast Riverine Flood Risk. https://doi.org/10.59490/jcrfr.2023.0005

Thieken AH, Bessel T, Kienzler S, Kreibich H, Müller M, Pisi S, Schröter K (2016) The flood of June 2013 in Germany: how much do we know about its impacts? Nat Hazards Earth Syst Sci 16(6):1519–1540. https://doi.org/10.5194/nhess-16-1519-2016

Tradowsky JS, Philip SY, Kreienkamp F, Kew SF, Lorenz P, Arrighi J, Bettmann T, Caluwaerts S, Chan SC, De Cruz L, De Vries H, Demuth N, Ferrone A, Fischer EM, Fowler HJ, Goergen K, Heinrich D, Henrichs Y, Kaspar F, Wanders N (2023) Attribution of the heavy rainfall events leading to severe flooding in Western Europe during July 2021. Clim Change 176(7):90. https://doi.org/10.1007/s10584-023-03502-7

Truedinger AJ, Jamshed A, Sauter H, Birkmann J (2023) Adaptation after Extreme Flooding Events: Moving or Staying? The Case of the Ahr Valley in Germany. Sustainability 15(2):1407. https://doi.org/10.3390/su15021407

Tunstall S, Tapsell S, Green C, Floyd P, George C (2006) The health effects of flooding: Social research results from England and Wales. J Water Health 4(3):365–380. https://doi.org/10.2166/wh.2006.031

Universität Zürich (2023) Trauma – Posttraumatische Belastungsstörungen (Traumatherapie). https://www.psychologie.uzh.ch/de/bereiche/hea/psypath/Psychotherapie/Trauma.html. Zugegriffen: 14. Sept. 2024

Wasserwirtschaftsverwaltung des Landesamts für Umwelt RLP (2024) *Onlinedienst Gewässernetz 2017*: https://wasserportal.rlp-umwelt.de/kartendienste [Daten], Creative Commons 4.0 „CC BY": https://creativecommons.org/licenses/by/4.0/. Zugegriffen: 14. Sept. 2024.

Weiss DS (2007) The impact of event scale: revised. In: Cross-cultural assessment of psychological trauma and PTSD. Springer US, Boston, S 219–238

Wiseman T, Foster K, Curtis K (2013) Mental health following traumatic physical injury: an integrative literature review. Injury 44(11):1383–1390. https://doi.org/10.1016/j.injury.2012.02.015

Zenker M-L, Bubeck P, Thieken AH (2024) Always on my mind: Indications of post-traumatic stress disorder among those affected by the 2021 flood event in the Ahr valley, Germany. Nat Hazards Earth Syst Sci 24(8):2837–2856. https://doi.org/10.5194/nhess-24-2837-2024

Zhong S, Yang L, Toloo S, Wang Z, Tong S, Sun X, Crompton D, FitzGerald G, Huang C (2018) The long-term physical and psychological health impacts of flooding: a systematic mapping. Sci Total Environ 626:165–194. https://doi.org/10.1016/j.scitotenv.2018.01.041

Zigmond AS, Snaith RP (1983) The hospital anxiety and depression scale. Acta Psychiatr Scand 67(6):361–370

Auswirkungen von Brücken auf das Abflussgeschehen 2021

Lisa Burghardt, Elena-Maria Klopries und Holger Schüttrumpf

J. Birkmann et al. (Hrsg.), *Die Flutkatastrophe 2021*, Sustainable Development Goals (SDG) – Umsetzung in Praxis, Lehre und Entscheidungsprozessen,
https://doi.org/10.1007/978-3-658-48688-4_10

Die Folgen des Katastrophenhochwassers 2021 sind in Rheinland-Pfalz und Nordrhein-Westfalen heute noch zu sehen. Die Schäden in Milliarden Höhe sind zu einem hohen Anteil an sogenannter kritischer Infrastruktur entstanden, die für die Versorgung und das Leben unentbehrlich sind (Krings 2020; Deutsches Komitee Katastrophenvorsorge e. V. 2022). Neben Wasser- und Energieversorgung zählen auch Transport und Verkehr zu den Sektoren der kritischen Infrastruktur (Pohl 2015). Durch den hohen Verlust an Straßen- und Bahnbrücken wurden ganze Stadtteile von der Verkehrsinfrastruktur abgetrennt und Hilfsmaßnahmen während des Hochwasserereignisses stark verzögert. Bis dato bilden Behelfsbrücken einen wichtigen Bestandteil der lokalen Infrastruktur, die nach und nach durch neue Brückenbauwerke ersetzt werden müssen. Doch der Wiederaufbau der Brücken sollte nicht ohne eine genaue Bewertung der Rolle von Brücken auf das vergangene Hochwasser geschehen, um das zukünftige Schadensausmaß an Brücken und in deren Umfeld möglichst gering halten zu können. Im Folgenden werden die Schäden an Brücken, hervorgerufen durch das Hochwasser im Jahr 2021, analysiert. Auf Basis von numerischen Modellierungen werden zudem die Auswirkungen von Brückenbauwerken auf das Hochwassergeschehen 2021 genauer untersucht. Hieraus lassen sich zahlreiche Empfehlungen für den Neubau von Brücken in den betroffenen Gebieten sowie Handlungsempfehlungen für den operationellen Hochwasserschutz ableiten.

10.1 Schäden an Brücken

An den vier deutschen Flüssen Ahr, Inde, Vicht und Erft wurde im Nachgang des Hochwassers eine Analyse der Brückenschäden durch Begehung vor Ort sowie auf Basis von Foto- und Videodokumentationen durchgeführt. Hierbei wurden angelehnt an Maiwald und Schwarz (2023) fünf Klassen für Brückenschäden unterschieden (siehe ◘ Tab. 10.1), die von verwitterungsähnlichen Schäden (Klasse D0) bis hin zum vollständigen Abriss des Bauwerks reichen (Klasse D4) (siehe ◘ Abb. 10.1). Eine Brücke erhält durch den Verlust des Geländers (Klasse D1) im Zuge der herkömmlichen Bauwerkprüfung aufgrund mangelnder Verkehrssicherheit die schlechteste Zustandsnote. Da allerdings ein Großteil der Bausubstanz noch vorhanden ist und auch gut erhalten sein kann, fallen die Instandsetzungskosten im Vergleich zu einem Neubau der gesamten Brücke sehr viel niedriger aus. Die hier aufgeführte Schadensklassifizierung konzentriert sich daher auf den Zustand des Bauwerks mit Blick auf Wiederaufbau- und Instandsetzungsaufwand.

Die höchste Schadensdichte wurde an der Ahr, besonders im Stadtgebiet Bad Neuenahr-Ahrweiler, verzeichnet. Von 114 Brücken vor der Flut, wurden knapp 40 Brücken als beschädigt (Schadensklasse D1 bis D3) und 40 Brücken als zerstört (Schadensklasse D4) klassifiziert. Gerade am Unterlauf der Ahr waren viele Brücken flachgegründet und erlitten aufgrund der hohen Wassermassen starke Schäden. Am nordrhein-westfälischen Fluss Inde traten deutlich weniger Brückenschäden auf. Rund 32 % der 77 frei zugänglichen Brücken wurden beschädigt oder zerstört, dabei wurden die meisten Brücken der Schadensklasse D1 zugeordnet. Vier Brücken wurden durch das Hochwasser vollständig zerstört. An der Vicht, die auf Höhe Stolberg in die Inde mündet, wurden 23 der 52 untersuchten Brücken als beschädigt eingestuft. Auch hier überwiegt der Anteil an leicht beschädigten Brücken und nur zwei Brücken wurden zerstört. Entlang der Erft wurden 15 % der 109 Brücken im hochwasserbetroffenen Gebiet zerstört. Brücken der Schadensklassen D1 und D2 erreichten den gleichen Anteil und fast die Hälfte der Brücken wurden bis zum Überbau angeströmt oder sogar überströmt.

Tab. 10.1 Schadensklassen für die Bewertung der Brückenschäden nach dem Hochwasser 2021. (Burghardt et al. 2024)

Schadensart	Schadensbeschreibung	Schadensklasse				
		Intakt	Beschädigt			Zerstört
		D0	D1	D2	D3	D4
Oberflächliche Schäden	Verwitterung	x				
	Feuchtegehalt der Bauteile	x				
	Verlust von Geländer, Leitplanke und Verzierungen		x			
	Kleinflächige Ablösungen (< 1 m^2) des Fahrbahnbelags		x			
	Verstopfung oder Abriss der Straßenentwässerung		x			
	Kleine Risse in der Bausubstanz (< 50 cm)			x		
	Leichte Freilegung der Bewehrung			x		
	Größere Risse in der Bausubstanz (> 50 cm)				x	
	Durchdringende Risse in der Bausubstanz (> 1 m)				x	
Strukturelle Schäden	Abriss der Zufahrtsrampen			x		
	Absinken, Verschieben oder Abreißen einzelner Brückenpfeiler				x	
	Abriss des Überbaus und einzelner Widerlager				x	
	Kolkbildung am Pfeiler				x	
	Großflächiger Abriss von Bausubstanz (> 1 m^3)				x	
	Vollständiger Abriss					x

Entlang der vier Flüsse zeigt sich, dass der Anteil an beschädigten oder zerstörten Brücken in dicht besiedelten Gebieten stark ansteigt. Brücken mit leichter Bauweise, wie beispielsweise für Wander- oder Radwegnutzungen, haben häufig höhere Schäden erlitten und wurden leichter von den Fluten mitgerissen. Die Öffnungsform der Brücke konnte nicht als signifikante Ursache für ein höheres Schadensausmaß an den Brücken identifiziert werden. An den Flüssen Ahr und Inde zeigte sich, dass mit größerer Öffnungsbreite und höherem Freibord an den Brücken tendenziell weniger Schäden aufgetreten sind. Besonders an Brücken entlang der Ahr wurden während und nach dem Hochwasser große Treibgutansammlungen und -rückstände beobachtet, die sich zum einen direkt auf die Belastung des Bauwerks und auch auf das Flutgeschehen selbst ausgewirkt haben.

10.2 Die Rolle von Brücken auf das Hochwassergeschehen 2021

Brücken stellen ein Abflusshindernis im Flussquerschnitt dar, die durch Pfeiler, Widerlager und bei hohen Wasserständen auch durch den Überbau den Abflussquerschnitt reduzieren. Bereits beim Umfließen des Brückenpfeilers kommt es zum sogenannten Pfeilerstau (siehe auch Patt und Jüpner 2020) und somit zu einer Erhöhung des Was-

Abb. 10.1 Beispielhafte Darstellung der Schadensmerkmale an Brücken nach dem Hochwasser 2021 an der Ahr. **a** Verlust des Geländers, **b** Verlust von Zufahrtsrampen, **c** Verlust des Überbaus, **d** vollständiger Abriss (**a**, **c** Burghardt et al. 2024; **b**, **d** eigene Abbildung)

serspiegels. Dieser hängt neben der Pfeilergeometrie auch von den hydraulischen Bedingungen, insbesondere Wassertiefe und Froude-Zahl ab (siehe auch Patt und Jüpner 2020). Sollte der Abflussquerschnitt zu gering sein, um den Abfluss eines Hochwassers abzuführen, wird das Wasser aufgestaut und es kommt je nach Form des Umlandes zum Umströmen oder sogar zum Überströmen der Brücken. Zusätzliche Verlegungen des Abflussquerschnitts durch Treibgut oder Geschiebe können bereits bei geringeren Wasserständen oder Froude-Zahlen zu einem Um- oder Überströmen der Brücke führen.

Im Zuge der Schadensanalyse wurden gerade in der Nähe der Brückenbauwerke starke Veränderungen des Umlandes durch Erosionsprozesse beobachtet. An einer Vielzahl von Brückenpfeilern kam es zur Kolkbildung und bei einigen Pfeilern durch die Unterspülung der Gründung auch zum Absinken. Auch die Widerlager vieler Brücken waren starken Belastungen ausgesetzt, wodurch diese freigelegt oder teilweise auch verschoben wurden. Dies zeigt, dass gerade an Brückenstandorten hohe Wasserstände und Fließgeschwindigkeiten erreicht wurden.

Mit Hilfe des hydrodynamischen-numerischen Programms HydroAS wurden im Nachgang des Hochwassers Berechnungen des Abflussgeschehens 2021 durchgeführt. Neben den geografischen Eingangsdaten, bereitgestellt durch das Landesamt für Umwelt Rheinland-Pfalz, dienten die Abflussberechnungen des Deutschen GeoForschungsZentrums als Randbedingungen für die Modellierung. Im Fokus lag hier die Auswirkung von Brückenbauwerken sowie Treibgutansammlungen auf das Flutgeschehen. Generell hängt der Einfluss einer Brücke auf einen möglichen Aufstau des Hochwassers stark von den angrenzenden Vorländern und beispielsweise angrenzender Bebauung ab. Nach den

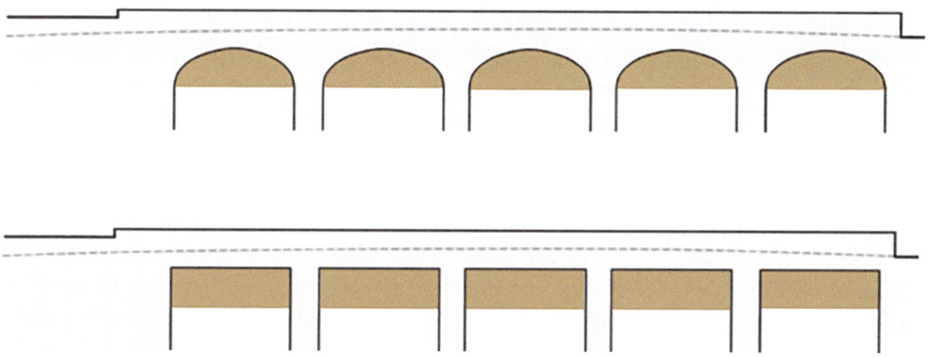

Abb. 10.2 Implementierung einer Verklausung der Brückenöffnung um 50 % (dargestellt in *braun*) gemessen von der Höhe der Öffnung in der numerischen Modellierung. (Eigene Abbildung)

numerischen Berechnungen hat die bloße Anwesenheit einer Brücke im Stadtgebiet Bad Neuenahr-Ahrweiler je nach Umland von entweder kaum merklichen Veränderungen der Wasserspiegellage am Brückenstandort bis hin zu über einem Meter Aufstau geführt.

Allerdings wurden auch große Mengen an Treibgut besonders an Brücken beim Hochwasser 2021 beobachtet. Die Verlegung des Brückenquerschnitts durch Treibgut hat zusätzlich zu einer Reduktion des Abflussquerschnitts geführt und sich demnach auch stark auf die Wasserspiegellagen des Flutgeschehens ausgewirkt. Die genauen Treibgutmengen lassen sich nur schwer rekonstruieren und sind bis heute unbekannt. Eine Verklausung der Brückenöffnung wurde mit Hilfe der Software HydroAS durch Absenken des Brückenüberbaus (siehe Abb. 10.2) um einen bestimmten Betrag dargestellt, da sich die Treibgutansammlungen, meist bestehend aus schwimmenden Materialien, von der Wasseroberfläche aus in Richtung Sohle ausbilden. Die Rekonstruktion des Abflussgeschehens 2021 mit Hilfe der numerischen Modellierung deutet jedoch darauf hin, dass eine Verlegung der Brückenöffnung von mehr als 50 % für die meisten Brücken nicht eingetreten sein wird. Mit steigendem Grad an Verklausung steigt auch der Einfluss des Brückenbauwerks auf Wasserstands- und Abflussbedingungen.

Unter der Annahme eines Verklausungsgrads von bis zu 50 % ergab sich im Modellgebiet Bad Neuenahr-Ahrweiler eine Erhöhung des Wasserstandes vor der Brücke um durchschnittlich 50 cm, maximal jedoch 1,16 m. Im oberstrom liegenden Gebiet Altenahr, das in einem engen Kerbtal ohne Retentionsflächen liegt, haben sich Treibgutansammlungen vor Brücken noch stärker ausgewirkt. Durch die Anwesenheit von Treibgut und einer schätzungsweise maximalen Verlegung des Brückenquerschnitts um 50 %, wurden Erhöhungen des Wasserstands um bis zu drei Meter berechnet (siehe Abb. 10.3).

Neben den Auswirkungen auf den Wasserstand führt die Verengung des Fließquerschnitts zusätzlich zu einer Erhöhung der Fließgeschwindigkeiten innerhalb der Brückenöffnung und in unmittelbarer Nähe des Bauwerks sowie unterstrom der Brücke. Dies birgt zum einen eine Gefahr für Personen, aber auch für die Stabilität des Bauwerks. Durch erhöhte Sohlschubspannungen werden Erosionsprozesse in Gang gesetzt oder beschleunigt. Die Ansammlung von Treibgut begünstigt dies zusätzlich, kann Kolkbildung verstärken und durch erhöhte hydrostatische Kräfte die Stabilität des Bauwerks in Gefahr bringen. Im Zuge der Schadensanalyse wurde zudem festgestellt, dass auch die Treibgutansammlung von Form und Beschaffenheit des Umlands beeinflusst wird und nicht immer, wie

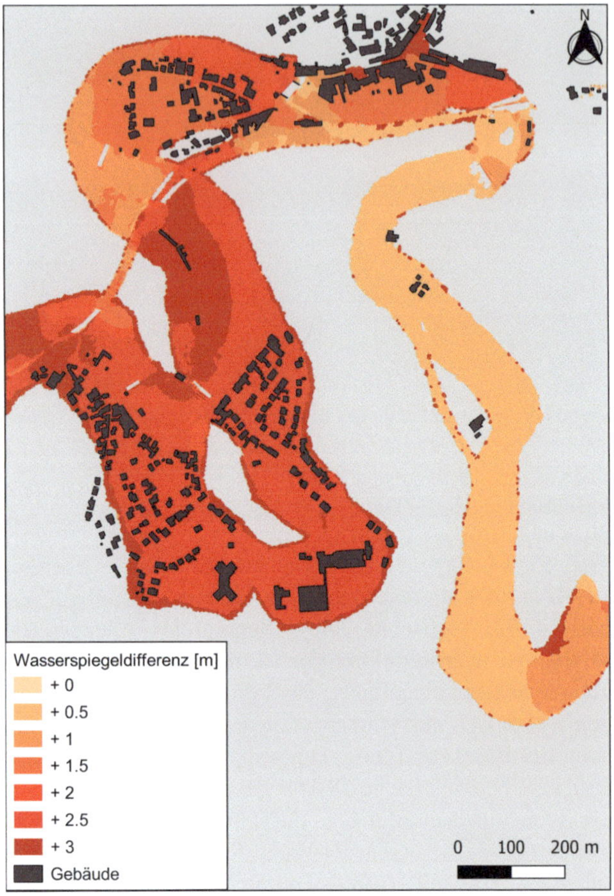

Abb. 10.3 Berechnung der Wasserspiegeldifferenzen in Altenahr für Berechnungsläufe mit und ohne Verklausungen. (Eigene Abbildung)

10

oben angenommen, symmetrisch verlaufen muss. In Flusskurven lagert sich das Treibgut in der Außenkurve ab, was zu erhöhten Fließgeschwindigkeiten in der Innenkurve führt, wo häufig dann größere Schäden am Bauwerk und stärkere Erosionen des Umlandes beobachtet wurden.

Aufgrund der höheren Lasten durch steigende Wasserstände könnten einige Brücken vor Erreichen eines hohen Verklausungsgrades schon versagt haben. Der Abriss von Brücken und den davor aufgestauten Treibgutansammlungen führt zur Ausbreitung von Schwallwellen, die durch höhere Fließgeschwindigkeiten und die Mitnahme von Treibgut ein hohes Schadenspotenzial bieten. Auch dies konnte mit Hilfe der Modellierung mit HydroAS untersucht werden. Durch den Abriss von Brückenbauwerken kam es zu einer Erhöhung der Fließgeschwindigkeiten um bis zu 2 m/s je nach Brückenstandort. Dies lässt sich aktuell nicht mit Messungen aus dem Hochwasserereignis belegen, Anwohnende berichteten jedoch, den Abriss und die Ausbreitung von Wellen durch das Tal beobachtet zu haben.

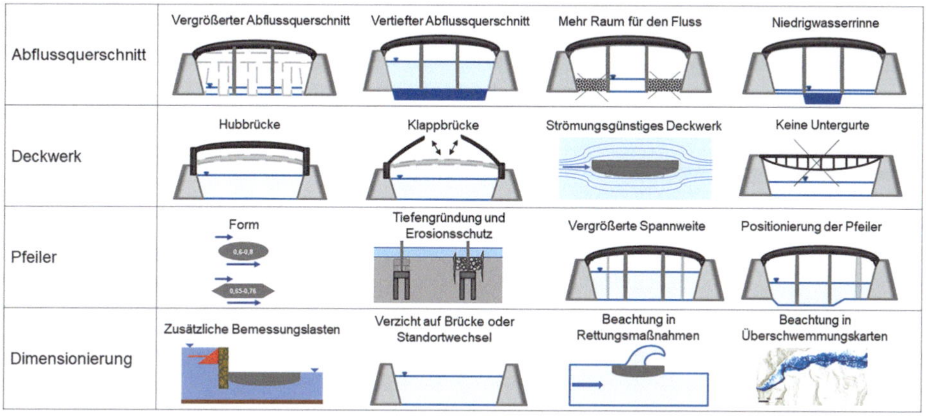

Abb. 10.4 Empfehlungen für den Wiederaufbau von Brücken. (Eigene Darstellung)

10.3 Empfehlungen für den Wiederaufbau

Aus den oben genannten Schadensbildern sowie Modellierungsergebnissen lassen sich bereits erste Empfehlungen für die Neuplanung der Brücken in den betroffenen Gebieten ableiten (siehe Abb. 10.4).

Hinweis für Entscheidungsprozesse

Als wichtigster Aspekt kann die Größe der Brückenöffnung genannt werden, die möglichst groß und frei von Hindernissen gehalten werden sollte. Neben der Brückenhöhe, die oft durch Anschlusspunkte festgelegt wird, spielen auch Öffnungsbreite und Anzahl der Pfeiler eine Rolle. Die Öffnungsbreite sollte ebenso wie die Höhe möglichst groß gehalten werden, um zunächst einmal einen Aufstau durch den Brückenüberbau aber auch gleichzeitig die Ansammlung von Treibgut zu reduzieren. Besonders die Anzahl der Pfeiler im Flusslauf und damit die Öffnungsbreite wirken sich stark auf die Verklausungswahrscheinlichkeit aus (siehe auch Gschnitzer et al. 2017), weshalb die Pfeilerzahl möglichst gering gehalten werden sollte. Besonders in eng bebauten Gebieten, die kaum Möglichkeit zum schadfreien Umströmen der Brücke lassen, können sich Hub- oder Klappbrücken anbieten, um Wasser- und Treibgutmassen abführen zu können. Auch ein möglichst dünner, stromlinienförmiger Überbau der Brücke kann sich negativ auf die Verklausungswahrscheinlichkeit auswirken und Treibgutansammlungen sowie den Aufstau vor der Brücke reduzieren. Der Einbau von klappbaren Geländern, die ab bestimmten Lasten nachgeben oder manuell abgelegt werden können, sind empfehlenswert, um Treibgutansammlungen vor Brücken und so einen Aufstau an Brückenstandorten zu minimieren.

Im Planungsprozess von Brücken an kritischen Standorten, die bereits häufiger von Hochwassern betroffen waren oder durch enge Flusstäler gekennzeichnet sind, sollten zusätzliche Lasten durch Aufstau und Überströmen in der Brückenstatik berücksichtigt werden. Auch Anpralllasten durch Treibgut sowie höhere Drücke aufgrund einer Verlegung des Brückenquerschnitts können für diese Standorte eine relevante Rolle in der Bemessung spielen. Tiefengründungen an Pfeilern sowie Erosionsschutz sind wichtige Hilfsmittel, um die Stabilität des Bauwerks im Hochwasserfall zu unterstützen. Generell sollten Abflusshindernisse in dicht bebauten Gebieten reduziert werden, um einen zusätzlichen Aufstau zu verhindern. Hier bietet sich beispielsweise die Zusammenlegung von Brücken unterschiedlicher Nutzung an. Zusätzlich kann durch Treibgutrechen die Ansammlung von Schwemmholz und weiteren Treibgutbestandteilen gesteuert und beispielsweise bereits oberstrom der Ortschaft verhindert werden.

Im Zuge der Modellierung hat sich gezeigt, dass Brücken einen starken Einfluss auf das Abschlussgeschehen, Wasserstände und Fließgeschwindigkeiten im Hochwasserfall haben können. Dies sollte in zukünftigen Warnprozessen, beispielsweise in Hochwassergefahren- und -risikokarten, beachtet werden, um neben Anwohnenden auch das Rettungspersonal schützen zu können. Mit Hilfe von Modellierungen für kritische Brückenstandorte können Alarm- und Einsatzpläne präzisiert werden, da beispielsweise der Ausfall von wichtigen Verbindungsbrücken aufgrund von Überströmen ab einem gewissen Wasserstand eingeplant werden kann. Auch eine frühzeitige Warnung kann eine entscheidende Rolle spielen, um beispielsweise große Treibgutmengen von Campingplätzen zukünftig verhindern zu können. An besonders engen Brückenstandorten in dicht bebauten Gebieten und für Brücken, die häufig bei Hochwasser an- und überströmt werden, ist es empfehlenswert, einen möglichen Rückstau durch die Brücke, durch Verklausung sowie die Folgen eines Brückenabrisses zu berechnen. Hierbei spielt auch die Qualität der Eingangsdaten, sowie die Annahme von Verklausungsszenarien für die Modellierung eine große Rolle, um die Abflussprozesse detailliert genug abbilden zu können.

10.4 Zusammenfassung

Das Hochwasser 2021 hat in Nordrhein-Westfalen und Rheinland-Pfalz zum Verlust einer Vielzahl von Brücken und damit dem Abriss der Verkehrsinfrastruktur geführt. Zudem haben sich die Bauwerke vor ihrem und durch ihr Versagen stark auf das Abflussgeschehen ausgewirkt. Besonders in dicht bebauten Gebieten kam es aufgrund der hohen Abflüsse und somit hohen Wasserstände in engen Flussquerschnitten bereits durch die Anwesenheit der Brückenbauwerke zu einem Aufstau der Wassermassen. Dies führte schnell zum Um- und Überströmen der Brücken und erhöhte somit die Lasten auf die Bauwerke. Gleichzeitig wurden Erosionsprozesse in Gang gesetzt, die Fundamente und Gründungen der Pfeiler sowie Widerlager von Brücken gefährdeten. Durch die Ansammlung von Treibgut wurden diese Effekte weiter verstärkt, Wasserstände unerwarteten Ausmaßes erreicht und ein Abriss der Bauwerke begünstigt. Der Abriss von bereits eingestauten Brückenbauwerken führte dann zur Ausbreitung von Schwallwellen, die durch ihre Fließgeschwindigkeit und das Mitreißen von Treibgut das Schadensausmaß erhöhten. Für die zukünftige Planung neuer Brückenbauwerke sollten demnach die Vielzahl an Lasten, die durch ein Hochwasser dieses Ausmaßes entstehen können, beachtet werden. Neben einer Vergrößerung des Abflussquerschnitts sollte auch der Treibgutrückhalt an besonders kritischen Brückenstandorten bedacht werden. Gleichzeitig sollte der Aufstau der Hochwasserwelle an Brückenbauwerken sowie das Um- und Überströmen dieser in Hochwassergefahren- sowie -risikokarten Beachtung finden, um Bevölkerung und Rettungskräfte besser schützen zu können. Durch einen hochwasserangepassten Neubau der Brücken kann auch im Hochwasserfall die Mobilität gewährleistet werden.

Zusammenfassung

- Das Hochwasser 2021 hat in Nordrhein-Westfalen und Rheinland-Pfalz zum Verlust einer Vielzahl von Brücken und damit dem Abriss der Verkehrsinfrastruktur geführt.
- Zudem haben sich die Bauwerke vor ihrem und durch ihr Versagen stark auf das Abflussgeschehen ausgewirkt.
- Der Abriss von bereits eingestauten Brückenbauwerken führte zur Ausbreitung von Schwallwellen, die durch ihre Fließgeschwindigkeit und das Mitreißen von Treibgut das Schadensausmaß erhöhten.
- Für die zukünftige Planung neuer Brückenbauwerke sollten demnach die Vielzahl an Lasten, die durch ein Hochwasser dieses Ausmaßes entstehen können, beachtet werden.
- Beim (Wieder-)Aufbau von Brückenbauwerken sollte neben einer Vergrößerung des Abflussquerschnitts auch der Treibgutrückhalt an besonders kritischen Brückenstandorten bedacht werden. Gleichzeitig sollte der Aufstau der Hochwasserwelle an Brückenbauwerken sowie das Um- und Überströmen dieser in Hochwassergefahren- sowie -risikokarten Beachtung finden, um Bevölkerung und Rettungskräfte besser schützen zu können. Durch einen hochwasserangepassten Neubau der Brücken kann auch im Hochwasserfall die Mobilität gewährleistet werden.

Danksagung Die Autorinnen und Autoren von ▶ Kap. 10 bedanken sich beim Bundesministerium für Bildung und Forschung (BMBF) und beim DLR Projektträger für die Unterstützung und Förderung des KAHR-Projekts. Diese Forschung wurde vom BMBF unter dem Förderkennzeichen 01LR2102A-M finanziert.

Literatur

Burghardt L, Klopries E-M, Schüttrumpf H (2024) Structural damage, clogging, collapsing: analysis of the bridge damage at the rivers Ahr, Inde and Vicht caused by the flood of 2021. J Flood Risk Manag. https://doi.org/10.1111/jfr3.13001

Deutsches Komitee Katastrophenvorsorge e. V. (DKKV) (2022) Die Flutkatastrophe im Juli 2021 in Deutschland. Ein Jahr danach: Aufarbeitung und erste Lehren für die Zukunft. DKKV-Schriftenreihe, Bd 62. DKKV, Bonn. Online verfügbar unter https://dkkv.org/wp-content/uploads/2023/02/DKKV__Schriftenreihe_Juli_2022__Webversion__-_2te_Version_August3.pdf

Gschnitzer T, Gems B, Mazzorana B, Aufleger M (2017) Towards a robust assessment of bridge clogging processes in flood risk management. Geomorphology. https://doi.org/10.1016/j.geomorph.2016.11.002

Krings S (Hrsg) (2020) 10 Jahre „KRITIS-Strategie“. Einblicke in die Umsetzung der Nationalen Strategie zum Schutz Kritischer Infrastrukturen. Deutschland. Stand: Februar 2020. Praxis im Bevölkerungsschutz, 21. Bundesamt für Bevölkerungsschutz und Katastrophenhilfe, Bonn

Maiwald H, Schwarz J (2023) Ermittlung von Hochwasserschäden unter Berücksichtigung der Bauwerksverletzbarkeit. Erweitertes EDAC-Hochwasserschadensmodell, 2. Aufl. Bauhaus-Universitätsverlag Weimar, Ilmtal-Weinstraße

Patt H, Jüpner R (2020) Hochwasser-Handbuch. Springer, Wiesbaden

Pohl R (2015) Brücken aus der Sicht des Wasserbauers. Bautechnik 92(7):461–468. https://doi.org/10.1002/bate.201500034

Wirkanalyse und Umsetzungsbewertung einzugsgebietsweiter Retentionspotenziale für die Ahr

Stefanie Stenger-Wolf, Holger Sauter, Jannis Wilkens, Maike Zimmermann, Jörn Birkmann und Holger Schüttrumpf

J. Birkmann et al. (Hrsg.), *Die Flutkatastrophe 2021*, Sustainable Development Goals (SDG) – Umsetzung in Praxis, Lehre und Entscheidungsprozessen,
https://doi.org/10.1007/978-3-658-48688-4_11

11.1 Anlass für eine einzugsgebietsweite Retentionspotenzialanalyse

Wasserstände von bis zu 10 Metern in Ortschaften entlang der Ahr machten Mitte Juli 2021 deutlich, dass lokale Hochwasseranpassungsmaßnahmen bei extremen Hochwassern alleine nicht ausreichen. Schon kurz nach dem Katastrophenhochwasser war klar: um auf ein solches Ereignis vorbereitet zu sein, müssen die Abflussmassen rückgehalten werden, bevor sie sich in der Ahr akkumulieren.

Das Hochwasser von Mitte Juli 2021 hat vielerorts die höchsten Aufzeichnungen der sich aktuell in Betrieb befindlichen Pegel bei Weitem übertroffen, und sogar das HQ_{Extrem} überschritten (Kisseler et al. 2023; Schüttrumpf 2021; Apel et al. 2022). Der zugrundeliegende Abfluss für die Berechnungen der Überflutungsgebiete bei einem HQ_{Extrem} ist nicht bekannt, lediglich, dass die Wiederkehrwahrscheinlichkeit geringer als einmal in 200 Jahren sein muss (FGG Rhein 2019). In der Regel wird an der Ahr dafür das HQ_{100} um den Faktor 1,26 erhöht (Vorogushyn et al. 2022). Am Pegel Altenahr entsprach der Hochwasserabfluss von Mitte Juli 2021 mit ca. 1000 m^3/s ungefähr dem 4-fachen des HQ_{100}, das sich zu der Zeit auf 241 m^3/s belief (Ludwig et al. 2023), und mittlerweile auf 500 m^3/s erhöht wurde (Hochwasservorhersagezentrale Rheinland-Pfalz 2024). Recherchen zu historischen Hochwassern ergaben allerdings, dass im Juli 1804 ein Rekordabfluss von um die 1200 m^3/s in Dernau erzielt wurde, und im Juni 1910 wurden die Abflüsse in Altenahr immerhin noch auf 500 m^3/s geschätzt (Roggenkamp und Herget 2014). Rückblickend erscheint eine Wiederholung eines vergleichbaren Hochwasser zu dem von Juli 2021 in absehbarer Zeit trotz seiner Extreme also als gar nicht so unwahrscheinlich (Roggenkamp und Herget 2024).

Für das zugrundeliegende Regenereignis von Mitte Juli 2021 wurde die Wiederkehrwahrscheinlichkeit auf einmal in vierhundert Jahren geschätzt (Tradowsky et al. 2023). Gleichzeitig steigt die Wahrscheinlichkeit für vergleichbare Starkniederschläge mit dem aktuellen klimawandelbedingten Temperaturanstieg um ca. 20 % (Tradowsky et al. 2023; IPCC 2023; Blöschl et al. 2019; Koks et al. 2022; Kreienkamp et al. 2021; Vautard et al. 2019). Es ist also durchaus wahrscheinlich für aktuelle Generationen, ein vergleichbares Hochwasser nochmal zu erleben.

11.2 Forschungsbedarf

Es gibt verschiedenste Möglichkeiten, die Retention in einem Einzugsgebiet zu verbessern, sodass weniger Niederschlag direkt zum Abfluss kommt, und die Hochwasserwelle abgeflacht wird. Zunächst kann Retentionsraum über technische Bauwerke vorgehalten werden. Weiterhin sind natürliche Rückhaltemaßnahmen eine weitere Option des Wasserrückhalts. In der Regel sind sowohl technische als auch natürliche Maßnahmen mit einem hohen Aufwand verbunden. Hierunter fällt neben den Aufwendungen für Planung, Bau, Betrieb und Unterhaltung auch der Aufwand durch eine veränderte Nutzung auf der beanspruchten Fläche. Die Flächenverfügbarkeit ist eine weitere Umsetzungshürde, denn optimale Standorte für eine Maßnahme sind nicht selten aufgrund der Besitzverhältnisse der Fläche oder naturschutzrechtlicher Belange zunächst nicht verfügbar. Des Weiteren soll durch Maßnahmen zur Erhöhung des Retentionsraums im Einzugsgebiet ein greifbarer, also zahlenmäßig erfassbarer Schutz erzielt werden, bei dem das Kosten-Nutzen-

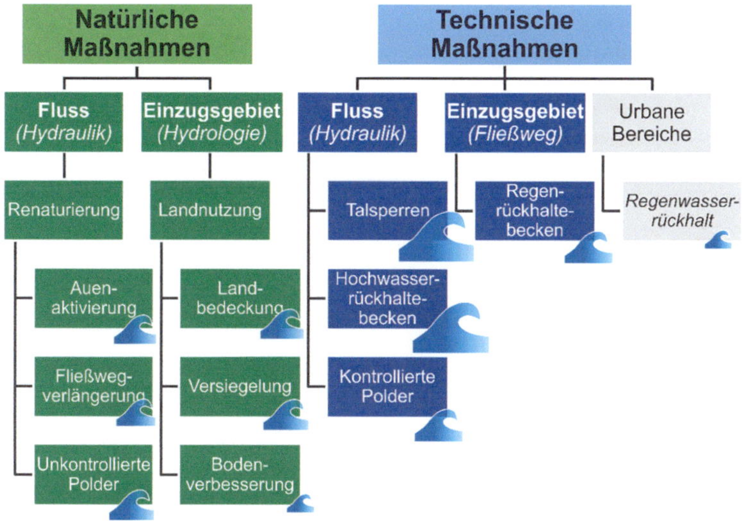

Abb. 11.1 Systematische Einteilung verschiedener natürlicher und technischer Retentionsmaßnahmen im Einzugsgebiet. Die Maßnahmen werden jeweils nach ihrer Lage und Wirkung unterschieden. So gibt es sowohl technische als auch natürliche Maßnahmen, die direkt am oder im Flusslauf ansetzen, und Maßnahmen, die im Einzugsgebiet in den hydrologischen Kreislauf eingreifen, also zum Ziele haben, den Direktabfluss zu verringern. Zudem gibt es Maßnahmen zum Regenwasserrückhalt im urbanen Raum. Die Größe der Flutwelle zeigt qualitativ das theoretische Retentionspotenzial an

Verhältnis positiv ist. Hierfür muss ein Schutzziel bekannt sein und in einen konkreten Zielwert formuliert werden.

Für die Retentionspotenzialanalyse wurden mögliche Retentionsraummaßnahmen in natürliche und technische Maßnahmen unterteilt. Beide Maßnahmentypen können am Fließgewässer selbst ansetzen und eine hydraulische Wirkung entfalten oder flächenmäßig im Einzugsgebiet wirken und in den hydrologischen Kreislauf eingreifen (Abb. 11.1). Natürliche Retentionsmaßnahmen am Gewässer selbst sind Renaturierungen oder naturnah gestaltete Polder. Bei Fließgewässerrenaturierungen wird über Auwaldaufforstung die Rauheit erhöht, was lokal zu einer Reduktion der Fließgeschwindigkeit und damit einem Abbremsen der Hochwasserwelle führt. Außerdem kann der Abfluss über eine Verlängerung des Gewässerlaufes weiter gebremst werden, wenn sich das Gewässer von einem begradigten Zustand zurück in einen mäandrierenden entwickeln kann. Über Geländeprofilierungen können zudem die Auen früher aktiviert werden, was dem Fluss mehr Raum gibt. Technischen Hochwasserrückhalt am Gewässer bieten Hochwasserrückhaltebecken und Talsperren. Während des Hochwassers von Mitte Juli 2021 pufferten die Talsperrensysteme in der Eifel und im Wupper-Einzugsgebiet große Wassermassen ab und verhinderten größere Schäden, obwohl sie ursprünglich nicht für einen solchen Hochwasserrückhalt ausgelegt waren (Wolf et al. 2024b). Die Steuerung von Talsperren muss in der Regel mehrere Ziele in Einklang bringen, in die sich der Hochwasserrückhalt neben der Bereitstellung von Trink- und Brauchwasser sowie der Energieerzeugung mit einreiht. Hochwasserrückhaltebecken hingegen dienen rein dem Hochwasserrückhalt und werden nicht im Dauerstau betrieben.

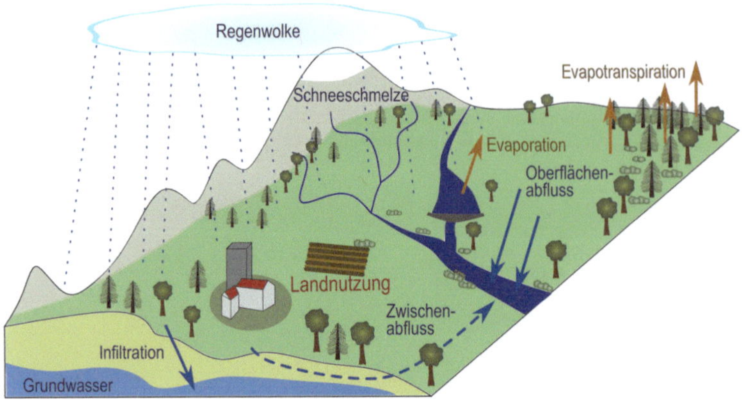

Abb. 11.2 Retentionsverbesserung im Einzugsgebiet durch angepasste Bodenbedeckung und Landnutzung. (Abbildung abgeändert nach Wolf (2023))

Im Einzugsgebiet kann über eine verbesserte Landnutzung bereits dort, wo der Niederschlag fällt, die Retention verbessert werden. Über eine angepasste Bodenbedeckung und Bewirtschaftung können die Infiltration und Evapotranspiration verbessert und so der Direktabfluss reduziert werden (Abb. 11.2). Auf versiegelten Flächen ist der Oberflächenabfluss („Surface Runoff" in Abb. 11.2) am höchsten, denn der Großteil des Niederschlags trägt zum Direktabfluss bei. Dichter gestufter Wald hingegen hat eine größere Oberfläche, was die Evapotranspiration begünstigt. Über einen durchwurzelten Boden wird zudem die Infiltration erhöht. Allerdings ist dieser Effekt bei Böden mit einem geringen Versickerungsvermögen sehr begrenzt. Eine hohe Hangneigung und Niederschlagsintensität reduzieren zudem die Effekte des natürlichen Flächenrückhalts. In vegetationsarmen Perioden, beispielsweise im Winter, ist der Rückhalteeffekt des Waldes ebenfalls reduziert.

Hochwasservorsorge ist in vielerlei Hinsicht aufwendig. Neben den Bau- und Unterhaltungskosten für Anpassungsmaßnahmen erfordern die zusätzlichen Aufgaben personellen Aufwand. Zudem ist die Flächenverfügbarkeit sehr oft eine große Herausforderung. Im Rahmen des vom BMBF geförderten KAHR-Projekts wurde eine Retentionsraumanalyse für das gesamte Ahr-Einzugsgebiet durchgeführt, um optimale Rückhaltemaßnahmen zu identifizieren. Hierbei wurden die Potenziale für verschiedene Maßnahmentypen grob abgeschätzt. Das Ziel war zunächst, vielversprechende Maßnahmentypen zu identifizieren. Dabei war das Erzielen einer hohen Genauigkeit zweitrangig, denn es ging um die Bestimmung der Größenordnung der erzielbaren Wirkung. Verglichen wurden natürliche Maßnahmen in Form von Landnutzungsanpassungen und technische Maßnahmen in Form von technischen Hochwasserrückhaltebecken.

11.3 Retentionspotenziale für das Ahr-Einzugsgebiet

Besonders im Mittellauf der Ahr ist das Tal steil und die Auenflächen sind sehr schmal. Die Bebauung in unmittelbarer Gewässernähe und die sonstige anthropogene Nutzung der Auen stellt ein hohes Schadenspotenzial und eine Herausforderung für den Hochwasserschutz dar (siehe beispielsweise Abb. 11.3). Um die Hochwassersituation im

Abb. 11.3 Die Skizze des Tals und der Ahr bei bordvollem Abfluss verdeutlicht die auf dem Foto gezeigte Situation: In engen Talabschnitten ist wenig Auenfläche vorhanden, sodass Wasserstände schnell ansteigen. An dem gezeigten betroffenen Café wurde eine Flutmarke von 9,24 m auf einem Transparent festgehalten. (Bildquelle: Wolf et al. (2024b))

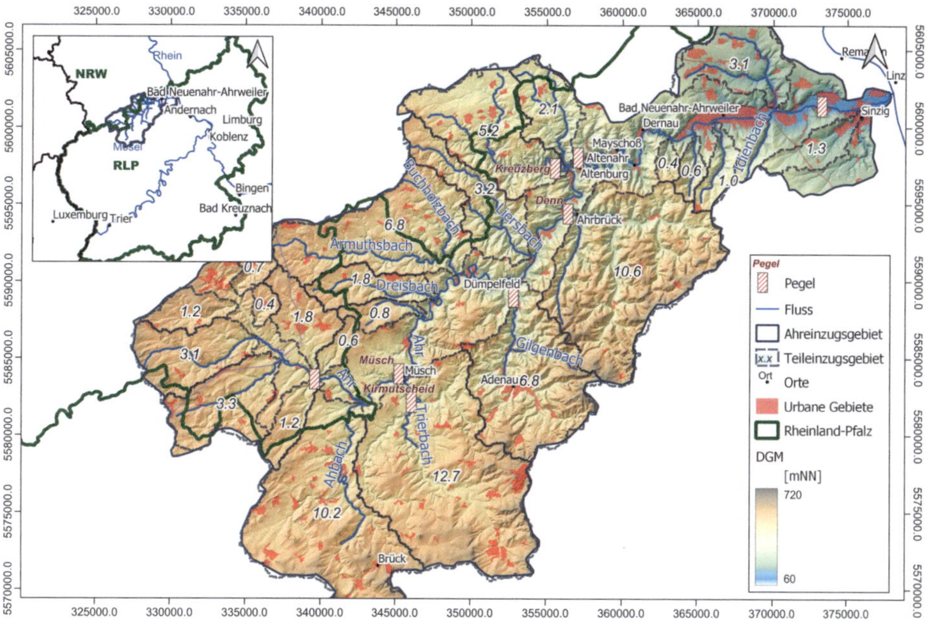

Abb. 11.4 (*oben rechts*) Lage des Einzugsgebiets der Ahr in Rheinland-Pfalz (RLP) und Nordrhein-Westfalen (NRW). (Hauptkarte) Geländehöhen des Ahrtals auf Grundlage des digitalen Geländemodells (DGM) und flächenmäßige Anteile der Zuflüsse am Gesamteinzugsgebiet. (Datenquellen: DGM: Bezirksregierung Köln (2023) und LVermGeoRP (2021), Gewässersystem, Städte und Bundeslandgrenzen: Geofabrik GmbH (2018a; 2018b), Urbane Bereiche: EEA (2016), Landesgrenze: ESTAT (2020), Teileinzugsgebiete basierend auf dem DGM25: European Environment Agency (EEA) (2016). Bildquelle: abgeändert nach Wolf et al. (2024a))

Ahrtal zu verbessern, bedarf es einer Retention der Wassermassen, bevor sich diese im Tal akkumulieren.

Abb. 11.4 schlüsselt die flächenmäßigen Anteile der Zuflüsse der Ahr am gesamten Einzugsgebiet basierend auf dem digitalen Geländemodell mit einer Auflösung von 25 m × 25 m auf. Insgesamt ist die Aufteilung sehr kleinteilig, und den größten Anteil am

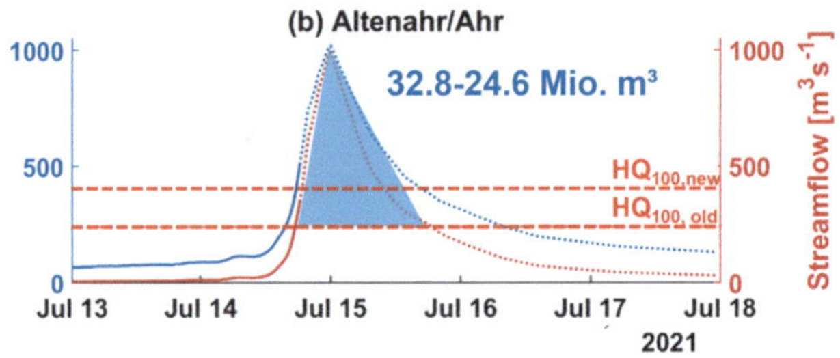

Abb. 11.5 Rekonstruierte Abflusskurve des Hochwassers von Mitte Juli 2021 an der Ahr von Mohr et al. (2023). Basierend auf der Rekonstruktion wurde abgeschätzt, welche Wassermassen mindestens rückzuhalten sind, unter der Annahme, dass ein HQ100 schadlos abgeführt werden kann. (Bildquelle: Wolf et al. (2024a). Alle Rechte vorbehalten)

11

Einzugsgebiet hat der Trierbach im Süden des Einzugsgebiets mit 12,7 %. Dies bedeutet, dass es keine einzelnen Zuflüsse gibt, die beispielsweise mit einem Hochwasserrückhaltebecken abgekoppelt werden könnten, um die Hochwassersituation zentral zu verbessern. Es wird Retention in allen Teileinzugsgebieten benötigt.

Zu Beginn der Analyse technischer und natürlicher Retentionspotenziale im Ahr-Einzugsgebiet wurde ein Rückhalteziel definiert. Das erklärte Ziel für das Ahrtal ist, ein Ereignis, wie es Mitte Juli 2021 auftrat, auf den Abfluss eines HQ_{100} zu reduzieren. Da die Pegel in den schwer betroffenen Gemeinden entlang der Ahr während des Hochwassers ausfielen, zerstört wurden oder ihr Messbereich überschritten wurde, wird hier auf Rekonstruktionen der Abflusswelle zurückgegriffen. Das Bemessungshochwasser HQ_{100} wurde nach dem Hochwasser von Mitte Juli 2021 nach oben korrigiert, wodurch sich eine Bandbreite für das rückzuhaltende Abflussvolumen ergibt (Abb. 11.5). Um sich vor einem Ereignis, wie dem von Mitte Juli 2021, zu schützen, liegt der Rückhaltebedarf bei rund 25 Mio. m^3 bis 33 Mio. m^3 (bei Annahme einer Scheitelkappung), wenn man davon ausgeht, dass die Gemeinden entlang der Ahr auf ein HQ_{100} vorbereitet sind. Da sich das nächste Starkregenereignis räumlich anders verteilen kann, muss tatsächlich ein unbekanntes Vielfaches des abgeschätzten Volumens vorgehalten werden.

11.3.1 Natürliche Rückhaltepotenziale durch Landnutzungsanpassung

Die natürlichen und technischen Retentionsraumpotenziale wurden über eine geoinformationssystem-gestützte Analyse bestimmt. Zur Bestimmung der natürlichen Potenziale wurde zunächst die Landnutzung im Einzugsgebiet der Ahr analysiert. So sind ca. 55 % des Einzugsgebiets heute bewaldet, und nur rund 7 % sind versiegelt (Abb. 11.6). Anfang des 19. Jahrhunderts waren nur ca. 30 % des Ahr-Einzugsgebiets bewaldet (Vélez Pérez et al. 2023), wodurch der natürliche Flächenrückhalt tendenziell schlechter war. Auf der anderen Seite war die Versiegelung auf den Flächen, die Mitte Juli 2021 überschwemmt wurden, deutlich geringer (Vélez Pérez et al. 2023). Das Schadenspotenzial im Überflutungsbereich ist also deutlich gestiegen. Zur Bestimmung der Größenordnung der natürlichen Retentionspotenziale durch Landnutzungsanpassungen wurden

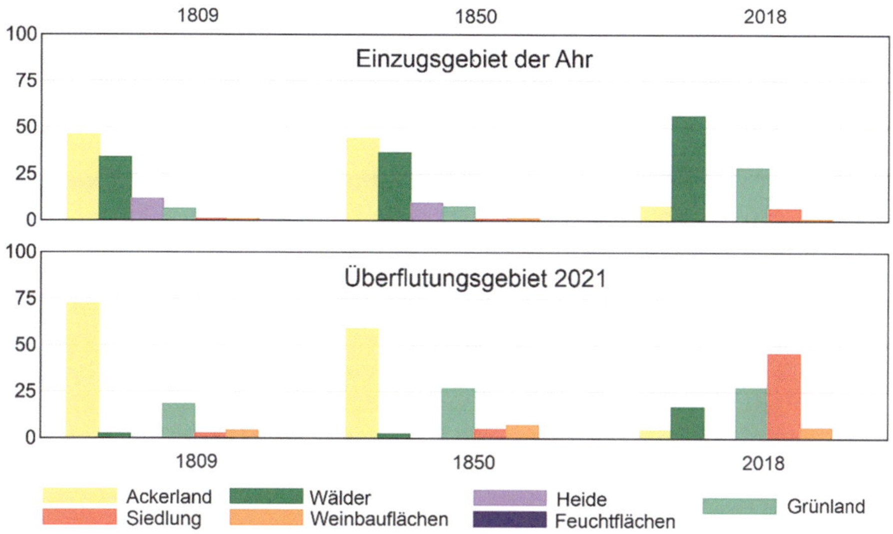

Abb. 11.6 Landnutzungsanalyse der letzten zwei Jahrhunderte im Ahrtal im Rahmen einer Masterarbeit am IWW (Vélez Pérez 2023; Vélez Pérez et al. 2023). Entwicklung der Landnutzung im Einzugsgebiet der Ahr von 1809 bis 2018 (*oben*) sowie Entwicklung der Landnutzung von 1809 bis 2018 im Überflutungsbereich des Hochwassers von Mitte Juli 2021 (*unten*). Abb. 11.6 wurde bereits mit gleicher Beschreibung in Wolf et al. (2024a) veröffentlicht

Siedlungsflächen, Waldflächen, landwirtschaftliche Flächen (Ackerflächen), Weinbauflächen und Wiesen bzw. Weiden betrachtet.

Vereinfacht kann über das Verfahren des Soil Conservation Service (SCS) über die sogenannte Curve Number (CN) abgeschätzt werden, welcher Anteil des kumulativen Niederschlags bei einer bestimmten Bodenbedeckung, Landnutzung, Bodengruppe und Vorfeuchte zum Abfluss kommt (Mockus 1965; Maniak 2016c; Gln. (11.1)–(11.3)):

$$A = \frac{(N - 0{,}2S)^2}{(N + 0{,}8S)} \qquad \text{(Abfluss)} \tag{11.1}$$

$$\text{mit} \quad S = \frac{1000}{CN} - 10 \qquad \text{(Rückhalt)} \tag{11.2}$$

$$\text{mit} \quad N > 0{,}2S \qquad \text{(Niederschlag)} \tag{11.3}$$

Dabei wird für die verschiedenen Landnutzungsarten unter Hinzuziehen der hydrologischen Bodengruppe die Curve-Number abgeschätzt. Je hoher die Curve-Number, desto höher ist der Direktabfluss. Die Curve-Number kann maximal 100 erreichen (Maniak 2016a, 2016c). In der vorliegenden Studie wurde die hydrologische Bodengruppe dem Datensatz „Global Hydrological Soil Groups" mit einer Auflösung von 250 m × 250 m entnommen (Ross et al. 2018). Die Landnutzung wurde der CORINE Land Cover (CLC) Klassifizierung entnommen (EEA 2016). Damit wurde je nach Landnutzung nach Maniak (2016b) eine mittlere Curve-Number abgeschätzt. Es wurde vereinfacht davon ausgegangen, dass die aktuelle Landnutzung im Ahr-Einzugsgebiet noch Verbesserungspotenziale bietet, aber weder hydrologisch sehr ungünstig, noch hydrologisch sehr günstig gestaltet ist. Folglich wurde zunächst mit einem mittleren Wert für die Curve-Number der

Direktabfluss basierend auf dem Niederschlag von Mitte Juli 2021 berechnet. Für die Berechnung wurde außerdem, dem Niederschlag der Vortage entsprechend, die Bodenfeuchteklasse II von drei möglichen Klassen angenommen.

Normalerweise findet das SCS-Verfahren in Niederschlags-Abfluss-Modellen Anwendung, der Modellaufbau für ein komplettes Einzugsgebiet kann allerdings mehrere Monate bis wenige Jahre dauern. Die erste Abschätzung der Größenordnung wurde deswegen manuell durchgeführt. Für den Niederschlag wurden stündliche Werte des DWD, bestehend aus einer Kombination von Radardaten und Bodenmessstationen (DWD 2021), zwischen dem 12.07.2021 und 19.07.2021 aufsummiert. Die angesetzte Niederschlagssumme ist somit über das Ahr-Einzugsgebiet räumlich differenziert und variiert von ca. 50 mm bis zu ca. 130 mm in einer Auflösung von 950 m × 950 m. Da die Analyse händisch durchgeführt wurde, wurde der gesamte Niederschlag als ein einziger Zeitschritt ausgewertet, und es wurde der Direktabfluss nach Gl. (11.3) abgeschätzt. Anschließend wurden optimale Werte für die Curve-Number abgeschätzt (■ Tab. 11.1). Hierbei wurde angenommen, dass überall ein dichter Mischwald hergestellt werden kann, Wald-Strauch-Übergangsstadien aufgeforstet werden, Felder als konservierendes Ackerland bewirtschaftet werden können, und eine Reduktion der durchgängigen Asphaltierung im urbanen Raum möglich ist. Dies sind zunächst rein theoretische Annahmen, die der Abschätzung der Größenordnung des Retentionspotenzials durch eine Landnutzungsanpassung dienen.

Wird nun der Abfluss einmal mit einem mittleren Wert für die Curve-Number und einmal mit einem optimalen Wert für die Curve-Number berechnet, und anschließend die Differenz gebildet, lässt sich die Größenordnung des Retentionspotenzials durch Landnutzungsänderungen abschätzen. Insgesamt wurde so ein ungefähres Rückhaltepotenzial von 7,2 Mio. m^3 auf allen Waldflächen im Ahrtal abgeschätzt, ein ungefähres Rückhaltepotenzial von 2,6 Mio. m^3 auf Ackerflächen, und ein ungefähres Rückhaltepotenzial von weniger als 0,5 Mio. m^3 für alle übrigen Flächen. Dieses Ergebnis ist in so weit nicht überraschend, da Waldflächen den größten flächenmäßigen Anteil am Ahr-Einzugsgebiet haben und urbane Gebiete nur einen kleinen Anteil (■ Abb. 11.6).

11

Im Rahmen einer Masterarbeit wurden fünf nach CORINE Land Cover als Laub- und fünf als Nadelwald klassifizierte Flächen genauer analysiert (■ Abb. 11.7; Zimmermann 2023). Es wurde die beschriebene Methodik angewendet, allerdings wurde ein pauschaler Niederschlag von 90,5 mm angesetzt (Zimmermann 2023). Dieser Wert entspricht dem durchschnittlichen Niederschlag, der dem Hochwasser von Mitte Juli 2021 zugrunde lag (DWD 2022). Auf den ausgewählten Waldflächen wurden die Annahmen zu der Curve Number mittels Digitaler Orthofotos verfeinert. Außerdem fanden die Baumwipfeldichte und das Wegenetz Eingang in die Abschätzungen (Zimmermann 2023).

Auf den fünf betrachteten Nadelwaldflächen von insgesamt 237 ha beträgt die mögliche Reduktion nach Zimmermann (2023) ca. 0,017 Mio. m^3, was einer Retentionssteigerung um 32 % im Mittel der fünf Flächen entspricht. Bezogen auf den Gesamtabfluss im Einzugsgebiet macht dies allerdings nur einen Unterschied von 0,06 % aus. Auf den fünf betrachteten Laubwaldflächen von insgesamt 317 ha kann der Direktabfluss potentiell um 0,015 Mio. m^3 reduziert werden. Rechnet man die Ergebnisse auf alle Waldgebiete hoch, ergibt sich ein theoretisches potentielles Retentionsvolumen von 4,5 Mio. m^3 (Zimmermann 2023). Dieser Wert weicht zunächst von den vorher bestimmten 7,2 Mio. m^3 ab, allerdings wurde insgesamt eine geringere Niederschlagssumme angesetzt. Die Abweichung liegt also nicht alleine an der verfeinerten Bestimmung der Curve-Number. Beide Werte bilden dennoch eine vergleichbare Größenordnung ab, die hilft, das theoretische

Tab. 11.1 Angesetzte Curve-Numbers nach im Ahr-Einzugsgebiet vorliegenden hydrologischen Bodengruppen, welche von A bis D die Infiltrationskapazität repräsentieren (C und C/D) für die verschiedenen Landnutzungsklassen (CORINE Land Cover Class (CLC)) sowie die angesetzten angenommenen Optimalwerte nach einer angepassten Landnutzung. Werte wurden abgeschätzt nach Maniak (2016b). Tab. 11.1 wurde mit gleicher Beschriftung bereits in Wolf et al. (2024b) veröffentlicht

	CLC Class	C	C/D	Optimalwerte	
				C	C/D
Durchgängig städtische Prägung	111	87,5	90	83,5	86
Nicht durchgängige städtische Prägung	112	83,5	86	83,5	86
Industrie- und Gewerbeflächen	121	92,5	93,5	83,5	86
Straßen-, Eisenbahnnetze und funktionell zugeordnete Flächen	122	95	95,5	83,5	86
Hafengebiete	123	98		95	
Flughäfen	124	80	83	80	83
Abbauflächen	131	80	83	80	83
Deponien	132	80	83	80	83
Baustellen	133	80	83	80	83
Städtische Grünflächen	141	80	83	80	83
Sport- und Freizeitanlagen	142	80	83	74	77
Nicht bewässertes Ackerland	211	89,5	92	73	77,5
Weinbau	221	79	80,5	79	80,5
Obst- und Beerenobstbestand	222	79	80,5	79	80,5
Wiesen und Weiden	231	78,5	82	77,5	82
Komplexe Parzellenstruktur	242	73	77,5	73	77,5
Landwirtschaftlich genutztes Land mit Flächen natürlicher Bodenbedeckung von signifikanter Größe	243	73	77,5	73	77,5
Laubwald	311	71,5	75	62	65,5
Nadelwald	312	75	78	62	65,5
Mischwald	313	62	65,5	62	65,5
Natürliches Grünland	321	71	78	62	65,5
Heiden und Moorheiden	322	80		66	71
Wald-Strauch-Übergangsstadien	324	77,5	83	62	65,5
Torfmoore	412	61	68	61	68
Wasserflächen	512	0	0	0	0

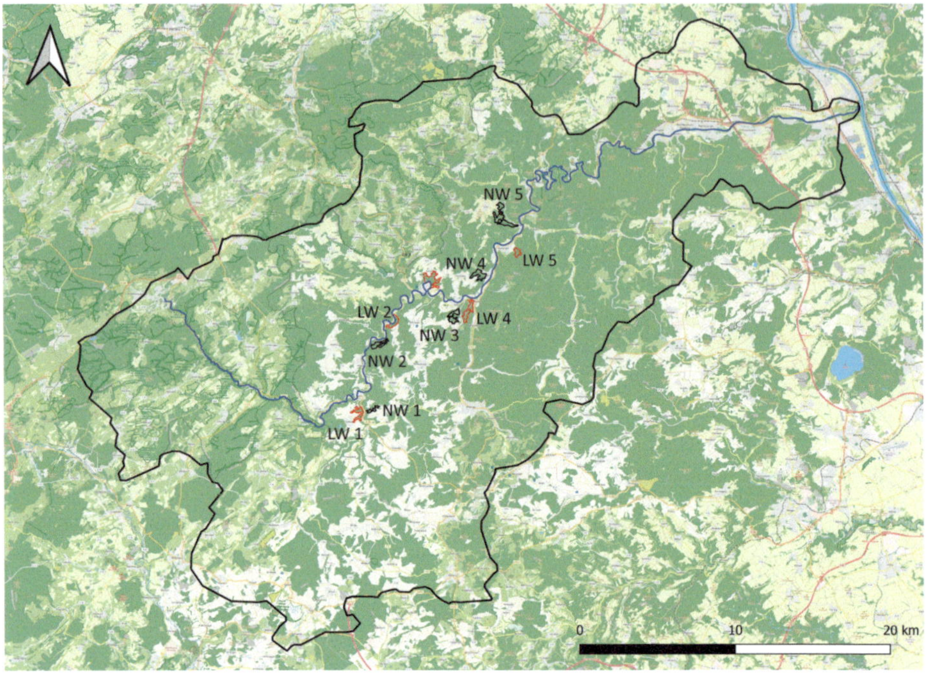

Abb. 11.7 Lage der im Rahmen einer Masterarbeit betrachteten ausgewählten Laub- und Nadelwaldgebiete. (Bildquelle: Zimmermann (2023))

natürliche Retentionspotenzial durch Landnutzungsanpassung mit anderen Maßnahmentypen zu vergleichen.

11.3.2 Natürlicher Rückhalt durch Gewässeranpassungsmaßnahmen

Die Auswirkung von Fließgewässerrenaturierungen lässt sich nicht händisch abschätzen, weswegen zwei ausgewählte Fließstrecken von jeweils zwischen 2 bis 2,5 km Länge systematisch in einer hydronumerischen Simulation untersucht wurden (Abb. 11.8). Die Ergebnisse ergaben hierbei, dass die Auwaldaufforstung im Vergleich zu einer Laufverlängerung und lokalen Geländeabsenkung eine deutlich größere Verzögerung des Spitzenabflusses bewirkt. Hierbei wurde eine Auwaldaufforstung über eine erhöhte Geländerauheit abgebildet, und die Laufverlängerung sowie lokale Geländeabsenkungen über Modifikationen des Geländemodells (Wilkens 2024). Für die beiden Abschnitte stellte sich bezogen auf ein HQ_{100} der Spitzenabfluss bei intensiver Auwaldaufforstung 70–75 min später ein. Die Fließgeschwindigkeit wurde in beiden Abschnitten am Ende des untersuchten Teilstücks um ca. 0,12 m/s bzw. 0,46 m/s reduziert (Wilkens 2024). Die Ergebnisse für eine lokale Geländeabsenkung und Laufverlängerung schwankten zwischen beiden Gebieten und lagen deutlich unter den Ergebnissen für eine Auwaldaufforstung (Wilkens 2024). Diese Ergebnisse sind lokal für die Hochwasservorsorge bedeutend, zeigen aber auch, dass die Wirkung von Renaturierungen für die Hochwasser-

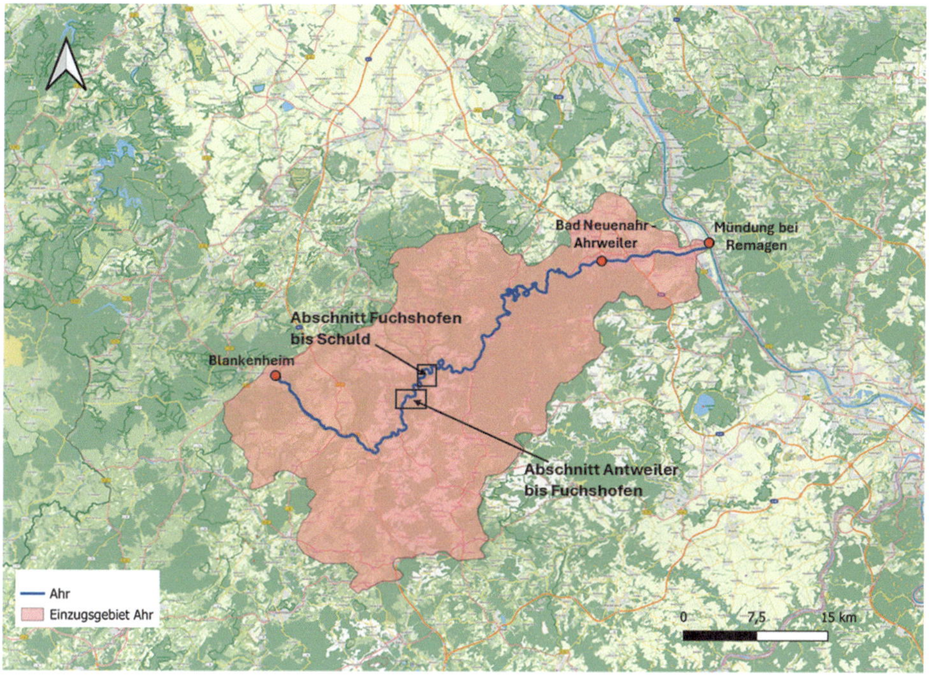

Abb. 11.8 Lage der beiden hydronumerisch simulierten Abschnitte zur Bestimmung des Rückhaltepotenzials durch eine Fließgewässerrenaturierung. (Bildquelle: Wilkens (2024))

vorsorge mithilfe hydronumerischer Simulationen des jeweiligen Abschnitts abgeschätzt werden muss.

11.3.3 Technische Rückhaltepotenziale

Insgesamt zeigen die ermittelten Werte (► Abschn. 11.3.1 und 11.3.2) allerdings, dass das zuvor ermittelte Rückhalteziel (vgl. Abb. 11.5) nicht alleine durch eine Landnutzungsanpassung und Fließgewässerrenaturierung erreicht werden kann. Daher wurde manuell das theoretisch maximale technische Rückhaltepotenzial abgeschätzt. Hierfür wurden in GIS auf Basis des digitalen Geländemodells mit einer Gitterweite von 1 m × 1 m (DGM1) zwischen Ortschaften Staudämme konstruiert, und anhand von Höhenlinien das Volumen abgeschätzt (Abb. 11.9). In die Analyse wurden nur Beckenvorschläge mit einem Mindestvolumen von 5 Mio. m^3 mit aufgenommen, um eine einzugsgebietsweite Einschätzung möglich zu machen. Zudem wurden nicht durchgängige Siedlungen von bis zu 25 Gebäuden nicht als Ausschlusskriterium gewertet, um möglichst große Stauvolumina generieren zu können (Wolf et al. 2024a, 2024b). Die Analyse diente zunächst nur der Abschätzung des theoretisch maximalen technischen Rückhaltepotenzials und nicht der Unterbreitung von konkreten Standortvorschlägen. Insgesamt wurden so 47 km^2 theoretisch potentielle Staufläche und ca. 1000 Mio. m^3 an theoretisch potentiellem Rückhaltevolumen ermittelt (Wolf et al. 2024a, 2024b).

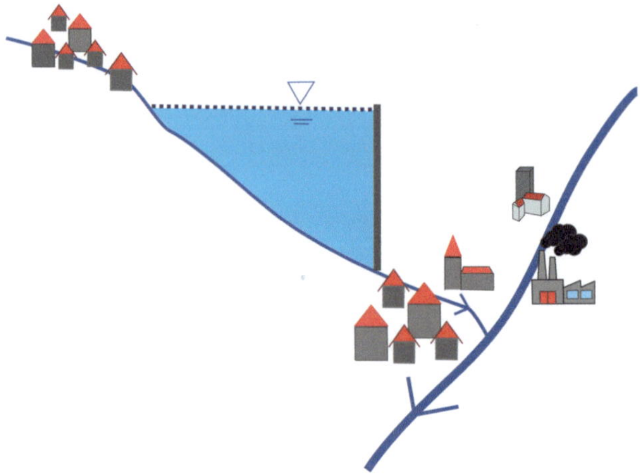

Abb. 11.9 Systematische GIS-gestützte Suche nach möglichen Standorten für Hochwasserrückhaltebecken im Ahrtal zur Abschätzung eines ungefähren technischen Rückhaltevolumens über das gesamte Einzugsgebiet. Abb. 11.9 wurde bereits mit gleicher Beschreibung in Wolf et al. (2024a) veröffentlicht

Hinweis für Entscheidungsprozesse

Auch wenn nicht unerhebliche Verbesserungen im Gebietsrückhalt über Landnutzungsanpassungen und Fließgewässerrenaturierungen erzielt werden können, ist der Bau von technischen Rückhalteräumen im Ahrtal unerlässlich, um sich auf große Hochwasser vorzubereiten.

Deswegen wird sich im Projekt „Plan zur Umsetzung und Weiterentwicklung von überörtlichen Maßnahmen zur Hochwasser- und Starkregenvorsorge für den Landkreis Ahrweiler und das Ahreinzugsgebiet unter Berücksichtigung der örtlichen Vorsorgekonzepte" („überörtlicher Maßnahmenplan" [üMP]) konkret mit der Priorisierung und Machbarkeit technischer und natürlicher Maßnahmen befasst (siehe auch ▶ Kap. 14).

Potentiell geeignete Flächen für einen technischen Hochwasserrückhalt sind mit einer hohen Wahrscheinlichkeit nicht frei verfügbar, sondern haben oft sogar mehrere konkurrierende Nutzungsansprüche.

11.4 Raumempfindlichkeit von Hochwasserrückhaltebecken

Die Flutkatastrophe vom Sommer 2021 hat zu einem hohen Engagement in Politik und Verwaltung geführt, die Schaffung von Retentionsraum durch natürliche und technische Maßnahmen der Hochwasservorsorge voranzutreiben. Der Wiederaufbauprozess im Ahrtal bietet hierfür die Möglichkeit damit auch langfristig einen Beitrag zur Steigerung der Hochwasserresilienz insgesamt zu leisten (Birkmann et al. 2023). Wie bereits im vorigen Unterkapitel dargestellt, liegen in der Verwirklichung von Maßnahmen des technischen Hochwasserschutzes erhebliche Rückhaltepotenziale für das Ahrtal. Die Suche nach möglichen Standorten für Hochwasserrückhaltebecken (HRB) erfordert jedoch neben der Berücksichtigung wasserwirtschaftlicher Aspekte auch die Auseinandersetzung mit raumplanerischen Vorgaben und die Berücksichtigung von Umweltbelangen. Durch eine Raumempfindlichkeitsuntersuchung (REU) soll die Standortsuche unterstützt werden und eine Grundlage zur Bewertung dieser potenziellen Flächen hinsichtlich dieser planerischen Restriktionen und Umsetzungshürden geschaffen werden. Dabei gilt es, gesetzliche und planungsrechtliche Vorgaben, die im Rahmen der Umsetzung konkreter Vorhaben greifen, bereits im Vorfeld in der Analyse durch die Auswahl geeigneter Kriterien zu berücksichtigen. Da die Rückhalteziele für das gesamte Einzugsgebiet, wie hier im Ahrtal, nur durch die Errichtung mehrerer kaskadierender HRB erreicht werden kann, ist von insgesamt vielen Jahren dauernden Planungs-, Genehmigungs- und schließlich Bauphasen zu rechnen. Deshalb ist auch eine frühzeitige Flächensicherung auf Ebene der Regional- und kommunalen Bauleitplanung sinnvoll, um konkurrierende Nutzungen in diesen Bereichen mittelfristig auszuschließen. Die Raumempfindlichkeitsuntersuchung zielt insgesamt darauf ab (a) einen Beitrag zur Standortfindung und -bewertung potenzieller HRB zu leisten, (b) eine Beurteilungsgrundlage für nachfolgende Prüf- und Genehmigungsverfahren zu schaffen und (c) abwägungsrelevante Informationen zur raumordnerischen Sicherung von geeigneten Potenzialflächen zusammenzustellen.

11.4.1 Planungsrechtlicher Rahmen

Gemäß § 14 des Bundesnaturschutzgesetzes (BNatSchG[1]) können HRB einen Eingriff in den Naturhaushalt darstellen. Erhebliche Beeinträchtigungen sind entsprechend § 13 BNatSchG zu vermeiden oder durch Ausgleichs- und Ersatzmaßnahmen zu kompensieren. Die frühzeitige Berücksichtigung dieser Vorgaben erfordert eine Abschätzung der möglichen Beeinträchtigungen auf den Naturhausalt. Zudem stellen größere HRB ab einem Einstauvolumen von 10 Mio. m^3 umweltverträglichkeitspflichtige Vorhaben nach dem Umweltverträglichkeitsgesetz (§ 6 UVPG[2]) dar, was eine Ermittlung der möglichen negativen Auswirkungen auf die in § 2 Abs. 1 UVPG aufgeführten Schutzgüter erfordert. Für kleinere HRB ist eine Vorprüfung des Einzelfalls (§ 7 Abs. 1 UVPG) erforderlich.

Die Schutzgüter nach UVPG sind:

1. Menschen, insbesondere die menschliche Gesundheit,
2. Tiere, Pflanzen und die biologische Vielfalt,
3. Fläche, Boden, Wasser, Luft, Klima und Landschaft,
4. kulturelles Erbe und sonstige Sachgüter sowie
5. die Wechselwirkung zwischen den vorgenannten Schutzgütern.

Bei überörtlich raumbedeutsam eingestuften Vorhaben (i. S. von § 1 Raumordnungsverordnung[3]) ist zudem die Raumverträglichkeit gemäß § 15 Abs. 1 des Raumordnungsgesetzes ROG[4] im Rahmen eines Raumordnungsverfahrens zu beurteilen. Auch hierbei sind die raumbedeutsamen Auswirkungen der Planung oder Maßnahmen abzuschätzen und die Übereinstimmung mit den Erfordernissen der Raumordnung zu prüfen.

Vor diesem Hintergrund berücksichtigt die vorliegende Analyse insbesondere auch die Inhalte des Regionalplans Mittelrhein-Westerwald (Planungsgemeinschaft Mittel-

1 Bundesnaturschutzgesetz vom 29. Juli 2009 (BGBl. I S. 2542), das zuletzt durch Artikel 3 des Gesetzes vom 8. Dezember 2022 geändert worden ist.

2 Gesetz über die Umweltverträglichkeitsprüfung, In der Fassung der Bekanntmachung vom 24.02.2010 (BGBl. I S. 94), zuletzt geändert durch Gesetz vom 08.05.2024 (BGBl. I S. 151) m. W. v. 16.05.2024.

3 Raumordnungsverordnung vom 13. Dezember 1990 (BGBl. I S. 2766), die zuletzt durch Artikel 12 des Gesetzes vom 22. März 2023 (BGBl. 2023 I Nr. 88) geändert worden ist.

4 Raumordnungsgesetz vom 22. Dezember 2008 (BGBl. I S. 2986), das zuletzt durch Artikel 1 des Gesetzes vom 22. März 2023 (BGBl. 2023 I Nr. 88) geändert worden ist.

rhein-Westerwald, Koblenz 2017), um mögliche Konflikte mit bestehenden Zielen und Grundsätzen der Raumordnung zu vermeiden bzw. darzustellen.

Gleichermaßen stellt die Regionalplanung auch einen Adressaten für die räumliche Sicherung von Flächen für die Hochwasservorsorge auf regionaler Ebene dar. Nach dem Raumordnungsgesetz ist dem vorbeugenden Hochwasserschutz bei der Abwägung mit konkurrierenden Raumnutzungen besonderes Gewicht beizumessen. Zur Flächensicherung kann die Regionalplanung dabei auf ein breites Instrumentarium zurückgreifen. Zusätzliches Gewicht erhält sie seit 2021 durch die Vorgaben des Bundesraumordnungsplans Hochwasserschutz[5], der weitergehende Regelungen in Bezug auf Retention enthält. Das Thema der Sicherung von Retentionsflächen wird sowohl für Renaturierungsmaßnahmen als auch für Maßnahmen des technischen Hochwasserschutzes in Grundsätzen des Bundesraumordnungsplans vorgegeben (vgl. 1.1.5 (G), II.1.6 (G)). So sollen auch aus wasserwirtschaftlicher Sicht geeignete Flächen u. a. für Hochwasserrückhaltebecken identifiziert und von entgegenstehenden Nutzungen freigehalten werden.

11.4.2 Raumempfindlichkeitsuntersuchung

11

Als Raumempfindlichkeit wird der Grad der Vereinbarkeit eines Projektes mit den Naturraumpotenzialen oder der Qualitätsminderung der Umweltgüter, die durch ein Vorhaben im betroffenen Untersuchungsraum zu erwarten sind, verstanden (Scholles und Fürst 2001). Die Raumempfindlichkeitsuntersuchung identifiziert auf Grundlage der möglichen Wirkungen des Vorhabens und der Empfindlichkeit relevanter Kriterien durch Überlagerung relativ konfliktarme Standorte und ist gerade in der Vorplanung eine wichtige Informationsbasis zur frühzeitigen Berücksichtigung umweltfachlicher Aspekte.

Grundsätzlich stellt der Vergleich von Standortalternativen ein Entscheidungsproblem dar, welches durch multikriterielle Bewertungsmethoden gelöst werden kann. Hierfür stehen zahlreiche formalisierte Verfahren zur Verfügung (Greene et al. 2011; Janssen 2001), die teils erhebliche Unterschiede in Bezug auf Komplexität und Nachvollziehbarkeit aufweisen. Für die Einstufung der Retentionseignung von Flächen (Vogel 2010) an der Elbe oder der Standortfindung für Hochwasserrückhaltebecken im Ahrtal (Eickelmann et al. 2022) liegen bereits Beispiele vor, an denen sich die vorliegende Arbeit insbesondere im Hinblick auf die verwendeten Bewertungskriterien orientiert. Die darin verwendeten kompensatorischen Verfahren scheinen jedoch im Kontext der Raumplanungsperspektive und Zielsetzungen dieser Arbeit weniger geeignet, weshalb methodisch auf die in der Planungspraxis hierzulande bekanntere Methode der Raumempfindlichkeitsuntersuchung zurückgegriffen wird (Diller 2010; Hoffmann 2016). Sie zeichnet sich durch eine hohe Transparenz und Nachvollziehbarkeit hinsichtlich der Bewertungskriterien und deren Aggregation zu einer Gesamtbewertung aus (Scholles und Fürst 2001) und kommt insbesondere häufig auf Ebene der Strategischen Umweltprüfung zum Einsatz.

Das schrittweise Vorgehen umfasst die Eingrenzung des Untersuchungsraumes, die Analyse möglicher Wirkungen von HRB auf die Schutzgüter, die Auswahl geeigneter räumlicher Bewertungskriterien und deren Gewichtung sowie die Aggregation

5 Verordnung über die Raumordnung im Bund für einen länderübergreifenden Hochwasserschutz (BRPHV) vom 19. August 2021, mit Länderübergreifender Raumordnungsplan für den Hochwasserschutz (Anlage zur Verordnung über die Raumordnung im Bund für einen länderübergreifenden Hochwasserschutz) vom 19. August 2021 (BGBl. I S. 3712).

der Schutzgutbewertungen zu Raumempfindlichkeitsklassen und die Erstellung einer Raumempfindlichkeitskarte. Die kriterienbasierte Analyse wird mit einem Geo-Informationssystem (GIS) durchgeführt. In den Bewertungsprozess wurden Akteure der Landes- und Regionalplanung in RLP, Vertreter der Verwaltung sowie des Naturschutzes und der Wasserwirtschaft des Landkreises Ahrweiler, private Ingenieurbüros sowie Vertreter aus der Wissenschaft (KAHR-Projektpartner) im Rahmen von zwei durchgeführten Workshops involviert. Dadurch konnten sowohl die Auswahl als auch die Gewichtung der Bewertungskriterien, die für die Raumempfindlichkeitsanalyse herangezogen wurden, durch Expertenwissen bestimmt werden.

11.4.3 Wirkungsanalyse und Bewertungskriterien

Die möglichen Wirkungen von Retentionsmaßnahmen sind abhängig von der Art und Intensität der Maßnahmen sowie der Empfindlichkeit der Schutzgüter. Bei der Errichtung von HRB ist im Bereich der Dämme und Sperrbauwerke meist von einem dauerhaften Verlust von Schutzgutfunktionen durch Überbauung auszugehen, während es in den Staubereichen nur im Ereignisfall zu temporären Überflutungen kommt. Hier ist die Dauer der Überflutung und die Einstautiefe für das Ausmaß der Wirkungen entscheidend. Gerade an den Zuflüssen und am Oberlauf der Ahr sind steile Hanglagen und tief eingeschnittene Bachtäler prägend. Dadurch ist bei der Errichtung von Hochwasserrückhaltebecken in diesen Bereichen mit teils sehr hohen Dämmen zu rechnen, um ausreichende Volumina zurückhalten zu können. Laut Experten der Wasserwirtschaft ist bei geplanten HRB aufgrund der nötigen Dimensionierung mit Einstauzeiten von bis zu ca. 21 Tagen zu rechnen.

Das Untersuchungsgebiet im Einzugsgebiet der Ahr im Landkreis Ahrweiler weist eine vergleichsweise hohe Struktur- und Artenvielfalt auf, die durch Biotopkomplexe von teils internationaler Bedeutung und zahlreiche Schutzgebiete nach Naturschutzrecht geprägt ist. Aufgrund der dargestellten Wirkungen kann vereinzelt mit Verlust und erheblichen Veränderungen von Biotopen und Habitaten sowie der Störung oder Individuenverlusten empfindlicher Arten gerechnet werden. Auch regionalplanerische Zielsetzungen für den Arten- und Biotopschutz bzw. -verbund können Konflikte mit der Errichtung und dem Betrieb von Hochwasserrückhaltebecken oder Poldern darstellen. Bei Inanspruchnahme besonders wertvoller Flächen können im Zuge der Umsetzungsplanungen umfangreiche Prüfungen, wie Zielabweichungsverfahren (sofern Ziele der Raumordnung entgegenstehen), FFH-Verträglichkeitsuntersuchungen oder spezielle artenschutzrechtliche Prüfungen notwendig werden.

Das Schutzgut Mensch und menschliche Gesundheit profitiert zunächst direkt durch die Anlage von Hochwasserschutzmaßnahmen durch den erhöhten Schutz der Unterlieger vor Hochwasser. Dennoch sind nachteilige Wirkungen zu berücksichtigen, die u. a. auf die Wohn- und Arbeitsverhältnisse, die Erholungsfunktion oder die menschliche Gesundheit abzustellen sind. Neben den Siedlungsbereichen werden auch anthropogene Infrastrukturen, wie Verkehrseinrichtungen oder Anlagen, der Ver- und Entsorgung mit besonderer Bedeutung für das Schutzgut Mensch berücksichtigt, da hier in der Regel bei Betroffenheit aufwendige und kostenintensive Umsiedelungen bzw. Verlagerung von Infrastruktur notwendig werden. Es ist bei der Errichtung von HRB auch zu beachten, dass von ihnen ein Risiko für die unterhalb liegende Bevölkerung ausgehen kann. Die Tendenz, dass unterhalb „hochwasserfreie“ Bereiche für die Siedlungsentwicklung nutzbar gemacht werden, ist insbesondere für sensible Nutzungen kritisch zu sehen – schließlich besteht Schutz nur bis zum jeweiligen Lastfall (vgl. Greiving et al. 2021).

Im Zuge der Bearbeitung wurden zunächst Geodaten, die als räumliche Bewertungskriterien den Schutzgütern zugeordnet wurden, zusammengestellt und für eine Überlagerung im Geo-Informationssystem aufbereitet. Tab. 11.2 gibt einen Überblick über die verwendeten Bewertungskriterien und die jeweiligen Datengrundlagen.

Für jedes Schutzgut wurden die potenziell betroffenen Kriterien und räumlichen Planungsziele in fünf Raumempfindlichkeitsklassen eingeteilt und Raumempfindlichkeitskarten erstellt. Die Einstufung erfolgte dabei unter Berücksichtigung der Ergebnisse des ersten Retentionsraumworkshops, bei dem die Empfindlichkeit unterschiedlicher Schutzgüter und planerischer Vorgaben gegenüber technischen Retentionsmaßnahmen bewertet wurde. Wesentlich für eine Zuordnung ist auch der Schutzstatus sowie die Vereinbarkeit

Tab. 11.2 Bewertungskriterien und Datengrundlagen für die Überlagerung

Schutzgut	Räumliches Kriterium	Datengrundlage
Arten und Biotope	NATURA 2000 (FFH/SPA-Gebiete)	Landschaftsinformationssystem der Naturschutzverwaltung Rheinland-Pfalz
	FFH-Lebensraumtypenkartierung	
	Naturschutzgebiet	
	Naturdenkmal	
	Naturpark	
	Geschützte Biotope (§ 30 BNatschG)	
	Biotopkomplexe	
	Bruthabitate	
	Landschaftsschutzgebiet	
	Ökokonto- und Kompensationsflächen (Kompensationsverzeichnis)	
	Vorranggebiet (VRG) Regionaler Biotopverbund	Regionalplanung Mittelrhein-Westerwald 2017
	VRG Ressourcenschutz	
	VRG Regionaler Grünzug	
	Vorbehaltsgebiet (VBG) Regionaler Biotopverbund	
	VBG Ressourcenschutz	
Mensch	Wohnbebauung, Siedlungs- und Gewerbeflächen	Regionalplanung Mittelrhein-Westerwald 2017
	Anlagen der Ver- und Entsorgung	Amtliches Liegenschaftskataster-Informationssystem (ALKIS) – Rheinland-Pfalz[a]
	Verkehrsinfrastruktur (Bahn-, Straßen- und Flugverkehr)	
	Freizeit- und Erholungseinrichtungen	
	Straßennetz: Autobahnen, Bundes-, Landes-, und Kreisstraßen	Landesbetrieb Mobilität Rheinland-Pfalz

11

Tab. 11.2 (Fortsetzung)

Schutzgut	Räumliches Kriterium	Datengrundlage
Kultur- und Sachgüter	Kulturgüter/Objekte, die nach Denkmalschutzrecht besonders geschützt sind	Amtliches Liegenschaftskataster-Informationssystem (ALKIS) – Rheinland-Pfalz
	Weinbaulagen	
	VRG Landwirtschaft	Regionalplanung Mittelrhein-Westerwald 2017
	VRG Forstwirtschaft	
	VRG Rohstoffabbau	
Landschaftsbild	Gesamtanlagen mit Fernwirkung	Regionalplanung Mittelrhein-Westerwald 2017
	VBG Erholung und Tourismus	
	Landesweit bedeutsame Kulturlandschaften, LEP IV, 2013	Abgrenzung der landesweit bedeutsamen historischen Kulturlandschaften nach Gutachten MWKEL 2013
Wasser	Trinkwasserschutzgebiete	Wasserschutzgebiete Wasserwirtschaftsverwaltung WWV RLP
	Mineralwassereinzugsgebiete	
	Heilquellenschutzgebiete	
	Vorrang- und Vorbehaltsgebiete für den Grundwasserschutz	Regionalplanung Mittelrhein-Westerwald 2017
	Grundwasserflurabstände	Grundwasserflurabstände: Landesamt für Geologie und Bergbau Rheinland-Pfalz (LGRB)
Boden	Ertragspotenzial Boden	Übersichtskarten zu Bodeneigenschaften und -funktionen (BFD200) Landesamt für Geologie und Bergbau Rheinland-Pfalz (LGRB)
Fläche	Verlust von Fläche kann nur nach Einzelfall bewertet werden und ist je nach Dimension der Bauwerke zu beurteilen. Ein dauerhafter Verlust entsteht nur durch die baulichen Anlagen	
Klima/Luft	Aufgrund mangelnder Geodaten zu diesem Schutzgut, konnten die Auswirkungen auf das lokale Klima nur verbal argumentativ nach Einzelfall eingeschätzt werden	

[a] Filterung der Daten nach Objektartenkatalog GeoInfoDok 6.0.

mit bestehenden raumordnerischen Zielsetzungen und Grundsätzen. Die räumliche Überlagerung der Kriterien wurde mit Relevanz- bzw. boolschen Entscheidungsbäumen, die mit UND/ODER – Verknüpfungsregeln zur jeweiligen Raumempfindlichkeitsklasse zugeordnet wurden, durchgeführt. Abb. 11.10 zeigt exemplarisch den Relevanzbaum für das Schutzgut Arten und Biotope. Im Rahmen der Klassifizierung der Raumempfindlichkeit für die Schutzgüter Landschaftsbild und Mensch wurden Abstände zu den jeweiligen Objekten berücksichtigt, wenn von mittelbaren Wirkungen auf die Strukturen auszugehen ist.

Durch Aggregation der schutzgutbezogenen Raumempfindlichkeiten wird eine abschließend zusammenfassende Raumempfindlichkeit gebildet, wobei der jeweils höchste

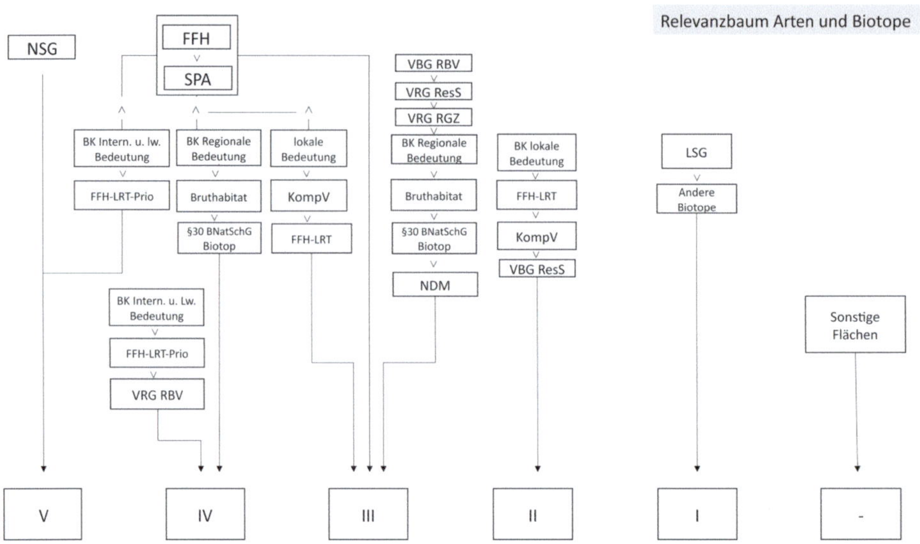

Abb. 11.10 Entscheidungsbaum (Relevanzbaum) für das Schutzgut Arten und Biotope. Die Kriterien sind durch boolesche UND (∧) bzw. ODER (∨) Operatoren verknüpft. Klassifizierung der Raumempfindlichkeit: *V* sehr hoch, *IV* hoch, *III* mittel, *II* gering, *I* sehr gering. Abkürzungen: *NSG* Naturschutzgebiet, *BK* Biotopkomplex, *intern.* internationaler, *lw.* landesweiter Bedeutung, *FFH* Flora Fauna Habitat – Gebiet, *SPA* Vogelschutzgebiet (Special Protected Area), *FFH-LRT* FFH-Lebensraumtyp, *VRG/VBG RBV* Vorrang- bzw. Vorbehaltsgebiet Regionaler Biotopverbund, *VRG/VBG ResS* Vorrang- bzw. Vorbehaltsgebiet Ressourcenschutz, *KompV* Kompensationsverzeichnis, *VRG RGZ* Vorranggebiet Regionaler Grünzug, *LSG* Landschaftsschutzgebiet

Wert eines Schutzgutes maßgeblich ist. Aus den gebildeten Raumempfindlichkeitsklassen lassen sich die Empfehlungen zur Standortfindung und Bewertung der Alternativen ableiten.

11.4.4 Raumempfindlichkeitskarte

Die Überlagerung der Raumempfindlichkeiten, die verschiedenen Schutzgütern zuzuordnen sind, erfolgt nach dem Maximalwertprinzip. Die höchste bei Überlagerung erreichte Raumempfindlichkeit eines Gebietes bestimmt die aggregierte Raumempfindlichkeitsklasse.

Durch die flächendeckende Darstellung in der Raumempfindlichkeitskarte (vgl. Abb. 11.11) für den Landkreis Ahrweiler können gegenüber den genannten Schutzzielen konfliktarme Bereiche identifiziert und geplante Standorte bewertet werden. In Gebieten, die mit hoch bis sehr hoch bewertet wurden, sind erhebliche Umweltauswirkungen zu erwarten die einem Vorhaben entgegenstehen können oder es ist von einem erhöhten Prüfungsaufwand in der Umsetzung auszugehen, da der Zulassung eines Vorhabens rechtlich verbindliche Schutznormen oder andere erhebliche Restriktionen entgegenstehen. Im Rahmen der Abwägung können die betroffenen Kriterien entscheidungserheblich sein. Es wird empfohlen, diese Bereiche bei der Planung von technischen Retentionsmaßnahmen möglichst zu umgehen bzw. frühzeitig die erkennbaren Konflikte

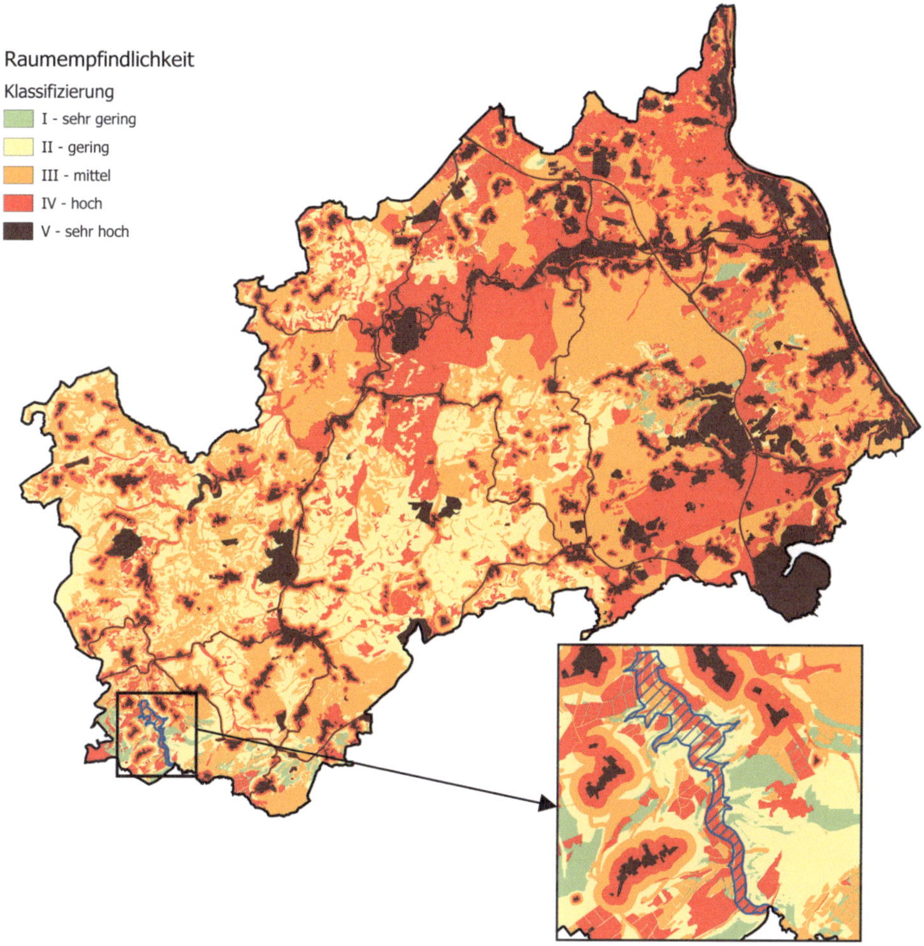

Abb. 11.11 Raumempfindlichkeitskarte für den Landkreis Ahrweiler, im Ausschnitt Überlagerung der Raumempfindlichkeitsklassen mit einer Potenzialfläche für HRB am Trierbach oberhalb Trierscheid. (Eigene Darstellung, Geobasisdaten: Landkreisgrenze Ahrweiler: © GeoBasis-DE/LVermGeoRP (2024), ▶ https://www.govdata.de/dl-de/by-2-0. Alle Rechte vorbehalten)

mit den zuständigen Behörden zu lösen. Gebiete mit überwiegend mittlerer Raumempfindlichkeit umfassen Sachverhalte oder Kriterien, die einem Vorhaben sachlich nicht zwingend entgegenstehen oder dessen nachteilige Auswirkungen durch Minimierungsmaßnahmen reduzierbar sind. Diese Bereiche sind für die Findung potenzieller Standorte eher geeignet, wenngleich Vorgaben bestehen, die im Sinne der Umweltvorsorge in die Abwägung einfließen sollten. Gebiete mit sehr geringer und geringer Einstufung umfassen Flächen, die gegenüber den Wirkungen von HRB relativ unempfindlich sind. Die Beurteilung der Raumempfindlichkeit der untersuchten Potenzialflächen kann durch die Flächenanteile der Raumempfindlichkeitsklassen (REK) abgeschätzt werden. Von den insgesamt auf 47 km^2 untersuchten Potenzialflächen befinden sich ca. 35 km^2 innerhalb des Landkreises Ahrweiler. Davon sind 16 % der Flächen als besonders unproblematisch anzusehen, da hier die Anteile der Raumempfindlichkeitsklassen IV und V insgesamt

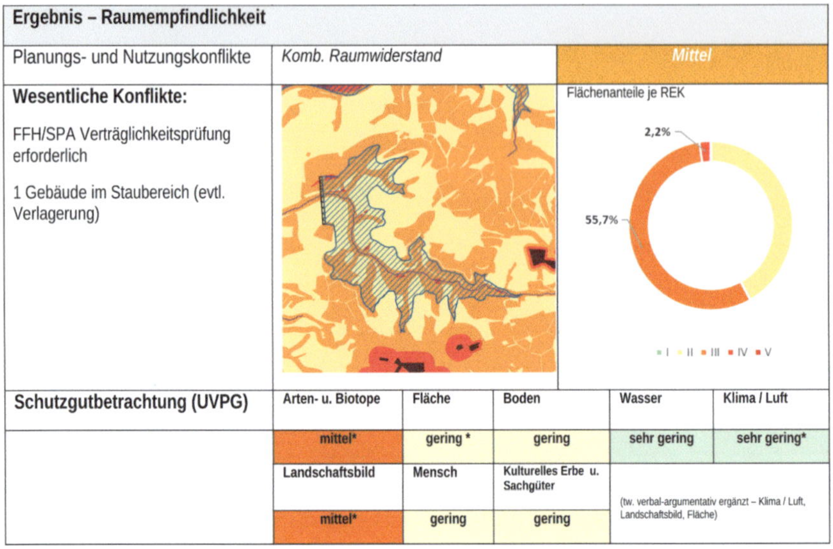

Ergebnis – Raumempfindlichkeit					
Planungs- und Nutzungskonflikte	*Komb. Raumwiderstand*				*Mittel*
Wesentliche Konflikte: FFH/SPA Verträglichkeitsprüfung erforderlich 1 Gebäude im Staubereich (evtl. Verlagerung)					Flächenanteile je REK 2,2% 55,7% I II III IV V
Schutzgutbetrachtung (UVPG)	Arten- u. Biotope	Fläche	Boden	Wasser	Klima / Luft
	mittel*	gering *	gering	sehr gering	sehr gering*
	Landschaftsbild	Mensch	Kulturelles Erbe u. Sachgüter	(tw. verbal-argumentativ ergänzt – Klima / Luft, Landschaftsbild, Fläche)	
	mittel*	gering	gering		

Abb. 11.12 Beispiel: Gebietssteckbrief (Ausschnitt) für die HRB-Potenzialfläche Limbach, Landkreis Ahrweiler

11

unter 10 % liegen. 19 % der untersuchten Fläche werden dagegen als besonders problematisch eingestuft, da hier die REK IV und V über 80 % der Fläche einnehmen.

Für ausgewählte Potenzialflächen für Hochwasserrückhaltebecken wurde zudem eine erste Einzelbetrachtung durchgeführt, deren Ergebnisse in Gebietssteckbriefen (vgl. Abb. 11.12) zusammengefasst wurden. Eine vergleichende Beurteilung der Standorte ist durch die enthaltene Schutzgutbetrachtung und die Darstellung der wesentlichen zu erwartenden Konflikte im Sinne der untersuchten Bewertungskriterien möglich. Die Gebietssteckbriefe erlauben die konkrete Auseinandersetzung mit den Konflikten der jeweiligen Standortpotenziale und ermöglichen die Identifikation von Lösungsmöglichkeiten.

11.5 Fazit und Ausblick

Insgesamt liefern die erarbeiteten Ergebnisse der Retentionsraumanalyse wertvolle Erkenntnisse, die in den Aktivitäten und Projekten auf lokaler und regionaler Ebene im Ahrtal verwertet werden können. Dies entspricht auch dem Anspruch des KAHR-Projektes, den Wiederaufbau und die Steigerung der Hochwasserresilienz unterstützend zu begleiten.

Es konnte gezeigt werden, dass erhebliche Retentionspotenziale durch angepasste Landnutzung sowie natürliche und technische Maßnahmen im Einzugsgebiet der Ahr bestehen, um in Kombination das angestrebte Rückhalteziel zu erreichen. Allein durch die langfristige Anpassung der Landnutzung könnten bis zu 10 Mio. m^3 Wasser in der Fläche zurückgehalten werden. Durch natürliche Maßnahmen, wie der Laufverlängerung am Fließgewässer oder Auwaldaufforstungen, kann zudem der Abfluss bei einem Hochwasserereignis deutlich verlangsamt werden. Die Ergebnisse zeigen aber auch, dass nur durch die Kombination mit technischen Maßnahmen, wie Hochwasserrückhaltebecken, für vergleichbar große Hochwasser, wie das vom Sommer 2021, ausreichend vorgesorgt werden kann.

Für alle genannten Maßnahmentypen des natürlichen und technischen Hochwasserschutzes stellen Flächenverfügbarkeit und Nutzungs- oder Zielkonflikte Hemmnisse bei der Umsetzung dar, die bei technischen Maßnahmen teils erheblich sein können. Insbesondere die umweltfachlichen Belange wurden exemplarisch für den Landkreis Ahrweiler durch eine flächendeckende Darstellung schutzgutbezogener Raumwiderstände bzw. -empfindlichkeiten gegenüber Hochwasserrückhaltebecken aufgezeigt.

Grundsätzlich ist festzuhalten, dass die Raumempfindlichkeit nur einen Aspekt darstellt, der bei Standortentscheidungen berücksichtigt werden muss. Daneben sind insbesondere Kriterien der Wirtschaftlichkeit sowie der technischen Machbarkeit zu berücksichtigen. Dennoch stellt das Ergebnis der Raumempfindlichkeitsuntersuchung eine wertvolle Informationsbasis für die Standortfindung und die Beurteilung von Potenzialflächen für Hochwasserrückhaltebecken dar. Damit unterstützen die vorliegenden Ergebnisse auch den „Plan zur Umsetzung und Weiterentwicklung von überörtlichen Maßnahmen zur Hochwasser- und Starkregenvorsorge für den Landkreis Ahrweiler und das Ahreinzugsgebiet unter Berücksichtigung der örtlichen Vorsorgekonzepte“ (üMP) des Landkreises Ahrweiler. Zudem kann die vorliegende Raumempfindlichkeitsuntersuchung als eine erste Grundlage zur Flächensicherung im Rahmen der Raumordnung dienen, wenngleich für diese Standorte noch weitere Untersuchungen und weitere Konkretisierungen nötig sind.

Darüber hinaus ergeben sich weitere Forschungsfragen, welche die vorliegenden Arbeiten zum Thema Retentionsraumpotenziale um weitere Aspekte sinnvoll ergänzen würden. Insbesondere das Thema des schadfreien Abflusses von Hochwasser und Starkregen innerhalb besiedelter Bereiche ist hier von Bedeutung. So könnten Freiflächen identifiziert werden, die langfristig von Bebauung freizuhalten sind, um das Schadenspotenzial nicht weiter zu erhöhen und Wassermassen schadlos und gezielt abzuleiten.

Zusammenfassung

- Um sich im Ahr-Einzugsgebiet auf ein Hochwasser wie das von Mitte Juli 2021 vorzubereiten, wurde ein Rückhaltevolumen von mindestens 25 bis 33 Mio. m^3 als Ziel definiert, wobei aufgrund der räumlichen Variabilität von Starkregenniederschlägen eigentlich ein Vielfaches hiervon vorgehalten werden muss.
- Erste Analysen, in denen das auf der Curve-Number basierende SCS-Verfahren händisch angewendet wurde, ergaben ein Potenzial durch Landnutzungsanpassungen von bis zu 10 Mio. m^3. Weitere natürliche Maßnahmen wie Auwaldaufforstungen und Laufverlängerungen von Fließgewässern würden den Abfluss bei Hochwasser zusätzlich verlangsamen.
- Technische Maßnahmen, wie der Bau von Hochwasserrückhaltebecken, sind jedoch unerlässlich, um auf extreme Ereignisse wie 2021 vorbereitet zu sein. Grobe GIS-Analysen ergaben hier ein theoretisches technisches Potenzial von bis zu 1000 Mio. m^3.
- Ein wesentlicher Teil der Retentionsraumpotenzialanalyse befasste sich mit der Raumempfindlichkeit von potenziellen Standorten für Rückhaltemaßnahmen. Dabei werden verschiedene Schutzgüter wie menschliche Gesundheit, Arten und Biotope, Boden, Wasser, Landschaftsbild und Kultur- und Sachgüter näher untersucht.
- Es zeigte sich, dass neben schutzgutbezogenen Konflikten Flächenverfügbarkeit und konkurrierende Nutzungsansprüche oft die größten Hindernisse für die Umsetzung dar-

stellen. Hier wird empfohlen, frühzeitig Flächensicherungsmaßnahmen in die Planung aufzunehmen, um langfristig eine wirksame Hochwasservorsorge zu gewährleisten.

Danksagung Die Autorinnen und Autoren von ▶ Kap. 11 bedanken sich beim Bundesministerium für Bildung und Forschung (BMBF) und beim DLR Projektträger für die Unterstützung und Förderung des KAHR-Projekts. Diese Forschung wurde vom BMBF unter dem Förderkennzeichen 01LR2102A-M finanziert.

11

Literatur

Apel H, Vorogushyn S, Merz B (2022) Brief communication – Impact Forecasting Could Substantially Improve the Emergency Management of Deadly Floods: Case Study July 2021 floods in Germany

Bezirksregierung Köln (2023) Digitales Geländemodell. https://www.bezreg-koeln.nrw.de/geobasis-nrw/produkte-und-dienste/hoehenmodelle/digitale-gelaendemodelle/digitalesgelaendemodell. Zuletzt geprüft am: 13.11.2023

Birkmann J, Schüttrumpf H, Handmer J, Thieken A, Kuhlicke C, Truedinger A, Sauter H, Klopries E-M, Greiving S, Jamshed A, Merz B, Solecki W, Kirschbauer L (2023) Strengthening resilience in reconstruction after extreme events – Insights from flood affected communities in Germany. Int J Disaster Risk Reduct 96:103965. https://doi.org/10.1016/j.ijdrr.2023.103965

Blöschl G, Hall J, Viglione A et al (2019) Changing climate both increases and decreases European river floods. Nature 573:108–111. https://doi.org/10.1038/s41586-019-1495-6

Diller C (2010) Methoden der Raumplanung in der Praxis. Universitätsverlag der TU Berlin

DWD (2021) Frei verfügbare RADOLAN-Daten und deren aktuelle Visualisierung; Tagessumme SF (rechts). https://opendata.dwd.de/climate_environment/CDC/grids_germany/daily/radolan/. Zugegriffen: 30. Dez. 2023

DWD (2022) RADKLIM-Bulletin; Projekt-Rundschau | Panorama und Werkstatt § Jährliche Fortschreibung und Erweiterung der Radarklimatologie § Jahres-Ergebnisse der Klimatologie § Extreme Niederschlagsereignisse des vergangenen Jahres. DWD (https://www.dwd.de/DE/fachnutzer/wasserwirtschaft/radarnieder schlag/radklim-bulletin/radklimbulletin2022download.pdf;jsessionid=7048FC7D1A9E83E1BD05D11D9 F4E4831.live11043?__blob=publicationFile&v=3. Zugegriffen: 18. September 2024)

EEA (2016) CLC 2018. European Environment Agency. https://land.copernicus.eu/pan-european/corine-land-cover/clc2018. Zugegriffen: 19. Okt. 2020

Eickelmann J, Engels S, Franzmann H, Gayko N, Hijazi A, Hölzner R, Hövel S, Kneilmann M, Knipper T, Niehues M, Paßlick O, Pilarczyk S, Schlott M, Schmidt M, Schultze-Schlutius T, Wrase T (2022) Identifizierung geeigneter Flächen für Hochwasserrückhaltebecken und Versiegelungsverbotszonen: Eine GIS-basierte Analyse im Ahrtal

ESTAT (2020) Countries, 2020 – Administrative Units – Dataset. https://giscoservices.ec.europa.eu/distribution/v2/countries/download/. Zuletzt geprüft am 27.10.2023

Geofabrik GmbH (2018a) gis_osm_waterways_free_1. CC BY-SA 2.0. https://download.geofabrik.de. Zuletzt geprüft am 19.10.2020

Geofabrik GmbH (2018b) gis_osm_places_a_free_1. CC BY-SA 2.0. https://download.geofabrik.de. Zuletzt geprüft am 19.10.2020

Greene R, Devillers R, Luther JE, Eddy BG (2011) GIS-based multiple-criteria decision analysis. Geogr Compass 5:412–432. https://doi.org/10.1111/j.1749-8198.2011.00431.x

Greiving S, Fleischhauer M, León CD, Schödl L, Wachinger G, Quintana Miralles IK, Prado Larraín B (2021) Participatory Assessment of Multi Risks in Urban Regions—The Case of Critical Infrastructures in Metropolitan Lima. Sustainability 13(5):2813. https://doi.org/10.3390/su13052813

Hochwasservorhersagezentrale Rheinland-Pfalz Pegel Altenahr / Ahr; Pegelkennwerte. https://www.hochwasser.rlp.de/flussgebiet/ahr/altenahr#pegelkennwerte. Zugegriffen: 24. Okt. 2024

Hoffmann A (2016) Methodeneinsatz in der räumlichen Planung: Methodenverständnis, Hemmnisse und Gelingenskriterien

IPCC Summary for Policymakers Climate Change (2023) Synthesis Report. A Report of the Intergovernmental Panel on Climate Change. Contribution of Working Groups I, II and III to the Sixth Assessment Report of the Intergovernmental Panel on Climate Change (36 pages)

Janssen R (2001) On the use of multi-criteria analysis in environmental impact assessment in The Netherlands. Multi Criteria Decis Anal 10:101–109. https://doi.org/10.1002/mcda.293

Kisseler E, Wolf S, Sauter H, Schüttrumpf H (2023) Reduktion der Vulnerabilität gegenüber Hochwasser – Empfehlungen zur Flächennutzung auf Basis von Flutmarken des Juli-Hochwassers 2021. Korrespondenz Wasserwirtsch 2023:498–506

Koks EE, van Ginkel KCH, van Marle MJE, Lemnitzer A (2022) Brief communication: critical infrastructure impacts of the 2021 mid-july western European flood event. Nat Hazards Earth Syst Sci 22:3831–3838. https://doi.org/10.5194/nhess-22-3831-2022

Kreienkamp F, Philip SY, Tradowsky JS et al (2021) Rapid attribution of heavy rainfall events leading to the severe flooding in Western Europe during July 2021

Ludwig P, Ehmele F, Franca MJ, Mohr S, Caldas-Alvarez A, Daniell JE, Ehret U, Feldmann H, Hundhausen M, Knippertz P, Küpfer K, Kunz M, Mühr B, Pinto JG, Quinting J, Schäfer AM, Seidel F, Wisotzky C (2023) A multi-disciplinary analysis of the exceptional flood event of July 2021 in central Europe – Part 2: Historical context and relation to climate change. Nat Hazards Earth Syst Sci 23:1287–1311. https://doi.org/10.5194/nhess-23-1287-2023

LVermGeoRP (2021) DGM 1 m Gitterweite ASCII (X,Y,Z) 114 Kacheln (1×1 km), Ahrtal Sonderbefliegung: 12./13.08.2021. Datenbereitstellung am 2. Februar 2022.

Maniak U (2016a) Einführung Hydrologie und Wasserwirtschaft. In: Maniak U (Hrsg) Hydrologie und Wasserwirtschaft. Springer, Berlin Heidelberg, S 1–19

Maniak U (2016b) Niederschlag-Abfluss-Modelle für Hochwasserabläufe. In: Maniak U (Hrsg) Hydrologie und Wasserwirtschaft. Springer, Berlin, Heidelberg, S 301–414

Maniak U (2016c) Niederschlag-Abfluss-Modelle für Hochwasserabläufe. In: Maniak U (Hrsg) Hydrologie und Wasserwirtschaft. Springer, Berlin, Heidelberg, S 301–414

Mockus V (1965) National Engineering Handbook; Section 4 Hydrology. http://www.irrigationtoolbox.com/NEH/Part630_Hydrology/neh630-ch21.pdf. Zugegriffen: 22. Juni 2022

Mohr S, Ehret U, Kunz M, Ludwig P, Caldas-Alvarez A, Daniell JE, Ehmele F, Feldmann H, Franca MJ, Gattke C, Hundhausen M, Knippertz P, Küpfer K, Mühr B, Pinto JG, Quinting J, Schäfer AM, Scheibel M, Seidel F, Wisotzky C (2023) A multi-disciplinary analysis of the exceptional flood event of July 2021 in central Europe – Part 1: Event description and analysis. Nat Hazards Earth Syst Sci 23:525–551. https://doi.org/10.5194/nhess-23-525-2023

Planungsgemeinschaft Mittelrhein-Westerwald (2017) Regionaler Raumordnungsplan Mittelrhein-Westerwald. Koblenz

Rath AV (2022) Vergleich von internationalen Hochwasserstatistiken auf Grundlage der International Disaster Database. Aachen (Bachelorarbeit)

Rhein FGG (2019) Bericht der FGG Rhein über die Überprüfung und Aktualisierung der Hochwassergefahren- und Hochwasserrisikokarten im deutschen Teil des Rheineinzugsgebiets nach § 74 WHG; Stand: 16. Dezember 2019. https://fgg-rhein.de/servlet/is/87526/. Zugegriffen: 24. Okt. 2024

Roggenkamp T, Herget J (2014) Reconstructing peak discharges of historic floods of the river Ahr, Germany. Erdkunde 68:49–59. https://doi.org/10.3112/erdkunde.2014.01.05

Roggenkamp T, Herget J (2024) Flood reconstruction – The unexpected rather frequent event at River Ahr in July 2021. Glob Planet Change 240:104541. https://doi.org/10.1016/j.gloplacha.2024.104541

Ross CW, Prihodko L, Anchang J, Kumar S, Ji W, Hanan NP (2018) Global Hydrologic Soil Groups (HYSOGs250m) for Curve Number-Based Runoff Modeling. https://daac.ornl.gov/SOILS/guides/Global_Hydrologic_Soil_Group.html. Zugegriffen: 30. Dez. 2023

Scholles F, Fürst D (2001) Handbuch Theorien+ Methoden der Raum- und Umweltplanung. Handbücher zur Umweltplanung-HzU, Bd 4. Dortmund

Schüttrumpf H (2021) Das Juli-Hochwasser 2021 in NRW-Ein erster Erfahrungsbericht. Wasser Abfall 23:14–17

Tradowsky JS, Philip SY, Kreienkamp F et al (2023) Attribution of the heavy rainfall events leading to severe flooding in Western Europe during July 2021. Clim Change. https://doi.org/10.1007/s10584-023-03502-7

Vautard R, Boucher O, van Oldenborgh GJ, Otto F, Haustein K, Vogel MM, Seneviratne SI, Soubeyroux J-M, Schneider M, Drouin A, Ribes A, Kreienkamp F, Stott P, van Aalst M (2019) Human contribution to the record-breaking July 2019 heat wave in Western Europe. https://www.worldweatherattribution.org/wp-content/uploads/July2019heatwave.pdf. Zugegriffen: 10. Aug. 2023

Vélez Pérez M (2023) Investigation of the impact of land use changes of the flood event 2021 at the Ahr River. RWTH Aachen University, Aachen (Masterarbeit)

Vélez Pérez M, Wolf S, Klopries E-M (2023) Quantifizierung des Einflusses der Landnutzung an der Ahr auf das Abflussverhalten. Korrespondenz Wasserwirtsch 2023(16):435–441. https://doi.org/10.3243/kwe2023.07.004

Vogel R (2010) GIS-basierte multikriterielle Bewertung der Retentionseignung von Auenflächen – am Beispiel der Elbe. Mensch und Buch Verlag, Berlin

Vorogushyn S, Apel H, Kemter M, Thieken AH (2022) Analyse der Hochwassergefährdung im Ahrtal unter Berücksichtigung historischer Hochwasser. Hydrol Wasserbewirtsch. https://doi.org/10.5675/HyWa_2021.5_2

Wilkens J (2024) Grenzen und Möglichkeiten des Hochwasserrückhaltes durch Renaturierungen entlang der Ahr. Aachen (Masterarbeit)

Wolf S (2023) The long-term-memory of a typical mid-European upland to lowland river : or why we struggle to reach a good ecological state for our rivers. Rheinisch-Westfälische Technische Hochschule, Aachen (Dissertation)

Wolf SS, Zimmermann M, Holste I, Schüttrumpf H, Bucholz O, Yörük A (2024a) Möglichkeiten und Grenzen von natürlichem oder technischem Hochwasserschutz am Beispiel der Ahr. Wasser Abfall 26(5):33–37. https://www.springerprofessional.de/moeglichkeiten-und-grenzen-von-natuerlichem-oder-technischem-hoc/27057246

Wolf SS, Steudtner F, Buchholz O, Yörük A, Schüttrumpf H (2024b) Natürlicher vs. Technischer Hochwasserschutz – Wo sind die Möglichkeiten und Grenzen? 57. Essener Tagung für Wasserwirtschaft, GWA 257, Aachen. Gesellschaft zur Förderung des Instituts für Siedlungswasserwirtschaft an der RWTH Aachen e. V., Aachen, S 61/1–61/17

Zimmermann M (2023) Identifikation von natürlichen Hochwasserrückhaltepotentialen an der Ahr. Aachen (Masterarbeit)

Hochwasserresilienz an Inde und Vicht

Felix Steudtner, Elena-Maria Klopries, Gerd Demny und Susanne Kozerke

J. Birkmann et al. (Hrsg.), *Die Flutkatastrophe 2021*, Sustainable Development Goals (SDG) – Umsetzung in Praxis, Lehre und Entscheidungsprozessen,
https://doi.org/10.1007/978-3-658-48688-4_12

Das außergewöhnliche Hochwasserereignis im Juli 2021 verdeutlichte die dringende Notwendigkeit nachhaltiger und resilienter Schutzmaßnahmen in hochwassergefährdeten Regionen. Besonders betroffen waren die Flüsse Inde und Vicht in Nordrhein-Westfalen, die erhebliche Schäden an Infrastruktur, Wohngebäuden und öffentlichen Einrichtungen hervorriefen. Städte wie Eschweiler und Stolberg erlebten massive Zerstörungen, was die Dringlichkeit eines umfassenden Hochwasserschutzkonzepts für das Einzugsgebiet unterstrich. In Reaktion auf das Ereignis entwickelten der Wasserverband Eifel-Rur (WVER) und das Institut für Wasserbau und Wasserwirtschaft (IWW) der RWTH Aachen den Masterplan „Hochwasserresiliente Stadt- und Gebietsentwicklung für das Einzugsgebiet von Inde und Vicht". Ziel des Plans ist es, geeignete Maßnahmen zu identifizieren, die den Hochwasserschutz erhöhen, die Resilienz der Region gegenüber solchen Ereignissen stärken und die Auswirkungen zukünftiger Hochwasserereignisse minimieren. Der Masterplan verfolgt einen ganzheitlichen Ansatz, der bauliche und nichtbauliche Maßnahmen kombiniert, um sowohl den gesetzlichen Standard des HQ_{100}-Schutzes sicherzustellen als auch nachhaltige, widerstandsfähige Strukturen zu schaffen. Im weiteren Verlauf werden der Masterplan sowie die erarbeiteten Maßnahmen zur Erhöhung der Hochwassersicherheit in der Region vorgestellt. Zudem wird die Potenzialanalyse der identifizierten Schutzmaßnahmen erläutert und die daraus gewonnenen Erkenntnisse dargestellt, die für die Umsetzung eines effektiven Hochwasserschutzkonzepts entscheidend sind.

12.1 Hochwasserschutzkonzept an Inde und Vicht

12

Das Hochwasserereignis im Juli 2021 führte an den Flüssen Inde und Vicht zu massiven Zerstörungen an der Infrastruktur. Besonders betroffen waren die Städte Eschweiler und Stolberg. Weitere Schadensschwerpunkte lagen in Kornelimünster an der Inde sowie in den Ortslagen Vicht, Zweifall, Mulartshütte und Roetgen an der Vicht.

Neben Schäden an Wohngebäuden wurden kritische Infrastrukturen, die für das Funktionieren des gesellschaftlichen Alltags unerlässlich sind, schwer getroffen (Krings 2020). Öffentliche Gebäude und Plätze, wie das Rathaus und das Stadtarchiv in Stolberg, aber auch Schulen und Kindertagesstätten, wurden beschädigt. Die Bewältigung der Notlage wurde durch die Tatsache erschwert, dass die Einsatzkräfte teils selbst von den Überschwemmungen betroffen waren. Zudem waren wichtige Verkehrswege, insbesondere Straßen und Brücken, aufgrund der Zerstörungen nicht mehr passierbar, was die Rettungs- und Hilfsmaßnahmen vor Ort erheblich erschwerte.

12.1.1 Masterplan „Hochwasserresiliente Stadt- und Gebietsentwicklung für das Einzugsgebiet von Inde und Vicht"

Kurz nach dem Hochwasserereignis 2021 wurde durch den Wasserverband Eifel-Rur (WVER) in Zusammenarbeit mit dem Institut für Wasserbau und Wasserwirtschaft (IWW) der RWTH Aachen das geförderte Vorhaben „Hochwasserresiliente Stadt- und Gebietsentwicklung für das Einzugsgebiet von Inde und Vicht" (Masterplan) initiiert. Das zentrale Ziel dieses Masterplans besteht darin, potenzielle Hochwasserschutzmaßnahmen zu identifizieren, die die Widerstandsfähigkeit gegenüber zukünftigen Hochwasserereignissen in der Region signifikant erhöhen können.

Im Rahmen des Masterplans wurden neben baulichen auch nichtbauliche Maßnahmenvorschläge entwickelt, um die Resilienz der Region ganzheitlich zu stärken. Dabei wurde besonders darauf geachtet, dass für bauliche Maßnahmen eine zusätzliche Wirksamkeit über den gesetzlichen Standard des HQ_{100}-Schutzes hinaus in Betracht gezogen wird, um auf zukünftige Hochwasserereignisse besser vorbereitet zu sein. Ziel war es, nicht nur den aktuellen Schutzstandard zu erfüllen, sondern auch langfristige Lösungen zu entwickeln, die die Region an Inde und Vicht besser auf Hochwasserereignisse vorbereitet.

Die Erarbeitung der Maßnahmenvorschläge erfolgte durch einen interdisziplinär aufgestellten Kreis von Fachleuten. Durch die Beteiligung von Expertinnen und Experten aus verschiedenen Fachgebieten – von Wasserbau und Wasserwirtschaft über Stadt- und Regionalplanung bis hin zum Katastrophenschutz – konnten in Workshops und Fachtreffen konkrete und praxisnahe Vorschläge entwickelt werden. Diese interdisziplinäre Zusammenarbeit ermöglichte es, eine Vielzahl an Erfahrungen zu berücksichtigen und innovative Lösungen zu erarbeiten.

Die Einbindung wesentlicher Akteure spielte eine entscheidende Rolle im gesamten Prozess. Besondere Bedeutung kam der Beteiligung von Vertreterinnen und Vertretern der betroffenen Gemeinden zu, die mit ihren technischen Expertisen und detaillierten Ortskenntnissen wertvolle Informationen und Einblicke lieferten. Diese lokale Expertise trug maßgeblich dazu bei, mögliche Konfliktpotenziale und Einschränkungen bereits in einem frühen Stadium zu identifizieren und in die Maßnahmenvorschläge einzubeziehen. Auf diese Weise wurden in dem Masterplan nur Maßnahmenvorschläge aufgenommen, die zunächst als potenziell wirkungsvoll und umsetzbar gelten.

Die abschließende Bewertung und Abnahme der Ergebnisse sowie die strategische Ausrichtung der weiteren Untersuchungen bzgl. Wirksamkeit der vorgeschlagenen Maßnahmen wurde durch einen Lenkungsausschuss vorgenommen. Dieser setzte sich aus Vertreterinnen und Vertretern des Landesumweltministeriums NRW (MULNV), der Bezirksregierung Köln, des WVER, des IWW sowie Beauftragten der betroffenen Kreise und Gemeinden zusammen. In enger Abstimmung gewährleisteten sie eine qualitativ hochwertige und praxisorientierte Fortschreibung des Maßnahmenplans (WVER 2024).

12.1.2 Identifizierte potenzielle Hochwasserschutzmaßnahmen

Die im Masterplan erarbeiteten Maßnahmenvorschläge im Einzugsgebiet der Inde und Vicht gliedern sich in insgesamt zehn unterschiedliche Maßnahmenkategorien, die sich übergeordnet in die drei Hauptgruppen ***Technische Maßnahmen***, ***Natürliche Maßnahmen*** und ***Lenkungsmaßnahmen*** einordnen lassen. Diese Kategorien beinhalten:

Technische Maßnahmen

- Bau von Hochwasserrückhaltebecken
- Hochwasserangepasster Objektschutz
- Errichtung von Treibgutfallen

Natürliche Maßnahmen

- Gewässermaßnahmen
- Umsetzung von Renaturierungsmaßnahmen
- Flächenumnutzung „Raum für den Fluss“
- Hochwasserangepasste Flächenbewirtschaftung im Einzugsgebiet

Tab. 12.1 Identifizierte Hochwasserschutzmaßnahmen an Inde und Vicht. (Eigene Darstellung)

Maßnahmentyp	**Gebiet**			
	Inde		**Vicht**	
	Identifiziert	**Im Modell umgesetzt**	**Identifiziert**	**Im Modell umgesetzt**
Bau von Hochwasserrückhaltebecken	24	24	17	17
Hochwasserangepasster Objektschutz für Einzelobjekte (besonders gefährdete, besonders kritisch)	16	5	13	1
Errichtung von Treibgutfallen	0	0	7	0
Gewässermaßnahmen	0	0	12	7
Umsetzung von Renaturierungsmaßnahmen	28	12	11	9
Flächenumnutzung „Raum für den Fluss"	8	3	9	7
Hochwasserangepasste Flächenbewirtschaftung im Einzugsgebiet	5	0	4	0
Leistungsfähigkeitsanpassung von Brücken und Durchlässen	8	6	5	1
Gezielte Lenkung von Hochwasser	13	5	9	3
Anpassung von Ufermauern, Böschungen und Deichen	16	10	6	5
Gesamt	**118**	**65**	**93**	**50**

12

Lenkungsmaßnahmen

- Leistungsfähigkeitsanpassung von Brücken und Durchlässen
- Gezielte Lenkung von Hochwasser
- Anpassung von Ufermauern, Böschungen und Deichen

Der Masterplan beinhaltet Schutzmaßnahmen, die kurzfristig umsetzbar sind, wie bspw. Gewässeraufweitungen, aber auch Maßnahmen für die eine Machbarkeitsprüfung, wie etwa bei Hochwasserrückhaltebecken (HRB), durchgeführt werden muss. Zudem unterscheiden sich die Maßnahmen von großflächigen baulichen Veränderungen bis hin zu kleineren infrastrukturellen Anpassungen am Gewässerlauf. Die zeitliche und bauliche Varietät der Maßnahmen ist für einen ganzheitlichen Ansatz und für die Gesamtstrategie zum Hochwasserschutz im Einzugsgebiet von besonderer Bedeutung. Die Wirksamkeit dieser Maßnahmenvorschläge muss mithilfe von numerischen Simulationen untersucht werden. Aufgrund der komplexen Modellierungstechnik sollten die Potenziale der drei Hauptgruppen zunächst einzeln evaluiert werden, um eine detaillierte und umfassende Bewertung der Maßnahmen sicherzustellen.

Im Rahmen des Masterplans wurden insgesamt über 200 Maßnahmenvorschläge für das gesamte Einzugsgebiet der Inde und Vicht erarbeitet. Diese Maßnahmen konzentrieren sich hauptsächlich entlang der Hauptgewässer, erstrecken sich jedoch auch auf Nebengewässer und landwirtschaftlich bzw. forstwirtschaftlich genutzte Flächen. Im Teilgebiet der Vicht wurden etwa 90 Maßnahmen identifiziert, während entlang der Inde rund 120 Maßnahmenvorschläge entwickelt wurden (siehe Tab. 12.1). Der Masterplan wird fortgeführt und bleibt dynamisch, sodass die Anzahl an potenziellen Maßnahmenvorschlägen variiert. Dies ermöglicht eine flexible Anpassung an neue Erkenntnisse und Entwicklungen im Einzugsgebiet.

12.2 Methodik der Wirksamkeitsuntersuchung von Hochwasserschutzmaßnahmen

Ein zentrales Ziel des Masterplans war die Durchführung einer umfassenden Potenzialanalyse aller erarbeiteten Maßnahmen. Diese Analyse soll die Wirksamkeit jeder einzelnen Maßnahme ermitteln und deren Beitrag zur Steigerung des Hochwasserschutzes sowie der Hochwasserresilienz bewerten. Durch diese systematische Untersuchung können Ressourcen optimal eingesetzt und die effektivsten Lösungen priorisiert werden.

Die Wirksamkeitsuntersuchung folgt einem strukturierten Vorgehen in vier Bearbeitungsschritten (siehe ◻ Abb. 12.1), um die wirksamsten Maßnahmen aus hydrologischer und hydraulischer Sicht zu identifizieren. Der erste Schritt ist die hydrologische Modellierung (Niederschlag-Abfluss-Modellierung), die die Form der Hochwasserwellen im Einzugsgebiet analysiert. Durch die Simulation verschiedener Abflussszenarien wird ermittelt, wie sich Hochwasserrückhaltebecken (HRB) auf die Abflussspitzen und Wellenform auswirken. Die Ergebnisse liefern die hydrologischen Eingangsdaten für die hydraulischen Modelle als Abflusswellen.

Im Anschluss erfolgt die hydraulische Modellierung, die die Auswirkungen der simulierten Abflussszenarien auf den Wasserstand, die Überflutungsflächen und Fließgeschwindigkeiten untersucht. Diese Methode berücksichtigt auch die im Masterplan entwickelten natürlichen und Lenkungsmaßnahmen. So können präzise Vorhersagen über den lokalen Wasserstand bei verschiedenen Szenarien getroffen und fundierte Entscheidungen für den Hochwasserschutz sowie die Stärkung der Resilienz getroffen werden. Des Weiteren können kritische Bereiche identifiziert werden, in denen gezielte Maßnahmen erforderlich sind, um Überflutungen zu minimieren.

Ein weiterer Schritt ist die Schadenspotenzialanalyse, die den Zusammenhang zwischen reduzierten Überflutungsflächen und geringeren Schäden quantifiziert. Durch die Simulation verschiedener Szenarien werden potenzielle Schäden an Infrastrukturen, Gebäuden und landwirtschaftlichen Flächen abgeschätzt. Eine reduzierte Überflutungsfläche führt in der Regel zu geringeren Schäden, was positive wirtschaftliche Auswirkungen hat.

Abschließend kann eine Kosten-Nutzen-Analyse durchgeführt werden, die monetäre Schäden durch Hochwasserereignisse mit den Bau- und Betriebskosten der Maßnahmen vergleicht. So lässt sich ermitteln, welche Maßnahmen nicht nur technisch, sondern auch wirtschaftlich sinnvoll sind.

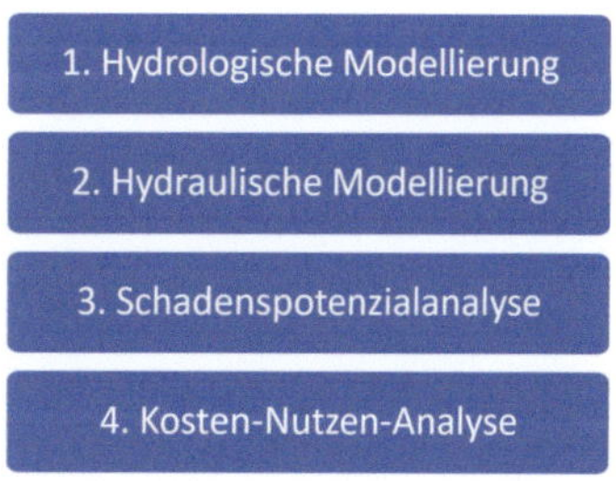

◻ **Abb. 12.1** Grundlegende Arbeitsschritte zur Wirksamkeitsuntersuchung von Hochwasserschutzmaßnahmen. (Eigene Darstellung)

Für die Ergebnisse, die im Folgenden vorgestellt werden (vgl. ► Abschn. 12.3 Potenzial der Maßnahmen), wurden die ersten beiden Bearbeitungsschritte durchgeführt.

12.2.1 Hydrodynamisch-numerische 2D-Simulation

Um die Wirksamkeit der identifizierten Maßnahmen aus dem Masterplan zu untersuchen, wurden für die Inde und die Vicht fluviale hydrodynamisch-numerische 2D-Modelle mit der Software HydroAs des Ingenieurbüros Hydrotec erstellt. Die Modellierung folgte den Standards für die 2D-hydraulische Modellierung von Fließgewässern für Hochwassergefahrenkarten (LfU Bayern 2018). Als Eingangsdaten dienten terrestrische Vermessungsdaten sowie LiDAR-Aufnahmen für die Flussbereiche, ergänzt durch das digitale Geländemodell (DGM) von Nordrhein-Westfalen und ALKIS- bzw. ATKIS-Daten aus dem GEOportal.NRW für das Umland.

Für die Vicht wurde ein Gesamtmodell mit einer Lauflänge von 23 km erstellt, das vom Grölisbach bei Petergensfeld bis zur Einmündung in die Inde bei Stolberg reicht und eine Fläche von 13,17 km^2 abbildet, was 12,7 % des Einzugsgebiets der Vicht (104 km^2) ausmacht. Die Inde wurde zur Reduzierung der Berechnungszeit in zwei Modelle unterteilt. Das Modell der oberen Inde erstreckt sich von Schmithof (Flusskilometer 48,7) bis Stolberg (Flusskilometer 27,7). Das Modell der unteren Inde umfasst den Abschnitt von Stolberg bis zur Einmündung in die Rur bei Kirchberg. Zusammen bilden die Teilmodelle der Inde eine Fläche von 26,43 km^2 ab, dies entspricht 7,1 % des Einzugsgebiets der Inde (374 km^2).

Für die Potenzialanalyse und den Vergleich der Wirksamkeit der einzelnen Maßnahmen wurden verschiedene Berechnungsnetze erstellt. Der Referenzzustand war jeweils das Netz des Ist-Zustands. Weitere Netzvarianten beinhalteten den Ist-Zustand plus natürliche Maßnahmen, Ist-Zustand plus Lenkungsmaßnahmen, Ist-Zustand plus alle Maßnahmen sowie Ist-Zustand ohne Brücken. Alle Berechnungsnetze wurden mit verschiedenen Abflussszenarien simuliert. Die Abflussszenarien setzen sich aus dem HQ_{100}, HQ_{1000}, HQ_{100_HRB28}, HQ_{1000_HRB28}, $HQ_{100\_HRB\_Potenzial}$ und $HQ_{1000\_HRB\_Potenzial}$ zusammen. Bei den Szenarien HQ_{100_HRB28} und HQ_{1000_HRB28} wurde das Retentionsvolumen der geplanten Hochwasserrückhaltebecken (HRB) an der Vicht berücksichtigt, während $HQ_{100\_HRB\_Potenzial}$ und $HQ_{1000\_HRB\_Potenzial}$ das Potenzial aller im Masterplan identifizierten HRB darstellt.

Für die hydraulische Simulation der Maßnahmen wurde eine fluviale Modellierung angewendet, bei der nur das Hauptgewässer abgebildet wurde. Die Abflüsse der Nebengewässer wurden durch die hydrologische Modellierung berechnet und als Abflusswellen ins Hauptgerinne eingespeist. Diese Vorgehensweise ermöglichte eine fokussierte Analyse des Hauptflusses, ohne durch Nebengewässer zusätzliche Komplexität zu erzeugen.

Aufgrund der modelltechnischen Einschränkungen konnten nicht alle Maßnahmen, wie in ◘ Tab. 12.1 dargestellt, aus dem Masterplan in den Modellen berücksichtigt werden. Einige Maßnahmen, wie die Treibgutfallen, wurden wegen der Beschränkung auf das Wasser, das ohne treibende Gegenstände simuliert wird, nicht berücksichtigt. Andere Maßnahmen lagen außerhalb des Modellgebiets oder konnten nur indirekt berücksichtigt werden, wie bei den HRB, die im Modell durch die Änderung der Abflusswellen als hydrologische Eingangsdaten berücksichtigt wurden. Insgesamt konnten 115 der identifizierten Maßnahmen (54,5 %) im Modell umgesetzt werden, davon 65 an der Inde (55,1 %) und 50 an der Vicht (53,8 %).

12.2.2 Ermittlung des Potenzials der identifizierten Maßnahmen

Das Potenzial der verschiedenen Hochwasserschutzmaßnahmen wurde anhand einer Vielzahl von Parametern untersucht und mit dem Ist-Zustand (Referenzzustand) verglichen. Dabei wurden sowohl quantitative als auch grafische Auswertungen herangezogen. Zu den quantitativen Parametern gehörten:

- Fläche des Überschwemmungsgebiets
- Fläche des bebauten Überschwemmungsgebiets
- Betroffene Gebäude
- Überströmte Brücken
- Maximalabfluss (Q_{max}) an der Mündung bzw. nach der Maßnahme
- Anlaufzeiten von MQ zu Q_{krit} und von Q_{krit} zu Q_{max}

Zusätzlich wurden graphische Auswertungen wie Abflussganglinien, Längs- und Querschnitte sowie Kartenmaterial (Überschwemmungsgebiete, Fließgeschwindigkeiten, Fließwege) genutzt, um das Potenzial der Maßnahmen umfassend zu bewerten.

Die Fläche des Überschwemmungsgebiets und die Anzahl der betroffenen Gebäude sind zentrale Parameter, die direkte Hinweise auf das Schadenspotenzial der verschiedenen Abflussszenarien geben. Diese Parameter verdeutlichen, wie effektiv die identifizierten Maßnahmen das Schadenspotenzial verringern können. Ein weiterer wichtiger Indikator ist der Maximalabfluss (Q_{max}) an der Mündung des untersuchten Flusses und unterhalb der jeweiligen Hochwasserschutzmaßnahme. Eine Senkung des Maximalabflusses reduziert die Hochwassergefahr signifikant.

Die Anlaufzeiten geben an, wie schnell der Hochwasserscheitel ansteigt und sind besonders wichtig für die Einsatzkräfte, die eine gewisse Vorwarn- und Reaktionszeit benötigen, um auf Notfallsituationen bestmöglich reagieren zu können. Längere Anlaufzeiten ermöglichen eine bessere Vorbereitung und effektive Einsatzmaßnahmen.

Abflussganglinien sowie Längs- und Querschnitte wurden zusätzlich verwendet, um den zeitlichen Verlauf der Hochwasserwelle und den Wasserstand anhand der topographischen Gegebenheiten detailliert zu analysieren und den Einfluss der Schutzmaßnahmen zu bestimmen.

Diese umfassende Herangehensweise ermöglicht eine gründliche Bewertung der Wirksamkeit und Effizienz der unterschiedlichen Hochwasserschutzmaßnahmen.

12.3 Das Potenzial der Maßnahmen

In den ◘ Tab. 12.2 und 12.3 sind die gekürzten Ergebnisse der Potenzialanalyse an der unteren Inde und an der Vicht dargestellt. Die vollständigen Ergebnisse sind zu umfangreich, um sie lückenlos darzustellen. Die angezeigten Ergebnisse spiegeln jedoch die allgemeinen Erkenntnisse wider, die im Folgenden näher erläutert werden.

Im Ist-Zustand bei einem HQ_{100}-Abflussszenario ist das Überschwemmungsgebiet (ÜSG) an der **Vicht** 1,29 km^2 groß, der Maximalabfluss an der Mündung beträgt etwa 94 m^3/s und 1482 Gebäude sind vom Hochwasser betroffen (siehe ◘ Tab. 12.2). Im Vergleich zum Referenzzustand führen sowohl die Lenkungs- als auch die natürlichen Maßnahmen zu keiner signifikanten Verbesserung des Hochwasserschutzes. Die Überschwemmungsfläche vergrößert sich leicht auf 1,31 km^2, und der Maximalabfluss bleibt mit etwa 93 m^3/s nahezu unverändert. Dies ist auf die gezielte Wasserlenkung durch

Tab. 12.2 Matrixtabelle der Potenzialanalyse an der Vicht für HQ_{100} und HQ_{1000}-Szenarien. (Eigene Darstellung)

Maßnahme	HQ-Szenario	ÜSG [km²]	Δ ÜSG		betroffene Gebäude	Δ betroffene Gebäude		Q_{max} Mündung	Δ Q_{max} Mündung	
			[km²]	[%]		n = [-]	[%]	[m³/s]	[m³/s]	[%]
Ist-Zustand		1,29	-	-	1.482	-	-	93,88	-	-
Lenkungsmaßnahmen	HQ_{100}	1,31	+ 0,02	+ 1,55	1.473	- 9	- 0,61	92,79	- 1,09	- 1,16
Natürliche Maßnahmen		1,31	+ 0,02	+ 1,55	1.426	- 56	- 3,78	93,20	- 0,68	- 0,72
Ist-Zustand + HRB28		0,49	- 0,80	- 62,02	115	- 1.367	- 92,24	63,41	- 30,47	- 32,46
Lenkungsmaßnahmen	HQ_{100_HRB28}	0,50	+ 0,01	+ 2,00	161	+ 46	+ 40,00	63,13	- 0,28	- 0,44
Natürliche Maßnahmen		0,59	+ 0,10	+ 20,41	121	+ 6	+ 5,22	61,71	- 1,70	- 2,68
Ist-Zustand		1,93	-	-	2.157	-	-	120,74	-	-
Lenkungsmaßnahmen	$HQ_{1.000}$	1,94	+ 0,01	+ 0,52	2.168	+ 11	+ 0,51	116,80	- 3,94	- 3,26
Natürliche Maßnahmen		1,93	+/- 0	+/- 0	2.092	- 65	- 3,01	115,16	- 5,58	- 4,62
Ist-Zustand + HRB28		1,62	- 0,31	- 16,06	1.787	- 370	- 17,15	100,17	- 20,57	- 17,04
Lenkungsmaßnahmen	$HQ_{1.000\_HRB28}$	1,63	+ 0,01	+ 0,62	1.809	+ 22	+ 1,23	98,58	- 1,59	- 1,59
Natürliche Maßnahmen		1,63	+ 0,01	+ 0,62	1.721	- 66	- 3,69	98,04	- 2,13	- 2,13

Tab. 12.3 Matrixtabelle der Potenzialanalyse an der unteren Inde für HQ_{100} und HQ_{1000}-Szenarien. (Eigene Darstellung)

Maßnahme	HQ-Szenario	ÜSG [km²]	Δ ÜSG		betroffene Gebäude	Δ betroffene Gebäude		Q_{max} Mündung	Δ Q_{max} Mündung	
			[km²]	[%]		n = [-]	[%]	[m³/s]	[m³/s]	[%]
Ist-Zustand		1,29	-	-	1.482	-	-	93,88	-	-
Lenkungsmaßnahmen	HQ_{100}	1,31	+ 0,02	+ 1,55	1.473	- 9	- 0,61	92,79	- 1,09	- 1,16
Natürliche Maßnahmen		1,31	+ 0,02	+ 1,55	1.426	- 56	- 3,78	93,20	- 0,68	- 0,72
Ist-Zustand + HRB28		0,49	- 0,80	- 62,02	115	- 1.367	- 92,24	63,41	- 30,47	- 32,46
Lenkungsmaßnahmen	HQ_{100_HRB28}	0,50	+ 0,01	+ 2,00	161	+ 46	+ 40,00	63,13	- 0,28	- 0,44
Natürliche Maßnahmen		0,59	+ 0,10	+ 20,41	121	+ 6	+ 5,22	61,71	- 1,70	- 2,68
Ist-Zustand		1,93	-	-	2.157	-	-	120,74	-	-
Lenkungsmaßnahmen	$HQ_{1.000}$	1,94	+ 0,01	+ 0,52	2.168	+ 11	+ 0,51	116,80	- 3,94	- 3,26
Natürliche Maßnahmen		1,93	+/- 0	+/- 0	2.092	- 65	- 3,01	115,16	- 5,58	- 4,62
Ist-Zustand + HRB28		1,62	- 0,31	- 16,06	1.787	- 370	- 17,15	100,17	- 20,57	- 17,04
Lenkungsmaßnahmen	$HQ_{1.000\_HRB28}$	1,63	+ 0,01	+ 0,62	1.809	+ 22	+ 1,23	98,58	- 1,59	- 1,59
Natürliche Maßnahmen		1,63	+ 0,01	+ 0,62	1.721	- 66	- 3,69	98,04	- 2,13	- 2,13

12

Hochwasserschutzmauern und die ungehinderte Ausbreitung des Wassers in Auenbereiche durch natürliche Maßnahmen zurückzuführen. Der Schutz einiger weniger Gebäude konnte erreicht werden, jedoch bleibt der Unterschied beim Maximalabfluss marginal. Ein ähnliches Bild ergibt sich im HQ_{1000}-Szenario (siehe Tab. 12.2), bei dem die Überschwemmungsfläche bei 1,93 bzw. 1,94 km^2 liegt und durch die natürlichen Maßnahmen 3 % der betroffenen Gebäude geschützt werden können. Der Maximalabfluss wird leicht abgesenkt, jedoch ist der Effekt der Maßnahmen begrenzt.

Ein deutlicherer Unterschied im Vergleich des HQ_{100}-Szenarios ist zwischen dem Referenzzustand und den Lenkungsmaßnahmen und den natürlichen Maßnahmen an der **unteren Inde** zu erkennen (siehe ◘ Tab. 12.3). Hier haben die Lenkungsmaßnahmen mit einer Abnahme der ÜSG um 5 % von 3,12 km^2 auf 2,96 km^2 und mit 25 % weniger betroffenen Gebäuden eine größere Wirkung. Auch die natürlichen Maßnahmen, obwohl sich die Überschwemmungsfläche um über 4 % auf 3,25 km^2 vergrößert, können mit knapp 17 % zusätzlich geschützten Gebäuden potenziell zum verbesserten Hochwasserschutz beitragen. Für das HQ_{1000}-Szenario nimmt die Wirkung der natürlichen Maßnahmen fast vollständig ab. Die Lenkungsmaßnahmen tragen hingegen weiter zum Hochwasserschutz bei, in dem die ÜSG um mehr als 10 % verringert wird und etwa 13 % weniger Gebäude betroffen sind als im Referenzzustand. Der Maximalabfluss in Eschweiler bleibt jedoch bei allen Abflussszenarien nahezu identisch.

Der Unterschied im Potenzial der Lenkungsmaßnahmen und der natürlichen Maßnahmen an der unteren Inde und an der Vicht ist geographisch zu erklären. Die Vicht befindet sich mit dem gesamten Flussverlauf im Mittelgebirge der Eifel und besitzt ein durchgehend enges Tal mit steilen Hängen. Zudem ist das Vichttal ab der Ortslage Vicht stark urbanisiert. Dadurch konnten nur vergleichsweise kleine Lenkungsmaßnahmen und natürliche Maßnahmen identifiziert werden, die aufgrund des limitierten Platzangebotes eine geringe Wirkung entfalten können. Im Vergleich dazu befindet sich die untere Inde ab Eschweiler im nördlichen Randbereich der Eifel, im Tiefland der Jülich-Zülpicher Börde. Hier ist der Fluss nicht durch ein enges Tal eingegrenzt und kann sich zu beiden Uferseiten ausbreiten. Daher konnten die Lenkungsmaßnahmen und natürlichen Maßnahmen an der unteren Inde großzügiger und effektiver gestaltet werden, wodurch auch das Potenzial der Maßnahmen zunimmt.

In Abflussszenarien, in welchen die Hochwasserrückhaltebecken in Rott und Mulartshütte an der Vicht mit einem Gesamtvolumen von 1,14 Mio. m^3 aktiv sind, zeigen sich sowohl an der Vicht als auch an der unteren Inde signifikante Verbesserungen. Das Potenzial solch großer Rückhalteräume ist im Vergleich zu den Lenkungsmaßnahmen und den natürlichen Maßnahmen nicht so stark von der geographischen Lage der Flussverläufe abhängig. Das Überschwemmungsgebiet an der **Vicht** reduziert sich bei einem HQ_{100} auf 0,49 km^2, was eine Reduktion von 62 % im Vergleich zum Ist-Zustand bedeutet. Zudem werden 1367 Gebäude (92 % der betroffenen Gebäude) geschützt, und der Maximalabfluss wird um ein Drittel auf etwa 63 m^3/s gesenkt. Diese HRB ermöglichen einen HQ_{100}-Schutz für Stolberg, wodurch Überschwemmungen im städtischen Bereich vermieden werden. Auch bei einem HQ_{1000} zeigt sich eine deutliche Verbesserung, obwohl die HRB nur für ein HQ_{100} dimensioniert sind. Das Überschwemmungsgebiet verringert sich auf 1,62 km^2, 370 Gebäude (17 %) werden geschützt, und der Maximalabfluss sinkt auf 100 m^3/s. Die Lenkungs- und natürlichen Maßnahmen haben an der Vicht in diesem Szenario keinen zusätzlichen Effekt.

Die zwei großen Hochwasserrückhaltebecken in Rott und Mulartshütte haben auch an der **unteren Inde** noch eine signifikante Wirkung. Bei einem HQ_{100} wird die ÜSG um knapp 28 % reduziert und 910 Gebäude (75 %) werden im Vergleich zum Referenzzustand geschützt. Der Maximalabfluss in Eschweiler wird um mehr als 20 % von 123 m^3/s auf 95 m^3/s gesenkt. Dieser Effekt ist auch beim HQ_{1000}-Szenario zu beobachten. Die Überschwemmungsfläche nimmt von 6,15 auf 4,78 km^2 um 22 % ab. Mehr als 30 % der

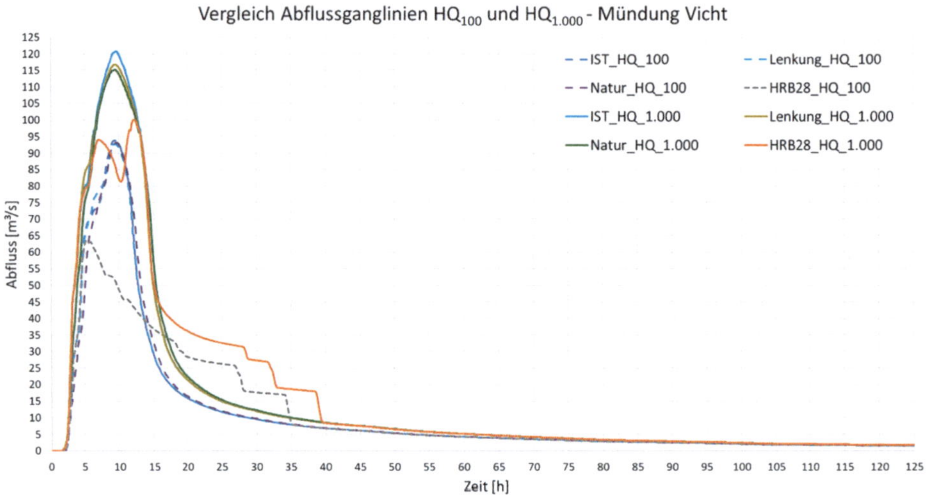

Abb. 12.2 Abflussganglinien für die Abflussszenarien HQ_{100} und HQ_{1000} an der Mündung der Vicht. (Eigene Darstellung)

12

betroffenen Gebäude können geschützt werden und der Maximalabfluss nimmt erneut um etwa 22 % ab. Eine weitere Verbesserung des Hochwasserschutzes kann mithilfe der Lenkungsmaßnahmen und natürlichen Maßnahmen erzielt werden. Bei einem HQ_{100}-Szenario können 113 bis 169 bzw. bei einem HQ_{1000} 350 bis 452 zusätzliche Gebäude geschützt werden. Jedoch haben auch hier die Lenkungsmaßnahmen und die natürlichen Maßnahmen keine signifikante Wirkung auf den Maximalabfluss. Dies unterstützt die Beobachtung, dass solche Maßnahmen an Flüssen im Tiefland ein höheres Potenzial besitzen als an Flüssen im Mittelgebirge.

Die im Masterplan identifizierten Potenziale der Hochwasserrückhaltebecken (nicht in den Ergebnistabellen angegeben) zeigen noch größere Verbesserungen hinsichtlich des Hochwasserschutzes. Am Beispiel der Vicht konnten mit 15 weiteren wirkungsvollen HRB das Rückhaltevolumen deutlich gesteigert werden. Im HQ_{100}-Szenario wird die Überschwemmungsfläche um weitere 20 % auf 0,39 km^2 reduziert, und nahezu alle Gebäude entlang der Vicht (bis auf 59) werden geschützt. Der Maximalabfluss sinkt auf unter 60 m^3/s. Auch im HQ_{1000}-Szenario kommt es zu einer erheblichen Verbesserung: Das Überschwemmungsgebiet wird auf 0,63 km^2 verringert, und 80 % der betroffenen Gebäude können geschützt werden. Der Maximalabfluss sinkt um ein Viertel auf etwa 75 m^3/s.

Allerdings erwiesen sich nicht alle im Masterplan vorgeschlagenen Standorte als wirkungsvoll. Die hydrologische Potenzialstudie zeigte, dass es an einigen Stellen durch das Zurückhalten der Hochwasserwellen zu ungünstigen Überlagerungen des Durchflusses kam, was in bestimmten Bereichen die Überflutungssituation sogar verschärfte. Diese Effekte verdeutlichen die Komplexität von Hochwasserschutzplanungen und unterstreichen die Notwendigkeit einer präzisen Analyse, um negative Auswirkungen auf benachbarte Bereiche zu vermeiden.

Die Erkenntnisse aus den Ergebnistabellen werden auch bei den graphischen Abflussganglinien der Vicht für das HQ_{100}-Szenario und HQ_{1000}-Szenario bestätigt (siehe Abb. 12.2). Die Abflussganglinien für die verschiedenen Szenarien mit Hochwasser-

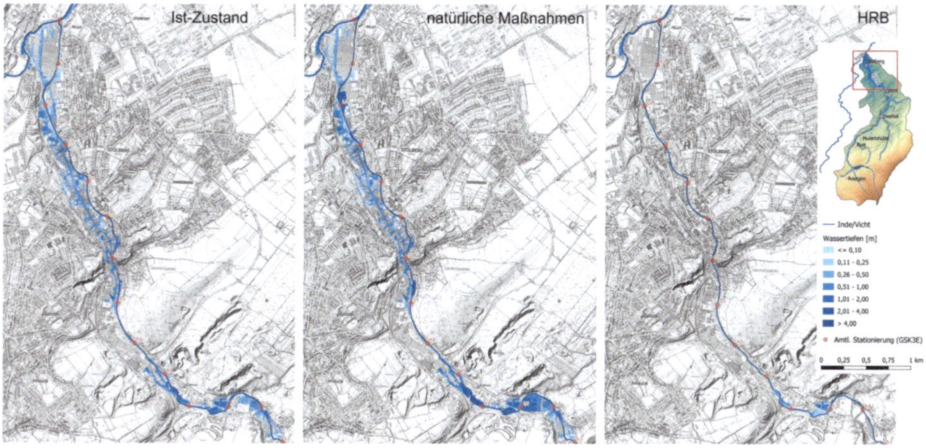

Abb. 12.3 Vergleich von Überflutungsflächen im Bereich der Stadt Stolberg. (Eigene Darstellung; Datengrundlage: DGM1 NRW (Bezirksregierung Köln 2022a), GSK3E (OpenGeodata.NRW 2022), DGK5 (Bezirksregierung Köln 2022b), ► https://www.govdata.de/dl-de/zero-2-0. Alle Rechte vorbehalten)

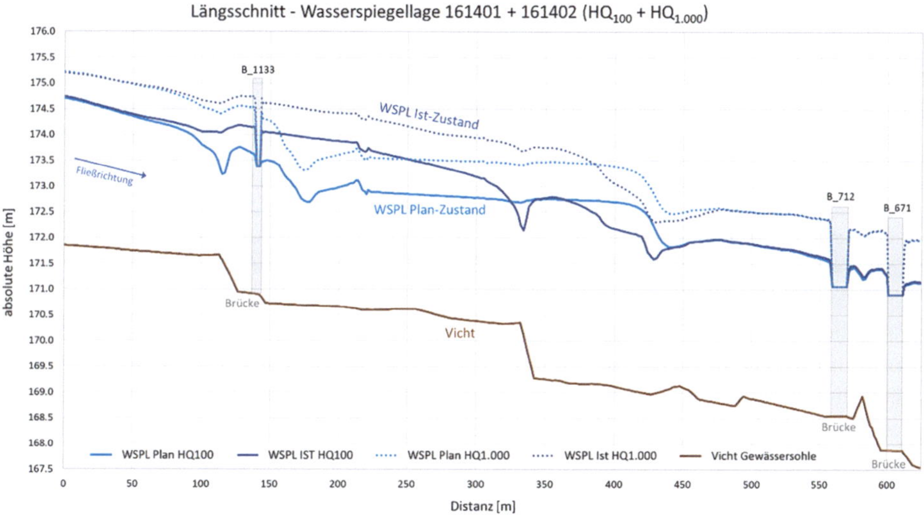

Abb. 12.4 Längsschnitt und Wasserspiegellage für die natürlichen Maßnahmen 161401 und 161402. (Eigene Darstellung)

schutzmaßnahmen verlaufen nahezu identisch, wobei die Szenarien mit den HRB in Rott und Mulartshütte eine Dämpfung der Hochwasserwelle bewirken, was zu einer veränderten Form und einem anderen Verlauf der Welle führt. Dies zeigt, dass die HRB den Flussverlauf unterstrom signifikant beeinflussen, während die Lenkungs- und natürlichen Maßnahmen nur eine begrenzte lokale Wirkung haben.

In Abb. 12.3 sind die Überflutungsflächen des Ist-Zustands, der natürlichen Maßnahmen und der zwei Hochwasserrückhaltebecken für den Bereich von Stolberg vergleichend gegenübergestellt. Es ist ersichtlich, dass zwischen dem Ist-Zustand und den natürlichen Maßnahmen kaum Unterschiede bestehen, außer in den Bereichen, in denen die identifizierten Hochwasserschutzmaßnahmen lokal wirken. Die HRB lassen einen deutlich sichtbaren Unterschied in den Überflutungsflächen erkennen, wobei die HRB das Überfluten des städtischen Bereichs von Stolberg verhindern, was den Hochwasserschutz deutlich stärkt.

Abschließend wird in ◘ Abb. 12.4 ein Längsschnitt einer natürlichen Maßnahme an der Vicht betrachtet. Bei den Maßnahmen 161401 und 161402 handelt es sich um eine Flächenumnutzung im Sinne des Raums für den Fluss. Für die Umsetzung wurde die Landnutzung geändert und die Flächen zur Reaktivierung der Aue abgesenkt. Die natürliche Maßnahme hat eine Länge von etwa 350 m, dies ist auch der Wirkungsbereich der Maßnahme, in dem der Wasserstand im Gerinne abgesenkt wird. Ober- und unterhalb der Maßnahme ist kein Effekt zu erkennen. Dadurch wird bestätigt, dass die Lenkungsmaßnahmen und die natürlichen Maßnahmen einen nur lokal begrenzten Wirkungsbereich haben und somit als ergänzende Maßnahmen neben den HRB zur Verbesserung des Hochwasserschutzes und der Resilienz gegen Hochwasser zu betrachten sind.

12.3.1 Fazit zum Potenzial der Maßnahmentypen

Die Potenzialanalyse zeigt, dass Hochwasserschutzmaßnahmen unterschiedlich stark zur Verbesserung des Hochwasserschutzes und der Resilienz gegenüber Hochwasserereignissen beitragen.

Hochwasserrückhaltebecken besitzen das größte Potenzial. An der Vicht ermöglichen zwei HRB an den Standorten Rott und Mulartshütte im Hauptlauf einen HQ_{100}-Schutz für die Stadt Stolberg im Unterlauf. Zudem wird der Hochwasserschutz an der unteren Inde signifikant gesteigert. HRB zeichnen sich durch ihren weitreichenden Wirkungsbereich aus und können je nach Anzahl und Volumen die Hochwassergefahr im gesamten Einzugsgebiet erheblich reduzieren. Durch die Speicherung von Wasser während eines Hochwasserereignisses wird die Hochwasserwelle effektiv abgeflacht, was sowohl die Wassertiefen als auch die Überschwemmungsflächen verringert und das Schadenspotenzial mindert. Langfristig führt die Verfügbarkeit von HRB zu einer Reduzierung der Überflutungswahrscheinlichkeit und einer gesteigerten Sicherheit für Anwohner und Anwohnerinnen sowie der Infrastruktur. Des Weiteren kann die Hochwasserresilienz weiter gesteigert werden, indem weitere HRB an Nebenflüssen oder im Hauptfluss selbst errichtet werden oder HRB mit einer größeren Dimensionierung als für ein HQ_{100}-Szenario gebaut werden. Diese müssen jedoch hydrologisch aufeinander abgestimmt werden, um eine Überlagerung von Abflusswellen und damit einer Verschärfung der Hochwassersituation im Einzugsgebiet zu verhindern.

Im Gegensatz dazu haben Lenkungsmaßnahmen und natürliche Maßnahmen eine lokal begrenzte Wirkung. Lenkungsmaßnahmen verändern den Fließweg und halten Wasser von schützenswerten Bereichen fern, können aber keine Wassermengen zurückhalten oder Abflüsse verringern. Natürliche Maßnahmen verbessern das Gewässerprofil und schaffen Raum für den Fluss, wodurch die Leistungsfähigkeit des Gewässers gesteigert wird. Ihre Wirkung ist jedoch stark auf den Bereich der Maßnahme begrenzt und hängt von der räumlichen Ausdehnung ab. Diese Maßnahmen sind besonders bei kleineren Hochwasserereignissen wirksam und haben bei extremen Abflüssen nur begrenzte Effekte. Wobei das Potenzial dieser Maßnahmen für einen Fluss im Tiefland größer ist als im Mittelgebirge, wo der Raum für solche Maßnahmen stark begrenzt ist. Sie eignen sich vor allem zur gezielten Absicherung von Siedlungsbereichen, Grundstücken und Gebäuden, tragen jedoch nicht wesentlich zur großflächigen Reduktion des Hochwasserrisikos bei. Vor allem für kritische Infrastrukturen sind dies geeignete Maßnahmen, um das Schutzziel zu erhöhen.

Die Potenzialanalyse verdeutlicht, dass eine Kombination verschiedener Maßnahmentypen erforderlich ist, um sowohl großflächige als auch lokale Schutzwirkungen optimal zu nutzen und darüber hinaus die Resilienz gegenüber Hochwasserereignissen zu stärken.

Hinweis für Entscheidungsprozesse

Während Hochwasserrückhaltebecken entscheidend für den allgemeinen Hochwasserschutz sind, ergänzen Lenkungs- und natürliche Maßnahmen diesen Schutz durch gezielte lokale Eingriffe.

12.4 Zusammenfassung

Das Hochwasserereignis im Juli 2021 verursachte im Einzugsgebiet der Flüsse Inde und Vicht in Nordrhein-Westfalen erhebliche Schäden an der Infrastruktur und den betroffenen Gemeinden, darunter Eschweiler und Stolberg. In Reaktion auf dieses extreme Ereignis wurde der Masterplan „Hochwasserresiliente Stadt- und Gebietsentwicklung für das Einzugsgebiet von Inde und Vicht“ entwickelt. Dieser Plan entstand in Zusammenarbeit zwischen dem Wasserverband Eifel-Rur (WVER) und dem Institut für Wasserbau und Wasserwirtschaft (IWW) der RWTH Aachen sowie weiteren relevanten Akteurinnen und Akteuren aus verschiedenen Fachrichtungen. Ziel ist es, die Widerstandsfähigkeit gegenüber künftigen Hochwasserereignissen zu stärken.

Der Masterplan identifiziert insgesamt über 200 potenzielle Hochwasserschutzmaßnahmen für die Flüsse Inde und Vicht. Die Maßnahmenvorschläge wurden in drei Hauptkategorien unterteilt: technische, natürliche und Lenkungsmaßnahmen. Der Masterplan bleibt dynamisch, sodass neue Maßnahmen bei Bedarf in das Konzept integriert werden können, um flexibel auf zukünftige Entwicklungen im Einzugsgebiet zu reagieren. Zur Untersuchung der Wirksamkeit von Hochwasserschutzmaßnahmen wurden Potenzialanalysen durchgeführt, bei der Niederschlags-Abfluss-Modelle, hydrodynamisch-numerische 2D-Modelle und Schadenspotenzialmodelle zum Einsatz kamen.

Besondere Bedeutung kommt den Hochwasserrückhaltebecken zu, die als die effektivsten Maßnahmen zur Reduktion von Hochwasserspitzen identifiziert wurden. Die geplanten Rückhaltebecken in Rott und Mulartshütte sollen bis 2028/29 fertiggestellt werden und ermöglichen es, die Überflutungsflächen in Stolberg um bis zu 62 % zu verkleinern und den Maximalabfluss um ein Drittel zu reduzieren. Zudem wird der Hochwasserschutz an der unteren Inde signifikant verbessert. Diese Maßnahmen bieten einen langfristigen Schutz und sind entscheidend für die Umsetzung des HQ_{100}-Schutzes.

Neben den technischen Maßnahmen spielen auch natürliche und Lenkungsmaßnahmen eine wichtige Rolle für die Erhöhung der Resilienz. Diese wirken lokal und können gezielt Grundstücke, Gebäude oder kritische Infrastruktur schützen.

Zusammenfassung

- Der Masterplan „Hochwasserresiliente Stadt- und Gebietsentwicklung für das Einzugsgebiet von Inde und Vicht" identifiziert insgesamt über 200 potenzielle Hochwasserschutzmaßnahmen für die Flüsse Inde und Vicht. Die Maßnahmenvorschläge wurden in drei Hauptkategorien unterteilt: technische, natürliche und Lenkungsmaßnahmen.
- Der Masterplan bleibt dynamisch, sodass neue Maßnahmen bei Bedarf in das Konzept integriert werden können, um flexibel auf zukünftige Entwicklungen im Einzugsgebiet zu reagieren.
- Zur Untersuchung der Wirksamkeit von Hochwasserschutzmaßnahmen wurden Potenzialanalysen durchgeführt, bei denen Niederschlags-Abfluss-Modelle, hydrodynamisch-numerische 2D-Modelle und Schadenspotenzialmodelle zum Einsatz kamen.
- Besondere Bedeutung kommt den Hochwasserrückhaltebecken zu, die als die effektivsten Maßnahmen zur Reduktion von Hochwasserspitzen identifiziert wurden.
- Neben den technischen Maßnahmen spielen auch natürliche und Lenkungsmaßnahmen eine wichtige Rolle für die Erhöhung der Resilienz. Diese wirken lokal und können gezielt Grundstücke, Gebäude oder kritische Infrastruktur schützen.

Danksagung Die Autorinnen und Autoren von ▶ Kap. 12 bedanken sich beim Bundesministerium für Bildung und Forschung (BMBF) und beim DLR Projektträger für die Unterstützung und Förderung des KAHR-Projekts. Diese Forschung wurde vom BMBF unter dem Förderkennzeichen 01LR2102A-M finanziert.

Literatur

Bezirksregierung Köln (2022a) Digitales Geländemodell. https://www.bezreg-koeln.nrw.de/geobasis-nrw/produkte-und-dienste/hoehenmodelle/digitale-gelaendemodelle/digitales-gelaendemodell. Zugegriffen: 24. Okt. 2025

Bezirksregierung Köln (2022b) Deutsche Grundkarte. https://www.bezreg-koeln.nrw.de/geobasis-nrw/produkte-und-dienste/topographische-karten/historische-topographische-karten/1937-2016. Zugegriffen: 24. Okt. 2025

Krings S (Hrsg) (2020) 10 Jahre „KRITIS-Strategie". Einblicke in die Umsetzung der Nationalen Strategie zum Schutz Kritischer Infrastrukturen. Deutschland. Stand: Februar 2020. Praxis im Bevölkerungsschutz, 21. Bundesamt für Bevölkerungsschutz und Katastrophenhilfe, Bonn

LfU Bayern – Landesamt für Umwelt Bayern (2018) Handbuch hydraulische Modellierung – Vorgehensweisen und Standards für die 2-D-hydraulische Modellierung von Fließgewässern in Bayern. Stand: Januar 2018. Pauli Offsetdruck, Oberkotzau

OpenGeodata.NRW (2022) Gewässerstationierungskarte. https://www.opengeodata.nrw.de/produkte/umwelt_klima/wasser/oberflaechengewaesser/gsk3e/; Informationen unter https://www.geoportal.nrw/?activetab=map#/datasets/iso/086e4f05-758c-4826-b3e6-33e82fdd4cba. Zugegriffen: 24. Okt. 2025

WVER – Wasserverband Eifel-Rur (2024) Hochwasserschutzkonzept an Inde und Vicht. https://www.hochwassergefahrenvorbeugen.de/inde-und-vicht/. Zugegriffen: 20. Sept. 2024

Wie die Bewältigung von Hochwasserereignissen gelingen kann – Optimierungsansätze im operativen Hochwasserschutz

Robert Jüpner, Selina Schaum, Max Beining und Alexandra Schüller

J. Birkmann et al. (Hrsg.), *Die Flutkatastrophe 2021*, Sustainable Development Goals (SDG) – Umsetzung in Praxis, Lehre und Entscheidungsprozessen,
https://doi.org/10.1007/978-3-658-48688-4_13

13.1 Einführung

Die Anpassung an die Folgen des Klimawandels stellt für die europäischen Staaten ein gemeinsames und prioritäres Ziel dar. Im Hinblick auf die Gefährdung durch Starkregen und Hochwasser liegt der Schwerpunkt dabei vorrangig auf Vorsorgemaßnahmen (Patt und Jüpner 2020). Diese Ausrichtung ist sinnvoll und zielführend, dennoch erfordert ein ganzheitlicher Ansatz auch die Berücksichtigung von Maßnahmen des operativen Hochwasserschutzes. Diese Maßnahmen bieten ein erhebliches Potenzial zur Schadensminderung während eines Ereignisses und sind insbesondere unerlässlich, um gefährdete Personen zu schützen und ihre Rettung sicherzustellen. Im Kreislauf des Hochwasserrisikomanagements, wie ihn die Bund/Länder-Arbeitsgemeinschaft Wasser (LAWA) formuliert hat (LAWA 2020), findet dieser Aspekt bislang nur geringe Beachtung (Schüller und Jüpner 2021).

Bis 2021 existierte keine einheitliche Begriffsdefinition für den operativen Hochwasserschutz. Schüller und Jüpner, 2021 haben diesen Begriff daher wie folgt definiert:

> „Der operative Hochwasserschutz umfasst alle einsatzvorbereitenden und durchführenden Planungen und Maßnahmen an der Schnittstelle zwischen Katastrophenschutz und Wasserwirtschaft mit dem Ziel, über die rein wasserwirtschaftliche Hochwasservorsorge hinaus Risiken vorzubeugen und Schäden durch Hochwasser und Starkregen zu reduzieren."

Die Autoren empfehlen, den operativen Hochwasserschutz nicht ausschließlich auf die Phase der Ereignisbewältigung zu konzentrieren, sondern für eine erfolgreiche und effektive Hochwasserbewältigung die „operative Hochwasservorsorge" und „operative Hochwassernachsorge" einzubeziehen. Am Beispiel der Hochwasservorhersage wird diese dreigliedrige Struktur in Abb. 13.1 dargestellt. Nicht nur während der Bewältigung können dabei konkrete Optimierungspunkte abgeleitet werden, auch in den Bereichen der Vorsorge und Nachsorge lassen sich entsprechende Ansätze identifizieren. So trägt die vorbereitende Erstellung komplexer Hochwasserszenarien und Datengrundlagen maßgeblich zu einem präzisen Lagebild während der Hochwasserbewältigung bei. Eine

13

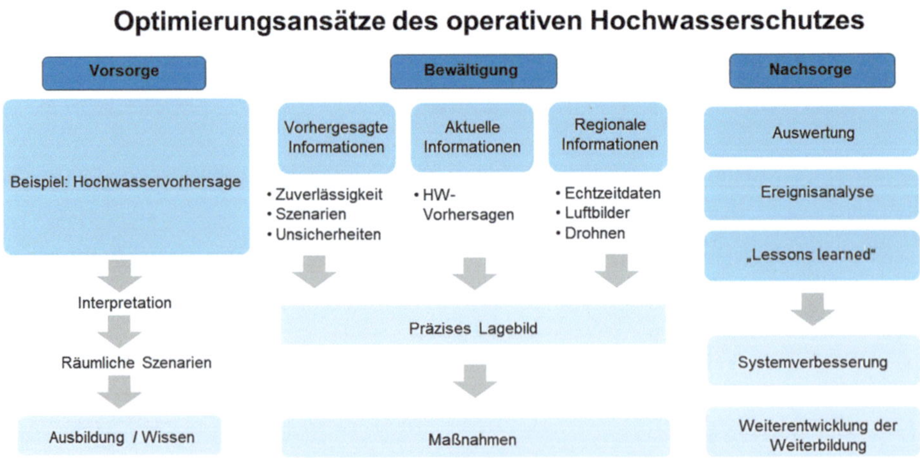

Abb. 13.1 Ansatz zur Optimierung mit dem Beispiel der Hochwasservorhersage, nach Definition des operativen Hochwasserschutzes von Schüller und Jüpner (2021)

detaillierte Auswertung nach einem Hochwasserereignis kann zur Systemverbesserung genutzt werden.

Die Ereignisanalysen der großen Hochwasser in den vergangenen Jahrzehnten in Deutschland und Europa zeigen fast durchweg das Potenzial durch eine effektivere und verbesserte Hochwasserbewältigung Menschenleben zu retten und Hochwasserschäden reduzieren zu können (siehe ► Abschn. 13.3). Dabei wird vielfach ein ausgeprägter Handlungsbedarf in der Hochwasservorsorge und Vorbereitung auf die Ereignisbewältigung formuliert (u. a. LHW 2014; Sächsische Staatskanzlei 2002).

Im Rahmen des KAHR-Projektes (siehe ► Kap. 2) bearbeitet das Fachgebiet Wasserbau und Wasserwirtschaft der Rheinland-Pfälzischen Technischen Universität Kaiserslautern-Landau das Teilprojekt „Optimierung des operativen Hochwasserschutzes". Aufbauend auf einer fundierten Analyse bestehender Strukturen des operativen Hochwasserschutzes in Rheinland-Pfalz (u. a. Schüller und Jüpner 2021; Jüpner 2022; Jüpner und Hoffmann 2022; Schüller 2022) sowie einer Ereignisanalyse des Juli-Hochwassers 2021 an der Ahr wird die Frage beantwortet, durch welche Vorsorgemaßnahmen zur Bewältigung großer und katastrophaler Hochwasserereignisse im Einsatzfall der Verlust von Menschenleben verhindert und Schäden wirksam reduziert werden können. Im Ergebnis werden u. a. Empfehlungen für strukturelle Optimierungen (vgl. ► Abschn. 13.2) als auch ein Weiterbildungskonzept für die Einsatzkräfte ausgearbeitet.

13.2 Herausforderungen bei der Hochwasserbewältigung

Hochwasser ist nicht gleich Hochwasser. Diese Aussage ist nicht nur durch die verschiedenen Ursachen von Hochwasser begründet, sondern zudem durch die unterschiedlichen Anforderungen an die Bewältigung dieser Ereignisse. Von Einsatzkräften wird vielfach betont, dass bei der Bewältigung auf Routinen und festgelegte Abläufe zurückgegriffen werden muss, unabhängig von der konkreten Einsatzaufgabe (Hochwasser, Waldbrand etc.). Dennoch erfordern Hochwasserereignisse oft spezifische Anpassungen und stellen sehr unterschiedliche Anforderungen an die Einsatzkräfte. So spielen beispielsweise die Vorwarnzeiten, Ressourcenbereitstellungen, lokale Fließgeschwindigkeiten oder Treibgutanfall bei verschiedenen Ereignissen eine unterschiedliche Rolle. Auf diese verschiedenen Herausforderungen (aber auch auf die Gemeinsamkeiten) wird nachfolgend eingegangen. Zudem wird eine Einordnung von Hochwasserereignissen in gewöhnliche und außergewöhnliche Hochwasserereignisse vorgenommen, die als Grundlage und Begriffsverständnis für die Optimierungsansätze dient.

13.2.1 Begriffsabgrenzung von Hochwasserereignissen im Hinblick auf deren Bewältigung

Die Hochwasserereignisbewältigung wurde hinsichtlich der Begriffsabgrenzung bisher nur unzureichend thematisiert. Zur Einordnung von Hochwasserereignissen existieren verschiedene Ansätze. In Carstensen et al. (2024) wird beispielsweise anlässlich des Hochwassers zum Jahreswechsel 2023/2024 eine Unterteilung in „planmäßige" und „nichtplanmäßige" Aspekte vorgenommen, wobei der letztgenannte Aspekt noch um die Beschreibung „nicht planmäßig, aber nicht überraschend" ergänzt wurde. Diese Unterteilung wurde hier auf ein Hochwasserereignis angewandt, ist jedoch grundsätzlich auf andere Hochwasserereignisse übertragbar. Als „planmäßig" sind hier alle Maßnahmen bezeichnet, die im Vorfeld geplant und im Ereignisfall umgesetzt wurden, wie z. B. das Öffnen des Pretziner Wehrs bei Magdeburg ab einem definierten Hochwasserabfluss (Carstensen et al. 2024). Die Aussage „Nicht planmäßig" hingegen beschreibt z. B. Schwachstellen an Deichen, die bereits unterhalb des Bemessungswasserstandes auftreten. Als „nicht planmäßig, aber auch nicht überraschend" sind nach dieser Herangehensweise Versagensfälle technischer Hochwasserschutzanlagen einzuordnen, bei denen im Hochwasserfall Bemessungsgrößen überschritten wurden und ein Deich durch Hochwasser überströmt wird (Carstensen et al. 2024).

Alternativ zur vorgeschlagenen Unterteilung in „planbare" und „nicht planbare" Ereignisse, ist die Bezeichnung „geplante" bzw. „nicht geplante" Ereignisse zur Klassifizierung denkbar, da es das Ziel der Hochwasservorsorge ist, nicht planbare Ereignisse grundsätzlich mitzudenken. Dieser Klassifizierungsansatz würde sowohl Bewältigungs- als auch Bemessungsaspekte technischer Hochwasserschutzanlagen berücksichtigen.

Eine naturwissenschaftlich begründete Klassifizierung beruht auf der Entstehung und den Eigenschaften verschiedener (Extrem-)Ereignisse. Fischer und Schumann (2023) bilden dabei typenspezifische Gruppen von Hochwasserereignissen, die durch Ereignisanalysen hinsichtlich meteorologischer Entstehungsprozesse und der Ganglinienform unterteilt werden. Genutzt wird diese Einteilung zur Ergänzung der bestehenden Hochwasserstatistik und somit für die Bemessungspraxis (Fischer und Schumann 2023). Bewältigungsaspekte werden dabei nicht berücksichtigt.

Die dargestellten Klassifizierungsansätze beantworten jedoch nicht die Frage, wann von einem „außergewöhnlichen Hochwasser" in Abgrenzung zu einem „gewöhnlichen Hochwasserereignis" gesprochen werden kann. Für eine erfolgreiche und effektive Hochwasserbewältigung ist diese Abgrenzung von praktischer Bedeutung. So ist z. B. die präzise Darstellung der Hochwasserlage – vor allem die zu erwartende räumliche Ausbreitung des Hochwassers – für die Auswahl operativer Hochwasserschutzmaßnahmen essenziell.

In der medialen Berichterstattung wird in der Ereignisdarstellung häufig von „Extremereignis" gesprochen. Dabei wird suggeriert, dass „das Gewöhnliche" überschritten wird. Das Adjektiv „extrem" bietet sich jedoch aus mehreren Gründen nicht als Bezeichnung bzw. Abgrenzungskriterium an: „Extrem" ist per Definition ein Superlativ und somit nicht steigerbar. Der Begriffsinhalt im Wortsinn beinhaltet, dass ein Zustand eingetreten ist, der schlimmer nicht sein kann. Die Nutzung des Begriffes „Extremhochwasser" im operativen Hochwasserschutz führt jedoch auch praktisch in die Irre. Wird zum Beispiel eine Hochwassergefahrenkarte mit der Darstellung des HQ_{extrem} so interpretiert, dass damit die Grenzen einer maximal möglichen Hochwasserausbreitung visualisiert werden, sind Fehlannahmen vorprogrammiert. Werden im Umkehrschluss die vermeintlich „hochwasserfreien Räume" für die Einrichtung von Materiallagerplätzen, Bereitstellungsräumen oder Evakuierungsbereichen genutzt, kann es zu erheblichen Hochwasserschäden kommen. Im Ahrtal wurde nach Auswertung des Hochwasserereignisses im Juli 2021 die ausgewiesene Überschwemmungsfläche in den HQ_{extrem}-Karten in der Realität deutlich überschritten (LfU RLP 2022).

Neben den Begriffen „extreme" und „außergewöhnliche" Hochwasserereignisse werden eine Reihe weiterer verbaler Größenbeschreibungen wie z. B. „kleines", „mittelgroßes", „großes" und „sehr großes Hochwasser" verwendet. Diese Einteilung wird u. a. zur Klassifizierung der Hochwassersituation mit Bezug zu Hochwassermeldepegeln verwendet, wobei sich die zugrundeliegenden Beschreibungen in den Bundesländern unterscheiden (LfU BY und LUBW 2024). In Hessen werden beispielsweise drei Meldestufen verwendet und die dritte Meldestufe, die mit dem „sehr großen Hochwasser" anderer Bundesländer gleichzusetzen ist, als „außergewöhnliches Hochwasser" bezeichnet (HLNUG 2024). Zudem existieren verschnittene Begriffskombinationen, wie in Thieken et al. (2021), in der z. B. das Hochwasser 2021 als „extrem außergewöhnliches" Ereignis eingestuft wird.

Auch bei der Beschreibung von Starkregen ist eine begriffliche Vielfalt gebräuchlich. Beim sogenannten Starkregenindex (SRI) liegt das außergewöhnliche Starkregenereignis mit dem SRI 6 & 7 unterhalb der Einstufung des extremen Ereignisses mit dem Stufen 8 bis 12 (Schmitt et al. 2018).

Zusammenfassend kann festgestellt werden, dass in der wasserwirtschaftlichen Praxis ein großes begriffliches Spektrum hinsichtlich der Beschreibung von Hochwasserereignissen existiert, die eine eindeutige Charakterisierung nicht ermöglicht und den Vergleich verschiedener Ereignisse erschwert.

Neben den wasserwirtschaftlichen Verwendungen unterschiedlicher Begrifflichkeiten sind im Bereich des Rettungswesens verschiedene Bezeichnungen üblich, die vor allem auf die Auswirkungen von Ereignissen abzielen. In DIN 13050 „Begriffe im Rettungswesen" wird ein Großschadensereignis mit einer großen Betroffenheit abgegrenzt. Die Katastrophe ist laut Norm ein noch größeres Ereignis, die mit eigenen Mitteln nicht bewältigt werden kann, wobei auch die Betroffenheit der Infrastruktur erwähnt wird (Deutsches Institut für Normung 2021). Im Niedersächsischen Katastrophenschutzgesetz (NKatSG 2022) wird ein außergewöhnliches Ereignis klar gesetzlich definiert. Hierbei handelt es sich, um ein „massives" Ereignis, das eine zentrale Unterstützung erfordert, wobei ein Katastrophenfall folgen kann (§ 1 Abs. 3 NKatSG). Der Begriff „außergewöhnlich" wird somit bereits verwendet, jedoch in einer geringeren Schwere des Ereignisses, wie auch beim SRI.

Somit lässt sich abschließend anhand der vorliegenden Literatur keine einheitliche Verwendung oder Abgrenzung feststellen, weshalb im vorliegenden Buchkapitel Herausforderungen bei der Bewältigung von Hochwasserereignissen auf Basis der eigenen nachfolgenden Abgrenzung abgeleitet werden.

13.2.2 Gewöhnliche und außergewöhnliche Hochwasserereignisse

Eine zielführende begriffliche Unterteilung, um die Herausforderungen im operativen Hochwasserschutz angemessen zu adressieren, ergibt sich aus der Betrachtung der grundlegenden Herangehensweise von Einsatzkräften bei der Bewältigung einer Hochwasserlage. Dabei kann eine wesentliche Unterscheidung als Ansatzpunkt dienen:

1. Als **gewöhnliche** Hochwasserereignisse werden alle Ereignisse verstanden, die
   - mit Routinemaßnahmen bewältigt werden können und im (vorliegenden) Alarm- und Einsatzplan abgebildet sind.
2. **Außergewöhnliche** Hochwasserereignisse dagegen sind Hochwasser, die
   - Bemessungsgrenzen übersteigen.
   - plötzliches Versagen von Hochwasserschutzanlagen beinhalten.
   - Überlagerungen beinhalten (z. B. Überlagerung von Zuflüssen).
   - kaskadenartige Reaktionen auslösen.
   - an außergewöhnliche Maßnahmen gekoppelt sind.
   - mit einem aufgetretenen Schaden verknüpft werden.

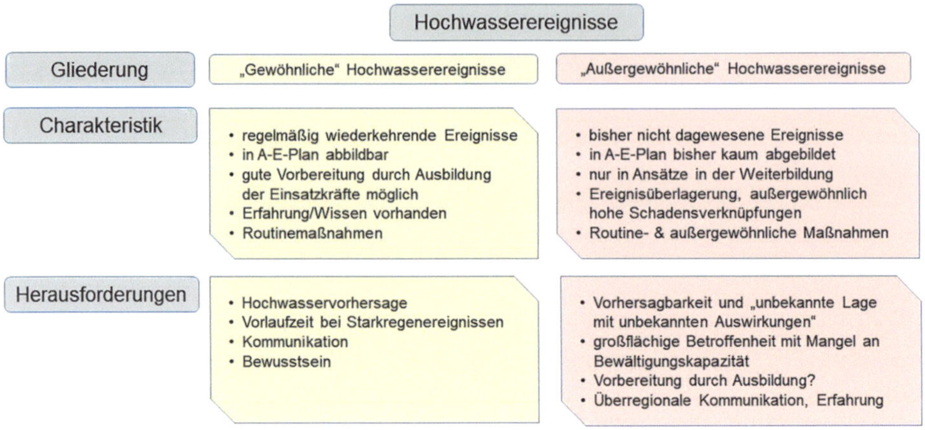

Abb. 13.2 Beispielhafte Ansätze der Charakteristik und Herausforderungen von „gewöhnlichen“ und „außergewöhnlichen“ Hochwasserereignissen

Die Ansätze der charakteristischen Unterscheidungen und spezifischen Herausforderungen sind in Abb. 13.2 gegenübergestellt und nachfolgend erläutert.

1. Gewöhnliche Hochwasserereignisse Ein in seinen Auswirkungen rechtzeitig vorhergesagtes und hinsichtlich seiner Ausprägung umfassend beschreibbares Hochwasserereignis bietet gute Rahmenbedingungen für eine erfolgreiche Bewältigung. Sowohl die Ausbildung der Einsatzkräfte als auch die Vorbereitung der erforderlichen Ressourcen (z. B. mobile Hochwasserschutzsysteme) können in einem planbaren Umfang genutzt werden. Die „gewöhnlich“ verfügbaren Bewältigungskapazitäten sind grundsätzlich ausreichend, um die Hochwasserlage erfolgreich bewältigen zu können.

13

Gewöhnliche Hochwasserereignisse sind in ihrer Charakteristik als regelmäßig wiederkehrende Ereignisse, die bereits in Alarm- und Einsatzplänen abgebildet sind, einzustufen. Dadurch, dass Erfahrungen und Wissen über diese Ereignisart vorhanden sind, ist eine Einbindung in die Weiterbildung der Einsatzkräfte gut möglich und wird üblicherweise auch so praktiziert. Gewöhnliche Hochwasserereignisse können grundsätzlich mit Routinemaßnahmen bewältigt werden. Da Hochwasserereignisse jedoch dynamisch sind, können gewöhnliche Ereignisse zu außergewöhnlichen Situationen führen, wenn z. B. ein zusätzlicher Ereignisparameter hinzu kommt. Die Einteilung ist somit als nicht stationär anzusehen.

Zudem gibt es trotz der eingeübten Routinen Herausforderungen, die in beiden Bereichen vorliegen. Eine möglichst präzise Hochwasservorhersage bildet die Grundlage für eine effiziente Bewältigung. Hier spielt insbesondere der Faktor Zeit eine Rolle, der bei Starkregenereignissen ein zentraler limitierter Faktor zur Bewältigung ist (Jüpner und Schüller 2025; Schüller et al. 2022b). Für eine effiziente Durchführung routinierter Abläufe ist eine zuverlässige Kommunikation erforderlich, die technisch und personell sichergestellt sein muss (DWA 2024).

Für die Praxis Die Vermittlung des nötigen Gefahrenbewusstseins ist auch bei häufig auftretenden Ereignissen eine Herausforderung, da die Gefahr besteht, dass Ereignisse in ihrer Wirkung unterschätzt werden.

2. Außergewöhnliche Hochwasserereignisse Hochwasserereignisse, die größer ausfallen als die verfügbare Bewältigungskapazität der technischen Hochwasserschutzanlagen sowie der verfügbaren Einsatzkräfte und -ressourcen im operativen Hochwasserschutz, können als „außergewöhnlich“ beschrieben werden.

Außergewöhnliche Hochwasserereignisse sind dadurch gekennzeichnet, dass von „bisher nicht dagewesenen“ Ereignissen gesprochen wird. Dabei ist üblicherweise gemeint, dass diese Ereignisse in der Erinnerung der Betroffenen nicht präsent sind, es aber durchaus in der Vergangenheit solche Ereignisse gegeben haben kann, so wie es auch im Ahrtal der Fall war (Vorogushyn et al. 2022). Durch die vielfältigen möglichen Szenarien ist eine Abbildung in Alarm- und Einsatzplänen grundsätzlich schwierig. Gleiches gilt für die Umsetzung dieser komplexeren Übungsszenarien in der Weiterbildungspraxis der Einsatzkräfte. Ein möglicher Ansatz wird in ▶ Abschn. 13.3.1 im Handlungsbereich 3 vorgestellt.

Zur Bewältigung außergewöhnlicher Hochwasserereignisse sind Routinemaßnahmen ebenfalls wichtig, müssen aber durch außergewöhnliche Maßnahmen, in Form von flexiblen und innovativen Bewältigungsoptionen (▶ Abschn. 13.3), ergänzt werden. Ein zielführender Ansatz ist dabei, die in der Praxis oder in Übungen getätigten Bewältigungsstrategien zu manifestieren, sodass in Zukunft darauf zurückgegriffen werden kann. Einschränkend muss jedoch gesagt werden, dass keine Hochwasserlage einer anderen gleicht. Dass Routinen als Grundlage für Improvisationen genutzt werden können, wurde beispielsweise in einer Studie mit dem Technischen Hilfswerk (THW) praktisch untersucht (Schröder und Geiger 2014).

Die besondere Herausforderung besteht in der Vorhersage. Eine vorhergesagte Abflussganglinie eines klassischen Flusshochwassers kann durch ein nicht geplantes Ereignis, wie etwa einen Deichbruch, eine unerwartete Änderung erfahren (Jüpner und Schüller 2025). Diese Besonderheiten müssen sich im Übungsgeschehen widerspiegeln. Durch den Charakter des Unbekannten ist die Einbettung in die klassischen Übungsszenarien eine weitere Herausforderung, die an dieser Stelle zu nennen ist. Insbesondere sehr großflächige Lagen, die einen außergewöhnlichen Ressourceneinsatz erfordern, bergen die Herausforderung einer überregionalen und auch überorganisationalen Kommunikation. Diese wird durch verschiedene bekannte Abläufe und Begriffsverständnisse erschwert.

Die vorgestellte grundlegende Unterteilung in „gewöhnliche“ und „außergewöhnliche“ Hochwasserereignisse berücksichtigt primär die bisher wenig berücksichtigte Sichtweise der Hochwasserbewältigung und die spezifischen Erfordernisse des operativen Hochwasserschutzes. Sie bietet eine gute Ausgangssituation für die nachfolgend aufgeführten Optimierungsansätze.

13.3 Optimierungsansätze

Die vorliegenden Ereignisanalysen des Juli-Hochwassers 2021 decken sich in ihren grundsätzlichen Einschätzungen hinsichtlich der notwendigen verbesserten Vorbereitung auf außergewöhnliche Hochwasserereignisse im Wesentlichen mit den Analysen vergangener Hochwasserereignisse an Elbe, Donau und anderen Gewässern. Die Notwendigkeit einer umfassenden Vorbereitung aller Akteure auf katastrophale Hochwasserereignisse sowie die präzise Kommunikation der Hochwasserrisiken werden insbesondere durch die Ergebnisse der Parlamentarischen Untersuchungsausschüsse zur Hochwasserkatastrophe 2021 in Nordrhein-Westfalen und Rheinland-Pfalz (Jüpner 2021) betont. Im Folgenden wird diese grundsätzliche Herangehensweise auf die drei Bereiche des operativen Hochwasserschutzes (vgl. ◘ Abb. 13.1) übertragen und in Form von konkreten Handlungsbereichen ausgeführt. Dabei werden neben den theoretisch abgeleiteten Optimierungspotenzialen und der Untersuchung bzw. Auswertung verschiedener Ereignisanalysen die einsatzspezifischen Erfahrungen der beteiligten Einsatzkräfte u. a. von Feuerwehr, THW und Bundeswehr aufgegriffen und berücksichtigt, siehe dazu auch ▶ Abschn. 13.4.

13.3.1 Operative Hochwasservorsorge

Im Bereich der operativen Hochwasservorsorge steht die fachliche und organisatorische Vorbereitung auf verschiedene Hochwasserereignisse und -szenarien im Mittelpunkt. Grundlage aller zielgerichteten Handlungen ist ein Vorhersagesystem, das die Hochwassersituation möglichst frühzeitig und präzise vorhersagt, räumlich darstellt und die Grundlage für regionale Alarm- und Einsatzpläne bildet (Jüpner und Schüller 2025; Schüller et al. 2022b).

Als **Handlungsbereich 1** kann die Verbesserung der **Hochwasservorhersage** abgeleitet werden. Dabei erscheinen folgende drei Zielstellungen besonders wichtig zu sein:

- Schneller verfügbare und präzisere Modellierungsergebnisse durch verbesserte Modelle auf breiterer Datenbasis (vgl. ▶ Kap. 4). Damit kann insbesondere die Vorwarn- und Reaktionszeit der Einsatzkräfte vergrößert werden.
- Ausweitung der Hochwasservorhersage auf eine größere Anzahl von Hochwassermeldepegeln. Dazu hat z. B. Nordrhein-Westfalen die Ausweitung des bestehenden Pegelmessnetzes um neue pegelstandorte beschlossen (Landtag NRW 2023).
- Allgemeinverständliche Darstellung charakteristischer Vorhersagewerte, wie zu erwartende pegelbezogene maximale Wasserstände oder Durchflüsse. Hierzu zählen der Bezug zu statistischen Werten wie Wiederkehrintervallen in Form eines HW_x oder HQ_x, die eine Interpretation der Hochwassergefahren- und Hochwasserrisikokarten (z. B. eines HQ_{100}-Szenarios) ermöglichen. Sinnvoll erscheint auch die Herstellung von Bezügen zu vergangenen Hochwasserereignissen, um die vorhandene Erinnerung an abgelaufene Hochwasserereignisse herstellen zu können. Diese Ideen sind z. B. in Rheinland-Pfalz umgesetzt worden, siehe ◘ Abb. 13.3. Außerdem gibt es beispielsweise für das Ahrtal nun sogenannte Anschlaglinien für bestimmte Pegelwerte (SGD NORD 2023)

13

Die Verbesserung der Hochwasservorhersage muss die gesamte Prozesskette von der Erhebung relevanter meteorologischer, hydrologischer und anderer Daten, deren Übertragung, Einbindung in Modelle, Berechnung und Interpretation berücksichtigen. Es ist deutlich zu erkennen, dass in den durch das Hochwasser 2021 betroffenen deutschen Bundesländern die politischen Beschlüsse gefasst und die praktische Umsetzung eingeleitet wurde. In Rheinland-Pfalz wird z. B. der Katastrophenschutz des Landes neu organisiert und strukturiert (LfU RLP 2024), vgl. Handlungsbereich 8.

Handlungsbereich 2 umfasst aufbauend auf der Hochwasservorhersage die detaillierte räumliche **Beschreibung des prognostizierten Hochwasserszenarios** und die absehbaren Konsequenzen. Dabei müssen vor allem

- die räumliche Überflutungsfläche ausgewiesen,
- die Betroffenheit kritischer und vulnerabler Infrastrukturen regionalspezifisch konkretisiert, sowie

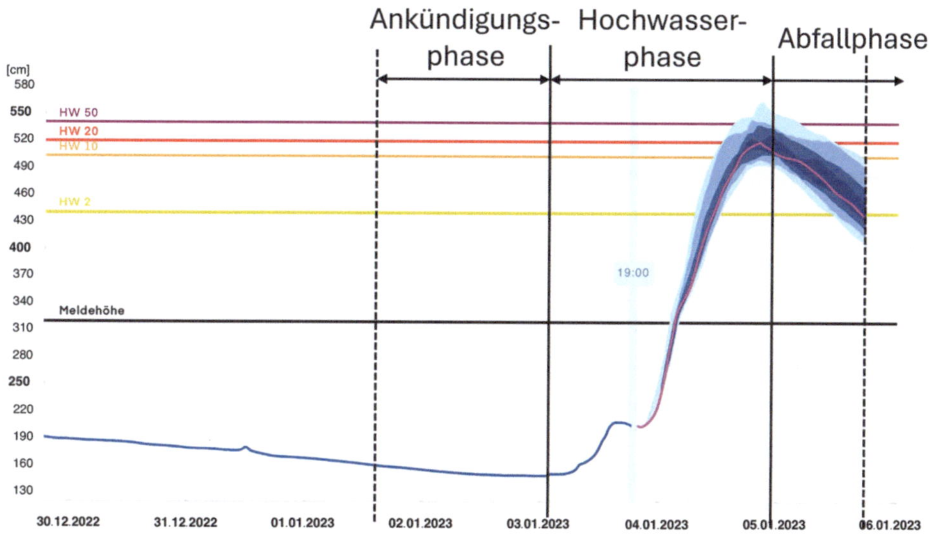

Abb. 13.3 Hochwasservorhersage für einen Hochwassermeldepegel in Rheinland-Pfalz. (Bildquelle: Jüpner und Schüller 2025)

- indirekte Folgen, z. B. durch Maßnahmen des operativen Hochwasserschutzes wie Straßensperrungen oder temporäre Stromausfälle, Lieferkettenunterbrechungen (Kaskadeneffekte) etc. für die betroffene Region sorgfältig analysiert und anschaulich dargestellt werden.

Eine derart durchdachte umfassende Beschreibung eines Hochwasserszenarios trägt wesentlich dazu bei, dass die Betroffenen die Hochwassergefahren erkennen und einschätzen und die Einsatzkräfte zielgerichtete Vorbereitungen treffen können.

Dieser Handlungsbereich scheint ein besonders hohes Optimierungspotenzial zu bieten. Bei der Analyse unterschiedlicher Ereignisanalysen des Hochwassers vom Juli 2021 sowie der Betrachtung von Reaktionen vieler Betroffener im operativen Hochwasserschutz fallen häufig Aussagen wie „noch nie dagewesene Situation“ oder „unvorstellbares Ausmaß“. Dies steht im Widerspruch zu den geltenden rechtlichen Grundlagen, wie sie z. B. im Wasserhaushaltsgesetz (WHG 2023), formuliert sind. Demnach ist die Vorbereitung auch auf die Bewältigung „extremer“ Hochwasserereignisse explizit gefordert. Ein detailliertes „Durchdenken“ der Folgen extremer Hochwasserereignisse sowie die resultierende Ableitung praxistauglicher Bewältigungsstrategien ist in der Praxis jedoch nach wie vor eher die Ausnahme als die Regel. Dabei sind die wesentlichen Informationen zur detaillierten Hochwassermodellierung grundsätzlich ebenso vorhanden, wie die Kenntnisse über die regionalen Auswirkungen. Das Bewusstsein der resultierenden negativen Konsequenzen von Hochwasserereignissen und der notwendigen Analyse als Teil eines umfassenden Vorsorgekonzeptes scheint jedoch zu wachsen.

Der **Handlungsbereich 3** baut auf den Handlungsbereichen 1 und 2 auf und ermöglicht die spezifische **Weiterbildung** von Einsatzkräften im operativen Hochwasserschutz hinsichtlich der Bewältigung außergewöhnlicher Hochwasserereignisse. Die derzeitige Ausbildungssituation von Einsatzkräften im operativen Hochwasserschutz – vor allem bei Feuerwehren, Technischem Hilfswerk und der Bundeswehr – ist dadurch charakterisiert, dass neben der „klassischen“ Katastrophenschutzausbildung sehr wenig Angebote in der Vermittlung von vertieftem Fachwissen zum Umgang mit „außergewöhnlichen Hochwassereinsatzlagen“ existieren (Jüpner und Schüller 2025). Als einzige Einsatzinstitution bietet das THW mit seinen Ausbildungsprofilen zum „Technischen Berater Hochwasserschutz und Naturgefahren“ ein den aktuellen Einsatzanforderungen adäquates Weiterbildungsprofil an. Eine Ausweitung und Weiterentwicklung in diesem Bereich zeigt ein hohes Potenzial für eine effektivere Hochwasserbewältigung, siehe auch Handlungsbereich 9.

Im Rahmen des KAHR-Projektes konnte durch die pilothafte Durchführung einer viertägigen Stabsrahmenübung die Bewältigung eines außergewöhnlichen Hochwasserereignisses getestet werden. Am Beispiel der Stadt Neuwied wurden verschiedene Szenarien eines katastrophalen Rhein-Hochwassers simuliert und unter realitätsnahen Bedingungen bearbeitet. Einsatzkräfte der Bundeswehr aus dem Landeskommando Rheinland-Pfalz sowie Technische Berater Hochwasserschutz und Naturgefahren des THW erstellten gemeinsam mit der RPTU Kaiserslautern-Landau die Übungsszenarien und führten diese aus. Die vorliegenden Auswertungen (Jüpner et al. 2024) zeigen die Wirksamkeit und die nachweisbaren Lerneffekte durch diese Art der praktischen Ereignissimulation.

13.3.2 Operative Hochwasserbewältigung

Die operative Hochwasserbewältigung bezieht sich auf die Phase der eigentlichen Ereignisbewältigung. Sie beginnt üblicherweise mit dem Ereigniseintritt ohne das eine klare zeitliche Abgrenzung möglich ist. Es ist unmittelbar einleuchtend, dass die Art und Weise der Hochwasserbewältigung die Hochwasserschäden direkt beeinflusst. Werden in einer betroffenen Region auf der Grundlage einer guten Vorbereitung, dem Vorliegen umfassender Alarm- und Einsatzpläne, mit gut ausgebildeten und geschulten Einsatzkräften und ausreichend Ressourcen die operativen Hochwasserschutzmaßnahmen rechtzeitig und zielgerichtet umgesetzt, können die Hochwasserschäden minimiert werden (Patt und Jüpner 2020).

Der **Handlungsbereich 4** betrifft ein schnell und kontinuierlich verfügbares, präzises und umfassendes **Lagebild**. Das erkennbare Optimierungspotenzial betrifft folgende wesentliche Aspekte

- ausreichende organisatorische und technische Vorbereitung bei den zuständigen Stäben bzw. Einsatzleitungen,
- Nutzung moderner softwarebasierter EDV-Lösungen und technischer Instrumente, die vor dem Ereignis vertieft geschult wurden,
- Einsatz automatisierter Software-Lösungen, z. B. für die Berichtserstellung,
- Kompatibilität mit anderen Einsatzstellen bzw. -organisationen,
- Einsatz von Entscheidungsunterstützungssystemen im Katastrophenschutz,
- Einsatz geeigneter Informations- und Managementsysteme, wie z. B. INGE – Interaktive Gefahrenkarten (Thieme 2023),
- Einsatz spezifischer Softwarelösungen zur Unterstützung operativer Hochwasserschutzmaßnahmen, wie z. B. Hochwasserlogistik-Softwaren.

Die vorliegenden Einsatzerfahrungen zeigen, dass insbesondere bei komplexen und großräumigen Hochwasserlagen eine präzise und umfassende Lagedarstellung und Lagebeurteilung zwingend erforderlich ist. Auch gehört zum Alltag der Stabsarbeit die

Simulation der Auswirkungen von möglichen operativen Hochwasserschutzmaßnahmen einschließlich deren Wechselwirkungen. Ohne die Möglichkeit, die – sich unter bestimmten Umständen rasch ändernde – Lage angemessen darstellen zu können, ist eine zielgerichtete und effiziente Ableitung und Umsetzung geeigneter operativer Hochwasserschutzmaßnahmen nur eingeschränkt möglich.

Ein grundsätzliches Optimierungspotenzial liegt in der Anwendung moderner IT und dem Einsatz geeigneter Softwarelösungen. In diesem Bereich gab es z. T. substanzielle Weiterentwicklungen, die grundsätzlich zu einer verbesserten Lagedarstellung im Hochwassereinsatz führen können. Beispielhaft seien an dieser Stelle KI-Anwendungen und automatisierte Datenauswertungs- und -analyseverfahren genannt (Mudersbach et al. 2022). Zu berücksichtigen ist dabei grundsätzlich ein möglicher Technikausfall durch Unterbrechungen von Stromversorgung, Internet oder Kommunikationswegen. Neben der Vorhaltung von Sicherungsmaßnahmen wie z. B. Notstromversorgungen sollten redundante Systeme als Alternativen genutzt werden können, siehe dazu u. a. (Fekete 2022).

Aufbauend auf dem Handlungsbereich 4 wird im **Handlungsbereich 5** großes Optimierungspotenzial im Einsatz von Instrumenten und Techniken zur **Echtzeitdaten**erfassung, -weiterleitung und -verarbeitung gesehen. Dazu zählen u. a.

- der Einsatz von Drohnen zur schnellen Visualisierung von Einsatzlagen, z. B. an verklausten Brücken,
- die Nutzung von Fernerkundungsdaten, um aktuelle Parameter des betroffenen Einsatzgebietes schnell, hoch aufgelöst und flächendeckend erheben zu können,
- Technologien der schnellen drahtlosen Datenübertragung von Einsatzabschnitten in den Einsatz- oder Führungsstab,
- die Datenaufnahme vor Ort mittels IT-Lösungen (z. B. konfigurierten Tablets), die eine unmittelbare Weiterverwendung dieser Daten erlauben.

Für die operativen Hochwasserschutzmaßnahmen wird durch den zielgerichteten Einsatz der o. g. Methoden und Technologien vor allem Zeit gewonnen.

Der **Handlungsbereich 6** umfasst praktisch umsetzbare und wirksame Maßnahmen zur **Hochwasserbewältigung**. Damit sind vor allem technologische Lösungen, wie z. B. mobile Hochwasserschutzsysteme zu bezeichnen, die eine „passgenaue" Bewältigung spezifischer Herausforderungen im Hochwassereinsatz ermöglichen. Auch hier sind verschiedene technische Weiterentwicklungen von bisher genutzten „klassischen Verteidigungsmaßnahmen" zu konstatieren. Betrachtet man beispielhaft die Verteidigung von Flussdeichen, so stehen heute neben den Sandsäcken eine Fülle von Sandsackersatzsystemen zur Verfügung. Diese sind Gegenstand wissenschaftlicher Untersuchungen und praktischer Erprobungen, vgl. Massolle et al. (2018) und Wagenhuber (2023).

13.3.3 Operative Hochwassernachsorge

Nach einem Hochwasser muss neben der *„Überwindung der Folgen für den Einzelnen und die Gesellschaft, Beseitigung von Umweltschäden"* (LAWA 2013) eine umfassende Ereignisanalyse erfolgen. Diese bietet die Möglichkeit der kritischen Reflexion der operativen Hochwasserschutzmaßnahmen mit dem Ziel, aus dem Ereignis zu lernen. Als beispielgebende Ereignisanalyse sei an dieser Stelle der „1. Kirchbach-Report" (Sächsische Staatskanzlei 2002) nachdrücklich empfohlen.

Der **Handlungsbereich 7** umfasst die Erarbeitung von detaillierten Ereignisanalysen und Ableitungen von **„Lessons Learned“**. Während traditionelle Ereignisanalysen von Staaten oder betroffenen deutschen Bundesländern die hydro-meteorologische Situation, den (hydrologischen) Ablauf des Hochwassers, die entstandenen Hochwasserschäden und (eher allgemein) die Hochwasserbewältigung thematisieren, besteht ein erkennbares Optimierungspotenzial für die Verbesserung des operativen Hochwasserschutzes in der kritischen Analyse der Wirksamkeit der durchgeführten Maßnahmen. Dazu erscheint es sinnvoll, dass möglichst alle relevanten Akteure im operativen Hochwasserschutz, die substanzielle Beiträge zur Hochwasserbewältigung leisten konnten, diese herausfordernde Aufgabe wahrnehmen. Leider ist in der Praxis immer wieder zu beobachten, dass für diese notwendige Reflexion und Analyse nur wenig Zeit und Ressourcen verfügbar sind, da die vielfältigen Anforderungen in der „Regenerationsphase“ nach dem Hochwasser die beteiligten Akteure in besonderem Maße fordern (Jüpner 2022).

Mögliche Optimierungsfelder im Bereich der operativen Hochwassernachsorge formulieren Jüpner und Schüller (2025) hinsichtlich der fachlichen Inhalte von Ereignisanalysen:

- Entstehung und Ablauf des Hochwassers: Darstellung der hydro-meteorologischen Situation, Ablauf und Besonderheiten der Hochwasserausprägung (Überlagerung von Hochwasserwellen, Rückstau etc.),
- Hochwasservorhersage: Dokumentation der Hochwasservorhersage, Genauigkeit der vorhergesagten und prognostizierten Werte, Ursachen der Abweichungen, aufgetretene Besonderheiten (Sedimenttransport, Verklausungen, Probleme der Datenerfassung etc.),
- Hochwasserbewältigung: Darstellung wesentlicher Gefahren- und Schadenslagen sowie Schwerpunktbereiche (Hotspots), Maßnahmen des operativen Hochwasserschutzes, Wirksamkeit und Effektivität ausgewählter Einzelmaßnahmen, Analyse außergewöhnlicher Situationen,
- Aussagen zur Organisation der Hochwasserbewältigung, Lagedarstellung, Kommunikation.

Als **Handlungsbereich 8** kann die Anpassung des gesamten **„Hochwasser-Bewältigungssystems“** formuliert werden. Dazu sei beispielhaft die politische Reaktion im Land Rheinland-Pfalz nach der Hochwasserkatastrophe vom Juli 2021 erwähnt. Das für den Katastrophenschutz zuständige Ministerium des Innern und für Sport Rheinland-Pfalz informiert im April 2024 unter der Überschrift „Meilenstein bei der Neuaufstellung des Katastrophenschutzes“ über die geplante Einrichtung eines neuen „Landesamtes für Brand- und Katastrophenschutz“ auf der Basis entsprechend geänderter gesetzlicher Grundlage (MdI RLP 2024). Damit wird aufbauend auf mehrjährigen und umfassenden (z. T. kontroversen) Diskussionen eine grundsätzliche Neuausrichtung des operativen Hochwasserschutzes auf den Weg gebracht. Ähnliche Entwicklungen konnte man z. B. nach dem Hochwasser vom August 2022 in Sachsen beobachten. In dessen Folge u. a. das Landeshochwasserzentrum Sachsen eingerichtet wurde.

Im **Handlungsbereich 9 „Anpassung des Weiterbildungsprogramms der Einsatzkräfte“** wird die Notwendigkeit, in möglichst vielen Szenarien zu denken, adressiert. Während der vergangenen Hochwasserereignisse sind in den letzten Jahren immer wieder besondere Einsatzlagen verzeichnet worden. Dabei stellt sich die Frage nach der Vorbereitung der Einsatzkräfte auf solche Szenarien. Als Beispiel ist etwa das drohende

Versagen der Steinbachtalsperre 2021 zu nennen. Hier ist es empfehlenswert, gewonnene Erkenntnisse systematisch aufzubereiten, um diese z. B. für Übungsszenarien zu nutzen.

Für die Lehre Im Zuge der fortschreitenden Digitalisierung ist die Nutzung von digitalen Formaten und neuen Technologien ein weiterer Optimierungspunkt, der bereits Eingang in die Weiterbildung gefunden hat. Beispielsweise wird im THW ein Virtual-Reality-Trainingssystem „Deichverteidigung“ genutzt, indem verschiedene Deichschäden in Echtzeit möglichst realitätsnah mit einer Virtual-Reality-Brille beübt werden (Szenaris 2024).

13.4 Vernetzung im operativen Hochwasserschutz – Beispiel H-Kat-Net

13.4.1 Aufgabenstellung des qualitativen Wissensaustauschs

Bedeutende Hochwasserereignisse wie das Juli-Hochwasser 2021 in Rheinland-Pfalz und Nordrhein-Westfalen sind Gegenstand zahlreicher Fachveröffentlichungen. Einsatzkräfte, die das Geschehen vor Ort miterlebt haben, können üblicherweise nicht die Möglichkeit nutzen, in Berichten ihre Einschätzungen einem breiten Publikum verfügbar zu machen. Auch werden interne Nachbesprechungen und Ereignisnachbereitungen meist als ausreichend angesehen.

Jedoch bietet der Blickwinkel der Einsatzkräfte eine besonders wichtige Perspektive, da sie die Abläufe, praktischen Herausforderungen und Maßnahmenumsetzungen miterlebt und vielfach mitgestaltet haben. Zudem bieten ihre aus der praktischen Bewältigung gewonnenen Einschätzungen Anknüpfungspunkte für vertiefte wissenschaftlich fundierte Weiterentwicklungen des operativen Hochwasserschutzes.

Der bei den Einsatzkräften vorhandene „Wissensschatz“ kann durch eine systematische Sammlung und Aufbereitung effektiv genutzt werden. Ein vielversprechender Ansatz ist die Etablierung von Netzwerken, die den gezielten Austausch von Erfahrungen und die fachliche Kommunikation fördern. Dadurch wird wertvolles Praxiswissen gebündelt und für die Weitergabe nutzbar gemacht.

Netzwerke ermöglichen die Verbindung von Wissenschaft und Praxis durch die Einbindung verschiedener relevanter Akteure im operativen Hochwasserschutz. Neben den sozialen Medien spielen interdisziplinäre Netzwerke eine wichtige Rolle für den Wissensaustausch und die Weiterentwicklung. Solche Netzwerke, auch als „Community of Practice“ bezeichnet, bringen gezielt Fachleute aus unterschiedlichen Disziplinen zusammen, um sich über gemeinsame Themen oder Herausforderungen auszutauschen und voneinander zu lernen. Als wissenschaftliche Herangehensweise ist die Etablierung, Förderung und Nutzung von (fachorientierten) Netzwerken lange bekannt und dokumentiert (vgl. u. a. Wenger 1998).

13.4.2 Regionales Katastrophenschutz-Netzwerk zur Bewältigung großer Hochwasserereignisse (H-Kat-Net)

Die Idee der Gründung eines Netzwerks zur gezielten Einbindung der Einsatzkräfte im operativen Hochwasserschutz in die wissenschaftliche Arbeit wurde gemeinsam von der

Arbeitsgruppe des Fachgebietes Wasserbau und Wasserwirtschaft (FWW) an der RPTU Kaiserslautern-Landau sowie engagierten und erfahrenen „Akteuren“ im operativen Hochwasserschutz entwickelt. Dabei wird der Schwerpunkt auf „große Hochwasserereignisse“ gelegt, wobei der Begriff „groß“ hier als Synonym für „außergewöhnliche Ereignisse“ verwendet wird. Darüber hinaus wurde die Fokussierung auf die vom Juli-Hochwasser 2021 besonders betroffenen Regionen in Rheinland-Pfalz und Nordrhein-Westfalen gelegt, ohne ein (regionales) Ausschluss-Kriterium zu formulieren.

Am 17.10.2023 erfolgte die Gründung des Netzwerks „H-Kat-Net“ durch Vertreterinnen und Vertreter der RPTU Kaiserslautern-Landau, die THW-Ortsverbände Sinzig und Euskirchen, dem Kreisverbindungskommando der Bundeswehr im Landkreis Bad Neuenahr-Ahrweiler sowie der Feuerwehr des Landkreises Ahrweiler. Die Gründung erfolgte dabei bewusst „niederschwellig“ als loses Netzwerk interessierter Personen ohne eine formale Struktur. Die organisatorische Betreuung erfolgt im Rahmen des KAHR-Projektes durch die RPTU Kaiserslautern-Landau.

Das Netzwerk veranstaltet quartalsweise Treffen in Präsenz, die wechselseitig von den Netzwerkpartnern organisiert werden. Dabei werden jeweils ausgewählte Fachthemen, wie z. B. Einsatzerfahrungen bei der Hochwasserbewältigung an der Steinbach-Talsperre im Landkreis Euskirchen, vorgestellt und diskutiert. Die Netzwerktreffen werden auf der Homepage des KAHR-Projektes angekündigt und über Ergebnisse berichtet.

Seit der Gründung ist der Kreis der Interessenten an der Netzwerkarbeit kontinuierlich gewachsen, und immer mehr Akteure aus verschiedenen Organisationen, wie dem Deutschen Roten Kreuz, verschiedenen Kommunalverwaltungen, Ingenieurbüros und der Polizei, nehmen an den Veranstaltungen teil. Das Netzwerk steht auch Interessierten offen, die nicht aus den von der Flut im Juli 2021 betroffenen Regionen stammen, und lädt dazu ein, Teil der Diskussion, des Austauschs und der Weiterentwicklung zu werden.

Nach einem Jahr der Arbeit des „H-Kat-Net“ können aus durchgeführten Evaluationen und Befragungen innerhalb des Netzwerkes folgende wesentliche Arbeitsergebnisse benannt werden:

1. Regelmäßige Treffen mit guter Resonanz fördern das „Netzwerken“
2. Die Fachdiskussionen zu verschiedenen Aspekten des operativen Hochwasserschutzes werden als konstruktiv wahrgenommen
3. Herausforderungen bei der Hochwasserereignisbewältigung können an praktischen Beispielen besonders gut verdeutlicht werden.
4. „Empfehlungen zur Optimierung des operativen Hochwasserschutzes“ in Form von neun Handlungsbereichen (vgl. ▶ Abschn. 13.3) wurden diskutiert und beschlossen.

13

Zusammenfassend lässt sich feststellen, dass die Arbeit innerhalb des H-Kat-Net als intensiv, teilweise mühsam, aber insgesamt betrachtet als sehr wertvoll angesehen wird. Der unmittelbare persönliche Austausch einschließlich der vertieften Betrachtung jeweils unterschiedlicher Perspektiven bietet einen überraschend großen Erkenntnisgewinn. Es bleibt dabei die zukünftige Herausforderung, die Arbeit der „Community of Practice“ in einem verlässlichen und praktikablen Rahmen zu etablieren und über den Projektzeitraum des KAHR-Projektes hinaus nutzbringend weiterzuentwickeln.

13.5 Zusammenfassung

Jedes Hochwasserereignis stellt hinsichtlich der Herausforderungen an die Bewältigung spezifische Anforderungen dar. Die Vielzahl an Begriffen (großes, extremes, seltenes Hochwasser etc.) berücksichtigt diese Aspekte der Bewältigung jedoch nur unzureichend, wodurch der Bedarf einer klaren begrifflichen Abgrenzung deutlich wird. Es wird vorgeschlagen, zwischen gewöhnlichen und außergewöhnlichen Hochwasserereignissen zu differenzieren, da diese unterschiedlichen Charakteristika und Herausforderungen hinsichtlich der Bewältigung aufweisen. Gewöhnliche Hochwasserereignisse können durch standardisierte Routinemaßnahmen bewältigt werden und sind in Alarm- und Einsatzplänen verankert (Schüller et al. 2022a). Außergewöhnliche Hochwasserereignisse hingegen überschreiten Bemessungsgrenzen, sind durch plötzliches Versagen von Hochwasserschutzanlagen, Ereignisüberlagerungen, oder z. B. kaskadenartigen Reaktionsfolgen geprägt und an außergewöhnliche Bewältigungserfordernisse und Schäden gekoppelt.

Um die Bewältigung von Hochwasserereignissen konkret zu optimieren, ist eine ganzheitliche Betrachtung des operativen Hochwasserschutzes erforderlich, die nicht nur die Phase während des Ereignisses umfasst, sondern auch Vor- und Nachsorgemaßnahmen einschließt. Für diese Phasen lassen sich neun konkrete Handlungsbereiche definieren. Dazu zählen u. a. die Nutzung von modernen Technologien, die detaillierte Beschreibung des gesamten Ereignisverlaufs sowie deren Schäden und die Weiterbildung der Akteure.

Wie Wissen weitergeben und Vernetzungen zwischen Akteuren gefördert werden kann, wird durch das Netzwerk H-Kat-Net verdeutlicht. Der wertvolle Erfahrungsschatz der Einsatzkräfte wird dabei systematisch aufgearbeitet, um bisherige Lücken in der Dokumentation zu schließen. Auf diese Weise leistet das Netzwerk einen wichtigen Beitrag zur Erhöhung der Resilienz gegenüber Hochwasserereignissen.

Zusammenfassung

- Jedes Hochwasserereignis stellt hinsichtlich der Herausforderungen an die Bewältigung spezifische Anforderungen dar.
- Um die Bewältigung von Hochwasserereignissen konkret zu optimieren, ist eine ganzheitliche Betrachtung des operativen Hochwasserschutzes erforderlich, die nicht nur die Phase während des Ereignisses umfasst, sondern auch Vor- und Nachsorgemaßnahmen einschließt.
- Für diese Phasen lassen sich neun konkrete Handlungsbereiche definieren. Dazu zählen u. a. die Nutzung von modernen Technologien, die detaillierte Beschreibung des gesamten Ereignisverlaufs sowie deren Schäden und die Weiterbildung der Akteure.
- Wie Wissen weitergegeben und Vernetzungen zwischen Akteuren gefördert werden kann, wird durch das Netzwerk H-Kat-Net verdeutlicht. Der wertvolle Erfahrungsschatz der Einsatzkräfte wird dabei systematisch aufgearbeitet, um bisherige Lücken in der Dokumentation zu schließen und einen wichtigen Beitrag zur Erhöhung der Resilienz gegenüber Hochwasserereignissen zu leisten.

Danksagung Die Autorinnen und Autoren von ▶ Kap. 13 bedanken sich beim Bundesministerium für Bildung und Forschung (BMBF) und beim DLR Projektträger für die Unterstützung und Förderung des KAHR-Projekts. Diese Forschung wurde vom BMBF unter dem Förderkennzeichen 01LR2102A-M finanziert.

Literatur

Bayerisches Landesamt für Umwelt, Landesanstalt für Umwelt Baden-Württemberg (2024) Länderübergreifendes Hochwasser-Portal (LHP), Hinweise zum Internetangebot, Klassifizierung der Hochwasser-Situation am Pegel, Bayerisches Landesamt für Umwelt (LfU) und Landesanstalt für Umwelt Baden-Württemberg (LUBW). https://www.hochwasserzentralen.de/info. Zugegriffen: 13. Sept. 2024

Carstensen D, Jüpner R, Müller U, Piroth K, Pohl R, Schüttrumpf H (2024) Expertenpapier der DWA. Hochwasser in Deutschland während des Jahreswechsels 2023/2024. GFA – Gesellschaft zur Förderung der Abwassertechnik e. V., Hennef

Deutsches Institut für Normung (2021) DIN 13050:2021-10, Begriffe im Rettungswesen. Beuth, Berlin

DWA – Deutsche Vereinigung für Wasserwirtschaft, Abwasser und Abfall (2024) Entwurf Merkblatt DWA-M 555 „Hochwasser – Risikokommunikation". GFA – Gesellschaft zur Förderung der Abwassertechnik e. V., Hennef

Fekete A (2022) Kritische Infrastruktur und Versorgung der Bevölkerung. Springer, Berlin https://doi.org/10.1007/978-3-662-65047-9

Fischer S, Schumann AH (2023) Berücksichtigung von Hochwassertypen in der Hochwasserstatistik. Hydrol Wasserbewirtsch 67(5):238–252. https://doi.org/10.5675/HyWa_2023.5_1

HLNUG – Hessisches Landesamt für Naturschutz, Umwelt und Geologie (2024) Hochwasserwarn- und meldedienst. https://www.hlnug.de/themen/wasser/hochwasser/warn-und-meldedienst. Zugegriffen: 13. Sept. 2024

Jüpner R (2021) Parlamentarischer Untersuchungsausschuss V (Hochwasserkatastrophe) des Landtages von Nordrhein-Westfalen. https://www.landtag.nrw.de/files/live/sites/landtag-r20/files/Internet/I.A.1/PUA/PUA_II/PUA-Gutachten-J%C3%BCpner-final-rev1.pdf. Zugegriffen: 9. Okt. 2024

Jüpner R (2022) Operativer Hochwasserschutz – Lernen aus der Hochwasserkatastrophe 2021? Wasserwirtschaft 113(10):60–66. https://doi.org/10.1007/s35147-022-1735-9

Jüpner R, Hoffmann H (2022) Operativer Hochwasserschutz – Schlussfolgerungen aus der Hochwasserkatastrophe 2021 in Deutschland. Wasser Abfall 24(11):18–21. https://doi.org/10.1007/s35152-022-1347-9

Jüpner R, Schüller A (2025) Operativer Hochwasserschutz. Springer. ISBN 978-3-658-46759-3.

Jüpner R, Fabisch M, Weber S, Aydin M, Liebscher T, Bailey M, Hoffmann H, De Koeijer T (2024) Training für die Bewältigung von Hochwasserkatastrophen – ein aktuelles Beispiel aus der Einsatzpraxis. Korrespondenz Wasserwirtsch 17(3):184–189

Landtag Nordrhein-Westfalen (2023) Zweiter Jahrestag der Flutkatastrophe: Umsetzungsstand des 10-Punkte-Arbeitsplans „Hochwasserschutz in Zeiten des Klimawandels". https://www.landtag.nrw.de/portal/WWW/dokumentenarchiv/Dokument/MMV18-1376.pdf. Zugegriffen: 9. Okt. 2024

LAWA – Bund/Länder-Arbeitsgemeinschaft Wasser (2013) Empfehlungen zur Aufstellung von Hochwasserrisikomanagementplänen. https://www.lawa.de/documents/empfehlungen_zur_aufstellung_von_hwrmpl_mit_anlagen_1552299352.pdf. Zugegriffen: 9. Okt. 2024

LAWA – Bund/Länder-Arbeitsgemeinschaft Wasser (2020) Einführung in den LAWA Klimawandel-Bericht 2020. https://www.lawa.de/documents/kompaktinfos_zum_lawa_klimawandel-bericht_2020_1637921187.pdf. Zugegriffen: 9. Okt. 2024

LfU RLP – Landesamt für Umwelt Rheinland-Pfalz (2022) Bericht. Hochwasser im Juli 2021. https://lfu.rlp.de/fileadmin/lfu/Startseitenbeitraege/2022/Hochwasser_im_Juli2021_1_.pdf. Zugegriffen: 9. Okt. 2024

LfU RLP – Landesamt für Umwelt Rheinland-Pfalz (2024) Hochwasservorhersagezentrale Rheinland-Pfalz. https://hochwasser.rlp.de/erlaeuterungen. Zugegriffen: 7. Jan. 2025

LHW – Landesbetrieb Für Hochwasserschutz und Wasserwirtschaft Sachsen-Anhalt (2014) Hochwasserbericht Hochwasserereignisse im Sommer 2002 in Sachsen-Anhalt. https://lhw.sachsen-anhalt.de/fileadmin/Bibliothek/Politik_und_Verwaltung/Landesbetriebe/LHW/neu_PDF/4.0/SB_Hochwasserschutz/Hochwasserbericht_2013.pdf. Zugegriffen: 9. Okt. 2024

Massolle C, Lankenau L, Koppe B (2018) Emergency Flood Control: Practice-Oriented Test Series for the Use of Sandbag Replacement Systems. Geosciences 8(12):45–48. https://doi.org/10.3390/geosciences8120482

MdI RLP – Ministerium für Inneres und Sport Rheinland-Pfalz (2024) Pressemitteilung „Meilenstein bei Neuaufstellung des Katastrophenschutzes". https://mdi.rlp.de/service/pressemitteilungen/detail/meilenstein-bei-neuaufstellung-des-katastrophenschutzes. Zugegriffen: 3. Sept. 2024

Mudersbach C, Leandro J, Reggiani P (2022) Verbesserung der Hochwasservorhersage für mittlere und kleine Fließgewässer insbesondere im Mittelgebirge von Nordrhein-Westfalen. https://www.landtag.nrw.de/files/live/sites/landtag-r20/files/Internet/I.A.1/PUA/PUA_II/Gutachten%20Prof.%20Leandro%2C%20Prof.%20Mudersbach%20und%20Prof.%20Reggiani.pdf. Zugegriffen: 9. Okt. 2024

NKatSG – Niedersächsisches Katastrophenschutzgesetz (2022) Niedersächsisches Katastrophenschutzgesetz, Letzte Änderung durch Art. 3 vom 26. August 2022 (Nds. GVBl. S. 405)

Patt H, Jüpner R (2020) Hochwasser-Handbuch, 3. Aufl. Springer Vieweg, Wiesbaden. ISBN 978-3-658-26742-1.

Sächsische Staatskanzlei (2002) Bericht der Unabhängigen Kommission der Sächsischen Staatsregierung – Flutkatastrophe 2002. 1. Kirchbach-Report. https://publikationen.sachsen.de/bdb/artikel/10825/documents/10951. Zugegriffen: 9. Okt. 2024

Schmitt T, Krüger M, Pfister A, Becker M, Mudersbach C, Fuchs L, Hoppe H, Lakes I (2018) Einheitliches Konzept zur Bewertung von Starkregenereignissen mittels Starkregenindex. Korrespondenz Wasserwirtsch 62(2):113–120. https://doi.org/10.3242/kae2018.02.002

Schröder A, Geiger D (2014) Routinen im Katastrophenmanagement. In: Grün O, Schenker-Wicki A (Hrsg) Katastrophenmanagement, Grundlagen, Fallbeispiele und Gestaltungsoptionen aus betriebswirtschaftlicher Sicht. Springer Gabler, Wiesbaden, S 153–175. ISBN 3-658-06172-3

Schüller A (2022) Möglichkeiten zur Weiterentwicklung des operativen Hochwasserschutzes – ein Beitrag aus wasserwirtschaftlicher Perspektive. Berichte des Fachgebietes Wasserbau und Wasserwirtschaft der Technischen Universität Kaiserslautern. Shaker Verlag (https://kluedo.ub.rptu.de/files/6918/_3-8440-8730-3_DOK.pdf (letzter Zugriff: 09.10.2024))

Schüller A, Jüpner R (2021) Operativer Hochwasserschutz im Wandel – Erste Erfahrungen aus der Hochwasserkatastrophe im Juli 2021. Korrespondenz Wasserwirtsch 11/2021:684–690

Schüller A, Gerz J, von Spiczak C, Illing C (2022a) Inhalte und Aspekte der Alarm- und Einsatzplanung Hochwasser. Korrespondenz Wasserwirtsch 2022(2):108–114. https://doi.org/10.3243/kwe2022.02.004

Schüller A, Meyer K, Gerz J (2022b) Alarm- und Einsatzplanung für Starkregenereignisse. Korrespondenz Wasserwirtsch 2022(3):170–175. https://doi.org/10.3243/kwe2022.03.004

SGD NORD – Struktur und Genehmigungsdirektion Nord (2023) Hochwasseranschlaglinien für Pegelstände an der Ahr veröffentlicht. https://sgdnord.rlp.de/service/pressemitteilungen/detail/sgd-nord-hochwasseranschlaglinien-fuer-pegelstaende-an-der-ahr-veroeffentlicht. Zugegriffen: 3. Jan. 2025

Szenaris (2024) Virtual-Reality-Trainingssystem „Deichverteidigung". https://www.szenaris.com/portfolio-items/virtual-reality-trainingssystem-deichverteidigung/. Zugegriffen: 5. Sept. 2024

Thieken A, Kemter M, Vorogushyn S, Berghäuser L, Sieg T, Natho S, Mohor GS, Petrow T, Merz B, Bronstert A (2021) Extreme Hochwasser bleiben trotz integriertem Risikomanagement eine Herausforderung, Potsdam, Germany. https://www.uni-potsdam.de/fileadmin/projects/extrass/Flut2021_StatementThiekenEtAl.pdf. Zugegriffen: 9. Okt. 2024

Thieme S (2023in) Hochwasserrisikomanagement mit INGE – Interaktive Gefahrenkarte für den kommunalen Hochwasserschutz. Tagungsband zur 14. Veranstaltung des Forums Hochwasserrisikomanagement, Augsburg, Aachen, 11. Mai 2023, S 117–127

Vorogushyn S, Apel H, Kemter M, Thieken AH (2022) Analyse der Hochwassergefährdung im Ahrtal unter Berücksichtigung historischer Hochwasser. Hydrol Wasserbewirtsch 66(5):244–254. https://doi.org/10.5675/HyWa_2022.5_2

Wagenhuber W (2023) Anforderungen an mobile Hochwasserschutz-Systeme. Wasser Abfall 98(9):32–33

WHG – Wasserhaushaltsgesetz (2023) Gesetz zur Ordnung des Wasserhaushalts, Letzte Änderung durch Art. 1 G vom 4. Januar 2023 (BGBl. I Nr. 5)

Wenger E (1998) Communities of Practice: Learning as a Social System. https://thesystemsthinker.com/communities-of-practice-learning-as-a-social-system/. Zugegriffen: 7. Okt. 2024

Interkommunale Kooperation zur Stärkung des regionalen Hochwasserrisikomanagements

Christian Stein, Daniel Bittner, Gerd Demny, Jens Hasse, Dietmar Jansen, Lothar Kirschbauer, Susanne Kozerke, Anton Schumann und Julian Struck

J. Birkmann et al. (Hrsg.), *Die Flutkatastrophe 2021*, Sustainable Development Goals (SDG) – Umsetzung in Praxis, Lehre und Entscheidungsprozessen,
https://doi.org/10.1007/978-3-658-48688-4_14

14.1 Einleitung

Die Flutkatastrophe vom Juli 2021 hat verdeutlicht, dass zur Reduktion von Hochwasserrisiken ein integriertes Hochwasserrisikomanagement (HWRM) notwendig ist, bei dem Akteure über Fachgrenzen und administrative Ebenen hinweg einzugsgebietsbezogen zusammenarbeiten. Im Rahmen des KAHR-Projekts wurden drei von der Flutkatastrophe besonders stark betroffene Teilregionen begleitet, um zusammen mit Praxispartnern einen Beitrag zur Stärkung der Hochwasserresilienz und zu einem klimaresilienten Wiederaufbau zu leisten. Diese Teilregionen mit ihren Gemeinden, kreisangehörigen Städten und Landkreisen sind in den Einzugsgebieten der Ahr in Rheinland-Pfalz sowie der Erft und Inde/Vicht in Nordrhein-Westfalen gelegen. In allen Teilregionen wurden nach der Flutkatastrophe neue Governance-Strukturen und -Prozesse auf- bzw. ausgebaut, um das Hochwasserrisikomanagement zu verbessern. Dieser Beitrag beschreibt, welche institutionellen Anpassungsprozesse in den Teilregionen eingeleitet wurden, und fasst erstmals Lösungsansätze sowie Herausforderungen („Lessons Learned") im Hinblick auf die regionale Kooperation beim Hochwasserrisikomanagement zusammen. Drei Fragestellungen werden in dem Beitrag behandelt:

1. Welche Strukturen und Prozesse zur Kooperation beim Hochwasserrisikomanagement wurden auf- bzw. ausgebaut?
2. Welche Lösungsansätze und Herausforderungen sind in den Teilregionen festzustellen?
3. Welche übergreifenden Erkenntnisse können aus den Erfahrungen abgeleitet werden?

Der Fokus des Beitrags liegt auf den Anpassungen in institutionellen bzw. Governance-Strukturen und -Prozessen, die für ein zukunftsfähiges Hochwasserrisikomanagement notwendig sind. Konkrete Maßnahmen zur Hochwasser- und Starkregenvorsorge, deren Wirkungen etc. werden in anderen Beiträgen vertiefend dargestellt (siehe ▶ Kap. 11 und 12). Eine vertiefende Betrachtung der Gemeinsamkeiten und Unterschiede zwischen den Teilregionen, u. a. hinsichtlich der institutionellen und gesetzlichen Rahmenbedingungen in Rheinland-Pfalz und Nordrhein-Westfalen, liefert wichtige Erkenntnisse dazu, welche Ansätze dazu beitragen können, das Hochwasserrisiko in (Mittelgebirgsregionen) in Deutschland und darüber hinaus zu reduzieren.

Die europäische Hochwasserrisikomanagementrichtlinie (EG-HWRM-RL) gilt rechtsverbindlich, wobei die regional angepasste Umsetzung auf Landes- und Kommunalebene der Verantwortung der Mitgliedsstaaten obliegt. Die EG-HWRM-RL schreibt die Risikobewertung, die Erstellung von Hochwasserrisiko- und -gefahrenkarten sowie die Aufstellung von Hochwasserrisikomanagementplänen vor, welche alle sechs Jahre überarbeitet werden müssen (Europäische Union [EU] 2007). Hierbei wird ein umfassendes Hochwasserrisikomanagement auf Ebene der Flusseinzugsgebiete angestrebt, welches über technische und bauliche Maßnahmen hinausgeht und die Einbeziehung aller direkt oder indirekt verknüpften Handlungsfelder und Akteure erfordert. Ergänzend formuliert die Länderarbeitsgemeinschaft Wasser (LAWA) Handlungsempfehlungen an die Bundesländer, die stärkere Nutzungsbeschränkungen in hochwassergefährdeten Gebieten, Maßnahmen des natürlichen Wasserrückhalts und technischen Hochwasserschutzes sowie die Ausweisung von Überschwemmungsgebieten vorsieht, deren Freihaltung angestrebt werden sollte. Für einen solchen integrativen Ansatz ist die Einbeziehung der verschiedenen Aspekte der Wasserwirtschaft, Raumplanung, Landnutzung, des Naturschutzes etc. auf internationaler, nationaler, regionaler und lokaler Ebene notwendig.

Durch die konsequente Umsetzung eines integrierten Hochwasserrisikomanagements werden alle Ziele und Vorgaben der EG-HWRM-RL abgedeckt (Merz et al. 2011; Müller 2010).

Da kein absoluter Schutz vor Hochwasserereignissen möglich ist und Risiken kontinuierlich neu evaluiert werden müssen, muss Hochwasserrisikomanagement als ein fortlaufender und iterativer Prozess verstanden werden (Müller 2010). Hierbei existieren verschiedene Konzepte bezüglich der Ausgestaltung eines HWRM-Zyklus. Ihnen gemeinsam ist die Einteilung in die Phasen der Bewältigung, Regeneration und Vorsorge wobei das Hochwasserereignis zwischen die Phasen der Vorsorge und Bewältigung fällt (Merz et al. 2011). Die Phase der Hochwasserbewältigung umfasst Maßnahmen, die der Begrenzung des Schadensausmaßes sowie der Dauer eines Ereignisses dienen. Während die Phase der Regeneration auf die Wiederherstellung der Voraussetzungen für den alltäglichen Betrieb abzielt, dient die Vorbeugung der Reduktion zukünftiger Vulnerabilität. In der Phase der Vorbeugung werden Maßnahmen ergriffen, die die Risikobewertungen vergangener Hochwasser berücksichtigen. Auf Grundlage der verbesserten Risikobewertung sollte sich das Hochwasserschutzniveau durch Maßnahmen der Vorbeugung positiv entwickeln (Patt und Jüpner 2020).

Das Hochwasser vom Juli 2021 sowie der anschließende Wiederaufbau (Regeneration) verdeutlichen die Notwendigkeit, die verschiedenen Phasen des HWRM-Zyklus integrativ zu betrachten. Das extreme Ereignis macht einen umfassenden Wiederaufbau notwendig. Der Fokus dieses Beitrags liegt auf den Phasen der Regeneration und der Vorbeugung, da der Wiederaufbauprozess ein Gelegenheitsfenster darstellt, in welchem die Resilienz gegenüber zukünftigen extremen Hochwasserereignissen gestärkt und so künftige Vulnerabilität minimiert werden können.

14.2 Aufbau neuer Governance-Strukturen und Prozesse zum Hochwasserrisikomanagement in den Teilregionen

14.2.1 Ahr-Einzugsgebiet

Das Einzugsgebiet der Ahr, ein Gewässer 2. Ordnung, liegt im nördlichen Teil von Rheinland-Pfalz und im südlichen Teil von Nordrhein-Westfalen (siehe ◘ Abb. 14.1). Es umfasst eine Größe von rd. 896 km^2, von denen rd. 214 km^2 (rd. 24 %) in Nordrhein-Westfalen liegen. Die Ahr entspringt in der Ortschaft Blankenheim (Nordrhein-Westfalen) der gleichnamigen Gemeinde und mündet nach rd. 85 km Fließweg bei Sinzig in den Rhein. Rd. 17 km (rd. 20 %) verlaufen dabei durch das südliche Nordrhein-Westfalen.

Der Landkreis Ahrweiler arbeitet seit der Flut vom Juli 2021 mit diversen Partnern an einer Optimierung der Hochwasservorsorge im Ahrtal. Das Hochwasserrisikomanagement für den Kreis Ahrweiler und das Ahr-Einzugsgebiet besteht aus vier zentralen Bausteinen: die Gewässerwiederherstellung (GWK), die Erarbeitung eines überörtlichen Maßnahmenplans (üMP), die Aufstellung, Fortschreibung und Umsetzung örtlicher Hochwasser- und Starkregenvorsorgekonzepte (öHSVK) durch die Kommunen, sowie die Zusammenarbeit von diversen Akteuren zu weiteren Bausteinen der Hochwasservorsorge im Rahmen der „Hochwasserpartnerschaft Ahr“.

Durch die Flutkatastrophe wurden im Ahr-Einzugsgebiet auch die Wasserläufe und Uferbereiche der Fließgewässer zum Teil schwer beschädigt. Im Rahmen des Wiederauf-

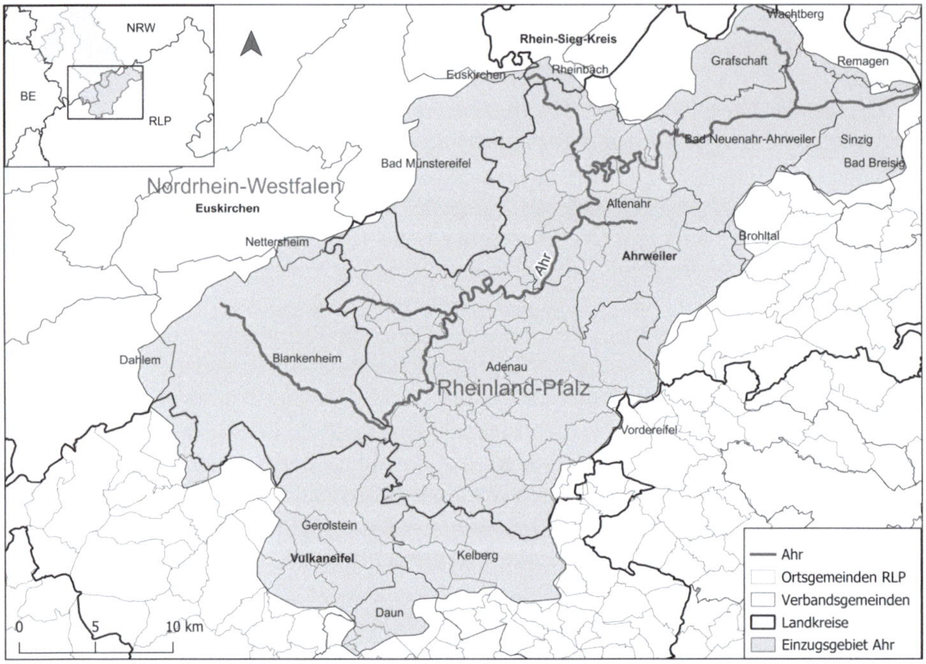

Abb. 14.1 Ahr-Einzugsgebiet mit kommunalen und Landesgrenzen. (Eigene Darstellung)

baus wurde vom Landkreis ein Gewässerwiederherstellungskonzept (GWK) für die Ahr und die Zuflüsse 2. Ordnung in Auftrag gegeben. Ziele des GWK sind, neben der Gefahrenbeseitigung, die Verbesserung der Gewässerstruktur und des Abflusses sowie die Schaffung von natürlichen Rückhaltefunktionen. Für die Erstellung des GWK wurden entlang der Ahr fünf Teilabschnitte entsprechend der kommunalen Grenzen gebildet, die sich an den jeweiligen Gewässereinheiten orientieren. Ingenieurbüros erarbeiteten für die Teilabschnitte Maßnahmenvorschläge, welche im Rahmen von öffentlichen Veranstaltungen in den jeweiligen Kommunen vorgestellt wurden. Eine besondere Herausforderung bei der Umsetzung der über 1000 festgestellten Maßnahmen sind die Abstimmungen mit den Kommunen und anderen Maßnahmenträgern sowie den Flächeneigentümer:innen. Zur effektiven Umsetzung des GWK wurde ein Projektsteuerungsbüro beauftragt, das aus dem Aufbauhilfefonds des Bundes und der Länder finanziert wird.

Der Landkreis Ahrweiler, die kreisangehörigen Kommunen sowie die im Ahr-Einzugsgebiet gelegenen Kreise Vulkaneifel (Rheinland-Pfalz) und Euskirchen (Nordrhein-Westfalen) mit der Gemeinde Blankenheim und der Stadt Bad Münstereifel arbeiten im Rahmen einer Kooperationsvereinbarung an einem überörtlichen Maßnahmenplan (üMP). Darin werden Hochwasserrisikovorsorgemaßnahmen mit überörtlicher Wirkung erarbeitet, spezifiziert und hinsichtlich ihrer Wirkung für die Hochwasser- und Starkregenvorsorge analysiert und bewertet. Ziel des üMP ist es, Maßnahmen zu identifizieren, die durch überörtliche Zusammenarbeit im gesamten Einzugsgebiet der Ahr über administrative Grenzen hinweg die örtlichen Hochwasser- und Starkregenvorsorgekonzepte (öHSVK) der Kommunen ergänzen und damit zu einer Risikominderung beitragen. Der üMP betrachtet das gesamte Einzugsgebiet und enthält technische sowie natürliche Hochwasserschutzmaßnahmen, die durch interkommunale Zusammenarbeit umgesetzt werden können und die lokalen Maßnahmen ergänzen. Für die Fortschreibung der Hochwasser- und Starkregenvorsorgekonzepte wurden vom Ministerium für Klimaschutz, Umwelt, Energie und Mobilität (MKUEM) des Landes Rheinland-Pfalz im Mai 2024 „Hinweise zur Fortschreibung eines örtlichen Hochwasser- und Starkregenvorsorgekonzepts" (Rheinland-Pfalz, Ministerium für Klimaschutz, Umwelt, Energie und Mobilität 2024a) veröffentlicht. Erkenntnisse aus dem KAHR-Projekt und Vorschläge aus der Bürgerschaft fließen in die Planung ein. Eine Lenkungsgruppe, bestehend aus Fachleuten der beteiligten Kommunen und der Kreisverwaltungen, trifft sich mehrmals im Jahr, um Sachstände und Zwischenergebnisse zu diskutieren. Weitere Akteure werden zu spezifischen Themen eingebunden.

Der überörtliche Maßnahmenplan und die Gewässerwiederherstellung werden in das Hochwasserrisikomanagement für das Ahrtal eingebunden. Weitere Bausteine des Hochwasserrisikomanagements werden im Rahmen der Hochwasserpartnerschaft Ahr bearbeitet. Diese besteht seit 2014 und ist ein freiwilliger Zusammenschluss von Städten und Gemeinden des Landkreises Ahrweiler sowie Kommunen aus angrenzenden Landkreisen mit dem Ziel, ganzheitliche Hochwasservorsorge unter Einbindung verschiedener Sektoren (Bauleitplanung, Land- und Forstwirtschaft, Katastrophenschutz, kritische Infrastruktur etc.) zu betreiben. Ein Austausch erfolgt im Rahmen regelmäßiger Workshops zu hochwasserrelevanten Themen, deren Häufigkeit nach der Flutkatastrophe vom Juli 2021 intensiviert wurde. In das überörtliche Hochwasserrisikomanagement werden auch die örtlichen Hochwasser- und Starkregenvorsorgekonzepte eingebunden, welche in den Kommunen vorliegen, allerdings nach der Flut 2021 teilweise angepasst werden müssen.

Um den überörtlichen Maßnahmenplan sowie Maßnahmen der Gewässerunterhaltung umzusetzen, befindet sich ein Gewässerzweckverband in der Gründung. Entsprechende Grundsatzbeschlüsse des Landkreises Ahrweiler und der kreisangehörigen Kommunen liegen vor. Verbandsordnung und Finanzierungschlüssel werden aktuell erarbeitet (Stand: September 2024).

14.2.2 Erft-Einzugsgebiet

Die Erft ist ein Gewässer 2. Ordnung, dessen Einzugsgebiet im südlichen Nordrhein-Westfalen liegt (siehe ◘ Abb. 14.2). Das Einzugsgebiet umfasst rd. 1918 km², wovon lediglich ein kleiner Teil des Teileinzugsgebiets Swist in Rheinland-Pfalz liegt. Die Erft entspringt in der Gemeinde Nettersheim in der Eifel und mündet in Neuss in den Rhein.

Die Interkommunale Hochwasserschutzkooperation Erft (hwsErft) wurde nach dem Juli-Hochwasser 2021 gegründet, um die Erstellung von aufeinander abgestimmten kommunalen Hochwasserschutzkonzepten (HWSK) zu koordinieren. Diese stellen die Basis eines ersten interkommunalen HWSK dar, in dem konkrete und priorisierte Maßnahmen zur langfristigen Verbesserung des Hochwasserschutzes benannt werden. Im September 2024 bilden 16 Kommunen, drei Kreise und der Erftverband die hwsErft. Langfristig soll diese das gesamte Erft-Einzugsgebiet umfassen. Der Erftverband, als einer der sondergesetzlichen Wasserverbände in NRW, ist für die Projektsteuerung und Koordination der interkommunalen Beiträge verantwortlich, die Kommunen erstellen kommunale HWSK.

Im Rahmen der Projektsteuerung behält der Erftverband die Terminübersicht und koordiniert die Arbeitspakete. Zudem finden Abstimmungen mit der Bezirksregierung Köln und dem Ministerium für Umwelt, Naturschutz und Verkehr des Landes Nordrhein-Westfalen bezüglich der HWSK statt, wodurch eine standardisierte Vorgehensweise für die

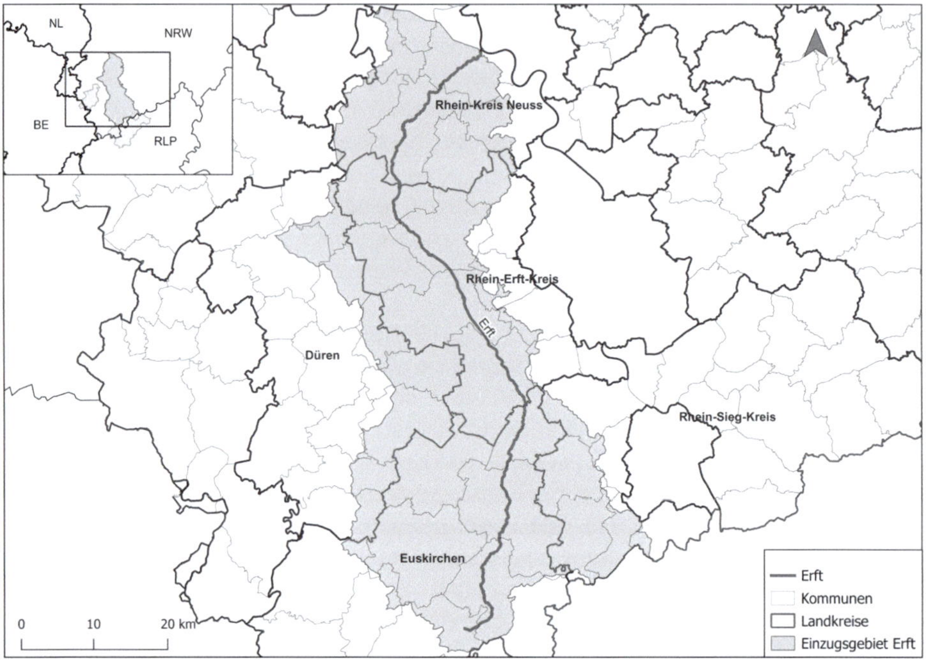

Abb. 14.2 Erft-Einzugsgebiet. (Eigene Darstellung)

hwsErft gewährleistet wird. Ein Beispiel hierfür sind die in enger Abstimmung erarbeiteten Musterförderanträge und -leistungsbeschreibungen für kommunale HWSK, die auch als Grundlage für künftige Leitfäden zur Erstellung (inter-)kommunaler HWSK in NRW dienen können. Die vereinheitlichten Vorgehensweisen haben die Förder- und Ausschreibungsprozesse für die Kommunen vereinfacht sowie eine schnellere Unterlagenprüfung durch die Bewilligungsbehörden und Qualitätssicherung gewährleistet. Der Erftverband unterstützt des Weiteren Aufgaben der Kommunikation und Öffentlichkeitsarbeit (Bittner et al. 2023).

Die interkommunalen Beiträge der hwsErft beinhalten die Schadenspotenzialermittlung, den Ausbau technischer Hochwasserschutzmaßnahmen und die Rückgewinnung von Überschwemmungsgebieten. Das beinhaltet die Planung und Umsetzung von Hochwasserrückhaltebecken, deren Auswirkungen sich signifikant auf das Abflussgeschehen auswirken und Abflussspitzen mindern. Die Rückgewinnung von Überschwemmungsgebieten durch technische Maßnahmen und Renaturierungen erhöht den temporären Wasserrückhalt in der Aue und mindert Fließgeschwindigkeiten. Die interkommunalen Beiträge ergänzen angestrebte kommunale Hochwasserschutzmaßnahmen (Bittner et al. 2023).

Die Kommunen erarbeiten die kommunalen HWSK. Die darin aufgezeigten Hochwasserschutzmaßnahmen sollen den Hochwasserschutz verbessern, ohne die Situation der Ober- und Unterlieger zu verschlechtern. Die Planung und spätere Umsetzung erfolgen auf Grundlage der Hochwassergefährdung und ermittelten Schadenspotenziale. Potenzielle Maßnahmen werden einer Restriktions- und Wirksamkeitsanalyse unterzogen und vom Erftverband bezüglich ihrer interkommunalen Wirksamkeit beurteilt (Bittner et al. 2023). Ein regelmäßiger Austausch der hwsErft wird in den Teilprojekten ermöglicht und stellt eine interkommunale Betrachtung und Abstimmung sicher. Das Arbeitspaket „Integration weiterer Konzepte“ ermöglicht eine integrative Betrachtung z. B. der Hochwasser- und Starkregenvorsorge und schafft die Grundlage zur ganzheitlichen Betrachtung wasserwirtschaftlicher Extrema. Die Einbindung der Bürger:innen erfolgt über Workshops und Ortsbegehungen, in denen Maßnahmenvorschläge aufgenommen werden, die auf persönlichen Vor-Ort-Erfahrungen beruhen. Über das WebGIS der hwsErft können Hochwasser- und Starkregengefahrenkarten sowie bestehende und geplante Hochwasserschutzmaßnahmen eingesehen und verfolgt werden (Erftverband o. J.).

Die koordinative Rolle des Erftverbands erweist sich im bisherigen Projektverlauf aufgrund der wasserwirtschaftlichen Expertise im Erft-Einzugsgebiet sowie des verfügbaren Fachpersonals als vorteilhaft. Auch die bereits vor dem Hochwasser 2021 bestehenden informellen und formellen Verbindungen des Erftverbandes erweisen sich als hilfreich, da dadurch frühzeitige und zielgerichtete Absprachen mit den Bezirksregierungen und den Ministerien ermöglicht wurden (Bittner et al. 2023).

14.2.3 Einzugsgebiet von Inde und Vicht

Das Einzugsgebiet von Inde und Vicht liegt in Nordrhein-Westfalen und ist durch Mittelgebirgsbäche sowie dichte Siedlungsstrukturen geprägt (siehe Abb. 14.3). Beide Flüsse sind in Teilen durch Siedlungen, Industriegebiete und Bergbau stark anthropogen geprägt (Reineke et al. 2006). Die Inde entspringt in Belgien, hat eine Gesamtlänge von

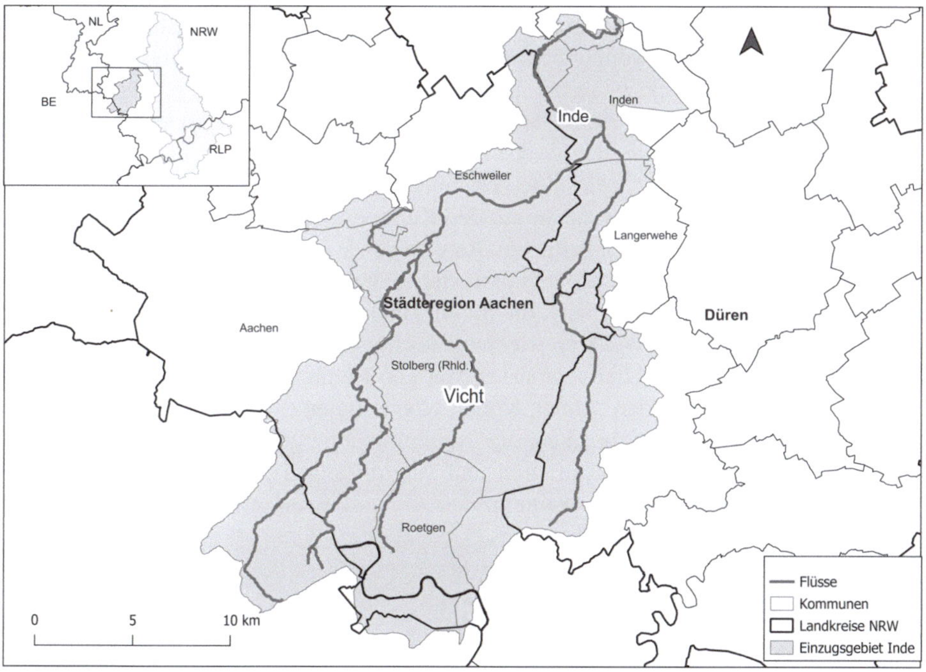

Abb. 14.3 Einzugsgebiet von Inde und Vicht. (Eigene Darstellung)

47 km und ein Einzugsgebiet von 344 km^2. Die Vicht mit 23 km Länge und einer Einzugsgebietsfläche von 104 km^2 entspringt in Roetgen (BRD) und fließt durch mehrere Gemeinden der Städteregion Aachen bis zur Mündung in die Inde (Wasserverband Eifel-Rur [WVER] 2024).

Unmittelbar nach der Flutkatastrophe wurde der Masterplan „Hochwasserresiliente Stadt- und Gebietsentwicklung für das Einzugsgebiet von Inde und Vicht" entwickelt (Kaleß et al. 2023). Zielsetzung war die Erhöhung der Hochwasserresilienz, um auf zukünftige extreme Ereignisse besser vorbereitet zu sein. Dazu wurden potenzielle Rückhaltemöglichkeiten und vorhandene technische Bauwerke aufgeführt und Maßnahmen in Anlehnung an das übergeordnete HWRM-NRW besprochen und bewertet. Der Wiederaufbau soll unter Berücksichtigung verschiedener Interessen hochwasserresilient und nachhaltig gestaltet werden, und Fehlinvestitionen sollen vermieden werden.

Insgesamt wurden für das Einzugsgebiet ca. 200 vielversprechende Maßnahmen identifiziert. In sechs Monaten sammelten Expert:innen im Rahmen von Workshops Maßnahmenvorschläge. Diese wurden im KAHR-Projekt sukzessive in Potenzialstudien auf Wirksamkeit überprüft (siehe ▶ Kap. 12). Mit der Schadenspotenzial- und der Kosten-Nutzen-Analyse lassen sich die wirkungsvollsten Projekte weiter in Richtung Umsetzung verfolgen.

Der Maßnahmenplan ist dynamisch und wird stetig angepasst. Projektideen und Maßnahmenvorschläge sind über ein Online-GIS-Portal des Wasserverbands Eifel-Rur (WVER) für die Öffentlichkeit zugänglich und werden regelmäßig aktualisiert (WVER o. J.). In den betroffenen Kommunen fanden öffentliche Veranstaltungen statt, um Maßnahmen zu erläutern und die Bevölkerung für die Möglichkeiten und Notwendigkeit der Eigenvorsorge zu sensibilisieren. Ein Ziel dieses Prozesses ist, die verfügbaren Flächen, wenn sie nicht für den Hochwasserschutz oder die Hochwasserresilienz wirksam sind, dann ggf. für das Starkregenmanagement oder die Klimaanpassung zu nutzen, mindestens aber frei von weiteren Schadenserwartungswerten zu halten. Dafür ist ein enger Austausch über viele Jahre mit allen Akteur:innen Voraussetzung, um diese Ziele langfristig allen ins Bewusstsein zu rücken und in ihr Handeln einzubinden.

Mit dieser Erkenntnis beschloss der Städteregionstag der Städteregion Aachen im März 2022 die Einrichtung einer interkommunalen Arbeitsgruppe „Regionales Hochwasserrisikomanagement" zusammen mit den Kommunen und dem WVER. Diese neue Organisationsstruktur soll dazu beitragen, Risiken sowie Präventivmaßnahmen zu ermitteln, zu bewerten und umzusetzen. Im Fokus stehen die Vernetzung und Koordination der vielfältigen Aktivitäten und Akteure, die für ein Hochwasserrisikomanagement erforderlich sind, und die ständige Weiterentwicklung des Masterplans.

Dafür sind Prozesse mit den regionsangehörigen Kommunen sowie anderen relevanten Akteur:innen (Ministerien, Ämter, Kreise, Hochschulen etc.) zu koordinieren, strategische Entscheidungen vorzubereiten, die Kommunikation gegenüber der Öffentlichkeit und politischen Gremien durchzuführen und Abgrenzungen gegenüber dem Katastrophenschutz und Klimaschutz zu ermitteln.

14.3 Erkenntnisse und Lösungsansätze für ein zukunftsfähiges Hochwasserrisikomanagement

14.3.1 Interkommunale und regionale Kooperation zur Abstimmung von HWRM-Konzepten im Einzugsgebiet

Ein effektives Hochwasserrisikomanagement (HWRM) bedarf einer ganzheitlichen überregionalen und interkommunalen Betrachtung, da sich das Abflussverhalten von Fließgewässern nicht an administrative Grenzen hält. Daher empfiehlt sich eine einzugsgebietsweite Betrachtung der Aufgaben im HWRM, wie u. a. im „10-Punkte Arbeitsplan Hochwasserschutz in Zeiten des Klimawandels" des Landes Nordrhein-Westfalen gefordert (Nordrhein-Westfalen, Ministerium für Klimaschutz, Umwelt, Landwirtschaft, Natur- und Verbraucherschutz des [MULNV] 2022). Ein interkommunales und integriertes Wassermanagement in Flusseinzugsgebieten, wie u. a. auch in der EU-Wasserrahmenrichtlinie gefordert, ist und bleibt jedoch eine Herausforderung, u. a. durch die verschiedenen Zuständigkeiten und Verwaltungsstrukturen (Länder, Ministerien, Bezirksregierungen, Kreise, Wasserverbände, Kommunen) sowie die Einbindung weiterer relevanter Akteure (Bürger:innen, Flächenbesitzer:innen etc.).

In allen drei Teilregionen wird sukzessiv daran gearbeitet, die regionale Kooperation für das gesamte (Teil-)Einzugsgebiet (EZG) auszubauen, um schlussendlich alle politisch-administrativen Gebietskörperschaften in das Hochwasserrisikomanagement einzubeziehen. Solche Strukturen, bei denen viele Akteure mit unterschiedlichen Verantwortlichkeiten zusammenarbeiten müssen, gewährleisten, „dass die formalen Zuständigkeiten erhalten bleiben, die Kooperationspartner sich jedoch [in Ihrem Bestreben] gegenseitig [. . .] unterstützen können" (Bittner et al. 2023, S. 279). Eine grundlegende Herausforderung besteht dabei darin, neue Wege und Ansätze zu finden, um kooperative, interkommunale Bestrebungen in bestehende Strukturen und -prozesse zu integrieren und komplementär zu ergänzen (Bodin 2017).

Im EZG der Ahr findet über die Hochwasserpartnerschaft Ahr seit 2014 ein Austausch über Gemeinde-, Kreis- und Landesgrenzen hinaus statt. Dieser wurde nach der Flut deutlich intensiviert und mittels Kooperationsvereinbarungen auch über die Kreis- und Ländergrenzen hinweg verbindlicher gemacht, da es in Rheinland-Pfalz, im Unterschied zu Nordrhein-Westfalen, keine sondergesetzlichen Wasserverbände gibt. Im EZG von Inde und Vicht wurden der Masterplan-Prozess, ausgehend von den am stärksten betroffenen Kommunen Stolberg und Eschweiler, auf weitere Kommunen im EZG ausgeweitet und das Regionale Hochwasserrisikomanagement der Städteregion Aachen mit Akteuren aus dem Kreis Düren verknüpft. Ähnliche Strategien wurden bereits auf die Einzugsgebiete Wurm, Urft/Olef, Obere und Untere Rur angepasst und auf die Randbedingungen der jeweiligen Zuständigkeiten und damit auf das gesamte Verbandsgebiet des WVER ausgeweitet. Im EZG der Erft wurden die bestehenden Strukturen im Verbandsgebiet des Erftverbandes nach dem Juli-Hochwasser 2021 weiter ausgebaut und im Jahr 2022 die Interkommunale Hochwasserschutzkooperation Erft gegründet.

14.3.2 Hochwasserrisikokommunikation und Beteiligung von Akteuren

Kommunikation ist ein wesentlicher Bestandteil des Hochwasserrisikomanagements, um Akteure für Hochwassergefahren und -risiken zu sensibilisieren und diese in Problemlösungen einzubeziehen (Kuhlicke et al. 2022). In einer Umfrage, welche im Rahmen einer KAHR-Publikation veröffentlicht wurde, gaben über 80 % der Befragten an, dass sie vor dem Hochwasser 2021 nicht wussten, dass sie in einem hochwassergefährdeten Gebiet wohnen (Truedinger et al. 2023). Ergebnisse wie diese veranschaulichen die Notwendigkeit einer stetigen Hochwasserrisikokommunikation. Erfolgreiche Risikokommunikation beinhaltet die Sensibilisierung für Gefahrenlagen und Eigenvorsorge. Hierbei muss Verständnis für das Zusammenspiel von staatlicher und privater Risikovorsorge erzeugt werden, da diese sich beeinflussen und im besten Fall ergänzen sollten. Gleichzeitig muss kontinuierlich verdeutlicht werden, dass ein absoluter Hochwasserschutz nicht möglich ist.

Hinweis für die Praxis

Für die Sensibilisierung und Partizipation der privaten Akteure haben sich in den Pilotregionen verschiedene Formate bewiesen, beispielsweise Ortsbegehungen, Bürger:innenworkshops oder interaktive Online-Karten (WebGIS). Die Partizipation der Bürger:innen hat ebenfalls positive Auswirkungen auf die Maßnahmengestaltung, da hierbei lokales Wissen der Bevölkerung bezüglich vergangener Hochwasser integriert werden kann. Solche Ansätze sollten idealerweise in Rahmen einer umfassenden Kommunikationsstrategie entwickelt, eingebettet und umgesetzt werden.

In der Praxis erweist sich auch die Kooperation mit Akteuren der Land- und Forstwirtschaft und den Naturschutzverbänden als vorteilhaft. Diese verfügen häufig über Flächen, die für die Umsetzung von Hochwasser- und Starkregenvorsorgemaßnahmen relevant sind, oder sollten wie Naturschutzverbände in die Planung einbezogen werden. Durch eine frühzeitige und kontinuierliche Einbindung können das Verständnis für die jeweiligen Perspektiven und Synergiemöglichkeiten bei der Umsetzung von Maßnahmen gestärkt und Konflikte vermieden werden. Gegebenenfalls ist dadurch die Kompromissbereitschaft für Mehrfachnutzungen z. B. bei Rückhalteflächen höher.

14.3.3 Adaptives Hochwasserrisikomanagement

Die Entwicklung und Umsetzung eines integrierten HWRM-Konzeptes ist ein komplexer und langfristiger Prozess. Für eine integrierte Planung und Steuerung im HWRM sind Prozesse und institutionelle Kapazitäten weiter auszubauen. Dabei müssen Pläne regelmäßig an sich verändernde Rahmenbedingungen und aufgrund von neuen Erkenntnissen angepasst werden. Beispielsweise können neue oder veränderte Brücken am Gewässer neue Modellierungen notwendig machen oder die Verfügbarkeit von Flächen Handlungsspielräume bei der Umsetzung von Maßnahmen verändern. Hochwasserrisiken und damit einhergehend die Möglichkeiten der Risikoreduktion sind keine Konstanten. Die Bedingungen von Hochwasserereignissen und -schäden verändern sich im Laufe der Zeit durch vielfältige Einflüsse wie Landnutzung, Klimawandel und Werteakkumulation in gefährdeten Flächen. Des Weiteren ist durch den Klimawandel damit zu rechnen, das klimatische Extremereignisse in ihrer Häufigkeit und Intensität zunehmen und sich das Niederschlagsregime verändert (Pörtner et al. 2022).

Hinweis für Entscheidungsprozesse

Vor dem Hintergrund von zunehmenden Unsicherheiten durch den Klimawandel und sich verändernden Rahmenbedingungen sind traditionelle Strategien zum Umgang mit Hochwasserrisiken zu überdenken und adaptive Managementansätze besonders erfolgversprechend, da diese vergleichsweise einfach an veränderte Rahmenbedingungen angepasst werden können (Grambow 2008; Merz et al. 2011). Solche adaptiven Ansätze betonen die Bedeutung von Flexibilität und kontinuierlichem Lernen im Umgang mit Unsicherheiten und zukünftigen Herausforderungen im HWRM. Das Ziel ist nicht die Erstellung eines abgeschlossenen HWRM-Plans, sondern die Initiierung eines Strategieprozesses, der aktiv verschiedene Akteur:innen einbindet und kontinuierlich weiterentwickelt wird. In den Teilregionen werden beispielsweise Maßnahmenpläne in Online-GIS-Plattformen gebündelt, diversen Akteur:innen zugänglich gemacht und kontinuierlich fortgeschrieben, so dass die Entwicklung von der Idee bis zur Umsetzung nachvollziehbar ist.

Der HWRM-Zyklus beinhaltet implizit einen adaptiven Ansatz, da die kontinuierlichen Risikobewertungen in der Phase der Vorbeugung auf vergangenen Ereignissen beruhen und die Berücksichtigung zukünftiger Entwicklungspfade der Umwelt und Gesellschaft erfordern (Grambow 2008; Greiving et al. 2013).

Adaptive Strategien können dazu beitragen, Synergien zwischen Klimaschutz-, Klimaanpassungs- und Nachhaltigkeitszielen herzustellen, indem sie Anpassungsmaßnahmen an sich verändernde Bedingungen flexibel gestalten (Birkmann 2023). Dazu zählen auch „No-Regret-Maßnahmen“ oder „Multi-Use-Ansätze“, deren Nutzen-Kosten-Untersuchungen häufig positiv ausfallen – unabhängig davon, wie sich die Wahrscheinlichkeit eines klimatischen (Extrem-)Ereignisses zukünftig entwickelt (Merz et al. 2011).

14.3.4 Potenzial von Modellierungs- und Visualisierungstools – in Planung und Kommunikation nutzen

Mit Hilfe von Visualisierungen können Ergebnisse hydrologischer und hydraulischer Berechnungen für die Bevölkerung sichtbar und verständlich dargestellt werden. Die Bürger:innen und auch die Hilfskräfte benötigen i. d. R. keine Abflussmengen, sondern Wasserstände, die dann auch für einzelne Objekte und Bereiche allgemeinverständlich dargestellt werden müssen. So reicht es für den Bürger und die Bürgerin nicht aus, dass am Pegel xy ein Wasserstand von x Metern erwartet wird, sondern sie/ihn interessiert der erwartete Wasserstand am eigenen Haus, um gezielt Maßnahme ergreifen zu können. In Rheinland-Pfalz wird hierfür das System „scenarify“ eingeführt, das mit Hilfe einer digitalen 3D-Hochwassersimulation die Wasserstände im gewünschten Bereich darstellt (VRVis o. J.). Hierzu wird ein digitales 3D-Simulations-Fachsystem für die Themen Sturzflut und Hochwasser aufgebaut (Rheinland-Pfalz, Ministerium für Klimaschutz, Umwelt, Energie und Mobilität 2023).

Die Verwendung neuer Algorithmen bietet zahlreiche Vorteile und Potenziale bei der Modellierung und Kommunikation, die sie zu sehr leistungsfähigen Werkzeugen im Bereich des Hochwasserrisikomanagements qualifizieren. Zu den Vorteilen gehört eine außerordentliche Schnelligkeit. So kann z. B. eine 60-Minuten-Dauerstufe KOSTRA-DWD für sehr große Einzugsgebiete im 2×2-Meter-Raster in wenigen Sekunden berechnet werden. Es sind benutzerfreundliche Editoren vorhanden, die schnell die Wirkung von Hochwasserschutzelementen berechnen, Ergebnisse überprüfen und exportieren sowie Eingangsparameter, digitale Geländemodelle (DGMs) und Zeitreihen austauschen können. Infiltrationsprozesse werden über Bodenparameter und Interzeption über Landnutzung abgebildet und parallel zur Ergebnisauswertung mitgerechnet. Hervorzuheben sind Möglichkeiten, Kanalnetzmodelle zu inkludieren, was Vorteile für das Starkregenmanagement bietet. Eine Verfeinerung des Berechnungsrasters kann automatisch während der Berechnung erfolgen.

Diese Simulationsmodelle bieten die direkte Visualisierung des Einzugsgebietes als 3D-Darstellung. Ein schnelles Durchspielen von Hochwasserschutzmaßnahmen ist vorteilhaft, da es die Konzepterstellung erheblich erleichtert. Niederschlagsraster können schnell durchgerechnet und so möglicherweise für Prognose-Szenarien eingesetzt werden. Das schnelle Berechnen und Darstellen der Auswirkungen von verschiedenen Szenarien kann beispielsweise im Katastrophenfall bei einem Deichbruch die Entscheidungsfindung beschleunigen und im Einsatzfall gezielt den Einsatz von Hilfskräften optimieren.

Für alle diese Aufgaben gilt es, Vertrauen in die Institution und deren fachlich begründete Entscheidungen (wieder-)aufzubauen; dies ist somit eine der Hauptaufgaben im Hochwasserrisikomanagement. Dafür haben sich die neuen Tools mit ihren Visualisierungsmöglichkeiten während Dialogveranstaltungen eindeutig als „Gamechanger" qualifiziert.

14.3.5 Hochwasserschutz- und Starkregen integrieren

14

Die Hochwasserkatastrophe 2021 war nicht nur ein Hochwasserereignis, sondern mit Niederschlagshöhen von über 200 mm am Tag ein extremes Starkregenereignis, das vielerorts auch fernab von Gewässer- und Bachläufen Überflutungsschäden verursacht hat. In Gewässernähe selbst konnte oftmals nicht unterschieden werden, ob die Überflutung aus dem Umfeld der betroffenen Siedlung oder aus dem Gewässer selbst kam, da beides zeitgleich auftrat. Das gleichzeitige Auftreten beider Phänomene ist insbesondere in kleinen Einzugsgebieten mit Flächengrößen kleiner 10 bis 25 km^2 zu beobachten (sog. Kopfgebiete). Fachlich wäre daher eine integrative Berechnung von Überflutungen aus Starkregen und Gewässerhochwasser erforderlich. Methodisch wird dies jedoch, z. T. aus historischen Gründen, mit unterschiedlichen Werkzeugen durchgeführt. In NRW erschwert zudem das derzeit bestehende Förderinstrumentarium die integrative Betrachtung beider Überflutungsursachen, da sie individuellen, voneinander getrennten, Förderrichtlinien unterliegen.

Fachlich können beide Überflutungsursachen über eine integrative Starkregen- und Hochwassersimulation zusammengeführt werden. Im Ergebnis wird dann nur noch ein Modell benötigt, um Betroffenheiten aus Starkregen und Flusshochwasser zu ermitteln und geeignete Schutzmaßnahmen zu entwickeln. Die integrative Maßnahmenplanung ist vor allem deshalb wichtig, weil Linienschutzmaßnahmen, wie z. B. Mauern und Deiche, den Starkregenabfluss in Richtung Gewässer behindern könnten. Umgekehrt werden sich Rückhaltemaßnahmen gegen Starkregen auf dem freien Feld auch günstig auf die Hochwasserentstehung im Bachlauf auswirken.

Die gemeinsame Betrachtung von Starkregen und Flusshochwasser ist nicht nur für kleine Kopfgebiete relevant. In mittelgroßen Einzugsgebieten treten Starkregen und Flusshochwasser i. d. R. zeitlich aufeinanderfolgend auf und sind somit hydrologisch entkoppelt. Aber auch hier können Linienschutzmaßnahmen entlang des Gewässerlaufes den Starkregenabfluss behindern und zu einer Verstärkung der Überflutung führen. Daher empfiehlt sich auch in diesen Fällen eine aufeinander abgestimmte Planung von Schutzmaßnahmen gegen Starkregen und Hochwasser.

Im Land NRW ist diese Problematik erkannt und in den zehn Punkte umfassenden Arbeitsplan „Hochwasserschutz in Zeiten des Klimawandels“ aufgenommen worden (MULNV 2022). Derzeit werden die oben beschriebenen Ansätze in einer Arbeitsgruppe ausgearbeitet mit dem Ziel, dies in den landesweiten Vorgaben für die Aufstellung von Hochwasserschutzkonzepten zu verankern.

14.3.6 Risikobasierter Ansatz: Anpassung der Bemessungsgrundlage für Hochwasserschutz Maßnahmen

In RLP wurde der „Zukunftsplan Wasser“ (Rheinland-Pfalz, Ministerium für Klimaschutz, Umwelt, Energie und Mobilität 2024b) aufgestellt, der für eine moderne, zukunftsfähige, klimaangepasste und energieeffiziente Bewirtschaftung der Wasserressourcen sieben Ziele aufweist:

1. Das Grundwasser soll nachhaltig geschützt und bewirtschaftet und die Trinkwassergewinnung langfristig gesichert werden.
2. Es sollen Strategien zum Management von Nutzungskonflikten, insbesondere aufgrund von Wasserknappheit, entwickelt werden.
3. Der naturnahe Landschaftswasserhaushalt soll regeneriert, stabilisiert und geschützt werden.
4. Es sollen saubere und klimaresiliente Gewässer geschaffen werden.
5. Es soll Vorsorge vor Extremereignissen für Mensch und Umwelt getroffen werden.
6. Kritische Infrastrukturen sollen geschützt werden.
7. Es soll eine energieeffiziente und leistungsfähige Wasserwirtschaft gestaltet werden.

Zur Erreichung dieser sieben Ziele werden zwölf Handlungsschwerpunkte bearbeitet, u. a. sollen die Information, Warnung sowie Risikokommunikation verbessert, der Wasserrückhalt in der Fläche gestärkt, Überflutungsflächen wiedergewonnen und Auen renaturiert und der technische Hochwasserschutz, die Flächenvorsorge und das Starkregenrisikomanagement vorangetrieben werden. In diesem Zusammenhang ist eine Erhöhung der Bemessungskennwerte um einen Klimaanpassungsfaktor bei allen neuen Hochwasserschutzbauwerken geplant. Für kritische oder sensible Infrastrukturen bzw. Nutzungen innerhalb von Risikogebieten sollen zusätzlich angepasste Vorgaben und Bemessungsgrundlagen entwickelt und dadurch angepasste Schutzniveaus vorgesehen werden.

In NRW orientiert sich die Festlegung des Hochwasserschutzes üblicherweise am Hochwasserabflusswert HQ_{100}. Auch wenn keine gesetzliche oder behördliche Festlegung dafür existiert, hat sich dieses Vorgehen in der Praxis durchgesetzt. Eine Ursache dafür ist die auf einem HQ_{100} basierende Festsetzung von Überschwemmungsgebieten: Die erfolgreiche Umsetzung einer Hochwasserschutzmaßnahme verringert das Überschwemmungsgebiet und eröffnet damit Entwicklungsmöglichkeiten der betroffenen Kommunen. Auch scheint die Wiederkehrzeit dieses Ereignisses in einem angemessenen Verhältnis zum aufzubringenden Aufwand für die Hochwasserschutzmaßnahme zu stehen.

Nach der Juli-Flut 2021 ergeben sich drei zentrale Herausforderungen:

1. Die bisher in den betroffenen Einzugsgebieten gültige Hochwasserstatistik ist durch das Auftreten des Ereignisses in Frage gestellt. Allerdings war das Ereignis so extrem (teilweise über $HQ_{10.000}$ nach bisheriger Statistik), dass die bislang üblichen statistischen Verfahren zur Einschätzung der Hochwasserwahrscheinlichkeit an ihre Grenzen kommen. Aber auch die Hochwasserstatistik der übrigen, nicht betroffenen Einzugsgebiete ist zu überdenken, da sich das Juli-Ereignis 2021 mit seinen Ausmaßen durchaus auch in anderen Landesteilen hätte ereignen können (siehe ▶ Kap. 4).
2. Der DWD schätzt, dass Hochwasserereignisse und Ereignisse wie das Tief „Bernd“ in 2021 durch den Klimawandel künftig häufiger auftreten werden. Hochwasserschutzbauten haben eine lange Lebensdauer (oftmals mehr als 80 Jahre) und sollten daher bereits bei der Planung an den Klimawandel angepasst werden (Deutscher Wetterdienst [DWD] 2021; Tradowsky et al. 2023).
3. An den Gewässern sind im Laufe der Jahrzehnte erhebliche Sachwerte entstanden, und der Trend geht auch weiterhin aufwärts. Ein Festhalten an einem „starren“ Schutzziel HQ_{100} wird diesem Umstand möglicherweise nicht mehr gerecht. Umgekehrt können örtliche Umstände die Auslegung auf das Schutzziel erschweren und zu „überteuerten“ Lösungen führen. Diese Probleme könnten mit einer risikobasierten Schutzzielfestlegung in Verbindung mit einer standardisierten Nutzen-Kosten-Betrachtung überwunden werden. Dabei dürfen aber nicht nur Sachwerte, sondern es müssen auch die Gefahren für Leib und Leben berücksichtigt werden.

Der zehn Punkte umfassende Arbeitsplan „Hochwasserschutz in Zeiten des Klimawandels“ des Landes NRW adressiert diese drei Aspekte einer risikobasierten Hochwasserschutzplanung. In einer Arbeitsgruppe der Hochwasserkommission NRW wurde des Weiteren beschlossen, landesweit Klimafaktoren bereitzustellen, die auf statistische Niederschlags- und Hochwasserabflussgrößen angewendet werden können. Damit wird es künftig möglich sein, die hydrometeorologischen und hydrologischen Auswirkungen des Klimawandels auf die Bemessung von Bauwerken der Siedlungswasserwirtschaft und des Hochwasserschutzes in NRW zu berücksichtigen.

Die Hochwasserschutzzielfestlegung soll künftig auf der Basis von Risikobetrachtungen und (volkswirtschaftlicher) Nutzen-Kosten-Untersuchungen durchgeführt werden. Die genaue Ausgestaltung der künftigen Vorgehensweise wird derzeit in einer Arbeitsgruppe, bestehend aus Expertinnen und Experten verschiedener Fachdisziplinen, u. a. aus der Wasserwirtschaft, der Raumplanung und dem Naturschutz sowie aus verschiedenen Institutionen wie Behörden, Hochschulen, Ingenieurbüros und Verbänden, erarbeitet. Die Untersuchungsergebnisse und Erkenntnisse aus dem KAHR-Projekt bilden dafür eine wertvolle Grundlage. Ein wesentliches Ziel ist es auch, Hochwasserschutzkonzepte künftig verstärkt flusseinzugsgebietsweit aufzustellen. Gerade bei kleinen und mittleren Einzugsgebieten bestehen kommunenübergreifende Zusammenhänge hinsichtlich der Hochwasserentstehungsgebiete und der Auswirkung möglicher Schutzmaßnahmen. So kann z. B. ein gemeinschaftlich errichtetes Hochwasserrückhaltebecken vorteilhafter sein als viele einzelne Linienschutzmaßnahmen an den betroffenen Ortslagen. Um diese interkommunale Zusammenarbeit zu initiieren und zu fördern, gilt es, die organisatorischen und behördlichen Voraussetzungen dafür in NRW weiter zu verbessern.

14.4 Fazit

Aus der Flutkatastrophe vom Juli 2021 wurde sowohl in NRW als auch in RLP die Lehre gezogen, dass nur mit Hilfe einer einzugsgebietsbezogenen interkommunalen Kooperation, auch über Ländergrenzen hinweg, eine Reduktion von Hochwasserrisiken möglich ist. Grundsätzlich werden in beiden Bundesländern vergleichbare Ansätze verfolgt mit dem Ziel eines integrierten Hochwasserrisikomanagements. Durch die sondergesetzlichen Wasserverbände, die jeweils für die gesamte Wasserwirtschaft in einem Gewässereinzugsgebiet zuständig sind und in denen alle Kommunen des jeweiligen Gewässereinzugsgebiets Mitglied sind, hat hier das Land NRW einen gewissen Vorteil. In RLP müssen entsprechende einzugsgebietsbezogene Strukturen in vielen Gebieten erst noch aufgebaut werden, z. B. in Form von Zweckverbänden.

Im Rahmen des Wiederaufbaus in den Flutgebieten hat sich insbesondere in RLP gezeigt, dass die ursprüngliche Ausweisung der Überschwemmungsgebiete auf der Grundlage eines HQ_{100} alleine nicht ausreichen. Hier wurden ergänzend Flächen markiert, in denen ein Wiederaufbau nicht zulässig ist. Außerdem wurde auf der Grundlage des Ereignisses vom Juli 2021 ein neuer Bemessungswert HQ_{100} ermittelt. In beiden Bundesländern wird daran gearbeitet, künftig durch die Bereitstellung von Klimafaktoren der Klimaveränderung Rechnung zu tragen. Durch die Flutkatastrophe wurde auch deutlich, dass es keinen absoluten Schutz vor einem Hochwasser gibt und die verschiedenen Phasen des HWRM-Zyklus integrativ zu betrachten sind.

In RLP werden die örtlichen Hochwasser- und Starkregenvorsorgekonzepte (öHSVK) fortgeschrieben und ergänzen im Einzugsgebiet der Ahr einen in Aufstellung befindlichen einzugsgebietsbezogenen überörtlichen Maßnahmenplan (üMP). In NRW werden die kommunalen Hochwasserschutzkonzepte (HWSK) von den Kommunen erstellt und im Erft-Einzugsgebiet in der Hochwasserschutzkooperation Erft durch den Erftverband als Projektsteuerer begleitet. Durch das Arbeitspaket „Integration weiterer Konzepte" kann eine integrative Betrachtung der Hochwasser- und Starkregenvorsorge ermöglicht werden.

Aus der intensiven Auseinandersetzung mit der Umsetzung des HWRM in den vergangenen drei Jahren lässt sich ableiten, dass eine gezielte, adressatenorientierte Hochwasserrisikokommunikation mit all seinen Facetten unerlässlich sein wird, um zum einen das Überflutungsrisiko als reales Risiko zu benennen, die Eigenverantwortung zu unterstützen und zum anderen das Vertrauen in die fachlichen Institutionen zu stärken. Alle Akteur:innen und Betroffenen sind zusammen verantwortlich, und nur so wird langfristig die Hochwasserresilienz erhöht werden können. Der regelmäßige Austausch in den regionalen und überregionalen Zusammenschlüssen bietet sehr gute Möglichkeiten, Probleme und Schwierigkeiten bei der Umsetzung von Maßnahmen zu benennen, zu diskutieren und lösungsorientierte Handlungen zu unterstützen.

Zusammenfassung

- Aus der Flutkatastrophe vom Juli 2021 wurde sowohl in NRW als auch in RLP die Lehre gezogen, dass nur mit Hilfe einer einzugsgebietsbezogenen interkommunalen Kooperation, auch über Ländergrenzen hinweg, eine Reduktion von Hochwasserrisiken möglich ist.
- Im Rahmen des Wiederaufbaus in den Flutgebieten hat sich insbesondere in RLP gezeigt, dass die ursprüngliche Ausweisung der Überschwemmungsgebiete auf der Grundlage eines HQ_{100} alleine nicht ausreichen. Außerdem wurde auf der Grundlage des Ereignisses vom Juli 2021 ein neuer Bemessungswert HQ_{100} ermittelt.
- In beiden Bundesländern wird daran gearbeitet, künftig durch die Bereitstellung von Klimafaktoren der Klimaveränderung Rechnung zu tragen.
- Durch die Flutkatastrophe wurde auch deutlich, dass es keinen absoluten Schutz vor einem Hochwasser gibt und die verschiedenen Phasen des HWRM-Zyklus integrativ zu betrachten sind.
- Eine gezielte, adressatenorientierte Hochwasserrisikokommunikation mit all seinen Facetten ist unerlässlich, um das Überflutungsrisiko als reales Risiko zu benennen, die Eigenverantwortung zu unterstützen und das Vertrauen in die fachlichen Institutionen zu stärken.
- Alle Akteur:innen und Betroffenen sind zusammen verantwortlich, und nur so wird langfristig die Hochwasserresilienz erhöht werden können.
- Der regelmäßige Austausch in den regionalen und überregionalen Zusammenschlüssen bietet sehr gute Möglichkeiten, Probleme und Schwierigkeiten bei der Umsetzung von Maßnahmen zu benennen, zu diskutieren und lösungsorientierte Handlungen zu unterstützen.

Danksagung Die Autorinnen und Autoren von ▶ Kap. 14 bedanken sich beim Bundesministerium für Bildung und Forschung (BMBF) und beim DLR Projektträger für die Unterstützung und Förderung des KAHR-Projekts. Diese Forschung wurde vom BMBF unter dem Förderkennzeichen 01LR2102A-M finanziert.

14

Literatur

Birkmann J (2023) Flutkatastrophe 2021 – Lehren und Lernen: Wiederaufbau und Klimaresilienz: Modellregion Ahrtal? Nachrichten Arl 53(1):45–49. https://doi.org/10.60750/23pd-2692

Bittner D, Struck J, Jansen D (2023) Die Interkommunale Hochwasserschutzkooperation Erft – Projektstruktur, Sachstand und Zwischenfazit. Korrespondenz Wasserwirtsch 16(11):728–731

Bodin Ö (2017) Collaborative environmental governance: achieving collective action in social-ecological systems. Science. https://doi.org/10.1126/science.aan1114

Deutscher Wetterdienst (DWD) (2021) Attributionsstudie: Klimawandel machte die Starkregenfälle wahrscheinlicher, die zu Überschwemmungen in Westeuropa führten. https://www.dwd.de/DE/presse/pressemitteilungen/DE/2021/20210824_attributionsstudie_starkregen.html (Erstellt: 24.08.) (Pressemitteilung)

Erftverband Interkommunale Hochwasserschutzkooperation Erft: Über das Projekt. https://hws-kooperation.erftverband.de/#Projekt

Europäische Union (2007) Richtlinie 2007/60/EG des europäischen Parlaments und des Rates vom 23. Oktober 2007 über die Bewertung und das Management von Hochwasserrisiken. Amtsbl Eur Union 50(L 288):27–34

Grambow M (2008) Wassermanagement: Integriertes Wasser-Ressourcenmanagement von der Theorie zur Umsetzung. Vieweg + Teubner https://doi.org/10.1007/978-3-8348-9435-9

Greiving S, Birkmann J, Diehl J (2013) Klimaanpassung: Handlungsfelder für die Raumentwicklung – vertiefend diskutiert am Beispiel Hochwasserrisikomanagement. In: Baumgart S, Terfrüchte T (Hrsg) Zukunft der Regionalplanung in Nordrhein-Westfalen, Bd 6. Akademie für Raumforschung und Landesplanung, S 67–75 (http://www.nbn-resolving.de/urn:nbn:de:0156-3816064)

Kaleß M, Reichert J, Demny G, Schüttrumpf H, Klopries E-M (2023) Entwicklung eines Masterplans für die Einzugsgebiete von Inde und Vicht zur Verbesserung der Hochwasserresilienz. In: Porth M, Schüttrumpf H, Ostermann U (Hrsg) Wasser, Energie und Umwelt. Springer Vieweg, S 25–34 https://doi.org/10.1007/978-3-658-42657-6_4

Kuhlicke C, Müller U, Heiland P, Hutter G, Illing C, Kutschera G, Scheibel M, Siekmann T, Tragner F, Assmann A, Pyka C (2022) Das neue DWA-Merkblatt Hochwasserrisikokommunikation. WasserWirtschaft. https://doi.org/10.1007/s35147-022-1727-9

Merz B, Bittner R, Grünewald U, Piroth K (Hrsg) (2011) Management von Hochwasserrisiken mit Beiträgen aus den RIMAX-Forschungsprojekten. Schweizerbart

Müller U (2010) Hochwasserrisikomanagement: Theorie und Praxis. Vieweg + Teubner

Nordrhein-Westfalen, Ministerium für Klimaschutz, Umwelt, Landwirtschaft, Natur- und Verbraucherschutz des (2022) Lernen aus dem Hochwasser – 10-Punkte Arbeitsplan Hochwasserschutz in Zeiten des Klimawandels. https://www.land.nrw/media/25821

Patt H, Jüpner R (2020) Einführung in die Thematik. In: Patt H, Jüpner R (Hrsg) Hochwasser-Handbuch: Auswirkungen und Schutz, 3. Aufl. Springer Vieweg, S 2–8

Pörtner H-O, Roberts DC, Tignor MB, Poloczanska E, Mintenbeck K, Alegría A, Craig M, Langsdorf S, Löschke S, Möller V, Okem A, Rama B (2022) Climate Change 2022 – Impacts, Adaptation and Vulnerability: Working Group II Contribution to the Sixth Assessment Report of the Intergovernmental Panel on Climate Change. https://doi.org/10.1017/9781009325844

Reineke T, Lehmkuhl F, Blümel H (Hrsg) (2006) Grenzüberschreitendes integratives Gewässermanagement. Naturschutz und Freizeitgesellschaft, Bd 7. Academia

Rheinland-Pfalz, Ministerium für Klimaschutz, Umwelt, Energie und Mobilität (2023) Katrin Eder: „VISDOM sensibilisiert für Starkregengefahren und stärkt die regionale Handlungskompetenz". https://mkuem.rlp.de/service/pressemitteilungen/detail/katrin-eder-visdom-sensibilisiert-fuer-starkregengefahren-und-staerkt-die-regionale-handlungskompetenz

Rheinland-Pfalz, Ministerium für Klimaschutz, Umwelt, Energie und Mobilität (2024a) Hinweise zur Fortschreibung eines örtlichen Hochwasser- und Starkregenvorsorgekonzepts [öHSVK]. https://hochwassermanagement.rlp.de/fileadmin/hochwassermanagement/Unsere_Themen/Was_macht_das_Land/Kompetenzzentrum_Hochwasservorsorge_und_Hochwasserrisikomanagement/Hinweise_fuer_die_Fortschreibung_der_oeHWSVK_formatiert_Stand_14.5.2024.pdf

Rheinland-Pfalz, Ministerium für Klimaschutz, Umwelt, Energie und Mobilität (2024b) Zukunftsplan Wasser. https://mkuem.rlp.de/themen/wasser/zukunftsplan-wasser-rheinland-pfalz

Tradowsky JS, Philip SY, Kreienkamp F, Kew SF, Lorenz P, Arrighi J, Bettmann T, Caluwaerts S, Chan SC, de Cruz Vries L, de Demuth H, Ferrone A, Fischer EM, Fowler HJ, Goergen K, Heinrich D, Henrichs Y, Kaspar F, Wanders N (2023) Attribution of the heavy rainfall events leading to severe flooding in Western Europe during July 2021. Clim Change 176:90. https://doi.org/10.1007/s10584-023-03502-7

Truedinger AJ, Jamshed A, Sauter H, Birkmann J (2023) Adaptation after extreme flooding events: moving or staying? The case of the Ahr valley in Germany. Sustainability 15:1407. https://doi.org/10.3390/su15021407

VRVis scenarify: Die Simulationssoftware. https://www.vrvis.at/produkte-loesungen/produkte-lizenzen/scenarify

Wasserverband Eifel-Rur (WVER) GIS-Portal: Das öffentliche Geoinformationssystem des WVER: Hochwasser 2021. https://gisportal-wver.hub.arcgis.com/

Wasserverband Eifel-Rur (WVER) (2024) Die Vicht. https://wver.de/fluss/die-vicht/

Risikobasierte Raumplanung für den Wiederaufbau

Stefan Greiving, Mark Fleischhauer, Alessa Trüdinger, Felix Othmer und Jörn Birkmann

J. Birkmann et al. (Hrsg.), *Die Flutkatastrophe 2021*, Sustainable Development Goals (SDG) – Umsetzung in Praxis, Lehre und Entscheidungsprozessen,
https://doi.org/10.1007/978-3-658-48688-4_15

15.1 Einleitung

Am 1. September 2021 ist der „Bundesraumordnungsplan Hochwasserschutz“ (BRPH) als Rechtsverordnung des Bundesministeriums des Innern und für Heimat (BMI) in Kraft getreten (BMI 2021). Damit bestehen erstmals innerhalb der Raumordnung einheitliche Regelungen für den planerischen Umgang mit Hochwasser im Binnenland und an der Küste. Eines der Ziele des BRPH ist die Einführung des Prinzips der risikobasierten Planung in die deutsche Raumplanung. Wodurch sich dieses Prinzip auszeichnet, welche Bedeutung es für den Wiederaufbau und die Bewertung der Analyse von Hochwasserrisiken und die Auswahl von Maßnahmen hat, wird in diesem Beitrag erörtert.

15.2 Risikobasierte Raumplanung

Risikobasierte Raumplanung stellt einen Paradigmenwechsel bezüglich der raumplanerischen Vorsorge dar, da nicht mehr nur der räumliche Umgriff einer potenziellen Gefährdung (z. B. das bei einem Bemessungsereignis überflutete Gebiet) zur Grundlage raumplanerischer Entscheidungen herangezogen wird. Vielmehr wird nunmehr sowohl nach verschiedenen Gefährdungskomponenten sowie nach den (geplanten) Nutzungen und somit nach allen Risikokomponenten differenziert.

15.2.1 Wodurch zeichnet sich risikobasierte Raumplanung aus?

Der Begriff der Raumplanung umfasst sowohl die überörtliche Raumordnung (Landes- und Regionalplanung) als auch die kommunale Bauleitplanung (insb. Flächennutzungs- und Bebauungspläne). Der Raumplanung kommt beim Management von Hochwasserrisiken eine vorsorgende Funktion zu. Hinsichtlich der gesetzlichen Lage bezüglich der Hochwasservorsorge sind im Bereich der Raumordnung besonders das allgemeine „Vorsorgeprinzip“ im Sinne des § 1 Absatz 1 Nummer 2 ROG „Vorsorge für einzelne Nutzungen und Funktionen des Raums“ sowie der ebenfalls im Raumordnungsgesetz verankerte Grundsatz § 2 Absatz 2 Nummer 3 Satz 4 ROG „Dem Schutz Kritischer Infrastrukturen ist Rechnung zu tragen“ hervorzuheben. Auch im BauGB sind unter § 1 Absatz 6 Nummer 12 die „Belange des Küsten- oder Hochwasserschutzes und der Hochwasservorsorge, insbesondere die Vermeidung und Verringerung von Hochwasserschäden“ benannt, die bei der Siedlungs- und Infrastrukturentwicklung sowie der Aufstellung von Bauleitplänen zu berücksichtigen sind. Allerdings ist die Bauleitplanung in diesem Fall auf Überschwemmungsgebiete (§ 76 Absatz 1 WHG) ausgerichtet, innerhalb derer nach § 78 Absatz 1 WHG weitgehende Bauverbote bestehen. Der Lastfall ist jedoch lediglich das im statistischen Mittel alle 100 Jahre auftretende Bemessungsereignis (HQ_{100}) – Gefährdungsintensitäten, Empfindlichkeiten und Schutzwürdigkeiten, ebenso wie die möglichen Auswirkungen zukünftiger Klimaänderungen wurden bislang nicht berücksichtigt.

Im Wasserrecht wird in § 73 Absatz 1 Satz 2 WHG der Begriff des Hochwasserrisikos als „Kombination der Wahrscheinlichkeit des Eintritts eines Hochwasserereignisses mit den möglichen nachteiligen Hochwasserfolgen für die menschliche Gesundheit, die Umwelt, das Kulturerbe, wirtschaftliche Tätigkeiten und erhebliche Sachwerte“ definiert, wobei implizit eine Berücksichtigung der Empfindlichkeit der potenziell betroffenen Infrastrukturen und Bevölkerungsgruppen enthalten ist.

Mit dem Inkrafttreten des BRPH wurde durch das Ziel I.1.1 das Prinzip der risikobasierten Planung in die deutsche Raumplanung eingeführt, wonach bei raumbedeutsamen Planungen und Maßnahmen einschließlich der Siedlungsentwicklung die Risiken von Hochwassern zu prüfen sind, was neben der Wahrscheinlichkeit des Eintritts eines Hochwasserereignisses und seinem räumlichen und zeitlichen Ausmaß auch die Wassertiefe und die Fließgeschwindigkeit umfasst. Gemäß § 4 Abs. 1 ROG sind alle öffentlichen Planungsträger einschließlich der Kommunen an Ziele der Raumordnung gebunden, und zwar auch ohne vorherige Anpassung der entsprechenden Landesraumordnungspläne und Regionalpläne.

15.2.1.1 Differenzierung der Gefährdung und Einbeziehung der Empfindlichkeit und Schutzwürdigkeit von Nutzungen

Ziel I.1.1 des BRPH erfordert eine differenzierte Beurteilung der Hochwassergefährdung und somit eine hochwasserangepasste Gestaltung von Siedlungs- und Infrastrukturen. Es verlangt auch, dass die unterschiedlichen Empfindlichkeiten und Schutzwürdigkeiten der einzelnen Raumnutzungen und Raumfunktionen in der planerischen Hochwasservorsorge berücksichtigt werden (BMI 2021). Folglich können weder die überörtliche Raumordnung noch die örtliche Bauleitplanung an der gängigen Praxis festhalten, Hochwasservorsorge allein am räumlichen Umgriff der gefährdeten Gebiete (d. h. Überschwemmungsgebiete gemäß § 76 WHG bzw. Risikogebiete nach § 78 b WHG) festzumachen.

„Empfindlichkeit" ist ein Parameter, der sich objektiv im Sinne einer Tatbestandsfeststellung durch Planungsträger ermitteln lässt. Dies gilt sowohl für empfindliche Raumnutzungen auf überörtlicher Ebene als auch für empfindliche Flächennutzungen auf kommunaler Ebene, wo sich die Empfindlichkeit für die jeweils nach Baunutzungsverordnung zulässigen baulichen Nutzungen ermitteln lässt. „Schutzwürdigkeit" hingegen ist ein politisch-normatives Konzept und erfordert eine dezidierte Auseinandersetzung mit Schutzzielen im jeweiligen Planungsverfahren.

Die Eignung von Raumnutzungen und die Zulässigkeit von baulichen Nutzungen im Rahmen einer risikobasierten Raumplanung kann – wie bereits angesprochen – nicht allein anhand der bestehenden Gefährdung beurteilt werden, sondern es kommt auch auf die jeweiligen Folgen an, so z. B. auf die Frage, ob sich eine überörtlich bedeutende kritische Infrastruktur im Gefährdungsbereich befindet oder im kommunalen Kontext, ob sich im Gefährdungsbereich Einrichtungen befinden, in denen sich regelmäßig Personen aufhalten und u. U. auch, ob es sich um Personen handelt, die sich im Ereignisfall nicht selbst helfen können, sondern evakuiert werden müssen.

15.2.1.2 Anpassung an sich ändernde Risiken

Angesichts des Klimawandels ist von einer Verstärkung von Hochwasser- und Starkregengefahren auszugehen. Diesem Umstand wird im BRPH durch das Ziel I.2.1 Rechnung getragen, wonach die Auswirkungen des Klimawandels im Hinblick auf Hochwasser- und Starkregenereignisse bei raumbedeutsamen Planungen und Maßnahmen einschließlich der Siedlungsentwicklung vorausschauend zu prüfen und nach Grundsatz I.2.2 raumbedeutsame Planungen und Maßnahmen zum Hochwasserschutz zu überprüfen und gegebenenfalls anzupassen sind. Im Sinne der Vorsorgeorientierung bedeutet eine risikobasierte Raumplanung somit auch die Anpassung an sich ändernde Risiken.

15.2.2 Welche Bedeutung hat risikobasierte Raumplanung für den Wiederaufbau?

In Wiederaufbaugebieten wie an der Ahr oder an der Erft besteht aufgrund der eingetretenen großflächigen Überflutungsereignisse mit der Folge erheblicher Zerstörungen und Schäden eine in vielerlei Hinsicht neue Situation, da viele Gebiete, deren Hochwassergefährdung bislang als gering oder gar nicht vorhanden eingeschätzt wurde, nach neueren Berechnungen zukünftig nicht mehr als sicher einzuschätzen sind (Land Rheinland-Pfalz, SGD Nord 2021). So stellt sich beispielsweise die Frage, ob am gleichen Ort wiederaufgebaut werden soll, und wenn ja, wie – oder ob ein Alternativstandort gesucht werden sollte. Im baulichen Bestand sind Maßnahmen einer hochwasserangepassten Bauweise und Maßnahmen zum Objektschutz sowie eine Nutzungsanpassung zu diskutieren. Flächenpotenziale für zukünftige Bebauung sind hinsichtlich ihrer Hochwassergefährdung neu und kritisch zu prüfen. Vor diesem Hintergrund kann eine risikobasierte Raumplanung an verschiedenen Punkten im Wiederaufbau ansetzen:

Differenzierung der Gefährdung Unter der Voraussetzung einer Neuberechnung der Hochwassergefährdung sind Raum- und Flächennutzungen an die Gefährdungen anzupassen. Beispielsweise kann in den am höchsten gefährdeten Bereichen gänzlich auf den Wiederaufbau verzichtet werden, wohingegen in anderen Bereichen – auch in Bestandsgebieten – Anpassungen der Nutzung oder eine Ertüchtigung der Bausubstanz im Sinne eines Planerfordernisses gemäß § 1 Abs. 3 BauGB geboten ist. Bei einer Überplanung eines vollständig oder teilweise zerstörten Plangebiets mit einem neuen Bebauungsplan wäre das Prinzip der risikobasierten Planung zu beachten. Baugenehmigungen und damit der Bestandsschutz gelten dabei nur fort, sofern das Gebäude als Ersatz und Fortführung des beschädigten Gebäudes erscheint. Wenn ein Gebäude jedoch zerstört ist, gilt nicht mehr der Bestandsschutz fort; dann sind für Wiederaufbau Einzelfallgenehmigungen notwendig (außer im Ahrtal in der „gelben Zone“, dort werden diese nicht genehmigt) (Deutscher Bundestag 2022, S. 1). Andernfalls, d. h. bei Nutzungsänderungen, Änderungen des Maßes der Bebauung oder der Stellung des Baukörpers wären auch beim Wiederaufbau die Festsetzungen des neuen Bebauungsplans zu beachten.

Bei Neuplanungen wäre im Sinne einer risikobasierten Planung in Baugebieten beispielsweise darauf zu achten, dass Festsetzungen zur baulichen Vorsorge und zum hochwasserangepassten Bauen auf Grundlage der Gefährdungsintensität getroffen werden.

15

Einbeziehung der Empfindlichkeit und Schutzwürdigkeit von Nutzungen Empfindliche oder schutzwürdige Raumnutzungen (vgl. z. B. ► Abschn. 11.4) bedeuten bei gleicher Gefährdung ein ungleich höheres Risiko. Eine risikobasierte Raumplanung hat beim Wiederaufbau dafür Sorge zu tragen, dass für empfindliche oder schutzwürdige Nutzungen wie z. B. Pflegeheime oder Stromversorgungsinfrastruktur, die in hochwassergefährdeten Bereichen (wieder) errichtet werden, eine entsprechende Flächen- oder bauliche Vorsorge getroffen wird oder dass sie gleich gänzlich außerhalb hochwassergefährdeter Bereiche errichtet werden (Truedinger et al. 2024).

Anpassung an sich ändernde Risiken Risikobasierte Raumplanung beim Wiederaufbau bedeutet auch, dass Flächen und Standorte – sowohl im Bestand, als auch bei Neuplanungen – im Hinblick auf eine zukünftige Veränderung der Risiken zu prüfen sind. Dies bezieht sich vor allem auf die Gefährdung durch zukünftige Hochwasserereignisse, die in ihrer Intensität und Häufigkeit zunehmen können. Zukünftige Veränderungen des Risikos können aber auch mit einer Veränderung der Empfindlichkeit einhergehen (Verschleiß der Bausubstanz bei Infrastrukturen oder Überalterung der Bevölkerung in gefährdeten Gebieten).

15.3 Analyse und Bewertung im Rahmen einer risikobasierten Raumplanung

Ein risikobasierter Planungsansatz in der überörtlichen Raumordnung und der kommunalen Bauleitplanung stellt neue Anforderungen an die Bereitstellung von Analyse- und Entscheidungsgrundlagen im Rahmen des Planungsprozesses. Zudem erweitert sich die Zielrichtung raumplanerischen Handelns von einem Beitrag zur reinen Gefahrenabwehr zu einem Beitrag zur Vermeidung und Verringerung raumrelevanter Risiken mit vielschichtigen Handlungs- und Gestaltungsoptionen.

15.3.1 Welche Informationen werden für die Risikobewertung in der Raumplanung benötigt?

Informationen zur Gefährdung Wichtigste Informationen für eine risikobasierte Raumplanung im Umgang mit Hochwasser sind spezifische Aussagen zum räumlichen Umgriff des Hochwassers, zur Wassertiefe und zur Fließgeschwindigkeit in Bezug auf das jeweilige Bemessungsereignis. Diese sollten nicht nur für Hochwasser-, sondern auch Starkregen- bzw. Sturzflutereignisse vorliegen und neben einem seltenen hundertjährlichen Ereignis (HQ_{100}) auch ein sehr seltenes Extremereignis (HQ_{extrem}) abbilden. Tatsächlich nehmen nur Teile der Überschwemmungsgebiete unmittelbar am Abflussgeschehen teil. In diesen Teilbereichen besteht die Funktion nicht nur in der Sicherung der Retention, sondern vor allem auch der Vermeidung von signifikanten Hochwasserschäden, die aufgrund der größeren Einstautiefe und höheren Fließgeschwindigkeiten auch Risiken für Leib und Leben auslösen.

Die Abgrenzung von hochwassergefährdeten Gebieten erfolgt anhand von Gefahren- und Risikokarten, die von den Ländern gemäß Wasserhaushaltsgesetz § 73 erstellt und veröffentlicht werden. In den Hochwassergefahrenkarten sind Wassertiefe und Fließgeschwindigkeit nach § 74 Abs. 2 WHG enthalten. Die Hochwasserrisikokarten zeigen, wo das Risiko durch die Gefährdung von empfindlichen Nutzungen, der menschlichen Gesundheit oder wirtschaftlicher Werte besonders groß ist. Während in allen Ländern in den Hochwassergefahren- und -risikokarten gemäß WHG das HQ_{100} als Bemessungsereignis dargestellt wird, unterscheiden sich die Bemessungsereignisse für ein extremes Hochwasser HQ_{extrem} zum Teil beträchtlich voneinander (Truedinger et al. 2024, S. 15).

Darüber hinaus legt auch die Raumordnung Überschwemmungsbereiche fest. Für die räumliche Abgrenzung und Reichweite der Ausschlusswirkungen für Nutzungen, die der Hochwasservorsorge entgegenstehen, ist ebenfalls das Prinzip der risikobasierten Planung zu beachten. Dafür sind drei unterschiedliche Vorgehensweisen denkbar, für die jeweils unterschiedliche fachliche Informationen zur Verfügung stehen bzw. erarbeitet werden müssen:

1. Berücksichtigung weiterer Parameter neben dem räumlichen Umgriff der gefährdeten Bereiche für die Abgrenzung von Überschwemmungsbereichen, um bspw. Bereiche, die unmittelbar am Abflussgeschehen teilhaben wie Abflussrinnen, auch außerhalb der HQ_{100}-Gebietskulisse einzubeziehen.
2. Einführung einer weiteren Vorranggebietskategorie mit eingeschränkter Ausschlusswirkung für den HQ_{extrem}-Bereich, die sich nur auf besonders empfindliche (z. B. soziale Infrastruktur) oder aus anderen Gründen besonders schutzwürdige Infrastruktur bezieht (Störfallbetriebe, Kritische Infrastrukturen).
3. Einführung eines Gefahrenzonenmodells, das die Zulässigkeit von Landnutzungen in Abhängigkeit vom gegebenen Hochwasserrisiko differenzierter steuert als bisherige Vorranggebiete.

Im Gegensatz zur Ermittlung und Kommunikation von Hochwassergefahren fehlen bislang flächendeckende Informationen im Zusammenhang mit Starkregenereignissen und deren Folgen, wie etwa Sturzfluten. Die Folgen von Starkregen für die Siedlungs- und Infrastrukturentwicklung werden hauptsächlich auf kommunaler Ebene diskutiert, da der Umgang mit Niederschlagswasser in den kommunalen Zuständigkeitsbereich fällt. Eine systematische, kommunale Risikomanagementstrategie für Starkregenereignisse würde zunächst die Erstellung von Gefahrenkarten erfordern. Diese Karten sind jedoch gegenwärtig keine verpflichtende Aufgabe auf kommunaler Ebene und liegen auch noch nicht für alle kommunalen Gebietskörperschaften (Kreise, Städte und Gemeinden) vor.

Auch die gemeinsame Betrachtung der Hochwasser- und Starkregen- bzw. Sturzflutgefährdung und die Erstellung von entsprechenden integrierten Gefahren- und Risikokarten wird noch nicht standardmäßig vorgenommen.

Informationen zur Vulnerabilität Im Rahmen der Terminologie des Sendai-Rahmenwerks wird Vulnerabilität verstanden als die durch physische, soziale, wirtschaftliche und ökologische Faktoren oder Prozesse bestimmten Bedingungen, die die Anfälligkeit eines Einzelnen, einer Gemeinschaft, von Vermögenswerten oder Systemen gegenüber den Auswirkungen von Gefahren erhöhen (UNDRR 2023). Gegenüber Starkregen und Hochwasser sind im Akutfall insbesondere Personen mit eingeschränkter Mobilität, Personen unter sechs Jahren und über 65 Jahren, Schwangere, Pflegebedürftige und Intensivpatientinnen und Patienten als besonders vulnerabel einzustufen (Der Paritätische Gesamtverband 2023), wodurch die Infrastrukturen, die von den genannten Personengruppen genutzt werden, aufgrund des erschwerten Evakuierungsmanagements als besonders empfindlich und vor dem Hintergrund des Schutzes von Leib und Leben als besonders schutzwürdig eingestuft werden müssen. Auch Personen mit psychischen Beeinträchtigungen und Erkrankungen, deren Erholung sich häufig langwieriger und schwieriger darstellt und die je nach Beeinträchtigung auch bereits bei der Bewältigung von Katastrophen schlechter gestellt sind, werden als besonders vulnerabel eingeschätzt. Außerdem sind beispielsweise aufgrund sozialer Umstände auch Wohnungs- und insbesondere Obdachlose, Flüchtlinge (zum Beispiel wegen Sprachbarrieren) und Sträflinge besonders vulnerabel. Für die Umsetzung eines risikobasierten Planungsansatzes müssen Planungsträger Zugang zu diesen Informationen erhalten oder sie selbst erheben.

Eine hohe Schutzwürdigkeit besteht auch für Infrastrukturen, die für die Funktionsfähigkeit moderner Gesellschaften von wichtiger Bedeutung sind und deren Ausfall oder Beeinträchtigung nachhaltige Störungen im Gesamtsystem zur Folge hat, also für kritische Infrastrukturen (KRITIS). Wesentlich ist dabei ihre besondere Schutzwürdigkeit als politisch-normatives Konzept gemäß dem bereits erwähnten § 2 Abs. 2 Nr. 3 Satz 4 ROG, wonach dem Schutz kritischer Infrastrukturen Rechnung zu tragen ist. Die Schutzwürdigkeit ist im Gesamtkontext der Infrastrukturnetze („Kritikalität") zu sehen und beschränkt sich nicht auf physische Elemente. Risikobasierte Planung im Zusammenhang mit kritischen Infrastrukturen bedeutet auch, dass den Planungsträgern auf regionaler und kommunaler Ebene bekannt ist, welche kritischen Infrastrukturen sich an welchem Standort im jeweiligen planerischen Zuständigkeitsbereich befinden und wie diese mit anderen (kritischen) Infrastrukturen in systemischer Hinsicht verflochten sind. Diese Informationen liegen auf regionaler und kommunaler Ebene in der Regel nicht flächendeckend vor, was unter anderem auch an Sicherheitsbedenken liegt. Gleichwohl besteht hier im Hinblick auf eine risikobasierte Raumplanung ein Informationsbedarf.

Informationen zum Umgang mit Unsicherheiten Im Hinblick auf Hochwasser- und Starkregengefährdungen führt der Klimawandel zu zusätzlichen Unsicherheiten bei der Abschätzung und Bewertung von Risiken. Da sich die möglichen Auswirkungen des Klimawandels nicht in Wahrscheinlichkeiten im Sinne von Wiederkehrintervallen („Jährlichkeiten"), sondern nur in szenariobasierten Bandbreiten ausdrücken lassen, kommt es hierbei auf eine sachgerechte Beurteilung der vorliegenden Informationen durch den Plangeber an. Um eine sachgerechte Beurteilung zu ermöglichen, werden zunächst die Ziele der Analyse und die Auswahl der Szenarien normativ gesetzt, um dann eine Tatbestandsfeststellung vorzunehmen. Dabei sind bestimmte Anforderungen und Grenzen zu beachten, wie das Hinzuziehen sämtlichen empirischen Materials, die Anwendung geeigneter Bewertungsverfahren, eine fachlich vertretbare und widerspruchsfreie Bewertung der Belange, und es ist zu ermitteln, ob zwischenzeitlich wissenschaftlich eindeutige Erkenntnisse vorliegen, die die weitere Anwendung der sogenannten Einschätzungsprärogative ausschließen. Mit der Ausübung der planerischen Einschätzungsprärogative entscheidet der Plangeber, ob er seine planerischen Festlegungen am oberen Rand des Möglichkeitsraums („Worst Case") oder an einer moderaten Entwicklung orientieren will (Greiving 2019, S. 71 f.).

15.3.2 An welchen Zielen sollte sich die Bewertung von Analysen und Maßnahmen orientieren?

Die im Rahmen einer risikobasierten Planung vorzunehmende Auswahl von Maßnahmen sollte auf Ziele ausgerichtet sein, die sich am Leitbild einer resilienten Raum- und Stadtentwicklung orientieren. Resilienz ist ein Begriff, der in einer Vielzahl von Zusammenhängen verwendet wird. Im Zusammenhang mit räumlicher Entwicklung lässt sich Resilienz definieren als die Fähigkeit, sich an Belastungen durch Gefahren anzupassen und sich schnell von deren Auswirkungen zu erholen (Henstra et al. 2004, S. 5). Godschalk (2003) betont, dass sich die Antworten der Raumentwicklung auf potenzielle Krisen im Sinne der Resilienz nicht nur allein auf die Gefahrenabwehr beschränken dürfen. Um katastrophenresiliente Städte und Regionen zu schaffen, leitet Godschalk Prinzipien resilienter Systeme ab, denen sich bestimmte Merkmale für deren Gestaltung und Verwaltung zuordnen lassen (Godschalk 2003, S. 5). Greiving et al. (2023) spezifizieren diese Prinzipien. Resiliente Systeme (Infrastrukturen) oder Räume (Städte und Regionen) lassen sich demzufolge anhand der folgenden Resilienzkriterien beschreiben:

- Fokus Raum- und Infrastrukturen:
  - Diversität – verschiedene, sich ergänzende Funktionen
  - Redundanz – gleiche, sich ersetzende Funktionen
  - Robustheit – Widerstandsfähigkeit, Stärke
  - Expositionsreduktion – gefährdete Standorte meiden
  - Effizienz – strategische und wirksame Allokation von Ressourcen
- Fokus Akteurinnen und Akteure sowie Institutionen:
  - Autonomie – Entscheidungen auf der geeigneten Ebene treffen
  - Kooperation – Mehrwert aus Zusammenarbeit erkennen und nutzen
  - Anpassungsfähigkeit – Flexibilität, auf neue Situationen zu reagieren und daraus zu lernen

Nach diesem Modell ist Resilienz eine Möglichkeit, mit Unsicherheiten umzugehen, da die Häufigkeit und das Ausmaß von Gefährdungen selten vorhergesagt werden können und die Anfälligkeit von infrastrukturellen und räumlichen Systemen vor einem Gefahrenereignis nicht vollständig bekannt ist. Daher müssen die räumlichen siedlungs- und infrastrukturellen Systeme so konzipiert sein, dass sie Gefahren widerstehen können und flexibel genug sind, um Extremsituationen zu bewältigen, ohne zu versagen, und dass sie so robust sind, dass sie sich von den Auswirkungen einer Katastrophe schnell erholen (Henstra et al. 2004).

Als wichtiger Schritt auf dem Weg zur Entwicklung von geeigneten Maßnahmen für die Gestaltung (Bestandsertüchtigung, Wiederaufbau, Neubau) resilienter Raumstrukturen steht die Definition von Schutzzielen, mit denen ein angestrebter Soll-Zustand operationalisiert werden kann. Dies greift die Empfehlung 10 „Neue Schutzstandards und Schutzziele für Kritische und Sensible Infrastrukturen müssen definiert und das Bewusstsein für ein unvermeidbares Restrisiko muss gestärkt werden.“ aus dem Empfehlungspapier „10 Empfehlungen aus Sicht der Wissenschaft zum Thema Wiederaufbau und Zukunftsfähigkeit der flutbetroffenen Regionen“ des KAHR-Projekts auf (KAHR 2022).

Unter einem Schutzziel versteht man den angestrebten Zustand eines Schutzguts, der bei einem Schadensereignis erhalten bleiben beziehungsweise nach einer Beschädigung wieder erreicht werden soll. Schutzgüter sind entweder aufgrund ihres ideellen oder ihres materiellen Werts vor einem Schaden zu bewahren. Ein Beispiel für ein Schutzziel ist die Versorgung der Bevölkerung in einer Krise mit Trinkwasser. Im Wassersicherstellungsgesetz werden für die Notversorgung der Bevölkerung im Verteidigungsfall (Schutzgut) beispielsweise 15 l Trinkwasser pro Person und Tag als Schutzziel vorgeschrieben (Gerhold et al. 2019).

Bei der Festlegung von Schutzzielen verlässt man die Sachebene und begibt sich auf die Wertebene, da hier politische, planerische, aber auch gesellschaftliche Wertentscheidungen mit einfließen. Dies ergibt sich auch schon aufgrund der Tatsache, dass unterschiedliche Akteurinnen und Akteure unterschiedliche Ziele verfolgen oder zumindest priorisieren. Jegliche Zielfestlegung sollte daher unter Beteiligung dieser relevanten Personen, Gruppen und Institutionen stattfinden, die Einbeziehung staatlicher oder kommunaler Entscheidungsbefugter ist notwendig, um die erforderliche politische Legitimation sicherzustellen.

15.4 Raumplanerische Maßnahmen und Instrumente

In diesem Abschnitt stehen raumplanerische Maßnahmen und Instrumente im Vordergrund, die einen Beitrag zu einer hochwasserresilienten Raumentwicklung für den Wiederaufbau, aber auch generell leisten können.

Die konkreten Planungsherausforderungen beim Wiederaufbau nach Naturkatastrophen liegen sowohl auf der lokalen als auch auf der regionalen Ebene und erzeugen einen hohen Kooperationsbedarf zwischen allen Akteurinnen und Akteuren in der Region sowie speziell zwischen räumlicher Gesamtplanung und Fachplanungen. Grundsätzlich stellt sich die Frage, ob der Wiederaufbau an gleicher oder an anderer Stelle erfolgen soll. Häufig stellt dabei allein die Suche nach verfügbaren, geeigneten Ersatz- bzw. Alternativstandorten eine Herausforderung dar, insbesondere, wenn die Region bereits dicht bebaut ist oder es aus topographischen Gründen nur wenig freies Bauland gibt. Um Ausfälle im Krisen- oder Störfall kompensieren zu können müssen kritische Infrastrukturen entweder außerhalb gefahrenexponierter Bereiche neu gebaut beziehungsweise dorthin verlagert werden oder durch bauliche Schutzmaßnahmen in ihrer Resilienz gestärkt oder die Redundanz erhöht werden. Hierbei können Zielkonflikte zwischen Nachhaltigkeit, Umweltschutz und Widerstandsfähigkeit als Leitprinzipien entstehen, die auch zwischen den verschiedenen Fachplanungen und der räumlichen Gesamtplanung fachlich und letztlich auch politisch ausgetragen und gelöst werden müssen.

Die nachfolgende Zusammenstellung greift in erster Linie neuere Ansätze zur planerischen Hochwasserrisikovorsorge auf. Für die bereits länger diskutierten und sich in der Umsetzung befindlichen Ansätze zur räumlichen Vorsorge gegenüber Hochwasser wird auf die einschlägigen Quellen verwiesen. Ein aktueller Überblick hierzu findet sich im Steckbrief „Hochwasservorsorge: Risikovorsorge gegenüber Hochwasser und Starkregenereignissen durch Raumordnung verbessern“ (BBSR 2024).

15.4.1 Welche raumplanerischen Maßnahmen und Instrumente tragen zu einer risikobasierten Raumplanung bei?

Um eine in Bezug auf Starkregen und Hochwasser resiliente Raumentwicklung zu ermöglichen, können verschiedene Maßnahmen ergriffen werden. Dabei gibt es nicht die *eine* Maßnahme, sondern es müssen abhängig von den das Risiko bestimmenden und weiteren Kontextfaktoren, wie der Exposition, der Schutzwürdigkeit, dem finanziellen Rahmen, der Verantwortlichkeiten usw., Maßnahmenbündel ergriffen werden.

Im Rahmen des KAHR-Projekts wurden im Juni 2022 im Rahmen eines Workshops mit der Landesplanung sowie den Regionalplanungsbehörden Nordrhein-Westfalens die Prinzipien der risikobasierten Planung und deren Umsetzung in der Regionalplanung diskutiert.

Dies greift die Empfehlung 7 „Hochwasser- und klimaresilientes Planen und Bauen muss auf allen Ebenen der räumlichen Planung integriert werden und alle Facetten der Klimawandelauswirkungen berücksichtigen.“ aus dem 10-Punkte-Papier des KAHR-Projekts auf (KAHR 2022).

15.4.1.1 Überörtliche Ebene: Gebietsfestlegungen durch die Regionalplanung

Auf überörtlicher Ebene kann die Regionalplanung Gebiete festlegen, in denen Siedlungsbereiche oder sensible oder kritische Einrichtungen besonders geschützt, hochwasserangepasst errichtet oder gar zurückgebaut werden müssen.

- **Retention und Abfluss**

Ein risikobasiertes Vorgehen erfordert eine räumliche Differenzierung der Überschwemmungsbereiche in die Funktionen Retention und Abfluss, um der erforderlichen Berücksichtigung der Gefährdungsintensität im Sinne von Ziel I.1.1 BRPH Rechnung tragen zu können. Dieses Vorgehen entspricht dem 2020 in Kraft getretenen Regionalplan Oberes Elbtal-Osterzgebirge (Sachsen). Ziel 4.1.4.2 lautet: „In Vorranggebieten vorbeugender Hochwasserschutz mit den Funktionen ‚Abfluss' bzw. ‚Herstellung Abfluss' sind alle raumbedeutsamen Planungen und Maßnahmen unzulässig, die den Abfluss von Hochwasser bzw. die Herstellung dieser Funktion beeinträchtigen können." In der Begründung heißt es: „Bemessen sind Vorrang- und Vorbehaltsgebiete vorbeugender Hochwasserschutz an den Anforderungen, die von extremen Hochwassern ausgehen. Es wurden Szenarien berücksichtigt, die dem im jeweiligen Einzugsgebiet größten anzunehmenden Hochwasser möglichst nahe kommen. Eine Orientierung am einhundertjährlichen Hochwasser ist im raumordnerischen Risikomanagement nicht ausreichend."

Daher sollte durch die Regionalplanung festgelegt werden, dass für die räumliche Abgrenzung derjenigen Teile der Überschwemmungsbereiche, die lediglich zur Erhaltung der Retentionswirkung dienen, weiterhin ein HQ_{100} zugrunde gelegt wird. Für vorhandene oder zu reaktivierende Abflussrinnen sollte hingegen ein HQ_{extrem} zu Grunde gelegt werden, wobei die vorhandenen Gefahrenkarten bzw. wasserwirtschaftlichen Grundlagen die erforderlichen Informationen zum Abflussgeschehen und damit zur Differenzierung zwischen Retentions- und Abflussfunktionen enthalten.

- **Siedlungsbereiche: Differenzierung nach Gefährdungsintensität**

15

In der Raumordnung wurde die Hochwasservorsorge bislang vorwiegend von der Sicherung von Überschwemmungsbereichen als Retentionsraum geprägt, was häufig den gänzlichen Ausschluss von Siedlungstätigkeit bedeutete. Die vollständige Freihaltung aller HQ_{extrem}-Flächen könnte zwar in Einzelfällen sinnvoll sein, würde aber Entwicklungsmöglichkeiten massiv einschränken. Eine pauschale Entscheidung, die weder die Gefährdungsintensität (mögliche Einstautiefen und Fließgeschwindigkeiten), noch die Empfindlichkeit und die Schutzwürdigkeit der verschiedenen Raum- bzw. Flächennutzungen berücksichtigt, entspricht nicht hinreichend den Prinzipien einer risikobasierten und adaptiven Planung. Eine weitere Differenzierung der Ansätze im Sinne einer risikobasierten Planung erscheint daher notwendig.

Ein Beispiel für eine stärkere Differenzierung hinsichtlich der Gefährdungsintensität ist das Vorgehen der Regionalplanungsbehörde Detmold beim aktuellen Regionalplan Ostwestfalen-Lippe. Hierzu hat die Bezirksregierung Detmold gemeinsam mit fachlichen Partnern eine differenzierte Prüfmethodik entwickelt, um das Hochwasserrisiko der geplanten Siedlungsbereichsfestlegungen zu ermitteln und zu bewerten. Neben der Betroffenheit von Überschwemmungsgebieten (HQ_{100}) wurden auch die Überlagerung von Flächen, die bei einem extremen Hochwasser (HQ_{extrem}) überflutet werden sowie die Parameter der Überschwemmungstiefe und Fließgeschwindigkeit ermittelt. In der Erläuterung dazu heißt es: „Soweit vorhandene Bebauung bzw. verbindlich ausgewiesene Siedlungsflächen im Regionalplan OWL als Siedlungsbereiche dargestellt und mit der zeichnerischen Festlegung als Überschwemmungsbereich überlagert sind, soll dies die Gefährdung hervorheben und zu angepassten Bauweisen und Schutzmaßnahmen anregen. Für die im Regionalplan OWL zeichnerisch nicht als Siedlungsbereich dargestellten Ortslagen gilt entsprechendes" (Bezirksregierung Detmold 2024, S. 231; 234).

▪ Siedlungsbereiche: Zielbindung auch im Bestand

Die Überlagerung der Überschwemmungsbereiche mit anderen Siedlungsraum-, Freiraum- und Infrastrukturfestlegungen ist ein wichtiger Beitrag zu einer risikobasierten Planung, da auf diese Weise der Handlungsbedarf für den Siedlungsbereich verdeutlicht wird. Dem trägt das Landesentwicklungsprogramm IV in Rheinland-Pfalz im Gegensatz zum Landesentwicklungsplan NRW bisher keine Rechnung.

Zudem sollte die Zielbindung auch für Umnutzung und Nachverdichtung bereits baulich geprägter Siedlungsstrukturen sowie der Standortsicherung bestehender Betriebe gelten. Damit kann die Regionalplanung eine wasserrechtliche Regelungslücke schließen, da § 78 WHG nur die erstmalige bauliche Inanspruchnahme von Überschwemmungsgebieten unterbindet. Als Ausnahmevoraussetzung für eine bauliche Inanspruchnahme wäre mindestens der Erhalt des Retentionsvolumens zu nennen. Gleichwohl bezieht sich diese Ausnahmeregelung allein auf die Hochwassergefahr, nicht jedoch das Hochwasserrisiko. Im Sinne der risikobasierten Planung wäre daher zumindest in der Begründung zu ergänzen, dass zu den Ausnahmevoraussetzungen neben dem Retentionserhalt auch eine an die spezifische Hochwassergefahr angepasste Bauweise und die Flächennutzung zählen.

▪ Kritische Infrastrukturen

Ein risikobasierter Planungsansatz in Bezug auf kritische Infrastruktur bedeutet neben einer möglichst nicht-exponierten Standortwahl sowie baulichen und organisatorischen Vorsorgemaßnahmen auch, dass kritische und sensible Infrastrukturstandorte nur durch redundante Strukturen außerhalb von Gebieten mit extremem Hochwasser durch wichtige Infrastrukturen wie Strom- und Wasserversorgung beliefert werden dürfen, um auch die indirekten Auswirkungen eines hochwasserinduzierten Infrastrukturausfalls zu berücksichtigen (vgl. ▶ Kap. 16). Solche Aspekte erhalten durch Vorbehaltsgebiete eine hohe Bedeutung und müssen im planerischen Abwägungsprozess berücksichtigt werden (Grundsatzcharakter). Darüber hinaus könnten solche Festlegungen in Form von Vorranggebieten einen Zielcharakter erhalten, sodass andere Planungsträger wie die kommunale Ebene oder Fachplanungen diese zwingend beachten müssen.

Beispiel aus der Praxis

Ein Beispiel hierfür ist der Regionale Planungsverband Leipzig-Westsachsen, der mit dem 2021 in Kraft getretenen Regionalplan Festlegungen mit Grundsatzcharakter zur Anpassung und Verlagerung vulnerabler Raumnutzungen, insbesondere von Siedlungsgebieten und kritischen Infrastrukturen, trifft. In Gebieten mit einem hohen Schutzbedarf gegenüber Hochwasser sollen sensitive Nutzungen und kritische Infrastrukturen geschützt werden. Darüber hinaus sollen in regionalen Schwerpunktbereichen für die Minderung bestehender Gefahrenpotenziale im Hochwasserfall hochwasserexponierte Anlagen zurück- oder umgebaut werden beziehungsweise der Neubau von Anlagen hochwasserangepasst erfolgen (RPV Leipzig-Westsachsen 2021, S. 132; 147 f.).

15.4.1.2 Kommunale Ebene: Ausschöpfung der Möglichkeiten des Städtebaurechts

- **Neubaugebiete: Risikobasierte Bauleitplanung**

Im Sinne der risikobasierten Bauleitplanung (sprich Bebauungs- und Flächennutzungsplanung) kann der Neubau von Wohn- und Gewerbenutzungen und Anlagen, die als kritische oder sensible Infrastrukturen anzusehen sind, in Gebieten verhindert werden, in denen ein erhöhtes Risiko für (Natur-)Gefahren existiert. In Bebauungsplänen kann die Freihaltung von Flächen sowie die hochwasserangepasste Bauweise von Gebäuden und Evakuierungsrouten festgesetzt werden. Leitungsgebundc Infrastrukturen können, sofern sie einer Baugenehmigung bedürfen, in gefährdeten Gebieten nicht gebündelt zugelassen werden, um die Redundanz beziehungsweise Versorgungssicherheit zu erhöhen.

Im Rahmen des KAHR-Projekts wurde im Januar 2023 in Erftstadt-Liblar ein Planspiel zum risikobasierten Bebauungsplan und einer hochwasserangepassten Siedlungsentwicklung durchgeführt, das Ansatzpunkte des Städtebaus zur Nutzungssteuerung und baulichen Gestaltung sensibler und kritischer Infrastrukturen aufgezeigt hat (Greiving et al. 2023). Formelle Instrumente wie der Flächennutzungsplan und Bebauungspläne bieten dabei vielfältige Möglichkeiten für eine hochwasserangepasste Siedlungsentwicklung und Bauweise, basierend auf fachlichen Grundlagen wie Hochwassergefahren- und -risikokarten (§ 74 WHG) sowie Überschwemmungsgebieten (§ 76 WHG) in Verbindung mit Bau- und Planungsverboten nach § 78 WHG. Im Detail wurden anhand eines Planentwurfs Festsetzungen zur Verringerung der Gefährdungsintensität (z. B. geringere Grundflächenzahl, multifunktionale Grünflächen), Festsetzungen zur Vermeidung von Schadenspotenzialen (z. B. Ausschluss vulnerabler oder gefährdender Nutzungen, Ausschluss von Kellergeschossen, Unzulässigkeit von Gebäudetechnik im Erdgeschoss) sowie Festsetzungen zur Stärkung des Reaktionsvermögens im Ereignisfall (z. B. Straßenprofil/Erhöhung der Fahrbahnmitte, bauliche Vorkehrungen zur Erleichterung der Evakuierung) diskutiert.

Eine wesentliche Grundlage für die Auswahl der vorgeschlagenen Festsetzungen war die Umweltprüfung, die in der Bauleitplanung nach § 2 Abs. 4 BauGB für die Berücksichtigung der Belange des Umweltschutzes durchzuführen ist. Dabei sind die voraussichtlichen, erheblichen Umweltauswirkungen zu ermitteln und in einem Umweltbericht zu beschreiben und zu bewerten, der nach den Vorgaben des § 2a S. 2 Nr. 2 BauGB und der Anlage 1 zum BauGB aufzustellen ist. Der Umweltbericht ist ein gesonderter Teil der Begründung von Bauleitplänen, der nach § 2a S. 3 BauGB das Ergebnis der Umweltprüfung zusammenfasst. Seit der BauGB-Novelle 2017 konkretisieren die Grundsätze der Bauleitplanung, dass bei der Aufstellung der Bauleitpläne unter anderem die Belange des Umweltschutzes und hier insbesondere Auswirkungen „[...] die aufgrund der Anfälligkeit der nach dem Bebauungsplan zulässigen Vorhaben für schwere Unfälle oder Katastrophen zu erwarten sind [...]“ zu berücksichtigen sind (§ 1 Abs. 6 Nr. 7 Buchstabe j BauGB). Dazu zählt auch eine Beschreibung und Bewertung der möglichen erheblichen Auswirkungen des geplanten Vorhabens infolge „der Risiken für die mensch-

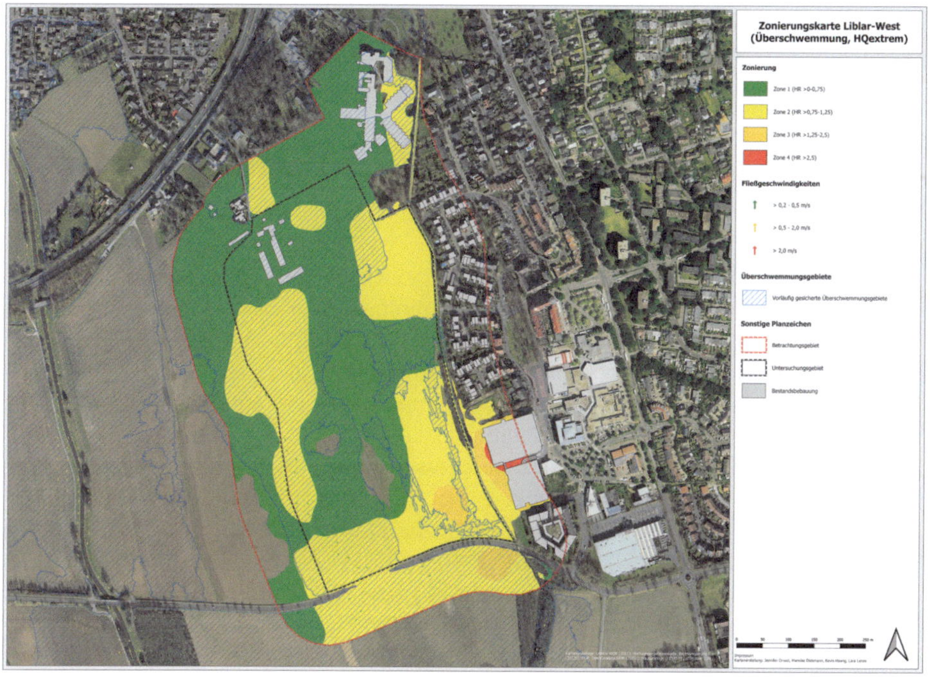

Abb. 15.1 Gefahrenzonenkarte Liblar-West (Überschwemmung, HQ_{extrem}). (Datenquellen: Bezirksregierung Köln (Hrsg.) (2022): Digitale Orthophotos (DOP). Verfügbar unter ▶ https://www.wms.nrw.de/geobasis/wms_nw_dop [abgerufen am 16.01.2023], LANUV NRW (Hrsg.) (2021): Hochwassergefahrenkarte. Verfügbar unter ▶ https://www.wms.nrw.de/umwelt/HW_Gefahrenkarte? [abgerufen am 16.01.2023]. OpenGeodata.NRW (Hrsg.) (2023): Hausumringe. Verfügbar unter ▶ https://www.opengeodata.nrw.de/produkte/geobasis/lk/akt/hu_shp/ [abgerufen am 16.01.2023])

liche Gesundheit, das kulturelle Erbe oder die Umwelt (zum Beispiel durch Unfälle oder Katastrophen)" (Anlage 1 Nr. 2b Doppelbuchstabe e) BauGB). Diese Wirkungen sind im Rahmen einer Risikoabschätzung bzw. -bewertung zu prüfen.

Die Risikobetrachtung im Umweltbericht verfolgt das Ziel, die potenziellen katastrophalen Risiken und ihre Auswirkungen zu ermitteln und zu bewerten. Eine integrierte Ermittlung und Bewertung der Anfälligkeit für Gefahren und Katastrophenrisiken zielt darauf ab, zu beurteilen, ob das Projekt tatsächlich für solche Ereignisse anfällig ist, und falls dies der Fall ist, Empfehlungen zur Vermeidung/Minimierung dieser Risiken und/oder notwendige Präventionsmaßnahmen (z. B. Festsetzungen im B-Plan) oder Alternativen (z. B. Standortwahl) (EC 2017) zu geben. Für den Umweltbericht im Rahmen des Planspiels wurde eine sog. „Gefahrenzonenkarte" erarbeitet, der die Parameter zur Differenzierung der Gefährdungsintensität in einen Indexwert überführt hat. Das Plangebiet wurde somit in Zonen eingeteilt, innerhalb derer entsprechende Einschränkungen der baulichen Nutzung im Testplan vorgeschlagen wurden (Abb. 15.1).

Im Ergebnis des Planspiels hat sich der Gefahrenzonenplan als geeignetes Instrument für eine Operationalisierung des Abwägungsbelangs Hochwasservorsorge in Risikogebieten erwiesen. Zudem konnte nachgewiesen werden, dass es möglich ist, innerhalb des geltenden planungsrechtlichen Rahmens risikobasiert zu planen (Greiving et al. 2023).

■ Bestandsgebiete: Anpassungen aktiv gestalten

Während bei der Neuentwicklung von Baugebieten ein großer Gestaltungsspielraum besteht, gilt es auch im Bestand, die Möglichkeiten des Städtebaurechts im Sinne einer risikobasierten Planung auszuschöpfen.

Im Rahmen des KAHR-Projekts wurden in einem Konsultationstermin mit ausgewählten Kommunalplanungen aus der StädteRegion Aachen im Mai 2024 u. a. Möglichkeiten und Grenzen für die Anpassung von Bestandsgebieten an die Hochwassergefährdung diskutiert.

Um den Leerstand in den Erdgeschossen der Seitenstraßen zu beseitigen und auch den Interessen der Gebäudeeigentümer entgegenzukommen, hat in einem der Beispielfälle die Stadt erwogen, die leerstehenden Geschäfte durch Wohnungen zu ersetzen und dafür den Bebauungsplan zu ändern. Problematisch war dabei jedoch, dass das Plangebiet fast vollständig im gesetzlich festgelegten Überschwemmungsgebiet liegt. Dennoch erwägt die Stadt, den Bebauungsplan zu ändern und offiziell Wohnnutzung im Erdgeschoss zuzulassen. Bislang war das Wohnen im Erdgeschoss im verbindlichen Flächennutzungsplan verboten. Aus wissenschaftlicher Sicht wurde dringend empfohlen, eine Wohnnutzung in den Erdgeschosszonen weiterhin auszuschließen. Die Gefahr für Leib und Leben ist bei einem erneuten Hochwasserereignis zu groß. Ein Alternativszenario sah den Abriss der Gebäude auf lange Sicht vor. Die frei werdenden Flächen könnten langfristig entsiegelt und begrünt werden, verbunden mit einer Prüfung, ob der Fluss an dieser Stelle renaturiert werden kann.

Als Instrument könnten hier städtebauliche Sanierungsmaßnahmen gemäß den §§ 136 ff. BauGB angewendet werden, mit denen in einem Gebiet bestehende städtebauliche Mängel oder funktionelle Defizite behoben werden können. Aufgrund dieser Regelung haben Kommunen die Möglichkeit, gezielt in bestehende Gebiete einzugreifen, um beispielsweise Maßnahmen zur Klimaanpassung umzusetzen. Dabei können die Belange der Klimaanpassung sowohl bei den Sanierungszielen als auch bei der Behebung städtebaulicher Mängel berücksichtigt werden. Gemäß § 136 Absatz 2 Satz 2c BauGB können auch Maßnahmen zur Förderung blau-grüner Infrastruktur ergriffen werden, um städtebauliche Mängel zu beheben. Zusätzlich besteht die Möglichkeit, im Rahmen einer städtebaulichen Sanierungsmaßnahme auch die lokal und regional bedeutende sensible und kritische Infrastruktur zu stärken oder zu verlagern, um sie letztendlich widerstandsfähiger zu machen.

Das Fallbeispiel hat bestätigt, dass die größte Herausforderung im Hochwasserrisikomanagement der Umgang mit bebauten Gebieten ist. Hier kollidieren die Interessen der Bewohner, des Einzelhandels, des Schutzes des kulturellen Erbes und des Hochwasserschutzes. Deshalb müssen Kompromisse eingegangen werden und der planerische und bauliche Hochwasserschutz muss angesichts der gegebenen Gefahr für Leib und Leben durch Verhaltensvorsorge und Notfallvorsorge ergänzt werden. Gefahrenzonenkarten, die Überflutungstiefen und Fließgeschwindigkeiten berücksichtigen, werden empfohlen, sind aber in Deutschland noch nicht weit verbreitet. Diese differenzierteren Karten sind wichtig, um den gesetzlichen Grundsatz der Verhältnismäßigkeit zu wahren. Planerische Vorsorgemaßnahmen, die private Eigentumsrechte einschränken, müssen daher immer in Abhängigkeit von dem jeweiligen grundstücksspezifischen Hochwasserrisiko festgelegt werden und nicht nur in Abhängigkeit von der Tatsache, dass das Grundstück hochwassergefährdet ist.

Der Konsultationstermin hat zudem gezeigt, dass insbesondere kleine und mittlere Kommunen mehr Unterstützung benötigen, um fundierte Entscheidungen treffen und ihrer Verantwortung für eine resiliente Stadtentwicklung gerecht werden zu können.

15.4.1.3 Weitere Elemente einer risikobasierten Raumplanung

- **Wiederaufbauparagraph**

Der neu eingeführte § 246c BauGB ermöglicht es den Ländern, Wiederaufbaugebiete festzulegen, innerhalb derer Vorhaben, die auf resilienten Wiederaufbau setzen, vorübergehend von den Regelungen der §§ 29–35 BauGB befreit sind. Von den Regelungen kann abgewichen werden, wenn diese oder vergleichbare Anlagen oder Einrichtungen bei Anwendung der genannten Vorschriften im Gebiet der Gemeinde, in der sie entstehen sollen, nicht oder nicht rechtzeitig zur Verfügung gestellt werden könnten. Dies bietet die Möglichkeit, nach Naturkatastrophen einen strategischen Transformationsprozess einzuleiten und die Katastrophenresilienz zu stärken, sei es durch eine hochwasserangepasste Bauweise am alten Standort oder durch einen vom bisherigen Standort versetzten Wiederaufbau.

Im Wiederaufbau kann § 246c BauGB genutzt werden, welcher zum Beispiel hochwasserangepasste Bauweisen und geringfügige Standortverlagerungen ermöglicht. Bei öffentlicher Infrastruktur sollten allerdings auch weitergehende Standortverlagerungen in Bereiche außerhalb des HQ_{extrem}-Gebiets geprüft werden.

Dies greift die Empfehlung 1 „Der Wiederaufbau nach dem Hochwasserereignis 2021 bietet auch eine Chance, einen strategischen Transformationsprozess einzuleiten und die Katastrophenresilienz zu stärken.“ aus dem 10-Punkte-Papier des KAHR-Projekts auf (KAHR 2022).

- **Nutzung informeller Instrumente**

In der Praxis kommen über die formellen Instrumente hinaus auch eine Vielzahl informeller Instrumente zur Anwendung. Integrierte Stadtentwicklungskonzepte (ISEKs) haben sich aufgrund ihrer querschnittsorientierten Ausrichtung für Themen wie das der Klimaanpassung etabliert und können zukünftig auch stärker Fragen einer resilienten kommunalen und regionalen Infrastrukturgestaltung einbeziehen. Sie können schließlich als Basis für Planungen und Entscheidungen sowie für die Ausgestaltung formeller Instrumente fungieren.

15.4.2 Wie lässt sich eine risikobasierte Raumplanung umsetzen?

Zunächst einmal ist eine Vermittlung der rechtlichen Erforderlichkeit für eine risikobasierte Raumplanung geboten, verbunden mit einer Darstellung der wesentlichen sich daraus ergebenden Anforderungen für Planungsverfahren. Angesichts der bereits gegenwärtig bestehenden Komplexität und Länge von Planungsverfahren sollte dabei nicht der Eindruck vermittelt werden, dass mit einem risikobasierten Vorgehen neue bzw. zusätzliche Anforderungen an Planung verbunden sein müssen. Es geht vielmehr darum, die prinzipiell vorhandenen Informationen insbesondere der Wasserwirtschaft zielgerichteter einzusetzen. Essenziell erscheint daher auch eine Zusammenarbeit des Planungsträgers mit der jeweils zuständigen Wasserwirtschaftsverwaltung.

Da Raumplanung mit seiner Bedeutung für Hochwasservorsorge und Wiederaufbau nicht den gesamten Hochwasserrisikokreislauf abbildet, bietet sich für ein effektives Hochwasserrisikomanagement zudem eine Abstimmung mit der jeweils zuständigen Bevölkerungsschutzbehörde an – in der Regel der entsprechende Landkreis – um sich auf gemeinsame Schutzziele zu verständigen. Auf diese Weise kann die Verhältnismäßigkeit und Erforderlichkeit der in der Raumplanung vorgesehenen Risikominderungsmaßnahmen besser beurteilt werden. So lässt sich die Frage nach der Notwendigkeit einer Standortverlagerung bzw. eines Wiederaufbaus an einem sicheren Alternativstandort nicht losgelöst von der Evakuierungsfähigkeit einer Einrichtung beantworten, wie das Beispiel der Levana-Schule (▶ Abschn. 16.3.1) verdeutlicht.

15.5 Fazit

Die konkreten Planungsherausforderungen beim Wiederaufbau nach Naturkatastrophen liegen sowohl auf der lokalen als auch auf der regionalen Ebene und erzeugen einen hohen Kooperationsbedarf zwischen allen Akteurinnen und Akteuren in der Region sowie speziell zwischen räumlicher Gesamtplanung und Fachplanungen. Grundsätzlich stellt sich die Frage, ob der Wiederaufbau an gleicher oder an anderer Stelle erfolgen soll. Häufig stellt dabei alleine die Suche nach verfügbaren, geeigneten Ersatz-/Alternativstandorten eine Herausforderung dar, insbesondere, wenn die Region bereits dicht bebaut ist oder es aus topographischen Gründen nur wenig freies Bauland gibt. Vor allem kritische Infrastrukturen müssen entweder außerhalb gefahrenexponierter Bereiche neu gebaut beziehungsweise dorthin verlagert, durch bauliche Schutzmaßnahmen in ihrer Resilienz gestärkt oder die Redundanz erhöht werden, um Ausfälle im Krisen- oder Störfall zu kompensieren. Hierbei können Zielkonflikte zwischen Nachhaltigkeit, Umweltschutz und Widerstandsfähigkeit als Leitprinzipien entstehen, die auch zwischen den verschiedenen Fachplanungen und der räumlichen Gesamtplanung ausgetragen und gelöst werden müssen.

Grundsätzlich zeigt sich aber, dass die Raumplanung in Form der Raumordnung und Bauleitplanung mit den bestehenden, aber auch den neuen rechtlichen Rahmenbedingungen und Vorgehensweisen einen bedeutenden, risikobasierten Planungsansatz implementieren und somit einen wichtigen Beitrag zu einem hochwasserresilienten Wiederaufbau leisten kann.

Zusammenfassung

- Durch den im Jahr 2021 in Kraft getretenen Bundesraumordnungsplan Hochwasserschutz (BRPH) bestehen erstmals innerhalb der Raumordnung einheitliche Regelungen für den planerischen Umgang mit Hochwasser im Binnenland und an der Küste. Eines der Ziele des BRPH ist die Einführung des Prinzips der risikobasierten Planung in die deutsche Raumplanung.
- Risikobasierte Raumplanung stellt einen Paradigmenwechsel bezüglich der raumplanerischen Vorsorge dar, da nicht mehr nur der räumliche Umgriff einer potenziellen Gefährdung zur Grundlage raumplanerischer Entscheidungen herangezogen wird, sondern nach verschiedenen Gefährdungskomponenten und den (geplanten) Nutzungen differenziert wird.
- Beim Wiederaufbau von Gebieten, die von Naturkatastrophen betroffen sind, kann risikobasierte Raumplanung durch die Differenzierung der Gefährdung, die Einbeziehung

der Empfindlichkeit und Schutzwürdigkeit von Nutzungen und die Anpassung an sich ändernde Risiken einen Beitrag zur Risikominderung leisten. Dazu sind neben Informationen zur Gefährdung auch Informationen zur Vulnerabilität und für den Umgang mit Unsicherheiten notwendig.

- Als raumplanerische Instrumente kommen auf überörtlicher Ebene u. a. Gebietsfestlegungen durch die Regionalplanung und auf kommunaler Ebene die Ausschöpfung der Möglichkeiten des Städtebaurechts zur Anwendung.
- Grundsätzlich zeigt sich, dass die Raumplanung mit den bestehenden, aber auch den neuen rechtlichen Rahmenbedingungen und Vorgehensweisen einen bedeutenden, risikobasierten Planungsansatz implementieren und somit einen wichtigen Beitrag zu einem hochwasserresilienten Wiederaufbau leisten kann.

Danksagung Die Autorinnen und Autoren von ▶ Kap. 15 bedanken sich beim Bundesministerium für Bildung und Forschung (BMBF) und beim DLR Projektträger für die Unterstützung und Förderung des KAHR-Projekts. Diese Forschung wurde vom BMBF unter dem Förderkennzeichen 01LR2102A-M finanziert.

Literatur

BBSR – Bundesinstitut für Bau-, Stadt- und Raumforschung (2024) Hochwasservorsorge: Risikovorsorge gegenüber Hochwasser und Starkregenereignissen durch Raumordnung verbessern. https://www.bbsr.bund.de/BBSR/DE/forschung/programme/moro/jahr/2022/klimawandel-energiewende-gestalten/downloads/steckbrief-hochwasservorsorge.pdf?__blob=publicationFile&v=4. Zugegriffen: 10. Jan. 2025

Bezirksregierung Detmold, Regionalplanungsbehörde (2024) Regionalplan OWL – Für den Planungsraum Ostwestfalen-Lippe. Stand: 16.04.2024 (Bekanntmachung im Gesetz- und Verordnungsblatt NRW). https://www.bezreg-detmold.nrw.de/system/files/media/document/file/3.32_regionalplan_owl_textteil2.pdf. Zugegriffen: 10. Jan. 2025

BMI – Bundesministerium des Innern und für Heimat (2021) Verordnung über die Raumordnung im Bund für einen länderübergreifenden Hochwasserschutz (BRPHV). https://www.bgbl.de/xaver/bgbl/start.xav#__bgbl__%2F%2F*%5B%40attr_id%3D%27bgbl121s3712.pdf%27%5D__1653916218508. Zugegriffen: 10. Jan. 2025

Der Paritätische Gesamtverband (2023) Bedarfe vulnerabler Bevölkerungsgruppen im Katastrophenfall. https://www.der-paritaetische.de/fileadmin/user_upload/171215_Positionspapier_Katastrophenfall_Bedarfe_vulnerable_Personengruppen_final.pdf. Zugegriffen: 10. Jan. 2025

Deutscher Bundestag (2022) Antwort der Bundesregierung auf die Kleine Anfrage der Fraktion der CDU/CSU – Drucksache 20/1384, 21.04.2022. Entschädigung für Grundstücke in den Flutgebieten. https://dserver.bundestag.de/btd/20/014/2001479.pdf. Zugegriffen: 10. Jan. 2025

EC – European Commission (2017) Environmental impact assessment of projects. Guidance on the preparation of the environmental impact assessment report (Directive 2011/92/EU as amended by 2014/52/EU). Luxembourg: Publications Office of the European Union. ISBN: 978-92-7974374-0.

Gerhold L, Hartmann J, Thiebes B, Thieken AH (2019) DESKRIS – Definition von Schutzzielen für Kritische Infrastrukturen – Forschungsstand, rechtlicher Rahmen und politische Entscheidungsfindung Bd 28. BBK: Forschung im Bevölkerungsschutz, Bonn (https://www.researchgate.net/profile/Lars-Gerhold-2/publication/349822466_Definition_von_Schutzzielen_fur_Kritische_Infrastrukturen/links/605b8c8b299bf17367685ba6/Definition-von-Schutzzielen-fuer-Kritische-Infrastrukturen.pdf, zuletzt geprüft am 10.01.2025)

Godschalk DR (2003) Urban hazard mitigation: Creating resilient cities. John Jay College, City University of New York, January 22–24. Plenary paper presented at the Urban Hazards Forum

Greiving S (2019) Analyse- und Bewertungskonzepte für Risiken im Vergleich. Inf Raumentwickl 4/2019:62–73

Greiving S, Fleischhauer M, Hartz A, Terfrüchte T (2023) Resiliente Raum- und Infrastrukturen gestalten: Neue Flächenansprüche in der Zeitenwende. Izr – Informationen Zur Raumentwickl 50(1):40–53

Henstra D, Kovacs P, McBean G, Sweeting R (2004) Background paper on disaster resilient cities. Prepaed by the Institute for Catastrophic Loss Reduction. Infrastructure Canada, Research and Analysis Division, Ottawa

KAHR – Klima-Anpassung, Hochwasser und Resilienz (2022) 10 Empfehlungen aus Sicht der Wissenschaft zum Thema Wiederaufbau und Zukunftsfähigkeit der flutbetroffenen Regionen. https://hochwasser-kahr.de/index.php/de/alle-produkte/10-empfehlungen. Zugegriffen: 10. Jan. 2025

Land Rheinland-Pfalz, SGD Nord (2021) Vorläufige Sicherstellung des Überschwemmungsgebietes Ahr. https://sgdnord.rlp.de/themen/wiederaufbau-ahr/ueberschwemmungsgebiet. Zugegriffen: 10. Jan. 2025

Regionaler Planungsverband Leipzig-Westsachsen (2021) Regionalplan Leipzig-Westsachsen. https://www.rpv-westsachsen.de/regionalplan-leipzig-westsachsen/. Zugegriffen: 10. Jan. 2025

Truedinger A, Birkmann J, Fleischhauer M, Greiving S, Kirschbauer L, Schäfer M, Burggraf C, Fabisch M, Jüpner R, Droste-Franke B, Apel H, Arenz H-T, Johann G, Wolf S, Schüttrumpf H, Dolfus F, Tholl L (2024) Praxishilfe – Stärkung der Resilienz der Stromversorgung gegenüber Starkregen und Hochwasser: Vorsorge, Anpassung, Wiederaufbau. Veröffentlichung im Rahmen des Vorhabens KAHR – Klima-Anpassung, Hochwasser und Resilienz (2021–2025), BMBF-Förderkennzeichen 01LR2102. Online verfügbar: https://hochwasser-kahr.de/images/produkte/Alle_Praxishilfe_Resilienz_Strom_KAHR_Westnetz.pdf. Zugegriffen: 10. Jan. 2025

UNDRR – United Nations Office for Disaster Risk Reduction (2023) Sendai Framework Terminology on Disaster Risk Reduction: Vulnerability. https://www.undrr.org/terminology/vulnerability. Zugegriffen: 10. Jan. 2025

Kritische und sensible Infrastrukturen: „Wieder- versus Neuaufbau“

Alessa Trüdinger, Jörn Birkmann, Mark Fleischhauer und Stefan Greiving

J. Birkmann et al. (Hrsg.), *Die Flutkatastrophe 2021*, Sustainable Development Goals (SDG) – Umsetzung in Praxis, Lehre und Entscheidungsprozessen,
https://doi.org/10.1007/978-3-658-48688-4_16

Um Hochwasserrisiken zu reduzieren und nach einem Ereignis das erneute Eintreten einer Flutkatastrophe wie vom Juli 2021 zu verhindern, ist es wichtig, resilienter (wieder-)aufzubauen und vorhandene Infrastrukturen umzugestalten, sodass ein Mehr an Resilienz geschaffen wird. Da die Auswirkungen von extremen Starkregen- und Hochwasserereignissen insbesondere auch von der Vulnerabilität und dem Vorbereitungsgrad auf solche Ereignisse abhängen und im Wiederaufbau die Ressourcen oftmals begrenzt sind, ist ein Fokus insbesondere auf vulnerable Bevölkerungsgruppen und kritische sowie sensible Infrastrukturen zu legen, sodass im Wiederaufbau die Exposition und Anfälligkeiten von diesen Infrastrukturen reduziert werden. Der Wiederaufbau bietet eine gute Möglichkeit, neue und resilientere Strukturen zu schaffen, doch gleichzeitig müssen auch die bestehenden Infrastrukturen – auch jenseits der betroffenen Gebiete – auf ihre Resilienz hin überprüft und gegebenenfalls angepasst werden. Zusätzlich führen veränderte Rahmenbedingungen durch neue Gesetze sowie Veränderungen aufgrund des Klimawandels zu neuen Anforderungen an die Gestaltung von Infrastrukturen.

16.1 Was zeichnet kritische und sensible Infrastrukturen aus?

Gerade im Hinblick auf Krisenfälle spielt die Funktionsfähigkeit von kritischen und sensiblen Infrastrukturen eine entscheidende Rolle, wobei sich in diesem Bereich Infrastrukturen gemäß ihrer Schutzwürdigkeit und Bedeutung für die Bevölkerung priorisieren lassen.

16.1.1 Infrastrukturtypen

Neben kritischen Infrastrukturen, kurz: KRITIS, gibt es noch weitere Infrastrukturtypen, die als besonders schutzwürdig angesehen werden können. Im Kontext der Raumplanung fällt der Begriff der Schutzwürdigkeit unter anderem in den Vorgaben des relativ neuen, länderübergreifenden Raumordnungsplanes für den Hochwasserschutz. Unter Schutzwürdigkeit ist ein politisch-normatives Konzept zu verstehen, welches in der Raumplanung angewendet wird, um bestimmte Nutzungen und Raumfunktionen in der Frage des Schutzes stärker zu gewichten, was entsprechende Konsequenzen für den Abwägungsprozess nach sich zieht (Hartz et al. 2020). Insgesamt gibt es laut Greiving et al. (2023) vier Typen von besonders schutzwürdigen Infrastrukturen, wobei manche Infrastrukturen, z. B. Krankenhäuser, auch mehreren Typen zugeordnet werden können:

Kritische Infrastrukturen (KRITIS) „sind Organisationen oder Einrichtungen mit wichtiger Bedeutung für das staatliche Gemeinwesen, bei deren Ausfall oder Beeinträchtigung nachhaltig wirkende Versorgungsengpässe, erhebliche Störungen der öffentlichen Sicherheit oder andere dramatische Folgen eintreten würden" (Bundesregierung Deutschland o. J.). Wichtig ist hierbei auch die Vermeidung von Funktionsverlusten außerhalb exponierter Bereiche und in anderen Infrastrukturbereichen (sog. „Domino- und Kaskadeneffekte"). Vom Bundesraumordnungsplan Hochwasserschutz werden KRITIS mit länder- oder staatsgrenzenüberschreitender Bedeutung sowie die nach der BSI-Kritisverordnung festgelegten Infrastrukturen erfasst (Truedinger et al. 2024b).

Sensible Infrastrukturen sind Organisationen und Einrichtungen, die von vulnerablen Gruppen, welche im Ereignisfall die Hilfe Dritter benötigen, genutzt werden und deren Beschädigung oder Ausfall zwar nicht das Funktionieren des staatlichen oder regionalen Gemeinwesens beeinträchtigt, aber zu erheblichen sozialen Schäden führen kann (Greiving et al. 2023). Da sich die vulnerablen Nutzerinnen und Nutzer, z. B. von Pflege- und Altenheimen, Schulen oder Kindergärten, in Gefahrensituationen vielfach nicht selbstständig retten können und bei einer Evakuierung die Hilfe Dritter benötigen, erfordern diese Infrastrukturen oft einen hohen Schutzaufwand. Im Bundesraumordnungsplan Hochwasserschutz werden diese Einrichtungen als „bauliche Anlagen, die ein komplexes Evakuierungsmanagement erfordern", bezeichnet (Truedinger et al. 2024b).

Infrastrukturen, von denen im Ereignisfall besondere Gefährdungen ausgehen. Diese Infrastrukturen sind als besonders schutzwürdig anzusehen, da sie das Potenzial haben, beispielsweise bei Überflutungsereignissen, die bereits entstandenen Schäden zu verschlimmern, insbesondere durch das Austreten gefährlicher Stoffe wie z. B. Treibstoff oder Chemikalien, was zur Kontamination von Grundwasser, Erdreich und Gebäuden führen kann (Greiving et al. 2023).

Infrastrukturen als Teil des Baukulturerbes. Die Beschädigung oder Zerstörung des baukulturellen Erbes kann eine Gesellschaft emotional stark belasten und langfristige psychologische Folgen nach sich ziehen – diese symbolische Kritikalität geht auf die kulturelle und identitätsstiftende Bedeutung zurück (Greiving et al. 2023).

Im Folgenden wird insbesondere auf die ersten beiden Infrastrukturtypen vertieft eingegangen.

16.1.2 Betroffenheit infolge der Flut 2021

Neben enormen Schäden an privaten und öffentlichen Gebäuden hat die Flutkatastrophe vom Juli 2021 auch zahlreiche besonders schutzwürdige Infrastrukturen direkt und indirekt getroffen. So wurden allein im Ahrtal beispielsweise etwa 40 Kindergärten und Kitas von der Flut verwüstet, hunderte Kilometer Stromleitungen beschädigt sowie mehrere Krankenhäuser und Kliniken direkt geschädigt oder in ihrer Funktionsausübung massiv behindert, u. a. durch den Ausfall der Strom- und Wasserversorgung (Landesregierung Rheinland-Pfalz 2022). Gerade die indirekte Betroffenheit, die infolge von Domino- und Kaskadeneffekten entsteht, darf nicht vernachlässigt werden. So können beispielsweise in weit entfernten, nicht direkt hochwasserexponierten Gebieten schwerwiegende Schäden und Funktionsverluste bei kritischen und sensiblen Infrastrukturen auftreten (s. ► Abschn. 16.1.3). Auch im Ahrtal konnten solche Effekte beobachtet werden. So musste nach der Flut beispielsweise die Augustinum Seniorenresidenz in Bad Neuenahr evakuiert werden, weil der Zeitpunkt der Wiederherstellung von Strom- und Wasserversorgung nicht absehbar war (Bugge 2021). Ebenfalls aufgrund der fehlenden Stromversorgung musste das Krankenhaus Maria Hilf in Bad Neuenahr-Ahrweiler, das vorübergehend mit den eigenen Notstromaggregaten betrieben wurde, nach einigen Tagen evakuiert werden. Über 100 Patientinnen und Patienten wurden in andere Krankenhäuser verlegt, da durch den endgültigen Ausfall der Stromversorgung die Wasserversorgung, Kühlung und Lüftung ausfielen (Truedinger et al. 2024b).

16.1.3 Kritikalität und Kaskadeneffekte

Die verschiedenen Sektoren, denen KRITIS und sensible Infrastrukturen zugeordnet werden können, wie etwa die Wasserversorgung, Energieversorgung und Gesundheitseinrichtungen, sind dabei nicht isoliert, sondern sowohl innerhalb ihres jeweiligen eigenen Sektors (Wasser, Energie, etc.) als auch sektorübergreifend miteinander verknüpft und daher häufig voneinander abhängig. Durch den Ausfall einer kritischen Infrastruktur wird oftmals deutlich, wie wichtig deren Leistungen für andere KRITIS sind. Diese Interdependenzen führen dazu, dass sich die Auswirkungen auf verschiedene Infrastruktursysteme und über administrative Grenzen hinweg mit zunehmender Schwere ausdehnen, was als **Kaskadeneffekt** bezeichnet wird (Truedinger et al. 2024b).

Die **Kritikalität** einer Infrastruktur ist das relative Maß für ihre Bedeutung in Bezug auf die Folgen, die ein Ausfall oder eine Störung der Infrastruktur für die Versorgungssicherheit der Gesellschaft mit wesentlichen Gütern und Dienstleistungen hat (BMI 2009). Bei kritischen und auch sensiblen Infrastrukturen sollte die Betrachtung damit nicht nur auf das Objekt selbst bezogen sein, sondern eben auch auf die Frage, wie verwundbar bestimmte (abhängige) Nutzergruppen dieser Infrastrukturen sind. Sensible Infrastrukturen und ihre Sensitivität zeichnen sich demzufolge insbesondere auch über die in der Infrastruktur untergebrachte besonders verwundbare Bevölkerungsgruppe aus, die im Ereignisfall die Hilfe Dritter benötigt, wie z. B. alte Menschen in Seniorenheimen oder Kinder mit körperlichen und geistigen Behinderungen in Förderschulen.

Störungen und Beeinträchtigungen von Leistungen kritischer und sensibler Infrastrukturen können also auch durch Kaskadeneffekte ausgelöst werden, indem eine betroffene Infrastrukturbranche Auswirkungen auf andere Bereiche hat (Schmitt 2019). Die hohe Bedeutung für zentrale gesellschaftliche Prozesse und das Funktionieren der Gesellschaft und Wirtschaft macht kritische Infrastrukturen besonders schutzwürdig, wie auch in § 2 Abs. 2 Nr. 3 S. 4 des Raumordnungsgesetzes hervorgehoben wird: „Dem Schutz kritischer Infrastrukturen ist Rechnung zu tragen." Ein Problem der bisherigen Risikobetrachtung in der Raumplanung besteht darin, dass Krisen und Risiken in der Regel nur im Hinblick auf den Raum und die spezifischen Gefährdungen analysiert werden, während kumulative und indirekte Effekte, die auch außerhalb exponierter Gebiete auftreten können, häufig nicht systematisch erfasst werden. Dies ist einerseits in der raumbezogenen Ausrichtung der Raumplanung begründet, wobei die Planungshoheit für den eigenen Planungsraum besteht, und andererseits im Auftreten von Kaskadeneffekten, die über kommunale oder Kreisgrenzen hinausgehen und daher oft unbeachtet bleiben. Darüber hinaus ist es in Teilen der kommunalen bzw. kreisweiten Zuständigkeit im Katastrophenschutz geschuldet, die den Blick über die eigenen Grenzen hinweg erschwert (Truedinger et al. 2024b).

Kaskadeneffekte verstärken also die Risiken innerhalb eines Systems und führen zu großen Herausforderungen bei Risikobewertung und -management, da hierzu oftmals kaum Informationen vorliegen und sie nur schwer abzuschätzen sind. Außerdem lassen sie sich aufgrund der (z. T. wechselseitigen) Abhängigkeiten weder durch ortsbezogene Managementkonzepte der Raumplanung noch durch technische Ansätze allein lösen, weswegen eine umfassende Risiko-Governance notwendig ist (Truedinger et al. 2024b). Hierzu gehören auch eine effektive Risikokommunikation mit klaren Verantwortlichkeiten, Aufgaben und Informationen zu Vorsorge- und Verhaltensmaßnahmen sowie die Beteiligung der lokalen und regionalen Akteurinnen und Akteure, um beispielsweise akteursbezogene Zusammenhänge oder Unsicherheiten zu erfassen (Truedinger et al. 2024b). Zudem ist es essenziell, Abhängigkeiten und daraus resultierende Kaskadeneffekte im Voraus zu identifizieren, was in Zusammenarbeit mit allen beteiligten Akteuren durch Diskussionsrunden, Planspiele, Simulationen oder Modellierungen von Abhängigkeiten und verschiedenen Ereignissen gelingen kann. Die Durchführung einer partizipativen Kritikalitätsanalyse, wie sie beispielsweise im Rahmen des BMBF-geförderten Projekts SysKa am Beispiel der Stadt Bochum durchgeführt wurde, dient der Identifizierung von Kaskaden und Kaskadeneffekten, sodass anschließend Maßnahmen und Ansatzpunkte zur frühzeitigen Unterbrechung der Kaskaden abgeleitet werden können (Truedinger et al. 2024b). Bei sensiblen Infrastrukturen wäre analog dazu der Grad der Sensibilität zu ermitteln.

16.1.4 Identifikation von schutzwürdigen Infrastrukturen

Die Identifikation schutzwürdiger Infrastrukturen stellt eine Herausforderung für Wissenschaft und Politik dar. Diese ist aber entscheidend für die Vor- und Nachsorge im Sinne des Risikomanagements. Da es sich bei Schutzwürdigkeit allerdings um ein politisch-normatives Konzept handelt, ist es nicht möglich, eine einheitliche, vollständige Liste aller schutzwürdigen Infrastrukturen für jedweden Raum vorzugeben.

KRITIS stehen bereits länger im Fokus von Politik, Wirtschaft und Wissenschaft – und die Notwendigkeit des Schutzes von KRITIS ist sowohl auf europäischer als auch auf Bundesebene gesetzlich verankert, z. B. im Raumordnungsgesetz. Auf Bundesebene gibt es außerdem durch die BSI-Kritisverordnung für verschiedene Sektoren Festlegungen und Schwellenwerte zur Bestimmung kritischer Infrastrukturen nach dem BSI-Gesetz. Als Voraussetzung und Schwellenwert wird hier oft angesetzt, dass die kritische Infrastruktur oder Dienstleistung essenziell für die Gesamtversorgung Deutschlands sein und mindestens 500.000 Personen versorgen muss, wodurch insbesondere in ländlichen Räumen viele KRITIS nicht von der BSI-Kritisverordnung und dem kommenden KRITIS-Dachgesetz erfasst werden. Demzufolge sind die KRITIS-Schwellenwerte auch für den eher ländlich geprägten Raum des Ahrtals und der flutbetroffenen Kreise in NRW nicht hinreichend passfähig.

D. h. es existieren – gerade in ländlichen Räumen mit geringerer Bevölkerungsdichte als in Ballungsräumen – zahlreiche weitere KRITIS mit regionaler oder lokaler Bedeutung, für die bislang keine spezifischen Operationalisierungen oder Schwellenwerte definiert wurden. Darüber hinaus gibt es KRITIS-Sektoren, die von der BSI-Kritisverordnung nicht abgedeckt sind, wie etwa der Bereich Staat und Verwaltung. Auch wenn regionale KRITIS oft unter den Schwellenwerten liegen, die im Anhang der BSI-Kritisverordnung festgelegt sind, besitzen sie eine übergeordnete Bedeutung, da sie häufig eine Versorgungsfunktion für benachbarte Gemeinden oder für andere KRITIS-Sektoren übernehmen. Zudem können solche KRITIS sowie sensible Infrastrukturen für besonders verwundbare Bevölkerungsgruppen in der Region eine zentrale Bedeutung haben, wie z. B. das Beispiel der Levana-Schule zeigt. Diese schutzwürdigen Infrastrukturen sollten zukünftig jeweils auf lokaler und regionaler Ebene in einem partizipativen Prozess definiert werden. Sie ergeben sich aus den Zuständigkeitsbereichen der relevanten regional und kommunal Handelnden (Kommunalverwaltung, Kreisverwaltung, Katastrophenschutz, regionale Energieversorger usw.) und können beispielsweise Pflegeeinrichtungen, Apotheken, Arztpraxen und landwirtschaftliche Großbetriebe umfassen (Truedinger et al. 2024b).

Die konkrete Identifizierung der schutzwürdigen Infrastrukturen und die Erstellung eines Katasters kann somit nur unter Berücksichtigung der Gegebenheiten vor Ort stattfinden. Eine Angabe von Schwellenwerten ist auf lokaler und regionaler Ebene schwierig, da in ländlichen Gebieten zum Teil auch deutlich kleinere Einrichtungen von besonderer Bedeutung sein können und die Schutzwürdigkeit in einem gesellschaftspolitischen Diskurs zu ermitteln ist. Eine wichtige Grundlage für diesen Diskurs sind allerdings wissenschaftliche Studien und Empfehlungen, die die Bedeutung und Schutzwürdigkeit der jeweiligen kritischen oder sensiblen Infrastruktur genauer beleuchten.

16.2 Herausforderung: Hochwasser und Starkregen

Für die Raumplanung sind nach § 78 Wasserhaushaltsgesetz (WHG) im Hinblick auf die Überflutungs-Thematik die festgesetzten Überschwemmungsgebiete zentral. Bei diesen sog. festgesetzten Überschwemmungsgebieten handelt es sich nach § 76 Abs. 2 Nr. 1 und 2 um Gebiete, die im statistischen Mittel einmal in hundert Jahren überschwemmt werden, und um Gebiete, die zur Hochwasserentlastung und Rückhaltung beansprucht werden. Hierbei besteht allerdings häufig das Problem, dass die Überschwemmungsgebiete nur für Gewässer I. und II. Ordnung in Hochwassergefahren- und -risikokarten festgehalten werden. So ist beispielsweise in Rheinland-Pfalz für die Festsetzung der Überschwemmungsgebiete an Gewässern I. und II. Ordnung die Struktur- und Genehmigungsdirektion Süd bzw. Nord als Obere Wasserbehörde zuständig und für die Festsetzung der Überschwemmungsgebiete an Gewässern III. Ordnung sind die Kreisverwaltungen bzw. Verwaltungen der kreisfreien Städte als Untere Wasserbehörden zuständig (Struktur- und Genehmigungsdirektion Süd RLP o. J.). Gemäß WHG müssen Überschwemmungsgebiete mindestens für die nach EU-Hochwasserrisikomanagementrichtlinie ermittelten Risikogewässer ausgewiesen werden (Struktur- und Genehmigungsdirektion Süd RLP o. J.), sodass es häufig keine Überschwemmungsgebiete und Karten für kleinere Gewässer gibt. Im Zuge der Ahrtal-Katastrophe und weiterer Ereignisse zeigte sich aber, dass nicht nur die Überschwemmungen von großen Flüssen hochproblematisch werden können, sondern gerade im Zusammenhang mit Starkregenereignissen auch die kleineren Gewässer sehr schnell anschwellen und ausufern können und somit ebenfalls ein hohes Gefahrenpotenzial bergen. Darüber hinaus können sich bei Starkregenereignissen auch abseits von Gewässern Fließwege mit beträchtlichen Wassertiefen und Fließgeschwindigkeiten ausbilden.

Hinweis für Entscheidungsprozesse

Ein weiteres Problem stellt die Datengrundlage, die zur Erstellung der Hochwassergefahrenkarten genutzt wird, dar. Zumeist können nur relativ kurze Zeitreihen berücksichtigt werden, historische Ereignisse fallen somit oft aus den Modellierungen heraus, sodass das HQ_{100}, also das (Abfluss-)Ereignis, das im statistischen Mittel einmal alle einhundert Jahre auftritt, oftmals unterschätzt wird. So gab es z. B. an der Ahr bereits 1804 und 1910 starke Hochwässer, diese sind aber nicht in die bisherige Erstellung der Karten eingeflossen. Auch etwaige Veränderungen und Unsicherheiten aufgrund des Klimawandels fließen nicht mit ein – wenngleich laut Attributionsforschung das Eintreten des Ereignisses vom Juli 2021 aufgrund des Klimawandels etwa 1,2–9-mal wahrscheinlicher gegenüber einem 1,2 °C kälteren Klima war und der Klimawandel die Intensität der maximalen 1-Tages-Regenfälle in der Sommersaison in der gesamten betroffenen Region bereits um 3–19 % erhöht hat (Tradowsky et al. 2023).

Eine weitere Unzulänglichkeit zeigt sich in Bezug auf das Thema Starkregen und Sturzfluten. Starkregen- bzw. Sturzflutgefahrenkarten haben bislang bundesweit keine rechtliche Verbindlichkeit und müssen weder flächendeckend noch nach bestimmten Vorgaben erstellt werden. Es gibt somit bislang keine bundesweit einheitliche Verpflichtung, die explizit auf die Erstellung von Starkregengefahrenkarten abzielt. Dennoch hat sich hier in den vergangenen Jahren bereits etwas getan. So gibt es beispielsweise gesetzliche Neuerungen, die die Berücksichtigung von Anpassungsnotwendigkeiten gegenüber klimawandelbeeinflussten Naturgefahren stärker einfordern, wie das in 2024 in Kraft getretene Klimaanpassungsgesetz (KAnG), welches die systematische Anpassung an die Folgen des Klimawandels fordert, z. B. durch die Verpflichtung zur Erstellung von Klimaanpassungskonzepten auf kommunaler Ebene gemäß § 12 KAnG und die Vorsorge vor und Minimierung von Klimarisiken, worunter auch Starkregen fällt (Truedinger et al. 2024b). Zusätzlich zur lokalen Ebene ist ein verstärktes überörtliches, vorsorgendes Risikomanagement für Starkregenereignisse erforderlich. Das Starkregen-induzierte Hochwasserereignis vom Juli 2021 hat gezeigt, dass die Auswirkungen von Starkregen auch eine bedeutende überörtliche Dimension haben, sodass auch auf der Ebene der Raumordnung raumplanerische Risikovorsorge erforderlich ist (Truedinger et al. 2024b). Nach § 8 KAnG müssen sowohl die bereits eingetretenen als auch die zukünftig zu erwartenden Auswirkungen des Klimawandels, einschließlich Starkregenereignissen und Sturzfluten berücksichtigt werden. Somit gibt es durchaus eine zunehmende rechtliche Notwendigkeit, Starkregenereignisse im Rahmen der Klimaanpassung in den Handlungsfeldern Hochwasservorsorge und Raumplanung zu berücksichtigen, was auch die Erstellung von Starkregengefahrenkarten und Starkregenrisikokarten umfasst. Einzelne Bundesländer, wie Rheinland-Pfalz und Nordrhein-Westfalen, haben darüber hinaus auch landesweite Sturzflutgefahren- und -risikokarten bzw. Hinweiskarten Starkregengefahren erstellt. Die Frage, ob dabei auch die Thematik der kritischen und sensiblen Infrastrukturen hinreichend aufgegriffen wurde und wird, ist allerdings noch nicht geklärt.

Beispiel aus der Praxis

In Rheinland-Pfalz wurden nach der Flut sowohl die festgesetzten Überschwemmungsgebiete in den betroffenen Regionen überarbeitet als auch die Starkregengefahrenkarten, die in überarbeiteter Form als sog. Sturzflutgefahrenkarten erstellt wurden. Diese bilden nun auch die Sturzflutgefahr mit Fließgeschwindigkeiten und Einstautiefen im besiedelten Gebiet ab, was bis dato nicht der Fall war, da bisher in den öffentlich verfügbaren Karten die Siedlungsgebiete ausgenommen waren. Mittlerweile gibt es in Rheinland-Pfalz die sog. Sturzflutgefahrenkarten für drei Szenarien: ein außergewöhnliches Starkregenereignis mit einer Regendauer von einer Stunde (SRI 7), je nach Region mit einer Regenmenge von ca. 40–47 mm (bzw. $1/m^2$) in einer Stunde, ein extremes Starkregenereignis mit einer Regendauer von einer Stunde (SRI 10), je nach Region mit einer Regenmenge von ca. 80–94 mm in einer Stunde, und ein extremes Starkregenereignis mit einer Regendauer von vier Stunden (SRI 10), je nach Region mit einer Regenmenge von ca. 112–136 mm in vier Stunden (Landesamt für Umwelt Rheinland Pfalz o. J.). Auch wenn es in Bezug auf Starkregenereignisse keine Jährlichkeiten in dem Sinne gibt, wie es bei Hochwasserereignissen der Fall ist, und sich hier Wiederkehrwahrscheinlichkeiten auch viel schwieriger festlegen lassen, kann das in Rheinland-Pfalz verwendete SRI 7 etwa mit einem hundertjährlichen Hochwasserereignis gleichgesetzt werden. Daher lautet in Rheinland-Pfalz auch die Empfehlung behördlicherseits, dass bei Planungen, z. B. im Kontext der Bauleitplanung, das SRI 7 als Bewertungsgrundlage herangezogen werden soll – und bei Planungen, die besonders schutzwürdige Infrastrukturen betreffen, das SRI 10 als extremes Szenario. Dennoch handelt es sich auch hier bislang nur um eine Empfehlung und keine rechtlich verbindliche Vorgabe.

16.3 Stärkung der Resilienz kritischer und sensibler Infrastrukturen – Fallbeispiele

Dass gerade bei kritischen und sensiblen Infrastrukturen ein besonderer Schutzbedarf besteht, deckt sich auch mit einer der „10 Empfehlungen aus Sicht der Wissenschaft zum Thema Wiederaufbau und Zukunftsfähigkeit der flutbetroffenen Regionen" des KAHR-Projektes (KAHR et al. 2022). So besagt Empfehlung Nr. 10, dass neue Schutzstandards und Schutzziele für kritische und sensible Infrastrukturen definiert werden müssen und das Bewusstsein für ein unvermeidbares Restrisiko gestärkt werden muss. Diese Empfehlung wurde auch bereits in der Praxis des Wiederaufbaus häufiger herangezogen. Beispielsweise wurde u. a. mit Bezug auf diese Empfehlung seitens des Ministeriums für Klimaschutz, Umwelt, Energie und Mobilität Rheinland-Pfalz die Finanzierung eines höheren Schutzstandards für das Wasserwerk Walporzheim (Schutz vor HQ_{extrem} anstatt vor HQ_{100}) durchgesetzt (Birkmann et al. 2024). Außerdem lehnte die SGD Nord, u. a. in Anlehnung an die 10. KAHR-Empfehlung, den geplanten Standort des neuen Feuerwehrgerätehauses der Feuerwehr Sinzig ab. Ein weiteres Beispiel für die gelungene Umsetzung dieser Empfehlung und Stärkung der Resilienz sensibler Infrastrukturen stellt die Levana-Schule dar.

Im Folgenden werden nun zwei konkrete Fallbeispiele, wie die Stärkung der Resilienz besonders schutzwürdiger Infrastrukturen gelingen kann, dargestellt. Sowohl die Thematik sensible Infrastrukturen/Levana-Schule als auch die Thematik Stromversorgung wurden im Rahmen des KAHR-Projektes intensiv betrachtet und begleitet. Anhand der Levana-Schule wird außerdem deutlich, warum auch sensible Infrastrukturen eine besondere Schutzwürdigkeit aufweisen und der Fokus, der bislang meist auf kritischen Infrastrukturen lag, aufgeweitet werden muss.

16.3.1 Fallbeispiel Levana-Schule

Die Levana-Schule gehört zu den besonders sensiblen und somit schutzwürdigen Infrastrukturen, da sie von einer besonders vulnerablen Personengruppe, und zwar von Kindern mit geistigen und teils auch körperlichen Behinderungen, genutzt wird, die im Ereignisfall auf die Hilfe Dritter angewiesen sind. Die Schule umfasst knapp 100 Schülerinnen und Schüler, wovon – Stand Schuljahr 2022/2023 – mindestens 30 eine Evakuierung nur mit besonderer Hilfestellung vollziehen können, da sie neben geistigen auch motorische Behinderungen haben. Alle Schülerinnen und Schüler weisen überdies geistige Beeinträchtigungen auf, weswegen sie laut Schulleitung einen Notfall kognitiv

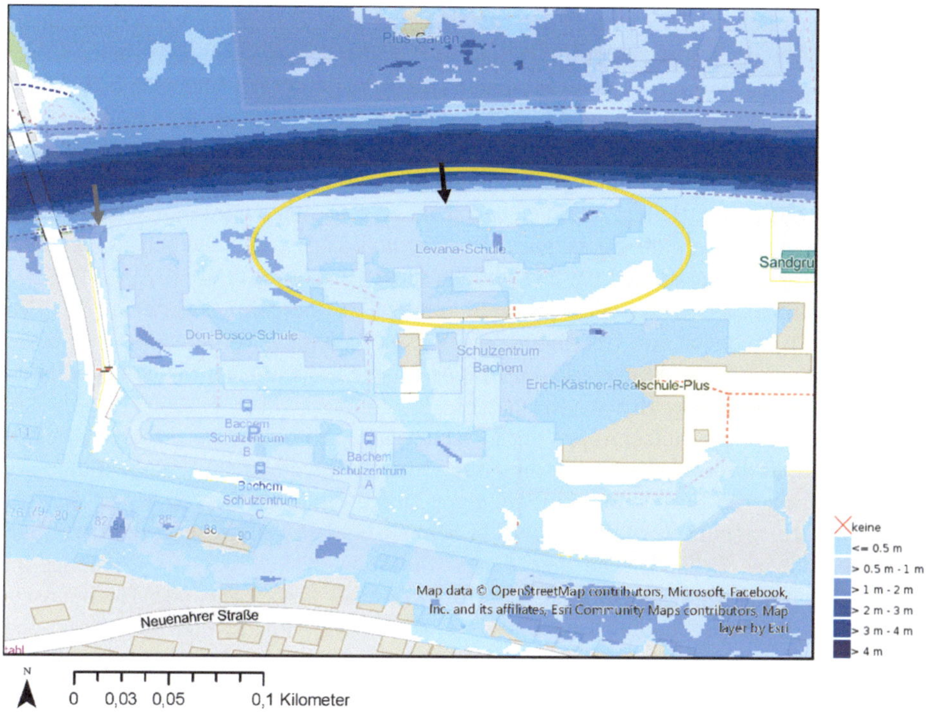

Abb. 16.1 Darstellung der HQ_{100}-Überflutungsflächen, überarbeitet vom Landesamt für Umwelt Rheinland-Pfalz (LfU RP) nach der Flut 2021, in der Umgebung der Levana-Schule (Hochwassergefahrenkarte Rheinland-Pfalz, © LfU RP, ▸ https://wasserportal.rlp-umwelt.de/kartendienste). (Bildquelle: Truedinger et al. 2024a)

kaum bis gar nicht umsetzen können. Dadurch ist der technische, personelle und organisatorische Aufwand bei einer Evakuierung sehr hoch einzuschätzen und gleichzeitig die eigenständige Evakuierung der Kinder faktisch ausgeschlossen.

Der bisherige Standort der Levana-Schule befindet sich in einem hochexponierten Gebiet, direkt an der Ahr – mit dem Hauptausgang zur Ahr hin (s. Abb. 16.1, schwarzer Pfeil).

Das Gebäude ist nur eingeschossig und weist damit auch keine Möglichkeit zur vertikalen Evakuierung von Personen und Gegenständen auf. Obschon die genauen Vorwarnzeiten aus bisherigen Studien nicht genau definierbar sind, ist davon auszugehen, dass selbst bei einer Vorwarnzeit von mehreren Stunden erhebliche Herausforderungen bei der Evakuierung der Personen sowie bei der Sicherung und Verlagerung des spezifischen Inventars bestehen. Die Erfahrungen aus dem Hochwasser 2016 und dem Hochwasser 2021 verdeutlichen, dass die Verlagerung des Inventars nicht hinreichend möglich ist (Birkmann et al. 2023). Zudem zeigt das Hochwasser 2021 auf, dass bei der Prognose erhebliche Probleme bestehen. So waren die Pegelprognosen im Juli 2021 lange Zeit viel zu niedrig und im späteren Verlauf des Ereignisses deckungsgleich mit der tatsächlichen Wasserhöhe am Pegel in Altenahr. Die Prognose hatte also für eine Vorplanung, wie z. B. für eine Evakuierung, keine prognostische Wirkung (Truedinger et al. 2024a).

Des Weiteren ist das Gebäude sowie das umgebende Gelände bereits bei einem HQ_{100} bis zu 1 m hoch überflutet und die Zufahrtswege zur Schule sowie der Sammelplatz, der sich zwischen Levana- und Don-Bosco-Schule befindet, sind in Teilen noch höher überflutet (s. ◻ Abb. 16.1, u. a. grauer Pfeil), sodass die sichere Erreichbarkeit bereits deutlich unterhalb des neuen HQ_{100} nicht mehr gegeben ist. So ist bereits bei einem Pegel in Altenahr von 410 cm ein Teil des Zufahrtsweges sowie des Sammelplatzes der Levana-Schule überflutet, wobei ein solcher Pegel im statistischen Mittel etwas seltener als alle 20 Jahre auftritt (nach den alten Berechnungen). Infolge des Klimawandels könnte ein solches Ereignis auch häufiger auftreten, wobei zu bedenken ist, dass massiv gebaute Schulgebäude eine Nutzungsdauer von 40–80 Jahren aufweisen und daher der Sicherheitsaspekt auch den zukünftigen Klimawandel berücksichtigen muss (Birkmann et al. 2023). Zukünftig kann es aufgrund des Klimawandels sowohl zu höheren Überflutungswerten als auch zu einer höheren Frequenz von Überflutungsereignissen kommen. Vor allem bei besonders schutzwürdigen Infrastrukturen wie der Levana-Schule besteht daher die Notwendigkeit der Berücksichtigung der HQ_{extrem}-Szenarien.

Da es sich bei der Nutzergruppe der Levana-Schule um besonders vulnerable Personen handelt, die im Ereignisfall auf die Hilfe Dritter angewiesen sind, ist die Frage der Evakuierbarkeit der Einrichtung ein wichtiger Punkt. Der Sammelplatz sowie die Zufahrtswege zur Schule werden bereits sehr schnell überschwemmt sein, d. h. vor dem Wassereintritt ins Gebäude. Die Zeit, die für die reine Evakuierung des Gebäudes im Brandfall mit 8–10 min angesetzt ist, wird im Falle eines Starkregen- oder Hochwasserereignisses außerdem deutlich länger sein, da auch Sammelplätze und Zufahrtsstraßen betroffen sein werden und das komplette Gelände evakuiert werden muss (Truedinger et al. 2024a). Aufgrund der eingeschossigen Bauweise ist eine vertikale Evakuierung ausgeschlossen – daher müssen die Schülerinnen und Schüler sowie Lehrkräfte zwangsweise das Schulgebäude sowie auch das Schulgelände verlassen. Da es sich allerdings um zahlreiche Kinder mit körperlichen und/oder geistigen Behinderungen handelt, ist eine solche Evakuierung hin zu einer hochwassersicheren Unterbringungsmöglichkeit mit erheblichem Zeit- und Personal- sowie Technikaufwand verbunden (Truedinger et al. 2024a). Einige der größeren Kinder müssten beispielsweise von vier Personen getragen werden, in einigen Fällen bräuchten einzelne Schülerinnen und Schüler sogar für sich allein einen kompletten Transportwagen, um liegend aus dem potenziellen Überflutungsbereich gefahren zu werden (Birkmann et al. 2023). Notfallkrankenwagen Typ B nach BBK-Norm von 2010 sowie die meisten anderen Transporter haben eine Wattiefe von ca. 30 cm und können daher durch eine Wassertiefe von maximal 30 cm fahren, sodass eine Erreichbarkeit der Schule bei einem HQ_{100} nach den neuen Berechnungen nicht mehr gewährleistet werden kann, da der Zufahrtsweg und der Sammelplatz in Teilen über 1 m tief überflutet sind. Auch Kinder können bei einer solchen Wassertiefe und höheren Fließgeschwindigkeiten diese Bereiche nicht mehr durchqueren. So werden laut Untersuchungen schwächere Personen, wozu Kinder allgemein sowie Menschen mit motorischen Behinderungen gehören, schon bei einer Wassertiefe von 20 cm (knöcheltief) und einer Fließgeschwindigkeit von rund 3 m/s umgerissen (MunichRE 2016). Laut einer 2019 erschienenen Arbeitshilfe aus Bayern ist die Personenflutsicherheit sogar schon bereits bei 2 m/s und 20 cm nicht mehr gegeben (Bayerisches Staatsministerium für Umwelt und Verbraucherschutz 2019). Zum Vergleich: In der Flutnacht betrug die Fließgeschwindigkeit der Ahr mehrere Meter pro Sekunde, Schätzungen zufolge etwa 3–5 m/s (RedaktionsNetzwerk Deutschland 2022) – in den bebauten Gebieten sind bis zu 2 m/s plausibel (Apel et al. 2022). Erschwerend hinzu kommen außerdem starke, unvorhersehbare Strömungen aufgrund von Verklausungen und Strömungshindernissen sowie Treibgut, z. B. in Form von Autos und Bäumen, welches ebenfalls Personen mit

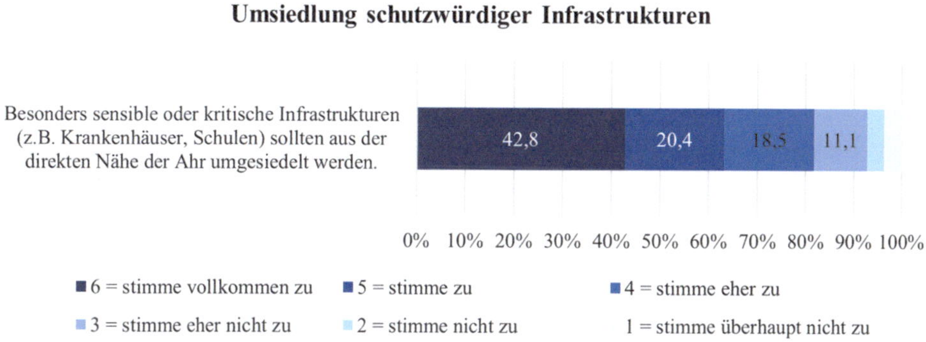

Abb. 16.2 Zustimmung zu einer Aussage in Bezug auf das Thema Umsiedlung schutzwürdiger Infrastrukturen auf einer sechsstufigen Likert-Skala mit n = 486 (Anm.: Werte unter 5 % sind nicht beschriftet) – die Daten stammen aus einer Haushaltsbefragung vom Sommer 2022. (Bildquelle: Truedinger et al. 2023 (verändert))

sich reißen kann. Darüber hinaus können beispielsweise auch hochgespülte Gullydeckel sowie dadurch freigelegte Kanaleinläufe, die eine große Sogwirkung entwickeln können, zur tödlichen Gefahr werden (Landesfeuerwehrschule Baden-Württemberg 2023).

Aktuelle Vorsorgekonzepte und gesetzliche Regelungen wie die des Bundesraumordnungsplans Hochwasserschutz empfehlen gerade für besonders empfindliche und sensible Nutzungen und somit besonders schutzwürdige Infrastrukturen – wie die Levana-Schule eine ist – auch die Betrachtung des HQ_{extrem} und möglicher Implikationen. Bei einem HQ_{extrem} wäre der Schulstandort in großen Teilen mindestens 2 bis 3 m überflutet. Damit wäre sowohl die Überlebenswahrscheinlichkeit im Gebäude als auch auf dem für die Evakuierung vorgesehenen Sammelplatz ausgeschlossen. Im Fall einer Hochwasser- und/oder Starkregen-Betroffenheit von Personengruppen, für die eine Evakuierung besonders aufwändig ist (z. B. Kita, Schule, Altenheim), wird vorrangig eine Prüfung eines alternativen Standortes empfohlen – selbst wenn vertikale Evakuierungsmöglichkeiten vorhanden sind (Bayerisches Staatsministerium für Umwelt und Verbraucherschutz 2019).

Auch eine größere Haushaltsbefragung von betroffenen Haushalten im Landkreis Ahrweiler, die im Rahmen des KAHR-Projekts durchgeführt wurde (s. ▶ Kap. 5), zeigt die besondere Bedeutung des Schutzes von sensiblen und kritischen Infrastrukturen als Ansatzpunkt zur Stärkung der Resilienz auf. Über 80 % der 486 Beantwortenden stimmten der Aussage, dass besonders sensible oder kritische Infrastrukturen, z. B. Krankenhäuser, Schulen, aus der direkten Nähe der Ahr umgesiedelt werden sollten, vollkommen bis eher zu (s. Abb. 16.2).

Der Wiederaufbau der Schule am gleichen Standort und in gleicher Weise ist daher aus unserer wissenschaftlichen Sicht nicht mit dem Vorsorgeprinzip und der Maßgabe, die Gefährdung von Leben und Gesundheit sowie erhebliche Hochwasserschäden zu vermeiden, (§ 1 Abs. 6 BauGB und § 78b WHG) vereinbar. Dem Schutz sensibler Infrastrukturen und Nutzungen bzw. Nutzergruppen – gerade von Kindern mit geistigen und körperlichen Behinderungen – ist nach neueren rechtlichen Vorgaben, aber auch im Kontext der Klimaresilienz und Klimaanpassung ein stärkeres Gewicht beizumessen. Dies erfordert den klimaresilienten Umbau sowie die Prüfung und nach Möglichkeit Durchführung einer Verlagerung des Standorts der Schule (Birkmann et al. 2023).

Auf Anfrage des Landkreises Ahrweiler wurde aus den aufgeführten Erkenntnissen und Argumenten im Rahmen des KAHR-Projekts eine wissenschaftliche Stellungnahme zur Unterstützung der Verlagerung der Levana-Schule verfasst. Diese diente als Diskussionsgrundlage für die grundsätzliche Förderung einer Verlagerung des Schulstandorts mit dem zuständigen Landesministerium. Aufgrund der wissenschaftlichen Stellungnahme wurde der Diskussionsprozess zwischen Land und Kreis fortgeführt und weitere Gutachten und Stellungnahmen nachgefordert. Schlussendlich zeigte sich das zuständige Ministerium positiv gegenüber der vollständigen Förderung einer Verlagerung, sodass im Herbst 2024 die politische Entscheidung im Landkreis fiel, die Levana-Schule zu verlagern und einen neuen Standort zu suchen (Au 2024). Dies zeigt, wie hilfreich wissenschaftliche, dynamisch anpassbare Begleitprojekte auch bei ganz konkreten, praktischen Fragestellungen vor Ort sein können (s. auch ▶ Kap. 22).

16.3.2 Fallbeispiel hochwasserresiliente Stromversorgung

Die Energieversorgung, und insbesondere die Branche Elektrizität, ist für einen resilienten (Wieder-)Aufbau von entscheidender Bedeutung, da viele andere schutzwürdige Infrastrukturen auf die Stromversorgung angewiesen sind. Fällt die Stromversorgung aus, entstehen oft zahlreiche Kaskadeneffekte.

Im Rahmen des KAHR-Projektes beschäftigten sich die Projektpartner gemeinsam mit dem assoziierten Partner Westnetz GmbH, einem regionalen Verteilnetzbetreiber für Strom und Gas, der bis zum Ende des Jahres 2023 eine unternehmensinterne Task-Force für die Bewältigung und den resilienten Wiederaufbau eingerichtet hatte, mit der Frage, wie die Stromversorgung resilienter gegenüber Hochwasser und Starkregen gemacht werden kann. Durch die Flut 2021 waren im Ahrtal und Umkreis nach Schätzungen der Westnetz GmbH rund 64.400 Haushalte von der Stromversorgung abgeschnitten und allein im Zuständigkeitsbereich der Westnetz GmbH waren in Rheinland-Pfalz im Mittel- und Niederspannungsbereich ungefähr 1300 km Kabel und Freileitung sowie 621 Ortsnetzstationen beschädigt worden (Truedinger et al. 2024b).

Um die Stromversorgung resilienter gegenüber Starkregen und Hochwasser zu machen, gibt es zahlreiche Maßnahmen, wobei meist die Kombination verschiedener Maßnahmen am erfolgversprechendsten ist. Neben Starkregen, Sturzfluten und Hochwasser gibt es außerdem noch weitere Gefahren wie z. B. Stürme, die es in Bezug auf eine umfassende Resilienzsteigerung ebenfalls zu betrachten gilt. Im Folgenden wird der Fokus jedoch weiterhin auf die Hochwasser- und Starkregenthematik gelegt. Die in ▶ Abschn. 15.3.2 vorgestellten Resilienzkriterien, mit denen sich resiliente Systeme oder Räume beschreiben lassen, können ganz konkret auf die Stromversorgung angewendet werden (s. ◘ Tab. 16.1). Hierdurch wird deutlich, was sich in der Praxis hinter Prinzipien wie Robustheit, Redundanz oder Expositionsreduktion verbirgt. So kann die Exposition des Stromversorgungssystems beispielsweise dadurch reduziert werden, dass Ortsnetzstationen und Mittelspannungsleitungen außerhalb der HQ_{extrem}-Gebiete geplant und gebaut werden.

Als Beispiel zur Planung eines Ortsnetzes unter Berücksichtigung der HQ_{extrem}-Überschwemmungsgebiete zur Reduzierung der Exposition (Prinzip Expositionsreduktion) dient der Ort Dernau (s. ◘ Abb. 16.3). Die Planung des neuen Stromnetzes sieht vor, dass die Mittelspannungsstationen außerhalb der HQ_{extrem}-Marke aufgestellt werden. Da aufgrund von technischen Gegebenheiten nicht alle Stationen aus diesem Bereich heraus

Tab. 16.1 Resilienzkriterien übertragen auf eine hochwasserresiliente Stromversorgung. (Tab. aus Truedinger et al. 2024b (verändert))

Prinzipien physischer Resilienz (Raum- und Infrastrukturen)	**Merkmale einer hochwasserresilienten Stromversorgung (Beispiele)**
Objektbezogene Perspektive	
Robustheit	Anlagen des Stromversorgungssystems werden so geplant und gebaut beziehungsweise ertüchtigt, dass sie auch den Auswirkungen extremer Ereignisse (zum Beispiel HQ_{extrem}, 500-jährlich wiederkehrendes Starkregenereignis) widerstehen können, ohne dass sie geschädigt werden. Selbst bei einer Überschwemmung kann die Stromversorgung aufrechterhalten werden, zum Beispiel durch die Gestaltung und Technologie der Anlagen (Auslegung der Betriebsmittel nach Schutzart IP68 [staubdicht, wasserdicht]). Durch die bauliche Gestaltung werden Schäden trotz Überschwemmung vermieden, zum Beispiel die Installation von Hausanschlüssen außerhalb von Kellern
Expositionsreduktion	Trassenführungen und Einzelstandorte werden so geplant, dass sie nicht gegenüber Gefahrenereignissen exponiert sind, sie beispielsweise also außerhalb des HQ_{extrem}-Hochwasserszenarios liegen oder außerhalb von potenziell einstaugefährdeten Bereichen. Die Überschwemmung reicht nicht an die Anlage heran, die Stromversorgung kann ungehindert aufrechterhalten werden, zum Beispiel durch Hochwasserschutzmaßnahmen (mobiler Hochwasserschutz, Deiche/Dämme, Standsicherung, Erhöhung der Anlage über den erwarteten (oder definierten) Hochwasserstand z. B. durch erhöhte Sockel, Verlagerung der Anlagen außerhalb der HQ_{extrem}-Marke usw.) Abschaltung eines Gebiets (automatisiert oder aus der Ferne), damit es nicht zu weiteren Folgeschäden kommt und Netzbereiche in überschwemmten Gebieten von nicht-betroffenen Bereichen getrennt werden können
Systemische Perspektive	
Effizienz	Stromversorgungssystem wird so geplant, dass im Krisenfall Wiederinbetriebnahme, Reparatur und Wiederaufbau schnell und möglichst kostengünstig geschehen kann (zum Beispiel zusätzliche Einspeisepunkte vorsehen, Modulbauweisen usw.)
Diversität	Unterschiedliche, sich ergänzende Funktionen innerhalb des Stromversorgungssystems (Produktion, Transformation, Transport/Leitungen, Speicherung) sind nicht an einem Ort gebündelt, sondern die Funktionen sind räumlich verteilt. Bei einer Beschädigung sind nicht alle Systemkomponenten betroffen; Reparaturen sind schneller möglich, Reparaturkosten geringer
Redundanz	Gleiche, sich ersetzende Funktionen innerhalb des Stromversorgungssystems werden so angeordnet, dass der Ausfall einer Komponente durch eine andere gleiche Komponente ersetzt werden kann. Bei der Beschädigung eines Standorts (Umspannwerk, Mittelspannungsstation) oder eines Leitungsabschnitts können andere Systemkomponenten (an anderen Standorten) diesen Ausfall kompensieren (relevant insbesondere bei isoliert gelegenen Anlagen oder Stichstrecken)

Tab. 16.1 (Fortsetzung)

Prinzipien physischer Resilienz (Raum- und Infrastrukturen)	Merkmale einer hochwasserresilienten Stromversorgung (Beispiele)
Organisatorische Perspektive	
Autonomie	Im Ereignisfall Einsatz von Ressourcen und Entscheidungsfindung durch dezentrale Einheiten vor Ort; setzt Ressourcenautonomie und Entscheidungsbefugnis sowie erprobte Kommunikations- und Entscheidungswege voraus; Mitteleinsatz kann schneller und effizienter erfolgen
Kooperation	Kooperation zwischen Infrastrukturbetreiber und Betreibern weiterer (kritischer) Infrastrukturen sowie den Behörden und Organisationen mit Sicherheitsaufgaben: Informationsaustausch zu Strategien, Ressourcen, aber auch Limitierungen; Nutzung von Synergien, Vermeidung von Konflikten, Identifikation von Interdependenzen/Ausfallkaskaden
Anpassungsfähigkeit	Kultur der Flexibilität, Zielorientierung (schnelle Wiederherstellung der Mindestversorgung), Offenheit für und Vorbereitung von pragmatischen, lösungsorientierten Ansätzen, Kultur des Lernens (beispielsweise im Rahmen von „Task-Forces" nach Katastrophen)

geplant werden können, werden in diesen Fällen besondere Schutzmaßnahmen (Prinzip Robustheit) umgesetzt – so werden diese Stationen auf fest verankerte Sockel gestellt, um zu verhindern, dass Wasser in die Station fließt.

Auch öffentliche Stellen und die Planung können zu einer resilienteren Stromversorgung beitragen. So ist es wichtig, dass auf kommunaler und regionaler Ebene KRITIS und weitere besonders schutzwürdige Infrastrukturen, wie insbesondere sensible Infrastrukturen, sowie deren Abhängigkeiten von der Stromversorgung (und weiteren Infrastrukturleistungen) identifiziert werden.

16

Im Rahmen der risikobasierten Raumplanung (Raumordnung und Bauleitplanung) kann der Neubau von Anlagen, die als schutzwürdige Infrastrukturen gelten, in Gebieten mit erhöhtem Risiko für (Natur-)Gefahren, welche zu einem Ausfall oder einer Funktionseinschränkung der Einrichtungen führen könnten, verhindert werden. Leitungsgebundene Infrastrukturen, die einer Baugenehmigung bedürfen, sollten in gefährdeten Gebieten nicht gebündelt zugelassen werden, um die Redundanz und die Versorgungssicherheit zu erhöhen. Darüber hinaus können in Bebauungsplänen Festsetzungen zur Freihaltung von Flächen sowie zur hochwasserangepassten Bauweise von Evakuierungsrouten und Gebäuden getroffen werden. Städtebauliche Sanierungsmaßnahmen können zur Ertüchtigung oder Verlagerung schutzwürdiger Infrastrukturen genutzt werden, insbesondere im Bestand, und durch Gestaltungssatzungen können Kommunen Vorgaben für die Entwicklung von Teilbereichen machen und so beispielsweise den Einsatz hochwasserangepasster Materialien fordern. Im Wiederaufbau kann außerdem § 246c BauGB angewendet werden, der unter anderem hochwasserangepasste Bauweisen und geringfügige Standortverlagerungen ermöglicht (s. ▶ Abschn. 15.4.1). Für öffentliche Infrastruktur sollten jedoch auch weitergehende Standortverlagerungen in Gebiete außerhalb des HQ_{extrem}-Gebiets in Betracht gezogen werden. Die Regionalplanung kann darüber hinaus Maßnahmen zur Risikoreduzierung für überschwemmungsgefährdete Gebiete festlegen, z. B. den

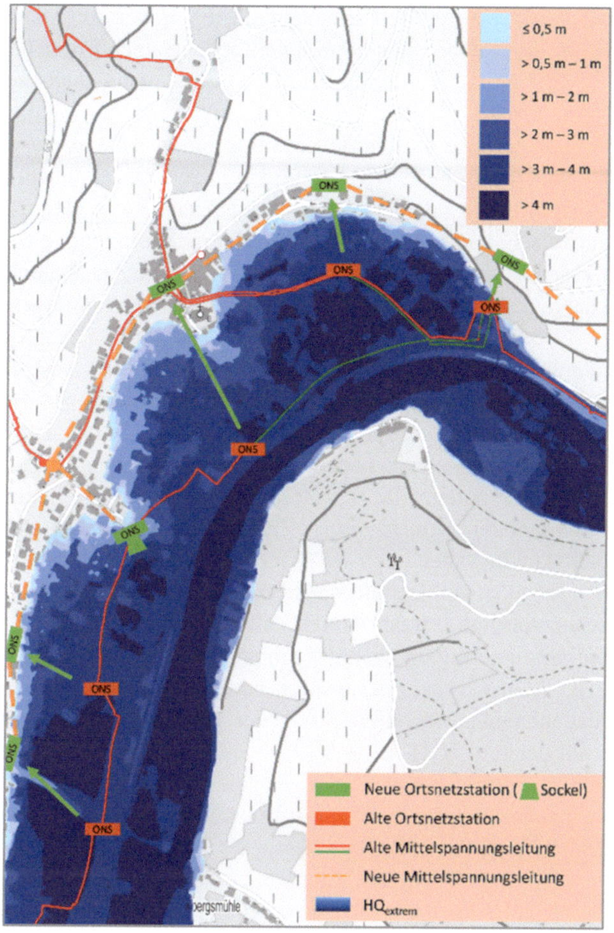

Abb. 16.3 Grundsätze zur Planung resilienter Energieinfrastrukturen – Konzeptplanung Dernau der Westnetz GmbH (erstellt auf Grundlage der ursprünglichen Gebietskulisse HQ_{extrem} vor der Neuberechnung). (Bildquelle: Truedinger et al. 2024b)

Ausschluss sensibler oder kritischer Einrichtungen in gefährdeten Bereichen. Darüber hinaus können durch die Regionalplanung spezielle Gebiete definiert werden, in denen solche Einrichtungen besonders geschützt, zurück- oder umgebaut oder hochwasserangepasst errichtet werden müssen, wie es im Regionalplan Leipzig-Westsachsen umgesetzt wurde (Regionaler Planungsverband Leipzig-Westsachsen 2021). Zusätzlich kann die Regionalplanung die Forderung umsetzen, dass kritische und sensible, stromabhängige Infrastrukturen nur durch redundante Strukturen oder Strominfrastruktur außerhalb der HQ_{extrem}-Gebiete versorgt werden dürfen. Dadurch wird das Problem der indirekten Betroffenheit durch Kaskadeneffekte adressiert (Truedinger et al. 2024b).

16.4 Fazit

Sowohl im Wiederaufbau nach Katastrophen als auch in der Vorsorge und somit im Bestand ist es wichtig, die Resilienz besonders schutzwürdiger Infrastrukturen zu stärken. Darunter fallen neben kritischen Infrastrukturen auch insbesondere sensible Infrastrukturen. Ein wichtiger Schritt auf dem Weg zur Resilienzstärkung ist dabei die Identifikation solcher Infrastrukturen, die auch auf regionaler und lokaler Ebene durchgeführt werden muss, da es neben den bundesweit bedeutsamen kritischen Infrastrukturen, die u. a. von der BSI-Kritisverordnung erfasst werden, auch noch viele lokal und regional bedeutsame, kritische und sensible Infrastrukturen gibt. Hierbei bietet sich eine partizipative Vorgehensweise an, bei der alle beteiligten Akteure aus Verwaltung, Wasserwirtschaft, Katastrophenschutz, Wirtschaft usw. in den Prozess integriert werden. Für kritische Infrastrukturen, die oft viele Interdependenzen mit weiteren schutzwürdigen Infrastrukturen aufweisen, kann beispielsweise auch eine partizipative Kritikalitätsanalyse durchgeführt werden, sodass im Anschluss frühestmögliche Ansatzpunkte zur Unterbrechung der Kaskaden identifiziert und entsprechende Maßnahmen umgesetzt werden können.

Neben der Hochwassergefahr gilt es außerdem auch die Starkregen- und Sturzflutgefahr zu betrachten, wobei die Flut 2021 gezeigt hat, dass diese Ereignisse nicht unbedingt getrennt voneinander auftreten. Gerade in Bezug auf Starkregen und Sturzfluten gibt es bislang kaum verbindliche Vorgaben – die Erstellung von Starkregen- und Sturzflutgefahrenkarten sowie von Hochwassergefahrenkarten auch für kleinere Gewässer hilft jedoch, die potenzielle Betroffenheit von schutzwürdigen Infrastrukturen einzuschätzen. Für besonders schutzwürdige Infrastrukturen sollten als Bewertungsgrundlage bei der Neuplanung, aber auch im Wieder- und Neuaufbau vorrangig Extremszenarien, z. B. die HQ_{extrem}-Gebietskulisse, herangezogen werden. Nach Möglichkeit sind besonders schutzwürdige Infrastrukturen außerhalb dieser Gebiete zu errichten – oder im Wiederaufbau zu verlagern, wie dies nun mit der Levana-Schule geschieht. Ist eine Verlagerung, z. B. aufgrund technischer Gegebenheiten, nicht möglich, sind Vorsorge- und Schutzmaßnahmen zu ergreifen. So hat der Verteilnetzbetreiber Westnetz GmbH beispielsweise nach der Flut 2021 seine Mittelspannungsstationen vorrangig in Gebiete außerhalb des HQ_{extrem}-Bereiches verlegt – war dies nicht möglich, wurde die entsprechende Mittelspannungsstation auf einen fest verankerten Sockel gebaut, sodass sie zumindest oberhalb der HQ_{extrem}-Anschlaglinie liegt.

Gerade in Bezug auf die strategische Berücksichtigung regional und lokal bedeutsamer KRITIS und sensibler Infrastrukturen hat die Raumplanung – sowohl die Raumordnung (beispielsweise die Regionalplanung) als auch die Bauleitplanung – eine wichtige Signal- und Impulsgeberfunktion für weitere Fachplanungen und die Kommunen. Im Sinne der risikobasierten und adaptiven Raumplanung kann der Neubau von Anlagen, die als schutzwürdige Infrastrukturen gelten, in Gebieten mit erhöhtem Risiko für (Natur-)Gefahren verhindert werden. Außerdem können in Bebauungsplänen Festsetzungen zur Freihaltung von Flächen sowie zur hochwasserangepassten Bauweise von Evakuierungsrouten und Gebäuden getroffen werden.

Die Befunde der KAHR-Forschung am Beispiel u. a. der Levana-Schule mit über 100 Kindern mit geistigen und körperlichen Behinderungen sowie die Empfehlungen aus der Zusammenarbeit mit dem Stromverteilnetzbetreiber Westnetz zeigen, dass die Stärkung der Resilienz und die besondere Bedeutung kritischer und sensibler Infrastrukturen auch nach Extremereignissen beim Neu- und Wiederaufbau Berücksichtigung finden kann bzw. muss, um besser auf zukünftige Ereignisse und Veränderungen im Kontext des Klimawandels vorbereitet zu sein.

Zusammenfassung

- Im Wiederaufbau nach Katastrophen und in der Vorsorge ist es entscheidend, die Resilienz besonders schutzwürdiger Infrastrukturen zu stärken.

- Die Identifikation besonders schutzwürdiger Infrastrukturen muss insb. auch auf regionaler und lokaler Ebene erfolgen. Eine partizipative Kritikalitätsanalyse kann z. B. helfen, Interdependenzen zwischen Infrastrukturen zu erkennen, und Ansatzpunkte zur frühzeitigen Unterbrechung von Kaskaden aufzeigen.
- Zusätzlich zur Hochwassergefahr müssen auch Starkregen und Sturzfluten berücksichtigt werden. Die Erstellung von Gefahrenkarten hilft, die potenzielle Betroffenheit zu beurteilen.
- Schutzwürdige Infrastrukturen sollten idealerweise in Risikogebieten, die über das HQ_{100} hinausgehen oder von extremen Starkregenereignissen betroffen sein könnten, vermieden oder aus diesen Bereichen verlagert werden.
- Die Raumplanung kann durch risikobasierte Bauleit- und Regionalplanung zur Resilienzstärkung beitragen, z. B. durch den Ausschluss bestimmter Nutzungen.

Danksagung Die Autorinnen und Autoren von ▶ Kap. 16 bedanken sich beim Bundesministerium für Bildung und Forschung (BMBF) und beim DLR Projektträger für die Unterstützung und Förderung des KAHR-Projekts. Diese Forschung wurde vom BMBF unter dem Förderkennzeichen 01LR2102A-M finanziert.

Literatur

Apel H, Vorogushyn S, Merz B (2022) Brief communication: Impact forecasting could substantially improve the emergency management of deadly floods: case study July 2021 floods in Germany. Nat Hazards Earth Syst Sci 22:3005–3014. https://doi.org/10.5194/nhess-22-3005-2022

Au B (2024) Weg ist frei: Levana-Schule bekommt neuen Standort im Kreis Ahrweiler, Rhein-Zeitung. https://www.rhein-zeitung.de/lokales/kreis-ahrweiler/weg-ist-frei-levana-schule-bekommt-neuen-standort-im-kreis-ahweiler_arid-2693591.html. Zugegriffen: 25. Sept. 2024

Bayerisches Staatsministerium für Umwelt und Verbraucherschutz (2019) Arbeitshilfe: Hochwasser- und Starkregenrisiken in der Bauleitplanung: Eine pragmatische Anleitung für Kommunen und deren Planer. https://www.stmuv.bayern.de/themen/wasserwirtschaft/hochwasser/doc/arbeitshilfe.pdf. Zugegriffen: 6. Jan. 2025

Birkmann J, Truedinger A, Sauter H, Greiving S (2023) Stellungnahme zur Levana-Schule als besonders sensible Infrastruktur (unveröffentlicht)

Birkmann J, Truedinger A, Schuettrumpf H (2024) Assessing the uptake and impact of scientific recommendations for flood and climate resilient reconstruction in the Ahr-Valley. J Flood Risk Manag 18(3):e70100. https://doi.org/10.1111/jfr3.70100

BMI – Bundesministerium des Innern (2009) Nationale Strategie zum Schutz Kritischer Infrastrukturen (KRITIS-Strategie). Bundesministerium des Innern, Berlin

Bugge F (2021) Fünf Altenzentren sind unbewohnbar: Ihre „idyllische Lage" wurde ihnen zum Verhängnis, RheinZeitung. https://www.rhein-zeitung.de/lokales/kreis-ahrweiler/fuenf-altenzentren-sind-unbewohnbar-ihre-idyllische-lage-wurde-ihnen-zum-verhaengnis_arid-2293849.html. Zugegriffen: 5. Jan. 2025

Bundesregierung Deutschland KRITIS-Definition der Bundesregierung. https://www.bmi.bund.de/DE/themen/bevoelkerungsschutz/schutz-kritischer-infrastrukturen/schutz-kritischer-infrastrukturen-node.html. Zugegriffen: 22. Nov. 2024

Greiving S, Kruse P, Othmer F, Fleischhauer M, Fuchs M (2023) Implementation of risk-based approaches in urban land use planning – the example of the city of Erftstadt, Germany. Sustainability 15(21):15340. https://doi.org/10.3390/su152115340

Hartz A, Saad S, Bächle S, Fleischhauer M, Greiving S, Gollmann C, Kirstein M, Nguyen B-H (2020) Vorsorgendes Risikomanagement in der Regionalplanung. Handlungshilfe für die Regionalplanung. Bundesinstitut für Bau-, Stadt- und Raumforschung (BBSR) im Bundesamt für Bauwesen und Raumordnung (BBR), Bonn

KAHR et al. (2022) 10 Empfehlungen aus Sicht der Wissenschaft zum Thema Wiederaufbau und Zukunftsfähigkeit der flutbetroffenen Regionen. https://hochwasser-kahr.de/index.php/de/alle-produkte/10-empfehlungen. Zugegriffen: 15. Okt. 2024

Landesamt für Umwelt Rheinland-Pfalz Sturzflutgefahrenkarten für Rheinland-Pfalz. https://wasserportal.rlp-umwelt.de/auskunftssysteme/sturzflutgefahrenkarten. Zugegriffen: 7. Jan. 2025

Landesfeuerwehrschule Baden-Württemberg (2023) Einsatztaktik für Fahrzeugführende: Hinweise für Unwettereinsätze. https://www.lfs-bw.de/fileadmin/LFS-BW/themen/taktik/sonderlagen/dokumente/Unwetter.pdf. Zugegriffen: 15. Okt. 2024

Landesregierung Rheinland-Pfalz (2022) Der Wiederaufbau in Rheinland-Pfalz nach der Naturkatastrophe vom 14./15. Juli 2021. https://wiederaufbau.rlp.de/fileadmin/wiederaufbau/2022/02-Februar/Der_Wiederaufbau_in_Rheinland-Pfalz_2021-2022.pdf. Zugegriffen: 15. Okt. 2024

MunichRe (2016) Unexpected, destructive, deadly: Flash floods. https://www.munichre.com/en/insights/natural-disaster-and-climate-change/flash-floods-unexpected-destructive-deadly.html. Zugegriffen: 6. Jan. 2025

RedaktionsNetzwerk Deutschland (2022) Landesamt warnte spät und uneindeutig vor Hochwasserkatastrophe im Ahrtal. https://www.rnd.de/panorama/gutachter-landesamt-warnte-spaet-und-uneindeutig-vor-ahr-hochwasser-LEXJJFQPQMVQGYZX2ZHIUIJAH4.html. Zugegriffen: 22. Nov. 2024

Regionaler Planungsverband Leipzig-Westsachsen (2021) Regionalplan Leipzig-Westsachsen, Amtlicher Anzeiger des Sächsischen Amtsblattes des Freistaates Sachsen

Schmitt HC (2019) Systemisches Kaskadenpotenzial von Kritis-Teilsektoren. Inf Raumentwickl 2019(4):48–61. https://www.researchgate.net/publication/339068518_Systemisches_Kaskadenpotenzial_von_KRITIS-Teilsektoren

Struktur- und Genehmigungsdirektion Süd RLP Festsetzung von Überschwemmungsgebieten. https://sgdsued.rlp.de/themen/wasserwirtschaft/ms-burgund/ueberschwemmungsgebiete. Zugegriffen: 22. Nov. 2024

Tradowsky JS, Philip SY, Kreienkamp F et al (2023) Attribution of the heavy rainfall events leading to severe flooding in Western Europe during July 2021. Clim Change 176:90. https://doi.org/10.1007/s10584-023-03502-7

Truedinger A, Birkmann J, Fleischhauer M, Ferreira C (2024a) Sensitive infrastructures and people with disabilities – Key issues when strengthening resilience in reconstruction. EGUsphere. https://doi.org/10.5194/egusphere-2024-1607

Truedinger A, Birkmann J, Fleischhauer M, Greiving S, Kirschbauer L, Schäfer M, Burggraf C, Fabisch M, Jüpner R, Droste-Franke B, Apel H, Arenz H-T, Johann G, Wolf S, Schüttrumpf H, Dolfus F, Tholl L (2024b) Praxishilfe – Stärkung der Resilienz der Stromversorgung gegenüber Starkregen und Hochwasser: Vorsorge, Anpassung, Wiederaufbau. BMBF-Förderkennzeichen 01LR2102, S 2021–2025 (Veröffentlichung im Rahmen des Vorhabens KAHR – Klima-Anpassung, Hochwasser und Resilienz)

Truedinger AJ, Jamshed A, Sauter H, Birkmann J (2023) Adaptation after extreme flooding events – Moving or staying? The case of the Ahr valley in Germany. Sustainability 15:1407. https://doi.org/10.3390/su15021407

Soziale Infrastrukturen bei Bewältigung und Wiederaufbau nach dem Hochwasser 2021

Zora Reckhaus, Marisa Fuchs und Christian Kuhlicke

J. Birkmann et al. (Hrsg.), *Die Flutkatastrophe 2021*, Sustainable Development Goals (SDG) – Umsetzung in Praxis, Lehre und Entscheidungsprozessen,
https://doi.org/10.1007/978-3-658-48688-4_17

17.1 Einleitung

Während des Wiederaufbaus steht meist dessen physisch-materielle Dimension im Mittelpunkt. Dies umfasst u. a. die Räumung von Geröll und Material, die Wiederherstellung von Gewässern, Gebäuden und Infrastrukturen bzw. die Ertüchtigung des technischen Hochwasserschutzes. Die Frage, wie eigentlich der soziale Wiederaufbau vonstattengeht und welche Unterstützungen dieser Bedarf, steht häufig hinten an und wird auf finanzielle Aspekte reduziert (Auszahlung von Versicherungen bzw. die schnelle Auflage von Wiederaufbauhilfen). Gunning und Rizzi (2022) meinen allerdings, dass der soziale Wiederaufbau oberste Priorität haben sollte, denn nur durch das Wiederherstellen von betroffenen sozialen Netzwerken und Beziehungen beginnt der eigentliche Bewältigungsprozess nach einem zerstörerischen Ereignis. Wird dieser Aspekt vernachlässigt, entstehen neue Probleme, wie die Verstetigung von Angst, Unsicherheit und Ungleichheit, was durchaus negativen Einfluss auf den gesamten Aufbauprozess haben kann. Immaterielle Aspekte wie beispielsweise Wohlbefinden, Identität, Erinnerung, Wissen und Beziehungen, aber auch Aspekte wie Engagement, Partizipation und der Aufbau von Netzwerken beeinflussen den materiellen Wiederaufbau in den betroffenen Gebieten und die Resilienz einer Gemeinschaft (ibid.).

Soziale Infrastrukturen (SI) nehmen beim sozialen, aber auch beim physisch-materiellen Wiederaufbau eine wesentliche Rolle ein. In diesem Beitrag wird argumentiert, dass sie wie eine Art gesellschaftlicher Resilienzverstärker wirken können. In Anlehnung an Klinenberg (2018, S. 4) sind hier SI als „die physischen Orte und Organisationen, die die Art und Weise, wie Menschen interagieren, gestalten“ zu verstehen. Damit haben SI eine materielle Dimension (das beinhaltet Räume und Gebäude, aber auch ganze Nachbarschaften, in denen sie wirken), eine personelle Dimension (z. B. Menschen, die in den SI arbeiten) sowie eine relationale Dimension (z. B. die Unterstützung und Gestaltung sozialer Interaktionen) (Otto et al. 2025). Damit liegt diesem Beitrag ein relativ weitgefasstes Verständnis von SI zugrunde, das auch die physischen Orte für die Entwicklung und Aufrechterhaltung sozialer Verbindungen sowie die persönliche Dimension umfasst. Treffpunkte wie Bibliotheken, Parks, Sporteinrichtungen, Schulen, Gemeindezentren und religiöse Einrichtungen sind für die Pflege des sozialen Gefüges von Städten von wesentlicher Bedeutung (Schubert 1995), da sie die Bildung und Aufrechterhaltung sozialer Netzwerke in Zeiten von Stress erleichtern (Aldrich 2023, S. 30). Damit schaffen SI Räume und Gelegenheiten, in denen Menschen zusammenkommen, interagieren und sinnvolle Verbindungen knüpfen (Latham und Layton 2019, S. 3; Wilson und Kershaw 2008). Diesem Verständnis folgend sind SI sowohl für die Resilienz von Kommunen und Städten wichtig als auch für einen besseren Aufbau nach einer Katastrophe.

17.2 Methodisches Vorgehen

Um die Rolle von sozialen Infrastrukturen bei der Bewältigung und dem Wiederaufbau nach dem Hochwasser 2021 zu analysieren, wurden Interviews mit verschiedenen sozialen Infrastrukturen in den betroffenen Pilotregionen Bad Neuenahr-Ahrweiler (RLP), Stolberg und Eschweiler (beides NRW) durchgeführt. Die Interviews wurden mit Sport- und Schwimmvereinen, Einrichtungen mit Kinder- und Jugendangeboten, Einrichtung für Begegnungen vor Ort für alle Bevölkerungsgruppen mit besonderem Fokus auf Senioren, Vereinen für Stadtentwicklung und Gewerbe, kirchlichen Einrichtungen und Einrichtungen für die Betreuung von Menschen mit Behinderungen geführt. Zusätzlich wurden neu entstandene Initiativen interviewt, die sich mit der Katastrophenbewältigung befassen. Die Interviews wurden einmal im Sommer 2022 und ein zweites Mal im Sommer 2023 durchgeführt, sodass Veränderungen und neue Entwicklungen erfasst werden konnten. Dabei wurde zum einen erfragt, welche sozialen, politischen, institutionellen, kulturellen und ökonomischen Faktoren aus Sicht der interviewten Akteure den physisch-materiellen aber auch den sozialen Aufbau hin zu einer klimaangepassten und resilienten Entwicklung unterstützen würden. Zum anderen wurde erfragt, welche Rolle die befragten SI einnehmen und wie sich diese im Laufe der Jahre entwickelt hat (für weitere Details: Otto et al. 2025).

Ergänzt wurde die Befragung durch empirische Erkenntnisse aus Interviews mit sozialen Infrastrukturen in der Stadt Hagen (NRW) sowie dort durchgeführten Erzählformaten mit Betroffenen und Helfenden angelehnt an den Erzählsalon von Richter und Rohnstock (2016). In Hagen wurden Interviews und Erzählbeiträge mit gemeinnützigen Organisationen, einer Kirchengemeinde sowie einer neu entstandenen Initiative geführt und ausgewertet, die jeweils in der Unterstützung betroffener Bürger:innen involviert, in Teilen aber auch selbst von der Flut betroffen waren.

Anhand der Interviews und Erzählformate wurde rekonstruiert, wie die sozialen Infrastrukturen zu urbanen Resilienz-Merkmalen beitragen. Als Untersuchungsschema werden die von Meerow und Stults (2016) auf Basis einer systematischen Analyse wissenschaftlicher Literatur zu Definitionen und Merkmalen urbaner Resilienz sowie einer Befragung von 134 Kommunalverwaltungen im amerikanischen Raum entwickelten 16 Merkmale urbaner Resilienz (vgl. ▫ Tab. 17.1) herangezogen.

17.3 Ergebnisse aus den Interviews

Trotz einer eigenen hohen Betroffenheit haben die interviewten SI eine starke *Robustheit* bewiesen. Häufig waren die Einrichtung, das Personal und auch die zu betreuenden Personen der SI vom Hochwasser betroffen. Trotz dieser dreifachen Betroffenheit (materiell, personell, relational) wurden schon wenige Tage nach dem Hochwasser Strukturen aufgebaut, die auf die neu aufkommenden, drängendsten Bedürfnisse der zu betreuenden Personen reagierten, wie z. B. das Einrichten von Notbetreuungen für Kinder, psychosoziale Betreuung von Betroffenen oder das Einrichten von Lager- und Verteilstationen für wichtige Güter. Durchaus entscheidend waren die schon vor dem Hochwasser etablierten Netzwerke. Denn häufig griffen die SI auf ihre bereits bestehenden Netzwerke zurück und setzten ihre Erfahrungen vor Ort ein, wobei sie auf Übergangs- und Zwischenlösungen setzten. Dabei wichen sie häufig auch von ihrer eigentlichen Kerntätigkeit ab – teilweise verbunden mit einem intensiveren Erlernen neuer Expertise und Fähigkeiten – und passten sich spontan an die sich geänderten Bedürfnisse und Gegebenheiten an. Dies trug auch zur kommunalen *Redundanz* bei, da das Personal neue Strukturen, die über ihren eigenen Aufgabenbereich hinausgehen, eingerichtet hat oder sich neue Strukturen mit einem Fokus auf die Katastrophenbewältigung gebildet haben. Diese ergänzten oder ersetzten Strukturen und Institutionen, die durch die Flut beeinträchtigt oder weggefallen waren. In Hagen kam ein für die stadtweite Koordination von Hilfsleistungen essentieller Verein temporär in Räumen der Stadtverwaltung unter, um dessen Funktion aufrecht zu erhalten.

Tab. 17.1 Merkmale urbaner Resilienz. (Eigene Tabelle nach Meerow und Stults (2016))

Resilienz-Merkmal	Kurzbeschreibung des Merkmals
Robustheit	Sicherstellung der Funktionsfähigkeit von städtischen Infrastrukturen und Organisationen und deren Widerstehen und schnelles Erholen nach Schocks
Redundanz	Reservesystem im Sinne des Bereithaltens von Strukturen, die ähnliche Funktionen übernehmen, sodass bei einem Wegfall einer dieser Strukturen, die Funktion durch eine andere Infrastruktur immer noch zur Verfügung gestellt werden kann
Diversität	Sicherstellung einer vielfältigen Wirtschaft, Infrastruktur und Ressourcenbasis zur Erhöhung der Widerstandsfähigkeit eines Systems, indem die unterschiedlichen Systemkomponenten einerseits verteilt sind, sodass sie nicht alle durch eine einzige Störung betroffen sind, und sich anderseits funktional ergänzen, sodass es mehrere Möglichkeiten gibt, kritische Bedürfnisse zu erfüllen
Dezentralisierung	Dezentrale Verteilung und Koordination von Dienstleistungen, Ressourcen und Entscheidungen auf unterschiedliche Stellen und Organisationen, sodass im Fall eines Wegfalls einer Institution nicht das gesamte System versagt
Integration	Integration von Strategien und Maßnahmen in die Handlungsabläufe der beteiligten Behörden und Organisationen, sodass ein gemeinsames Verständnis der Umsetzung und Zielsetzung besteht
Inklusion	Zugang zu städtischen Infrastrukturen und Dienstleistungen für alle Bewohner:innen einer Stadt, inklusive der Beteiligungsmöglichkeit an Entscheidungen
Gerechtigkeit	Gewährleistung, dass sich die Vor- und Nachteile von Strategien und Maßnahmen gerecht auf die beteiligten und betroffenen Bewohner:innen auswirken
Reflexive Prozesse	Strukturen, die der Nutzung von Rückmeldungen und Lehren aus den Erfahrungen dienen, um zukünftige Strategien entsprechend zu verbessern
Rückkopplung	Mechanismen, die es ermöglichen Personen in Entscheidungspositionen schnell und effektiv Rückmeldung zu geben, sodass aktuelle Entwicklungen und Bedürfnisse in der Entscheidungsfindung berücksichtigt werden können
Transparenz	Sicherstellung, dass alle Prozesse und Entscheidungen öffentlich und transparent gemacht werden, wodurch die Rechenschaftspflicht gegenüber Bürger:innen gesichert und das Vertrauen zwischen Bürger:innen und entscheidenden Institutionen gestärkt wird
Flexibilität	Gewährleistung von flexiblen und anpassungsfähigen (städtischen) Plänen und Entscheidungsablaufen, um auf Unvorhersehbarkeiten und spontane Gegebenheiten und Bedürfnisse reagieren zu können, damit sinnvolle und wirksamen Maßnahmen umgesetzt werden können
Zukunftsorientiertes Denken	Integration von Informationen über zukünftige Entwicklungen (z. B. Bevölkerungsentwicklung, Klimawandel, Digitalisierung) in Planungs- und Entscheidungsabläufe
Anpassungskapazität	Sicherstellen von Ressourcen für alle Bewohner:innen einer Stadt, um sich an plötzlich auftretende Ereignisse und schleichende Veränderungen anpassen zu können
Vorhersehbarkeit	Angepasste Systeme, sodass sie im Störfall auf vorhersehbare und sichere Weise ausfallen und dementsprechend vorbereitet auf diese Ausfälle reagieren können
Umweltschutz	Schutz von natürlichen Systemen und ihren Ökosystem-Dienstleistungen
Effizienz	Effiziente Operation lokaler Regierungen

Zum Resilienzmerkmal *Diversität* trugen soziale Infrastrukturen auf unterschiedliche Art und Weisen bei. Soziale Infrastrukturen, die in einer Stadt aktiv sind, sind von Natur aus bereits divers, da es sehr viele unterschiedliche Initiativen, Organisationen und Vereine mit je sehr verschiedenen Angeboten und unterschiedlichen Klientelen gibt. Auch wenn räumliche Schwerpunkte (bestimmter) SI innerhalb einer Stadt erkennbar sind, sind deren Standorte insgesamt über die gesamte Stadt verteilt. Dies bietet den Vorteil, dass gegebenenfalls nicht alle sozialen Infrastrukturen (mit ähnlichen Funktionen) im gefährdeten Bereich liegen. Die nicht direkt betroffenen sozialen Infrastrukturen in den Pilotregionen nutzten diese Gegebenheit, um z. B. benötigte Räume anzubieten und um stärker betroffene soziale Infrastrukturen zu unterstützen. Aufgrund ihres diversen Netzwerkes mit unterschiedlichen Bevölkerungsgruppen konnten sie zusätzlich die spezifischen Bedürfnisse dieser Gruppen abfragen und sie auch mobilisieren.

Die hohe Diversität der SI trägt auch stark zum Merkmal *Dezentralisierung* von Dienstleistungen und Entscheidungen bei. Wegen ihrer vielfältigen Angebote wie z. B. Kinderbetreuung, Jugendarbeit, Altenpflege sowie ihrer Selbständigkeit und Unabhängigkeit von staatlichen Strukturen bieten sie eine Vielzahl von dezentralen Anlaufstrukturen an. Dadurch konnten sie zu einem gewissen Grad Lücken füllen, die durch die Überforderung der verantwortlichen Institutionen entstanden sind, Entscheidungen dort fällen, wo es notwendig war, und dadurch andere, staatliche Strukturen entlasteten.

Zu dem Merkmal *Integration* haben die sozialen Infrastrukturen nur in begrenztem Maße beitragen können, denn dieses Merkmal adressiert in dem stark top-down organisierten Wiederaufbausystem vor allem die übergeordneten Politik- und Verwaltungsebenen, auf der die sozialen Infrastrukturen keine tragende Rolle spielen. Die Verantwortung für die Integration von Plänen und Maßnahmen in die Handlungsabläufe von Behörden und Organisationen liegt in den Entscheidungs- und Verwaltungsstrukturen der betroffenen Stadt, des Kreises oder des Bundeslandes. Soziale Infrastrukturen verfügen nicht über diese Entscheidungsgewalt und Position. Insgesamt konnte festgestellt werden, dass soziale Infrastrukturen wenig Einfluss darauf haben, wie der allgemeine physische Wiederaufbau übergeordnet gesteuert wird. Dennoch gibt es auf lokaler Ebene kleinere Formen der Integration, bei denen SI als Vertreterinnen bestimmter Bevölkerungsgruppen einbezogen werden. Ein Beispiel hierfür ist ein Austausch mit der Hochwasserhilfe im Rahmen der aktuellen Erarbeitung eines neuen Hochwasserschutzkonzepts in Hagen, der erste Ansätze für eine Einbindung solcher Akteure erkennen lässt.

Während bei der Integration ein eher begrenzter Einfluss festzustellen ist, gibt es großes Potenzial bei dem Merkmal *Inklusion*. Eine wichtige Eigenschaft von sozialen Infrastrukturen ist, dass sie häufig als Schnittstelle zwischen Bürger:innen und Behörden fungieren und zum Beispiel eine direktere Verbindung zur Stadt oder anderen verantwortlichen Institutionen haben. Dadurch entsteht ein Informationsfluss über beschlossene Entscheidungen und Strategien beispielsweise von der Stadt zu den Bürger:innen, z. B. durch gemeinsame Informationsveranstaltungen. Dies ist besonders wichtig, wenn sich Informationskanäle über traditionelle Massenmedien – wie es z. B. in Hagen im Zusammenhang mit der Förderrichtlinie des Landes der Fall war – als unzureichend erweisen. In den Interviews wurde deutlich, dass die sozialen Infrastrukturen eine große eigene Verantwortung in dieser Kommunikationsfunktion sehen und sich auch vorstellen können, diese Rolle in Zukunft verstärkt, auch bezüglich Themen wie Risikovorsorge, einzunehmen. Es ist aber wichtig zu betonen, dass es bisher keine Einschätzung von außen z. B. von Bürger:innen gibt, wie diese Angebote angenommen werden oder in Zukunft genutzt würden. Außerdem konnte nur anhand vereinzelter Beispiele gezeigt werden, dass die

Bürger:innen dadurch auch eine bessere Möglichkeit haben, sich an Entscheidungsprozessen zu beteiligen.

Der *Gerechtigkeitsanspruch*, den das Resilienzkonzept mit sich bringt, ist zentral für die Arbeit vieler sozialer Infrastrukturen. Zum einen bringen sie ein großes Bewusstsein für soziale Ungleichheit mit und sind explizit darauf ausgerichtet, Ungleichheiten zu überwinden und Zugänge für alle Gesellschaftsgruppen zu schaffen, vor allem für Gruppen, die sonst vom sozialen Leben eher ausgeschlossen werden. Beispiele dafür sind Einrichtungen für politische Bildung in benachteiligten Vierteln, Angebote für Menschen mit Behinderung oder Senior:innen. Im Kontext des Hochwasserereignisses konnten SI aufgrund ihrer Erfahrungen mit ihrer Klientel die Bedürfnisse schnell erfassen und bedarfsgerechte Angebote schaffen. So wurden z. B. bedarfsgerechte Zugänge zu Informationen oder aufsuchende Hilfe zur Unterstützung unterschiedlicher Belange eingeführt. Es wurde aber auch deutlich, dass der Bedarf so groß und individuell ist, dass die sozialen Infrastrukturen auch mit ihren Kapazitäten an ihre Grenzen stoßen. Zudem mussten einige SI im Zuge der Flutbewältigung selbst auch zusätzlich in die Beziehungsarbeit zu einigen Personen/-gruppen investieren, die sozial eher isoliert sind. Eine Unterstützung durch die öffentliche Hand ist also von großer Notwendigkeit.

Durch den engen Kontakt zwischen Bürger:innen und sozialen Infrastrukturen und den sozialen Infrastrukturen untereinander wurden *reflexive Prozesse* ermöglicht, die dazu beigetragen haben, dass Angebote verbessert und angepasst worden sind. Außerdem wurden durch soziale Infrastrukturen Bedarfe von Betroffenen und Verbesserungsvorschläge gebündelt an Entscheidungsträger:innen weitergeleitet. Der Eindruck der sozialen Infrastrukturen war aber, dass diese Vorschläge selten beachtet wurden; bisher gab es dazu aber keine weiteren Untersuchungen. Daher wird vor diesem Hintergrund auch das Merkmal *Rückkopplung* nur begrenzt erfüllt. Es gibt Mechanismen und Anlaufstellen, die es ermöglichen, Anliegen an Entscheidungsträger:innen weiterzuleiten oder Rückmeldungen zu geben. Soziale Infrastrukturen haben über diese Stellen häufig einen Überblick und sind motiviert, diese Mechanismen und Anlaufstellen für die Belange ihrer Klientel zu nutzen. Wie effektiv das ist und wie mit diesen Informationen weiter verfahren wird, bleibt aber erstmal unklar.

Durch den engen Kontakt zu Bürger:innen, die Kommunikation über Entwicklungen im Wiederaufbau sowie die Unterstützung z. B. bei der Antragstellung und anderen bürokratischen Verfahren kann in begrenztem Maße Transparenz über die vor Ort ablaufenden Prozesse ermöglicht werden. Trotzdem besteht auch beim Merkmal *Transparenz* eine Abhängigkeit von übergeordneten Politik- und Verwaltungsebenen wie der Gemeinde, dem Kreis oder dem Land. Wenn diese nicht entscheiden, transparent über Prozesse zu berichten, kann dies durch SI kaum aufgefangen oder ausgeglichen werden. Insbesondere soziale Infrastrukturen, die sich als Notfallhilfe nach der Flut gegründet haben, fordern *Flexibilität* in Prozessen ein. Viele weitere (auch etablierte) soziale Infrastrukturen betonen die Schwierigkeiten, die sich aus unflexiblen bürokratischen Prozessen und Abläufen ergeben. Sie fordern für sich und im Namen ihrer Klientel mehr Flexibilität, um effizienter und effektiver auf die Bedürfnisse vor Ort unkompliziert reagieren zu können. Soziale Infrastrukturen selbst beweisen ihre Flexibilität im Wiederaufbauprozess und gestalten ihre Angebote oftmals so, dass sie auf die veränderten Bedürfnisse reagieren können. Allerdings sind einige, formalisierte SI wie z. B. Organisationen der Wohlfahrtspflege zu einem gewissen Grad auch gegenüber ihren Finanzquellen rechenschaftspflichtig oder – wie das Nachrangigkeitsprinzip gemäß § 2 SGB XII im Zusammenhang mit Spendengeldern für den Wiederaufbau zeigt – von formalen Strukturen abhängig.

Der Wunsch nach *zukunftsorientierten Veränderungen* ist in den betroffenen Regionen sichtbar und auch soziale Infrastrukturen haben eine Vorstellung, wie diese aussehen kann. Häufig ist das mit positiven Veränderungen für ihr Klientel verbunden. Da soziale Infrastrukturen jedoch oft nicht gut in übergeordnete Kommunikationsprozesse eingebunden sind, bei politischen Entscheidungen außer Acht gelassen werden und Verantwortlichkeiten oftmals unklar sind, bleibt dieser Wunsch oft unerfüllt. Ausnahmen bilden sehr behördennahe soziale Infrastrukturen, die häufig näher an den Entscheidungen und Kommunikationsprozessen angebunden sind. Inwieweit diese Vorschläge aber in den Wiederaufbauprozess eingebunden werden, kann bisher nicht nachgewiesen werden. Das Merkmal *Anpassungsfähigkeit* wird durch soziale Infrastrukturen umgesetzt. Der Wegfall von Infrastrukturen und Dienstleistungen wird durch neu geschaffene Angebote unterschiedlicher Ressourcen abgepuffert. Diese Angebote umfassen u. a. die direkte Beratung von Betroffenen bei Förderanträgen, Informationen insbesondere zur Hochwasser- und Starkregenvorsorge, die Sensibilisierung für Überflutungsgefahren oder sogar die direkte finanzielle Unterstützung durch Spenden.

In den Interviews gab es keine Hinweise darauf, dass das Merkmal *Vorhersehbarkeit* durch soziale Infrastrukturen umgesetzt werden kann. Die Systeme von SI waren bisher nicht so angelegt, dass sie im Störfall auf vorhersehbare und sichere Weise ausfallen. Abgesehen von der Notwendigkeit einer verstärkten Hochwasservorsorge wurden keine Strategien genannt, die eine solche Vorhersehbarkeit adressieren. Das Merkmal *Effizienz* wird durch soziale Infrastrukturen nicht erfüllt, was aber weniger an ihrer Motivation liegt, sondern eher an ihrer fehlenden Einbindung durch übergeordnete Politik- und Verwaltungsebenen. Auch das Thema *Umweltschutz* wird nur bedingt in den Interviews thematisiert. Es ist den SI klar, dass dies ein relevantes Thema ist, das im Wiederaufbau beachtet werden muss, aber es fehlt ein konkretes, unmittelbar verbindliches, übergeordnetes und übergreifendes Nachhaltigkeitskonzept oder eine Vision für die betroffenen Regionen, zu denen soziale Infrastrukturen beitragen können. Vereinzelt, z. B. bei Einrichtungen der politischen Bildung, gibt es einen inhaltlichen Fokus auf den Bereich Umwelt- und Klimaschutz; meist liegt der Schwerpunkt jedoch auf rein sozialen Themen.

Im Fall des Hochwasserereignisses 2021 wird durch die Interviews deutlich, dass soziale Infrastrukturen zu vielen Dimensionen von Resilienz ganz entscheidend beitragen. Gleichzeitig sind soziale Infrastrukturen jedoch noch wenig in entscheidende Prozesse eingebunden, sodass der soziale und physische Wiederaufbau noch nicht ausreichend verknüpft ist – oder besser verknüpft werden könnte. Vor diesem Hintergrund wurden drei Thesen entwickelt, die im folgenden Abschnitt diskutiert werden.

17.4 Zusammenfassende Thesen zu der Rolle von sozialen Infrastrukturen im Wiederaufbau

Basierend auf den Ergebnissen sind in diesem Kapitel die wesentlichen Erkenntnisse entlang weniger Thesen zusammengefasst.

Erste These: Soziale Infrastrukturen sind im Wiederaufbau nach einer Katastrophe ein wichtiger Faktor für einen besseren Wiederaufbau. Sie wirken wie ein Resilienzverstärker während des Aufbaus.

Soziale Infrastrukturen ermöglichen soziale Interaktionen und stellen Räume für den Aufbau und die Verfestigung sozialer Beziehungen zur Verfügung. Sie sind besonders sensibilisiert für besonders verwundbare Personengruppen und haben einen Überblick über die unterschiedlichen Bedarfe vor Ort, vor allem wenn sie, z. B. durch lange Erfahrungen, starke und spezifische Ortskenntnisse haben. Soziale Infrastrukturen haben nach der Hochwasserkatastrophe 2021 bewiesen, dass sie in der Lage sind, flexibel auf sich ständig und schnell ändernde Bedarfe zu reagieren und Lösungen für aufkommende Probleme zu finden. Sie unterstützen vor allem die Wiederherstellung bzw. die Erneuerung von Netzwerken und Beziehungen. Hervorzuheben ist in diesem Zusammenhang auch ihre Fähigkeit zur sozialen Innovation und das Einführen von unbürokratischen Zwischenlösungen. Es wurden neue Begegnungsformate ausprobiert, Räume umfunktioniert und Menschen zusammengeführt, die vor der Hochwasserkatastrophe keine Berührungspunkte hatten. Zusätzlich halfen die nach dem Hochwasser etablierten Strukturen, auch auf andere Krisen schnell reagieren zu können. Der Angriffskrieg von Russland auf die Ukraine im Februar 2022 führte dazu, dass ukrainische Bürger:innen nach Deutschland kamen. Diesen wurden dann auch prioritär durch Sportvereine, kirchliche Einrichtungen und spontan geschaffenen Unterstützungsnetzwerken geholfen. Die Einrichtungen berichteten, dass sie durch die Erfahrungen aus dem Hochwasserereignis schnell auf diesen Ausnahmezustand reagieren konnten. In der gegenwärtigen gesellschaftlichen Situation, in der mehrere Krisen gleichzeitig auftreten, gewinnt diese Eigenschaft eine besondere Bedeutung. Durch die Aktivitäten von sozialen Infrastrukturen werden Resilienzmerkmale wie Inklusion, Flexibilität und auch zukunftsorientiertes Denken adressiert. Es ist zu beobachten, dass diese Entwicklungen und Aktivitäten auch zwei Jahre nach dem Hochwasserereignis bestehen bleiben. Damit diese auch langfristig etabliert werden und die Resilienz kurzfristig und auch langfristig vor Ort gestärkt wird, müssen die sozialen Infrastrukturen allerdings verstärkt unterstützt werden, worauf in den Interviews immer wieder hingewiesen wurde.

Zweite These: SI waren für viele Betroffene ein wichtiger Anlaufpunkt nach dem Hochwasser, insbesondere von Personengruppen, die keinen oder nur sehr schweren Zugang zu öffentlich bereitgestellten Ressourcen hatten.

In den Interviews wurde immer wieder darauf hingewiesen, mit welchen Barrieren viele Betroffene bei der Beantragung von finanzieller Wiederaufbauhilfe bzw. Fördermitteln konfrontiert sind. Zusätzlich besteht auch drei Jahre nach dem Hochwasser noch immer ein großer Bedarf an Informationen, wie Privatpersonen sich selbst vor zukünftigen Extremwetterereignissen schützen können. Um die Situation zu verbessern, gilt es Sprach- und Mobilitätsbarrieren abzubauen, um den Zugang zu digitalen und analogen Angeboten zu gewährleisten. Die unterschiedlichen und spezifischen Bedürfnisse und Barrieren müssen adressiert werden, damit die Betroffenen eine Chance haben, den Wiederaufbau zu bewältigen und eine resiliente Transformation umzusetzen. Soziale Infrastrukturen haben in den letzten Jahren verschiedene Strukturen aufgebaut, um unterschiedliche Bevölkerungsgruppen zu erreichen. Es gibt unter anderem Beratungsangebote, Informationspunkte, Shuttleservices sowie aufsuchende Beratung. Soziale Infrastrukturen fungieren oftmals als Schnittstelle zwischen den kommunalen Verwaltungen und den Menschen vor Ort. Sie ermöglichen Zugang zu den wesentlichen Ressourcen z. B. durch die Unterstützung bei der Beantragung von finanziellen Mitteln, das Weiterleiten von Spenden, der Vermittlung von Wissen oder den Aufbau von sozialem Kapital. In diesem Zusammenhang stellt sich die Frage, ob es notwendig ist, soziale Infrastrukturen besser vorzubereiten und zu schulen und sie so in das Krisenmanagement zu integrieren, dass sie bei Eintritt eines Extremwetterereignisses diese Rolle schnellstmöglich wieder einnehmen können.

Dritte These: Dem Wert von sozialen Infrastrukturen im Wiederaufbauprozess wird während des Wiederaufbaus zu wenig Beachtung und Anerkennung geschenkt.

In den Diskussionen über den Wiederaufbau reagieren Planer:innen und Politker:innen auf Katastrophen häufig mit einer Fokussierung auf den physischen Wiederaufbau. Der soziale Wiederaufbau und vor allem die bedeutende Rolle von SI werden dabei häufig politisch und finanziell deutlich weniger unterstützt. Das heißt nicht, dass es grundsätzlich keine politische oder finanzielle Unterstützung oder Anerkennung gibt. Ganz im Gegenteil, die Solidarität und unterstützende Rolle von sozialen Infrastrukturen wird häufig anerkannt, dies manifestiert sich aber nicht in einer mit Ressourcen unterlegten Unterstützung, zumindest legen das die Ergebnisse unserer Interviews nahe. Bei der Suche nach neuen Standorten oder bei der Verstetigung von neuen Prozessen sind soziale Infrastrukturen auf politische und finanzielle Unterstützung angewiesen. Diese wird aber häufig als nicht ausreichend empfunden. Des Weiteren haben SI zwar enorme Kenntnisse über die sozialen Begebenheiten und Problemlagen vor Ort, sie werden aber häufig nicht in die Schutz- und Vorsorgeentscheidungen des Hochwasserrisikomanagements eingebunden. Es gibt nur selten die Ambition, soziale Infrastrukturen in diese Prozesse mit einzubinden und dort, wo es geschah, wird der Gestaltungs- und Entscheidungsspielraum von SI als relativ begrenzt wahrgenommen.

17.5 Fazit

Dieser Beitrag verdeutlicht, dass soziale Infrastrukturen eine zentrale Rolle im Wiederaufbau nach Katastrophen spielen und wesentlich zur urbanen Resilienz beitragen, indem sie soziale Interaktionen ermöglichen, Netzwerke wiederherstellen und flexible Lösungen für aufkommende Probleme von Betroffenen bieten. Ihre besondere Sensibilität gegenüber sozialen Verwundbarkeiten und ihre tiefen Ortskenntnisse erlauben es ihnen, schnell auf sich verändernde Bedarfe zu reagieren, was sie zu unverzichtbaren Akteuren in Katastrophensituationen macht. Trotz ihrer wichtigen Beiträge, insbesondere für Bevölkerungsgruppen, die nur schwer Zugang zu öffentlichen Ressourcen haben, fehlt es jedoch häufig an ausreichender politischer und finanzieller Unterstützung und Einbindung. Während sich der physische Wiederaufbau im Fokus der politischen Aufmerksamkeit befindet, bleibt der soziale Wiederaufbau oft unterbewertet. Angesichts des Beitrags der sozialen Infrastrukturen in der Bewältigungs- und Wiederaufbauphase und ihres Potenzials für die Mitwirkung an Vorsorge- und Anpassungsprozessen ist unbestritten, dass soziale Infrastrukturen einen neuen Stellenwert im Katastrophenmanagement und in der Risikoprävention benötigen.

Die Bewältigungs- und Wiederaufbaustrukturen nach dem Hochwasser 2021 wurden ad hoc und unvorbereitet aufgebaut, was einen beeindruckenden, aber auch strapaziösen Kraftakt darstellt. Diese Strukturen langfristig zu etablieren und vorzubereiten, wäre ein erster Schritt, damit sie im Ernstfall schnell und den Umständen entsprechend unkompliziert reagieren können. Dabei ist zu beachten, dass den SI auch ausreichend Ressourcen

– sei es Personal, Geld oder Wissen – zur Verfügung gestellt werden müssen. Darüber hinaus sollten soziale Infrastrukturen aufgrund ihrer für uns offensichtlichen Schutzwürdigkeit einen besonderen Hochwasserschutz bereitgestellt bekommen, damit sie im Katastrophenfall geschützt sind und bei einem Hochwasserereignis ihre wichtigen Funktionen (schnell) weiter erfüllen können.

Hinweis für Entscheidungsprozesse

Um Resilienz und dementsprechend den Building Back Better-Ansatz zu verwirklichen, muss der psychosoziale Wiederaufbau genauso sichtbar werden, wie der physische Wiederaufbau. Dafür müssen soziale Infrastrukturen abgefragt und ihre lokale Expertise, ihre Netzwerke sowie ihre Beziehungen zu ihren Klienten sinnvoll eingesetzt werden, sodass auch die breite Zivilgesellschaft erreicht und bedarfsgerecht aufgebaut werden kann. Die Umsetzung dieses neuen Verständnisses hat auf Grund des Klimawandels eine gewisse Dringlichkeit und sollte zu den Grundprinzipien jedes Vorsorgekonzepts gehören und entsprechend mit Ressourcen unterstützt werden.

Zusammenfassung

- Soziale Infrastrukturen spielen eine zentrale Rolle im Wiederaufbau nach Katastrophen und tragen wesentlich zur urbanen Resilienz bei, indem sie soziale Interaktionen ermöglichen, Netzwerke wiederherstellen und flexible Lösungen für aufkommende Probleme von Betroffenen bieten.
- Angesichts des Beitrags der sozialen Infrastrukturen in der Bewältigungs- und Wiederaufbauphase und ihres Potenzials für die Mitwirkung an Vorsorge- und Anpassungsprozessen benötigen soziale Infrastrukturen einen neuen Stellenwert im Katastrophenmanagement und in der Risikoprävention. Die Umsetzung dieses neuen Verständnisses hat auf Grund des Klimawandels eine gewisse Dringlichkeit und sollte zu den Grundprinzipien jedes Vorsorgekonzepts gehören und entsprechend mit Ressourcen unterstützt werden.
- Trotz ihrer wichtigen Beiträge, insbesondere für Bevölkerungsgruppen, die nur schwer Zugang zu öffentlichen Ressourcen haben, fehlt es jedoch häufig an ausreichender politischer und finanzieller Unterstützung und Einbindung.
- Die ad hoc und unvorbereitet aufgebauten Bewältigungs- und Wiederaufbaustrukturen langfristig zu etablieren, vorzubereiten und mit ausreichend Ressourcen auszustatten, wäre ein erster Schritt, damit sie im Ernstfall schnell und den Umständen entsprechend unkompliziert reagieren können.
- Darüber hinaus sollten soziale Infrastrukturen aufgrund ihrer Schutzwürdigkeit einen besonderen Hochwasserschutz bereitgestellt bekommen, damit sie im Katastrophenfall geschützt sind und bei einem Hochwasserereignis ihre wichtigen Funktionen (schnell) weiter erfüllen können.
- Um Resilienz und dementsprechend den Building Back Better-Ansatz zu verwirklichen, muss der psychosoziale Wiederaufbau außerdem genauso sichtbar werden, wie der physische Wiederaufbau.

Danksagung Die Autorinnen und Autoren von ▶ Kap. 17 bedanken sich beim Bundesministerium für Bildung und Forschung (BMBF) und beim DLR Projektträger für die

Unterstützung und Förderung des KAHR-Projekts. Diese Forschung wurde vom BMBF unter dem Förderkennzeichen 01LR2102A-M finanziert. Die Autor:innen danken allen Interview-Partner:innen für ihr Engagement und ihre wertvollen Beiträge.

Literatur

Aldrich D (2023) How social infrastructure saves lives: a quantitative analysis of Japan's 3/11 disasters. Japanese J Polit Sci. https://doi.org/10.1017/S1468109922000366

Gunning L, Rizzi P (2022) Introduction: the invisible reconstruction project and its aims. In: Gunning L, Rizzi P (Hrsg) Invisible Reconstruction. Cross-disciplinary responses to natural, biological and man-made disasters. UCL Press, London.

Klinenberg E (2018) Palaces for the people: how social infrastructure Can help fight inequality, polarization, and the decline of civic life, first edition. Crown, New York

Latham A, Layton J (2019) Social infrastructure and the public life of cities: studying urban sociality and public spaces. Geogr Compass. https://doi.org/10.1111/gec3.12444

Meerow S, Stults M (2016) Comparing conceptualizations of urban climate resilience in theory and practice. Sustainability 8:701. https://doi.org/10.3390/su8070701

Otto D, Reckhaus Z, Kuhlicke C (2025) Caring, coping and rebuilding: the role of social infrastructure during and after the 2021 flood event in Germany. J Flood Risk Manag. https://doi.org/10.1111/jfr3.70007

Richter R, Rohnstock N (2016) Der Erzählsalon als Verfahren strategischen Erzählens. Konzeptionelle Voraussetzungen und empirische Gestalt am Beispiel des Projekts Lausitz an einem Tisch. DIEGESIS 5:2

Schubert H (1995) Soziale Infrastruktur. In: ARL – Akademie für Raumforschung und Landesplanung (Hrsg) Handwörterbuch der Raumordnung. Akademie für Raumforschung und Landesplanung, Hannover, S 847–851

Wilson SL, Kershaw MA (2008) Caring for young children after a hurricane: Florida's childcare workers reflect on support and training needs. Child Youth Environ. https://doi.org/10.1353/cye.2008.0058

Post-Katastrophen Wiederaufbau in benachteiligten Quartieren

Marisa Fuchs und Daniela Michalski

J. Birkmann et al. (Hrsg.), *Die Flutkatastrophe 2021*, Sustainable Development Goals (SDG) – Umsetzung in Praxis, Lehre und Entscheidungsprozessen,
https://doi.org/10.1007/978-3-658-48688-4_18

18.1 Einleitung

Extremwetterbedingte Katastrophen erweisen sich auch als soziale Katastrophen und fordern die urbane Gemeinschaft heraus (Heidbrink et al. 2010). Während einige Quartiere ein Katastrophenereignis schnell bewältigen, benötigen andere deutlich länger für die Bewältigung, die Erholung und den Wiederaufbau (Aldrich 2012). Die soziale Konstruktion von Katastrophen und soziale Ungleichheiten beeinflussen die Vulnerabilität und Resilienz einzelner Menschen, Bevölkerungsgruppen sowie Quartiere. Daher sind Katastrophen immer auch im Zusammenhang mit den gesellschaftlich geschaffenen Bedingungen von Vulnerabilität zu verstehen (Browning et al. 2006; O'Brien et al. 2009).

Beim Wiederaufbau nach einer Katastrophe geht es nicht nur um den physischen Wiederaufbau beschädigter Gebäude und Infrastrukturen nach dem Prinzip des *build back better* (UNDRR 2015). Wiederaufbaupolitik und -planung müssen ganzheitlich sein (Jha et al. 2010) und auch Fragen der Gerechtigkeit, des Zugangs zu Ressourcen, der Chancengleichheit für benachteiligte Quartiere sowie der Verringerung der Anfälligkeit von Quartieren gegenüber Gefahren berücksichtigen (Lewis 1999). Somit wird der Wiederaufbau zu einem hochpolitischen Ereignis und zu einer Frage der angemessenen Governance.

Dieser Beitrag beleuchtet die besonderen Herausforderungen und Anforderungen der Katastrophenbewältigung – insbesondere des Wiederaufbaus nach Katastrophen – in und für benachteiligte Quartiere. Die Forschungserkenntnisse beruhen auf Fallstudien in benachteiligten Quartieren der Städte Hagen und Stolberg (beides in Nordrhein-Westfalen [NRW], Deutschland), die vom überregionalen Starkregen- und Hochwasserereignis infolge des Tiefdruckgebiets *Bernd* 2021 betroffen waren. In Hagen lag der Forschungsschwerpunkt auf der sozialen Resilienz betroffener Nachbarschaften, während in Stolberg der Einfluss der gebauten Umwelt auf die soziale Resilienz im Vordergrund stand. Im Rahmen der Fallstudien wurden teilnehmende Beobachtungen bei Veranstaltungen in den betroffenen Gebieten, informelle Gespräche sowie halbstrukturierte Interviews mit Akteur:innen verschiedener behördlicher und gemeinnütziger Institutionen sowie betroffenen Quartiersbewohner:innen durchgeführt. Zudem wurden formelle und informelle Planungsgrundlagen gesichtet. Die Erkenntnisse der Fallstudien werden darüber hinaus in den internationalen Forschungskontext eingebettet.

18.2 Übersicht der Fallstudien

Beide Fallstudien befinden sich im Wiederaufbau-Governance-Setting von NRW (s. Abb. 18.1). In den Tagen vom 12.–15. Juli 2021 kam es in der Südhälfte von NRW örtlich zu Sturzfluten und Ausuferungen kleinerer Fließgewässer.

18.2.1 Fallstudienquartier in Hagen

Hagen liegt im südöstlichen Teil des Ruhrgebiets und hat rund 190.000 Einwohner:innen (EW). Mit einer EW-Dichte von rund 3235 EW/km^2 an Siedlungs- und Verkehrsfläche (SuV) ist die Stadt Hagen deutlich dichter besiedelt als der NRW-Landesdurchschnitt (= 2233 EW/km^2 SuV) (IT.NRW 2023a). Hagen gehört zu den vergleichsweise ärmeren Städten in NRW, sowohl was den öffentlichen als auch die privaten Haushalte

Abb. 18.1 Fallstudiengebiete. (Eigene Darstellung)

(Ø 20.862 €/EW verfügbares privates Einkommen, Rang 374 von 396 in NRW) angeht (IT.NRW 2023a).

Vier Flüsse und etwa 360 kleinere Bäche durchqueren die Stadt. Viele davon verlaufen auf Privatgrundstücken und sind kanalisiert. Das Hochwasserereignis 2021 betraf mehrere Tausend Menschen und erforderte über 1500 Hilfseinsätze. Allein die Schäden an der öffentlichen Infrastruktur beliefen sich auf rund 83,5 Mio. € (Stadt Hagen 2023). Zwei Jahre nach dem Hochwasser haben die Bürger:innen bereits rund 20,9 Mio. € an öffentlichen Mitteln für den privaten Wiederaufbau in Hagen beantragt (ohne Versicherungsleistungen und laufende Anträge) (Stadt Hagen 2023).

Das Fallstudienquartier Eckesey hat rund 6000 EW und zeichnet sich durch eine vergleichsweise hohe Bevölkerungsdichte und eine eher fluktuierende Bevölkerungszusammensetzung aus. Die Sozialstruktur ist zum Zeitpunkt der Flutkatastrophe im Vergleich zur Gesamtstadt überdurchschnittlich jung, migrantisch geprägt und finanzschwach (Stadt Hagen 2020c). Das Quartier weist mitunter einen der höchsten Jugendquotienten sowie der höchsten Betroffenheitsquoten[1] und höchsten Anteilen an Bewohner:innen mit Migrationshintergrund[2] (ca. 70 %) in Hagen auf. Laut einer Wohnungsmarktstudie des Instituts für Raumforschung & Immobilienwirtschaft (IRI 2016) dominierten unter den migrantischen Milieus im Quartier zum Studienzeitpunkt ein weitgehend isoliertes, religiös verwurzeltes Milieu[3], gefolgt von einem durch junge Menschen geprägten,

---

1 Anzahl der arbeitslosen SGB II und SGB III-Empfänger:innen im Verhältnis zu je 100 Personen im erwerbsfähigen Alter (18–59 Jahre).

2 In dieser Statistik zählen dazu alle Nichtdeutschen, Eingebürgerten, im Ausland geborenen Deutschen sowie Aussiedler:innen, ebenso Kinder unter 18 Jahren, von denen mindestens ein Elternteil einen Migrationshintergrund hat.

3 Das archaische, patriarchalisch geprägte Milieu, sozial und kulturell isoliert und in vormodernen Mustern sowie religiösen Traditionen verankert, zeigt starke Tendenzen zur Abgrenzung und pflegt weiterhin die (Familien-)Traditionen des Herkunftslandes (SINUS Markt- und Sozialforschung GmbH 2024).

hedonistisch-subkulturellen Milieu[4], das sich u. a. auch durch geringeres Engagement für die Quartiersentwicklung auszeichnet (eine Aktualisierung dieser Untersuchung liegt bislang nicht vor). Insgesamt konzentrieren sich verschiedene soziale Problemlagen im Fallstudienquartier (Stadt Hagen 2020b).

Städtebaulich ist das rund 87 ha große Quartier durch räumliche Zäsuren sowie durch eine unattraktive Lage zwischen Verkehrs- und Gewerbeflächen geprägt (▣ Abb. 18.2; Stadt Hagen 2020b). Die Siedlungsstruktur besteht überwiegend aus geschlossener Blockrand- und einer Reihenhausbebauung mit Mehrfamilienhäusern, aber auch Einfamilienhäusern (IRI 2016). Viele der Wohnungen gehören einer regionalen Wohnungsgenossenschaft. Angesichts der geringen finanziellen Ressourcen lag 2015 die Mietbelastungsquote[5] im Quartier bei rund 20 % (IRI 2016). Schon vor der Flutkatastrophe wies das Quartier eine signifikant unattraktive Bausubstanz, diverse Problemimmobilien, einen Investitionsstau und eine hohe Leerstandsquote auf (Stadt Hagen 2020b). Zudem mangelt es an Grünflächen, Freiraumvernetzung und Durchlässigkeit (Stadt Hagen 2020b, 2020a). Es fehlen ausreichend attraktive öffentliche Plätze für die breite Bewohnerschaft, wenngleich verschiedene soziale Infrastrukturen wie eine Grundschule, ein Jugendzentrum und örtliche Vereine (z. B. Sportverein, Kulturverein, Freiwillige Feuerwehr) spezifische Möglichkeiten der Begegnung bieten.

Mit dem Integrierten Stadtentwicklungskonzept kündigte die Stadt Hagen 2020 bereits vor dem Flutereignis 2021 das Ziel der Stabilisierung und Aufwertung benachteiligter Quartiere und für das Fallstudienquartier ein Teilräumliches Handlungskonzept im Zuge eines Programmgebiets der Städtebauförderung an – allerdings mit der niedrigsten Prioritätseinstufung. Diese Priorisierung könnte sich jedoch im Zuge der aktuellen strategischen (Neu-)Ausrichtungen der Stadtentwicklung nochmals verschieben.

18.2.2 Fallstudienquartier in Stolberg

Die Stadt Stolberg liegt im äußersten Südwesten von NRW und bildet mit ihren gut 56.000 EW die drittgrößte Gemeinde der Städteregion Aachen. Mit einer EW-Dichte von 2555 EW/km^2 SuV ist Stolberg etwas dichter besiedelt als der NRW-Landesdurchschnitt (IT.NRW 2023b). Ähnlich wie Hagen gehört Stolberg zu den ärmeren Städten in NRW (Ø 22.018 €/EW verfügbares privates Einkommen, Rang 352 von 396 in NRW) (IT.NRW 2023b). Bis heute ist Stolberg durch den technisch-wirtschaftlichen Strukturwandel der rohstoffverarbeitenden Industrie geprägt.

Stolberg liegt im hügeligen Vennvorland am nördlichen Rand des linksrheinischen Schiefergebirges. Die Innenstadt erstreckt sich entlang des Zusammenflusses von Inde und Vichtbach. Aufgrund der schmalen Talsohle im Innenstadtbereich kam es infolge des Tiefdruckgebiets Bernd zu Überflutungen mit hohen Einstautiefen und Fließgeschwindigkeiten durch den Vichtbach, insbesondere in den oberhalb des Zentrums gelegenen Quartieren. Allein der Wiederaufbauplan für öffentliche Infrastrukturen umfasst ein Finanzvolumen von 225 Mio. € (Stadt Stolberg 2023).

4 Das junge, freizeitorientierte Unterschichtmilieu mit schwachem Selbstbewusstsein und Underdog-Mentalität sucht nach Unterhaltung und Konsum, während es sich den Leistungs- und Anpassungserwartungen der Mehrheitsgesellschaft entzieht (SINUS Markt- und Sozialforschung GmbH 2024).

5 Anteil der Warmmiete an der Haushaltskaufkraft.

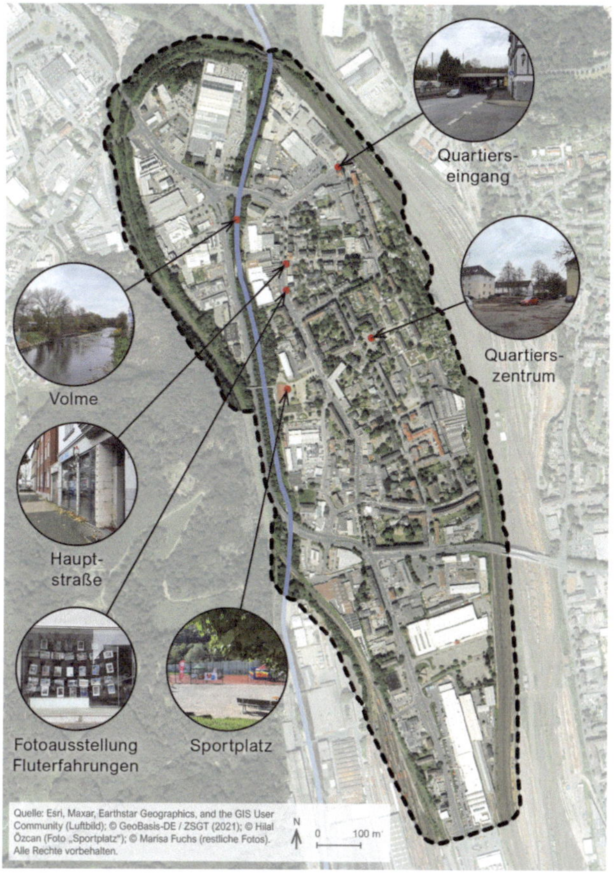

Abb. 18.2 Luftbild des Fallstudienquartiers in Hagen (Fotoaufnahmen von 11/2022 und 07/2023). (Eigene Darstellung)

Das Fallstudienquartier liegt innerstädtisch und umfasst Teile der Stadtgebiete Oberstolberg und Unterstolberg (siehe Abb. 18.3). Rund 10.300 Menschen leben in den beiden Stadtgebieten (Städteregion Aachen 2022). Im Vergleich zur Gesamtstadt ist die Sozialstruktur überdurchschnittlich jung und finanzschwach. Die Quartiersbevölkerung besteht zu etwas mehr als einem Drittel aus ausländischen Menschen[6]. Die Arbeitslosenquote ist in beiden Stadtteilen hoch (Oberstolberg: 11,5 % Arbeitslosenquote, 31,5 % SGB II-Bedarfsgemeinschaften; Unterstolberg: 23,3 % Arbeitslosenquote, 27,6 % SGB II-Bedarfsgemeinschaften), ebenso die Kinderarmutsquote (Oberstolberg knapp 52 %, Unterstolberg knapp 41 %) (Städteregion Aachen 2022).

Oberstolberg ist mit ca. 36 ha der räumliche, kulturelle und politische Stadtmittelpunkt von Stolberg. Hier finden sich das Rathaus sowie zentrale Einkaufs- und Versorgungsstraßen (Rathausstraße und Steinweg), eine Grundschule, zwei Kitas und Großtagespflegestätten, ein Gesundheitszentrum und weitere Dienstleister der Gesundheitswirtschaft. Unterstolberg ist insgesamt ca. 80 ha groß und bildet den nördlichen Eingang in die Innenstadt. Der Stadtteil stammt aus der Entstehungszeit ab Mitte des 19. Jahrhundert

6 Als ausländische Personen werden in dieser Statistik alle Personen gezählt, die im Sinne des Art. 116 Abs. 1 des Grundgesetzes nicht Deutsch sind.

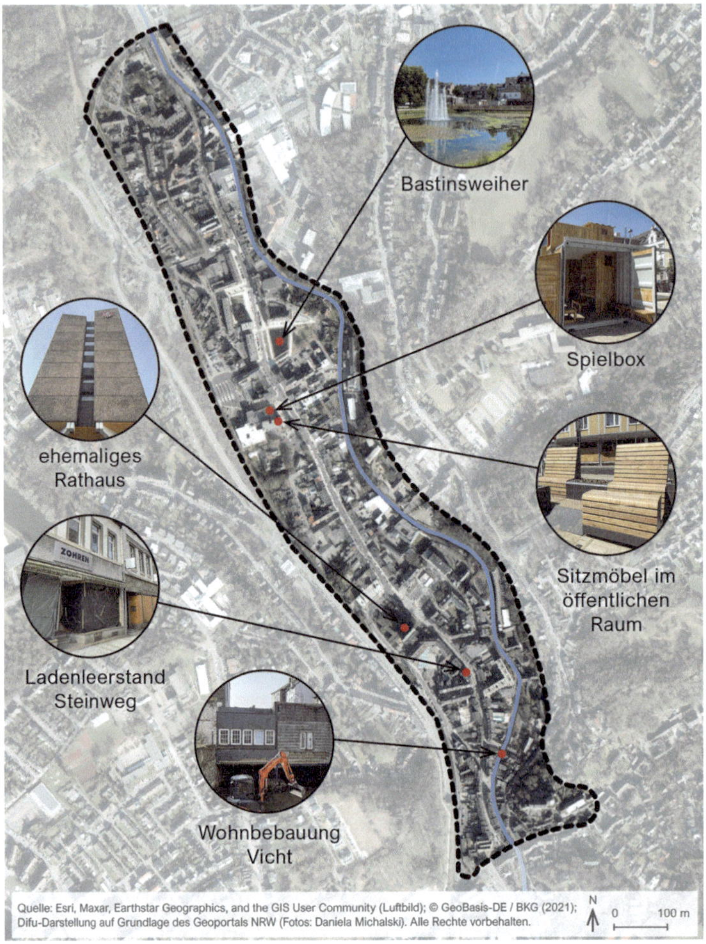

Abb. 18.3 Luftbild des Fallstudienquartiers in Stolberg. (Eigene Darstellung)

und weist kleinteiliges Gewerbe und Handwerk (Ladenlokale im EG, darüber Wohnraum) sowie Grundschule, Kirche und mehrere Moscheevereine auf.

Beide Stadtteile sind geprägt von überwiegend geschlossener Blockrandbebauung. Die Grundstücke in Oberstolberg sind aufgrund der räumlichen Enge am Vichtbach eher klein und stark überbaut. Der Bereich *Oberer und Unterer Steinweg* bildet die zentrale Geschäftsstraße, ist jedoch durch Leerstand und Mindernutzungen geprägt. Zahlreiche Gebäude stehen hier unter Denkmalschutz, ein Teil davon sanierungsbedürftig. Unterstolberg weist z. T. einen erheblichen Sanierungsstau auf, der Leerstand erstreckt sich mitunter auf ganze Gebäude.

Aufgrund der hohen baulichen Dichte gibt es nur wenige Grün- und Freiflächen in der Innenstadt. In Oberstolberg finden sich u. a. drei größere Plätze (Kaiserplatz, Heinrich-Böll-Platz und Willy-Brandt-Platz). In Unterstolberg ist der Kaplan-Josef-Dunkel-Platz zentral und wichtige Anlaufstelle für die Bewohner:innen (u. a. Vorplatz des Jugendzentrums Westside, des Josefshauses, der Musikschule), außerdem gibt es zwei weitere Stadtteilplätze (Mühlener Markt und Jordanplatz) sowie den Bastinsweiher, eine zentral gelegene Parkanlage. Die Vicht verläuft mitten durch die beiden Stadtteile. Der Bach ist kanalisiert, die Uferbereiche sind überwiegend nicht zugänglich.

Für Teilbereiche von Ober- und Unterstolberg wurde im Jahr 2011 das Entwicklungskonzept *Talachse Innenstadt* beschlossen. Außerdem liegt ein integriertes Handlungskonzept *Berg- und Talachse – Miteinander für Münsterbusch, Ober- und Unterstolberg* aus dem Jahr 2018 vor. Seit dem Jahr 2016 ist die Talachse Innenstadt ein förmlich festgelegtes Sanierungsgebiet, der Geltungsbereich wurde im Jahr 2021 aktualisiert. Beide Stadtteile sind Teil des Programmgebiets im Städtebauförderungsprogramm *Sozialer Zusammenhalt*.

18.3 Ursachen und Folgen sozialer Segregation im stadtstrukturellen Kontext

Die Verteilung der Bevölkerung nach verschiedenen sozialstrukturellen Merkmalen und somit auch soziale Ungleichheit ist seit langem in der Raumstruktur europäischer Städte erkennbar und nachweisbar (Baum 2007). Seit den Arbeiten der Chicagoer Schule in den 1920er Jahren werden die Ursachen und Ausprägungen einer ungleichen räumlichen Verteilung sozialer Gruppen in der sozialwissenschaftlichen Stadtforschung als *soziale Segregation* bezeichnet (Park et al. 1925). Soziale Segregation kann sowohl freiwillig als auch unfreiwillig entstehen. Dies zeigt sich deutlich am Beispiel von reichen und armen Bevölkerungsgruppen und einem auf spezifische Nachfragegruppen ausgerichteten, räumlich differenzierten Wohnungsangebot: einkommensstarke Haushalte haben weniger Restriktionen in der Wohnstandortauswahl, können ihren Wohnort weitgehend nach ihren persönlichen Präferenzen auswählen und wohnen meist freiwillig segregiert in privilegierten Lagen (Farwick 2007; Baum 2007). Arme Haushalte – ebenso wie Haushalte, die verstärkt sozialer Diskriminierung ausgesetzt sind – wohnen oft unfreiwillig segregiert in günstigen Wohnlagen oder in Wohnungen des sozialen Wohnungsbaus (Farwick 2007; Baum 2007).

In benachteiligten, sozial segregierten Quartieren fehlt es häufig an strukturellen Rahmenbedingungen, z. B. einer institutionellen und normativen Einbindung in Organisationen, Strukturen und Handlungskontexte (Baum 2007). Soziale Segregation ist daher eng mit der Integration und somit auch Exklusion von Menschen verbunden (Baum 2007). Sozialräumliche Verteilungsprozesse werden dann problematisch, wenn die Bevölkerungszusammensetzung, eine mangelnde Infrastruktur und institutionelle Einbindung zu einer Kumulation individueller Problemlagen führen, die die ohnehin deprivierte Struktur so verstärkt, dass es zu einer ökonomischen, sozialen, kulturellen und/oder auch politischen Ausgrenzung oder gar Exklusion der Quartiersbewohner:innen kommt (Baum 2007). In diesem Zusammenhang sprechen u. a. Häußermann und Siebel (1987; 1995) auch von der *Spaltung der Stadt*, bei der sich unterschiedliche soziale Schichten innerhalb einer Stadt so weit voneinander abkoppeln, dass sie sich nicht mehr gegenseitig beeinflussen.

So kommt es zu räumlichen Konzentrationen benachteiligter Bevölkerungsgruppen in Quartieren, in denen negative Quartierseffekte zusätzlich zu den individuellen Problemlagen benachteiligend wirken. Dies trägt zu einer Verfestigung der ohnehin prekären sozialen Lage der Bewohner:innen bei (Farwick 2007). Oftmals kommen Stigmatisierungsprozesse aufgrund der Wohnadresse hinzu, was zu kollektiven Zuweisungen diskreditierender Merkmale und Vorurteilen führt (Baum 2007; Hamedinger 2002). Allerdings können benachteiligte Quartiere auch positive Quartierseffekte haben, wie eine stärkere Quartiersbindung. Diese entsteht aus einer Wechselbeziehung von Ausgrenzung[7] und Integration[8] (Baum 2007).

18.4 Soziale Vulnerabilität und Resilienz

Im Gegensatz zum allgemeinen Resilienzbegriff befasst sich das Verständnis der *sozialen Resilienz* explizit mit Fragen des menschlichen Handelns, sozialer Praktiken, Machtbeziehungen, Institutionen und Diskursen (Keck und Sakdapolrak 2013). Unter Einbeziehung von als notwendig erachteten Kompetenzen und Fähigkeiten für den Umgang mit Störungen, Unsicherheiten und Wandel definieren Obrist et al. (2010, 289) soziale Resilienz als

> „the capacity of actors to access capitals in order to – not only cope with and adjust to adverse conditions (that is, reactive capacity) – but also search for and create options (that is, proactive capacity) and thus develop increased competence (that is, positive outcomes) in dealing with a threat".

Sozialforscher:innen wie Lorenz (2013) betonen, dass soziale Resilienz nicht nur von Einzelpersonen oder Gemeinschaften abhängt, sondern auch von gesellschaftlichen Faktoren, die beeinflussen, ob und wie Menschen auf Ressourcen zugreifen oder Fähigkeiten und Kompetenzen erwerben können. Somit wird soziale Resilienz nicht als vorgegebenes Potenzial verstanden, sondern als soziale Leistung oder als Effekt der sozialen Eigenschaften einer Gemeinschaft (Bonß 2015). Das Konzept der sozialen Resilienz ist daher ein Zusammenspiel zwischen den Fähigkeiten von Personen oder Gemeinschaften und sozialen Strukturen (Bohle et al. 2009; Bonß 2015).

Das Resilienzkonzept wird als *urbane Resilienz* auch in der Stadtforschung und -praxis immer stärker aufgegriffen – insbesondere im Zusammenhang mit Klimaanpassung und Katastrophenschutz (siehe z. B. Kuhlicke et al. 2024). Auf dem Weg hin zu einer robusten, adaptiven und zukunftsfähigen Stadt soll es als normatives Leitbild dienen, wie urbane Räume weiterzuentwickeln, Entscheidungen und Prioritäten zu treffen sowie Prozesse und Planungen aufzusetzen und bei Bedarf anzupassen sind.

Das Konzept der Vulnerabilität beschreibt die Verwundbarkeit von Mensch-Umwelt-Systemen gegenüber Stress, Gefahren und Risiken. Das Vulnerabilitäts- und Risikokonzept des IPCC (2023) betrachtet Vulnerabilität zusammen mit Gefahren (Hazards) und Exposition als zentrale Komponente des Katastrophenrisikos. Diesem Konzept folgend wird Vulnerabilität definiert als

> „the propensity or predisposition to be adversely affected and [it] encompasses a variety of concepts and elements, including sensitivity or susceptibility to harm and lack of capacity to cope and adapt [. . .]."

Dabei wird anerkannt, dass auch soziale und kontextuelle Determinanten die Vulnerabilität beeinflussen und nicht alle Personen oder Gruppen in einem bestimmten Kontext

---

7 Durch begrenzte Zugangsmöglichkeiten zur Stadt und mangelnde Handlungschancen außerhalb des Quartiers.

8 Rückzug in das unmittelbare Wohnumfeld, wo man den Normen genügt und genügen möchte; letzteres in Anlehnung an Dubet und Lapeyronnie (1994) und den *devianten Konformismus*.

gleich vulnerabel und resilient sind (IPCC 2023). Daher spricht man von *sozialer* Vulnerabilität (Bohle 2002; Birkmann 2007; Cutter 2006; Cutter und Finch 2008). Morrow (2008, S. 4) fasst es so zusammen:

> „Social vulnerability is a catchall phrase that has become part of the discussion related to how social and cultural conditions place some at higher risk to environmental impacts such as climate change or natural hazards. Simply stated, social vulnerability occurs when unequal exposure to risk is coupled with unequal access to resources."

18.5 Der Einfluss der gebauten Umwelt auf Bewohner:innen von benachteiligten Quartieren

Die Qualität der gebauten Umwelt hat Auswirkungen auf das Wohlbefinden und die Gesundheit des Menschen. Aus der Architekturpsychologie ist bekannt, dass salutogenes, also gesundheitsförderndes und gesundheitserhaltendes Wohnen unter anderem Rückzugs- und persönliche Bereiche, einen Ausblick in eine Umwelt mit erholsamen Qualitäten und ein Netz positiver Sozialkontakte beinhaltet (IWAP 2024; Flade 2020).

In benachteiligten Quartieren müssen die Bewohner:innen aufgrund beengter Wohnverhältnisse oft auf persönliche und individuelle Rückzugsbereiche verzichten. Zudem mangelt es gerade in benachteiligten Quartieren häufig an einer Umwelt mit erholsamen Qualitäten, stattdessen sind hier umweltbedingte Mehrfachbelastungen festzustellen (z. B. Lärm, Luftschadstoffe, mangelnder Zugang zu Grünflächen, bioklimatische Belastungen) (Hornberg und Maschke 2017).

Qualitätsvolle Frei- bzw. öffentliche Räume sind jedoch u. a. eine Voraussetzung dafür, dass Quartiersbewohner:innen ein Netz an positiven Sozialkontakten knüpfen können. Laut Adli haben öffentliche Plätze sogar einen „zentralen Public-Health-Auftrag", da sie zum Rausgehen auffordern und damit die Bewegung und den sozialen Austausch fördern (Roloff 2019; Adli 2017). Zudem wird in der Gesundheitsforschung davon ausgegangen, dass das Vorhandensein von Grün in erreichbarer Nähe der eigenen Wohnung einen Anteil daran hat, ob Stress in einer Depression resultiert oder nicht (Adli 2017, 214). Es besteht also ein direkter Zusammenhang zwischen Umweltbelastungen und -ressourcen, Gesundheit/Krankheit und der sozialen Lage (Hornberg und Maschke 2017).

Menschen, die in benachteiligten Quartieren leben, sind sowohl sozialen und gesellschaftlichen Belastungen ausgesetzt (dazu gehören u. a. Armut, Diskriminierung oder Benachteiligung), als auch physischen Stressoren (dazu gehören u. a. Platzmangel, mangelhaftes Freiflächenangebot, klimatische Belastungen) (Metzler-Amlacher 2024; Böhme 2022).

18.6 Höheres Katastrophenrisiko für benachteiligte Quartiere

In benachteiligten Quartieren kommen bereits im Vorfeld einer Katastrophe komplexe Problemlagen zusammen, die die Vulnerabilität der Bevölkerung erhöhen. Die Wechselwirkungen von sozialen Ungleichheiten und Problemlagen mit Einflüssen der gebauten Umwelt, die bereits im Vorfeld von Katastrophen bestehen, kommen auch zum Ausdruck, wenn sich die Katastrophen ereignen (Tierney 2007). Neben der höheren Vulnerabilität führt eine tendenziell häufigere und/oder stärkere Exposition gegenüber Überflutungen zu einem insgesamt tendenziell höheren Katastrophenrisiko für benachteiligte Quartiere.

Auch in Hagen korrelieren sozioökonomische Probleme mit der Betroffenheit gegenüber Klimawirkungen und Umweltbelastungen (Stadt Hagen 2018). Eine unzureichende Hochwasser- und Starkregenvorsorge zeigt sich in der Hagener Fallstudie u. a. darin, dass vielen Betroffenen das Wissen um die allgemeine Gefahren- und konkrete Gefährdungslage fehlte oder sie diese falsch einschätzten. Unzureichendes Wissen über Überflutungsrisiken sowie geringere finanzielle Ressourcen führten dazu, dass kaum Vorsorge- und Schutzmaßnahmen ergriffen wurden. Häufig fehlte bzw. fehlt es auch an Wissen über geeignete Schutz- und Vorsorgemaßnahmen und möglichen Peer-Effekten (siehe z. B. Galster 2011). So hatten zum Beispiel nur die wenigsten Betroffenen einen ausreichenden Versicherungsschutz.

Das Stolberger Fallstudiengebiet liegt teilweise in einem festgesetzten Überschwemmungsgebiet. Weniger als ein Fünftel der Gebäudeeigentümer:innen in Ober- und Unterstolberg ist im Besitz einer Eigentumswohnung, überwiegend handelt es sich also um Eigentümer:innen von Wohn- und Geschäftshäusern, die vermietet werden. Die Flutkatastrophe hat nicht zuletzt deshalb einen beträchtlichen Schaden angerichtet, da sich die Gebäude bereits vor der Katastrophe in einem schlechten Sanierungszustand befanden. Im Jahr 2016 hatte die Stadt auf den offenkundigen Sanierungsbedarf im Quartier reagiert und für die *Talachse Innenstadt* eine Sanierungssatzung zur Behebung der städtebaulichen Missstände beschlossen. Eine Umfrage der Stadt Stolberg unter den Eigentümer:innen der Gebäude im selben Jahr ergab jedoch, dass nur 10 % der Eigentümer:innen beabsichtigte, Sanierungsmaßnahmen vorzunehmen. 36 % der befragten Personen gaben an, kleinere Maßnahmen zu planen, während eine Mehrheit von 54 % antwortete, keine Sanierungsmaßnahmen zu planen (DSK 2016).

Die in sozial segregierten Quartieren durch erzwungene Prozesse entstandene ungünstige Häufung von Vulnerabilität, Benachteiligung und städtebaulichen Missständen unterstreicht die Notwendigkeit umfassender Unterstützung für diese Quartiere und verdeutlicht die Komplexität der Anforderungen. Die Bewältigung einer Katastrophe in benachteiligten Quartieren erfordert besondere Beachtung der Governance-Strukturen, die als Reaktion auf Krisen oder Gefährdungen entstehen (Schott 2012) und in Antizipation potenzieller Bedrohungen entwickelt werden (Medd und Marvin 2005). Es gilt also, die soziale Konstruktion von Vulnerabilität und Resilienz in deren gesamten Wirkungsbereich zu berücksichtigen.

18.7 Benachteiligte Quartiere in Wiederaufbausystemen

18

Wiederaufbausysteme bestehen zumeist aus einer Kombination bestehender sowie in Reaktion auf das Katastrophenereignis neu entwickelter Steuerungs- und Unterstützungsstrukturen. Da die Auswirkungen insbesondere überregionaler Katastrophen die Wiederaufbaukapazitäten (z. B. Fachwissen, Management, Finanzen) der lokalen öffentlichen Verwaltung, einzelner Sektoren und Gemeinschaften schnell überschreiten, sind häufig eine Vielzahl an Akteur:innen aus Politik, Verwaltung und Privatwirtschaft sowie Nichtregierungsorganisationen und der Zivilgesellschaft in unterschiedlichem Maße involviert (Cho 2014).

So entwickeln sich ergänzend zu größeren staatlichen Wiederaufbaustrukturen häufig auch lokale Initiativen aufstrebender Gruppen[9], insbesondere um Lücken zu füllen, wenn die Bedürfnisse der Quartiersbewohner:innen von den bestehenden Organisationen nicht erfüllt werden (Auf der Heide 1989) oder die (üblichen) Rollen und Strukturen nicht funktionieren (Stallings und Quarantelli 1985). Diese Initiativen organisieren sich außerhalb formaler Strukturen, um nach einer Katastrophe die Kontrolle über ihre (eigenen) Wiederherstellungsbemühungen zu übernehmen (Drabek und McEntire 2003) und/oder um einen wirksamen Wiederaufbauprozess zu ermöglichen (Aldrich 2012; Barton 1969; Fischer 1998; Gilligan 2018).

18.7.1 Überblick über Wiederaufbaustrukturen und -instrumente im Fallstudien-Setting

Eines der wesentlichen Instrumente, das auch in anderen Katastrophenkontexten häufig zum Einsatz kommt (siehe z. B. Fothergill und Peek 2004), sind öffentliche Fördermittel für den Wiederaufbau. Nach dem Flutereignis im Juli 2021 richtete die Bundesregierung innerhalb von drei Monaten einen nationalen Wiederaufbaufonds in Höhe von 30 Mrd. € ein, von denen 12,3 Mrd. € Privathaushalten und Unternehmen der Wohnungswirtschaft in NRW zur Verfügung stehen (MHKBG NRW 2022). Die Förderbestimmungen und -leistungen sind in der entsprechenden Förderrichtlinie des Landes NRW festgehalten. Mithilfe der öffentlichen Fördermittel können Betroffene bis zu 80 % ihrer Kosten für die Schadensbeseitigung und den Wiederaufbau finanzieren.

Die Klärung der finanziellen Ansprüche sowie die Beschaffung und Verwendung von Mitteln sind mit erheblichem bürokratischem Aufwand verbunden. Zudem können Betroffene öffentliche Fördermittel ausschließlich online beantragen. Darüber hinaus erfordert die Inanspruchnahme von öffentlichen Förderungen – wie auch etwaigen Versicherungsleistungen – die Einholung von Gutachten zur Schadensbeurteilung, mindestens drei verschiedene Kostenvoranschläge von Handwerksbetrieben und im Fall von öffentlichen Fördermitteln auch einen Nachweis, wie und für welchen Zweck die Mittel verwendet wurden. Nicht nur für benachteiligte Bevölkerungsgruppen kann dies schnell zu einer besonderen Heraus- bzw. Überforderung führen.

Um den Betroffenen bei der Bewältigung des Wiederaufbaus zu helfen, entwickelten sich in Hagen zwei Arten von Beratungsstrukturen. Einerseits berät seit dem Flutereignis eine kommunale Fachberatung Hochwasser Bürger:innen und Unternehmen kostenfrei zu geeigneten Hochwasserschutz- und Anpassungsmaßnahmen. Bisher führte diese stadtweit über 500 Einzel-Beratungsgespräche durch (Stand: Februar 2025). Neben dem kommunalen Beratungsangebot besteht auch die Möglichkeit der kostenpflichtigen Beratung durch private Unternehmen.

Andererseits richtete – neben einem stadtweit agierenden Hochwasserhilfe-Verbund (bestehend aus vier gemeinnützigen Organisationen und dem kommunalen Fachbereich Jugend und Soziales) – eine lokal tätige gemeinnützige Organisation im Hagener Quartier

9 Aufstrebende Gruppen können informelle Ansammlungen von Personen oder formelle Organisationen sein, darunter Helfende, Anwohnerinnen, außerörtliche Freiwillige, örtliche Unternehmen, gemeinnützige Organisationen, Kirchengemeinden und Regierungsbehörden. Sie übernehmen im Vergleich zu ruhigen Zeiten neue Rollen oder Funktionen oder gehen ungewohnte Beziehungen ein (Drabek und McEntire 2003; Stallings und Quarantelli 1985; Perry 2007).

ein eigenes Hochwasserhilfe-Büro ein, um den Betroffenen kostenfrei bei der Beantragung öffentlicher Fördermittel zu helfen. Das Büro hat bis Juli 2023 bereits über 100 Betroffene bei der Beantragung öffentlicher Mittel beraten und auch fortlaufend unterstützt[10]. Damit unterstützen sie deutlich mehr als ein Drittel der Personen, die bis dato erfolgreich öffentliche Fördermittel beantragten. Bis zum Januar 2025 wurden im Hagener Fallstudienquartier bereits 308 Anträge auf öffentliche Fördermittel mit einer Gesamtsumme von 6,68 Mio. € bewilligt (MHKBD NRW 2025).

Neben der Hochwasserhilfe durch die lokale Organisation haben sich im Hagener Quartier auch Teile der Bewohner:innen in kleinen Gruppen organisiert, um den Wiederaufbau zu unterstützen. So wurde das Vereinsheim des lokalen Fußballvereins durch Vereinsmitglieder und deren Angehörige mit Geld und handwerklichen Fähigkeiten wiederaufgebaut. Darüber hinaus spielen informelle Nachbarschaftsnetzwerke im Quartier eine entscheidende Rolle bei der Weitergabe von Informationen und der situativen Unterstützung während des Wiederaufbaus, etwa bei alltäglichen Aufgaben, um Ressourcen für die Organisation der Wiederaufbauarbeiten freizusetzen.

Die Stadt Stolberg reagierte vielfältig auf den Beratungsbedarf der betroffenen Quartiersbewohner:innen. So wurde u. a. das Aufgabenportfolio eines Quartiersarchitekten erweitert. Dieser war kurz vor dem Flutereignis mit der Umsetzung des Hof- und Fassadenprogramms beauftragt worden. Mit dem Programm sollten private Gebäudeeigentümer:innen bei der Aufwertung ihrer Fassaden und Freiflächen finanziell unterstützt werden. Nach dem Flutereignis übernahm der Architekt mit seinem Team auch die Beratung zu Gebäudeschäden und zum Einsatz von Fördermitteln auf Grundlage der Wiederaufbaurichtlinie. Ziel war, in Einzelberatungen mit Gesprächsangeboten in einem temporär aufgestellten Container sowie mit von der Stadt organisierten Informationsveranstaltungen möglichst viele Menschen in ihrer Notsituation bedarfsgerecht zu unterstützen. Zusätzlich begleitete der Quartiersarchitekt die Infomobile des Wasserverbands Eifel-Rur (WVER) durch das betroffene Quartier. Die Infomobile wurden von den Betroffenen mit über 100 Beratungsgesprächen pro Einsatz stark frequentiert. Hinzu kamen mehrere hundert Gespräche im temporär aufgestellten Container oder auf der Straße. An den insgesamt fünf Informationsveranstaltungen nahmen jeweils bis zu 150 Personen teil. In Anspruch genommen haben die Beratungsangebote zunächst vor allem Eigentümer:innen, die ihre Immobilien für eigene Zwecke nutzen. An zweiter Stelle und zeitlich etwas verzögert suchten Eigentümer:innen von Mehrfamilienhäusern das Gespräch.

18.7.2 Zugang zu notwendigen Wiederaufbau-Ressourcen in benachteiligten Quartieren

Im Wiederaufbau erweisen sich einkommensärmere Haushalte als wesentlich vulnerabler, was auch zahlreiche andere internationale Studien bereits belegten (siehe z. B. Bolin und Bolton 1986; Bolin und Stanford 1998; Hewitt 2014; Tierney 1988). Diese Haushalte sind alternativlos auf externe Geldmittel angewiesen. Die staatlichen Wiederaufbauhilfen, die nach dem Flutereignis im Juli 2021 speziell für den physischen Wiederaufbau bereitgestellt wurden, berücksichtigen soziale Verwundbarkeiten,

18

10 Der stadtweit agierende Hochwasserhilfe-Verbund hat bis Juli 2023 642 Haushalte beraten und 526 fortlaufend unterstützt (Stadt Hagen 11.07.2023). Damit tragen beide Hochwasserhilfen zu etwa der Hälfte der 1349 bis dahin gestellten privaten Förderanträge bei.

benachteiligende Standortbedingungen und lokale Resilienzkapazitäten jedoch nur unzureichend, da deren Adressierung von Einzelpersonen auch den Zugang von den jeweiligen Einzelpersonen und ihren Fähigkeiten bedingt. Daraus ergeben sich Herausforderungen in den Bereichen: (1) Informationen über verfügbare Hilfen und Ressourcen, (2) Beschaffung von Ressourcen und (3) Verteilung von Ressourcen zwischen verschiedenen Quartieren.

Ein zentrales Problem ist zunächst die Verbreitung von Wissen über verfügbare Ressourcen. Besonders sozial isolierte Bewohner:innen und Gruppen sind gefährdet, da sie weniger in Informationsnetzwerke eingebunden sind. Im Hagener Quartier erwiesen sich die traditionellen Massenmedien zudem als unzureichend. Viele Betroffene erfuhren erst spät durch Familie, Freund:innen, Nachbarschaftsnetzwerke oder das örtliche Hochwasserhilfe-Büro von den öffentlichen Fördermitteln. Auch zwei Jahre nach der Einrichtung des Wiederaufbaufonds war dies einigen Betroffenen im Hagener Quartier noch immer nicht bekannt oder ihnen war nicht bewusst, dass sie anspruchsberechtigt sind.

Insbesondere mit Blick auf den Zugang zu Informationen über verfügbare Hilfen und Ressourcen übernimmt auch der öffentliche Raum eine wichtige Funktion, da er für alle zugänglich und damit ein geeigneter Ort für Begegnungen ist (Flade 2020). Gerade wenn die Wohnverhältnisse beengt – oder wie nach der Flutkatastrophe ggf. nur eingeschränkt nutzbar – sind, die technischen Zugangsmöglichkeiten zu Informationsquellen im Internet fehlen oder Sprachbarrieren bestehen, ermöglicht der öffentliche Raum unkompliziert den Kontakt und den Wissensaustausch mit anderen Betroffen. Treffpunkte im öffentlichen Raum werden jedoch nur dann angenommen, wenn sie gut erreichbar, kleinräumig und unverwechselbar sind. Vor allem aber müssen sie mit Sitzgelegenheiten ausgestattet sein (Flade 2020), damit sie zum Verweilen einladen.

Im Hagener Quartier mangelte es bereits vor der Flut an ausreichenden qualitätvollen öffentlichen Begegnungsräumen, während im Stolberger Quartier der öffentliche Raum durch die Flutkatastrophe 2021 stark zerstört wurde, was den Zugang zu Informationen über soziale Kontakte erschwert. Die Umsetzung von Maßnahmen, die zur Wiederherstellung des öffentlichen Raums und seiner Nutzbarkeit notwendig sind, erfordern Zeit. Um den Aufenthalt an öffentlichen Plätzen in der Innenstadt wieder zu ermöglichen, hat die Stadt Stolberg im Zuge des Wiederaufbaus schnellstmöglich Sitz- und Spielmöglichkeiten an einem zentralen Platz in der Innenstadt errichten lassen. Auch die Maßnahmen zur schnellen Wiederherstellung und Nutzbarkeit des Freiraums rund um den Bastinsweiher in Unterstolberg haben dazu beigetragen, den Aufenthalt im öffentlichen Raum schnell wieder möglich zu machen.

Auch die Beschaffung und Inanspruchnahme von Ressourcen offenbart soziale Verwundbarkeiten. Trotz eines Leitfadens des Staates haben viele Betroffene in beiden Fallstudienquartieren aus verschiedenen Gründen Schwierigkeiten mit der Beantragung öffentlicher Mittel, sei es aufgrund fehlender Vertrautheit mit der bürokratischen Sprache oder mangelnder technischer Fähigkeiten. Ältere Menschen haben oft keinen erforderlichen PC-Zugang oder E-Mail-Adressen, um die Fördermittelanträge online einzureichen. Auch Betroffene mit eingeschränkten Deutschkenntnissen sind benachteiligt. Folglich kann der bürokratische Antragsprozess entmutigend wirken und Betroffene davon abhalten, die ihnen zustehenden Mittel zu beantragen und Wiederaufbaumaßnahmen einzuleiten. Ähnliche Herausforderungen wurden auch in anderen internationalen Studien nach Katastrophen dokumentiert (siehe z. B. Fothergill 2004; Rovai 1994). Im Gegensatz zu wohlhabenderen Betroffenen waren ärmere Betroffene weniger begabt in der Handhabung des bürokratischen Systems, sodass sie erst gar keine Mittel beantragten (Fothergill 2004; Rovai 1994).

Die Verteilung von Ressourcen zwischen verschiedenen Quartieren ist ebenfalls entscheidend für den Wiederaufbau in benachteiligten Quartieren. Die Ressourcenverteilung ist eng verbunden mit der institutionellen, politischen und medialen Aufmerksamkeit, die benachteiligte Quartiere erhalten. Im Hagener Quartier haben viele Bürger:innen das Gefühl, dass ihnen bei staatlichen Unterstützungsbemühungen weniger Aufmerksamkeit oder Priorität zuteilwird als anderen Quartieren, aber auch als anderen betroffenen Städten bzw. Regionen. Dies gilt auch für die allgemeine öffentliche Unterstützung jenseits des Wiederaufbaus.

18.7.3 Verschärfung von Ungleichheiten im Wiederaufbau

Internationale Untersuchungen zeigen, dass der Wiederaufbau von benachteiligten Quartieren nach Naturkatastrophen nicht gleichmäßig im Vergleich zur insgesamt betroffenen Region verläuft, sondern asymmetrisch. Ein Grund dafür ist u. a., dass Pläne und Programme des Wiederaufbaus nur bedingt auf diejenigen städtebaulichen und sozialen Rahmenbedingungen sowie spezifischen Charakteristiken eingehen, die sich in der Katastrophe als verwundbar herausstellten. Auch die Orientierung des Wiederaufbaus am Status quo ante trägt weder dazu bei, die Vulnerabilität betroffener Bevölkerungsgruppen zu verringern, noch das wünschenswerte Leitbild des *build back better* umzusetzen. Waren Quartiere vor einer Katastrophe benachteiligt und vulnerabel, sind sie es während und nach dem Wiederaufbau entsprechend weiterhin. Zu diesem Schluss kommt Mark Kammerbauer (2014) aufgrund seiner Forschungen zu den Auswirkungen des Orkans Katrina in New Orleans.

Zeit ist bei Überschwemmungsschäden von entscheidender Bedeutung und beeinflusst sowohl die Reparaturkosten als auch zusätzliche Ausgaben wie vorübergehende Unterbringung, Transport oder den Verlust wirtschaftlicher Aktivitäten. Darüber hinaus ist der Zeitfaktor ausschlaggebend für die Verfügbarkeit knapper Ressourcen wie Bauleitungen, Handwerker:innen und Materialien. Die Asymmetrie im Wiederaufbauprozess zeigt sich nach der Flutkatastrophe 2021 in NRW jedoch bereits jetzt an den unterschiedlichen Geschwindigkeiten, mit denen sich betroffene Quartiere allein mit Blick auf die Stadtstrukturen und das Stadtbild von den Folgeschäden der Flut erholen. Denn der private Wiederaufbau schreitet kontinuierlich ungleichmäßig voran: einige betroffene Bewohner:innen haben den Wiederaufbau bereits abgeschlossen, während andere die berechtigten Ressourcen noch nicht in Anspruch genommen oder noch unter den Folgen der Katastrophe zu leiden haben.

Der Zeitfaktor spielt außerdem eine wichtige Rolle für die mentale Gesundheit der Betroffenen (► Kap. 9). Verzögert sich der Wiederaufbau von Gebäuden und Infrastrukturen, entfällt auch der positive Einfluss, den die gebaute Umwelt auf den Menschen haben kann. Im Umkehrschluss verstärken sich die physischen Stressoren und damit die Stressbelastung der Betroffenen insgesamt. Es kann also angenommen werden, dass je länger der physische Wiederaufbau dauert, desto höher ist die gesundheitliche Belastung der ohnehin vorbelasteten Menschen, die in benachteiligten Quartieren leben.

18

> „Stress wirkt nach, körperliche und emotionale Stresssymptome klingen nicht sofort ab, es braucht Zeit, bis der Mensch wieder sein normales Gleichgewicht gefunden hat. Nach ho-

> hen Stressphasen kann das sogar mehrere Wochen dauern. Sind die Erholungsphasen nicht mehr ausreichend, geraten sogar normale Alltagsbelastungen zum Stressfaktor … Stress schädigt auf Dauer die Gesundheit. Aus Befindlichkeitsstörungen werden Erkrankungen." (Metzler-Amlacher 2024)

Herausforderungen und Verzögerungen beim physischen Wiederaufbau und beim Zugang zu notwendigen Ressourcen belasten schutzbedürftigere Bürger:innen in benachteiligten Quartieren demnach unverhältnismäßig stark. Auf lange Sicht führt dies zu größerer sozialer Ungleichheit innerhalb und zwischen den Quartieren einer Stadt.

18.8 Schlussfolgerungen und Empfehlungen

Die Fallstudien zeigen, dass staatliche Wiederaufbausysteme – insbesondere auf übergeordneten Verwaltungs- und Entscheidungsebenen – oftmals nicht ausreichend auf den sozialen Kontext und die Vulnerabilität benachteiligter Quartiere eingehen. Selbstverständlich ist der Wiederaufbaufonds eine wesentliche, wenn nicht die entscheidende Grundvoraussetzung für den dortigen Wiederaufbau. Allerdings obliegt der Zugang zu diesen Mitteln den Einzelpersonen und deren Fähigkeiten. Indes zeigt diese Studie auch, dass viele Betroffene nicht in der Lage waren, ohne Unterstützung auf die notwendigen Ressourcen zuzugreifen. Dies führt zu zwei wesentlichen Erkenntnissen: einerseits steigt dadurch die Gefahr einer Verschärfung von sozialen Ungleichheiten und andererseits wird die hohe Bedeutung lokaler Unterstützungsstrukturen, wie etwa durch soziale Infrastrukturen (s. ▶ Kap. 17), hervorgehoben.

Auf kommunaler Ebene bestehen verschiedene Möglichkeiten, benachteiligte Quartiere im Wiederaufbauprozess zu unterstützen und der Gefahr sozialer Verschärfung entgegenzusteuern. Aufgrund von Mobilitätsarmut und/oder Herausforderungen z. B. in der Kinderbetreuung ist jede Form von aufsuchender Beratung und Begleitung nach einem Katastrophenereignis zentral. Dies untermauert auch die Studie von Dash et al. (1997) nach dem Hurrikan Andrew. Mobile Beratungsteams können ebenfalls einen direkten Kontakt zur Bevölkerung herstellen. Sofern vor Ort bereits etabliert, können auch Quartiers- oder Sanierungsmanager:innen eine Anlaufstelle sein. Sie müssen aber gegebenenfalls für die besonderen Rahmenbedingungen sensibilisiert werden und einige Zeit in eine Beziehungsarbeit investieren. Vor Ort sollten gegebenenfalls Dolmetscher:innen eingebunden werden und schriftliche Informationen sollten mehrsprachig und in leichter Sprache aufbereitet sein.

Auch wenn die soziale Resilienz nicht explizit Gegenstand der Wiederaufbauhilfen sind, trägt doch jede Investition in den physischen Wiederaufbau – wenngleich dies allein nicht ausreicht – indirekt auch wesentlich zur Stabilisierung der Nachbarschaften bei und kommt somit den Bewohner:innen benachteiligter Quartiere zugute. Jedoch sind ergänzende Maßnahmen erforderlich, um eine umfassende soziale Resilienz zu gewährleisten. Aufgrund meist beengter Wohnverhältnisse übernimmt der öffentliche Raum in benachteiligten Quartieren immer schon eine wichtige Funktion für Freizeit, Erholung und Sozialkontakt. Nach einem Flutereignis wird der öffentliche Raum noch einmal wichtiger, damit Nachbarschaftskontakte bestehen bleiben und Hilfsangebote die Bevölkerung erreichen. Beliebte und bekannte Treffpunkte im Quartier sollten daher kurzfristig so wiederhergestellt werden, dass ein Aufenthalt möglich ist. Mobile Elemente leisten ebenso wie dauerhaftes Stadtmobiliar schnell einen sichtbaren und wichtigen Beitrag. Zusätzlich zur Wiederherstellung der Aufenthaltsqualitäten im öffentlichen Raum ist es für die Bewohner:innen eines benachteiligten Quartiers zentral, dass die sozialen Einrichtungen und Begegnungsstätten wiederaufgebaut werden bzw. wiedereröffnen. Gebäude und Infrastrukturen zu reparieren und instandzusetzen ist ohnehin wichtig, um das Quartier schnellstmöglich wieder funktionsfähig zu machen. Mit entsprechenden Maßnahmen signalisiert die Kommune den Bewohner:innen nicht zuletzt auch die Hoffnung auf eine schnelle Verbesserung der Rahmenbedingungen im Quartier.

Es liegt nahe, den Wiederaufbauprozess in benachteiligten Quartieren gezielt für die Aufwertung der Wohnsituation und des Wohnumfeldes im Sinne von *Build Back Better* zu nutzen. Angesichts der Rechtslage, der vielen städtebaulichen Missstände und sozialen Unterstützungsbedarfe in benachteiligten Quartieren kann dies nur durch eine Bündelung unterschiedlicher Fördermittel gelingen, die auf viele Jahre prioritär in den betroffenen Quartieren zum Einsatz kommen sollten.

Aufgrund der zahlreichen Vorbelastungen ist anzunehmen, dass die zusätzlichen Belastungen durch die Flutkatastrophe 2021 die Bewohner:innen benachteiligter Quartiere härter getroffen haben als in anderen Quartieren. Welche Auswirkungen Naturkatastrophen im Allgemeinen und das Flutereignis 2021 im Besonderen auf benachteiligte Quartiere haben, ist jedoch bislang wenig erforscht. Auch besteht eine Forschungslücke zu der Frage, welche Bedeutung ein Fluss als Erholungsraum für Betroffene und/oder vorbelastete Personen nach einer Flutkatastrophe überhaupt noch haben kann.

Zusammenfassung

- Extremwetterbedingte Katastrophen sind immer auch im Zusammenhang mit den gesellschaftlich geschaffenen Bedingungen von Vulnerabilität und Resilienz zu verstehen.
- Während einige Quartiere Katastrophen schnell bewältigen, benötigen andere deutlich mehr Zeit für den Wiederaufbau.
- Zwei Studien in benachteiligten Quartieren der Städte Hagen und Stolberg, die von der Flutkatastrophe im Juli 2021 betroffen waren, zeigen, dass größere, übergeordnete Wiederaufbausysteme oft nicht ausreichend auf die sozialen Bedingungen benachteiligter Quartiere eingehen.
- Der Zugang zu Wiederaufbauhilfen hängt stark von den Fähigkeiten der Quartiersbewohner:innen und der institutionellen Einbindung ab, wodurch soziale Ungleichheiten verstärkt werden können. Mobile Beratungsteams und leicht zugängliche Informationen können helfen, diesen Ungleichheiten entgegenzuwirken.
- Investitionen in den öffentlichen Raum und soziale Infrastrukturen sind essenziell, um die Resilienz der Quartiere zu stärken.

Danksagung Die Autorinnen von ▶ Kap. 18 bedanken sich beim Bundesministerium für Bildung und Forschung (BMBF) und beim DLR Projektträger für die Unterstützung und Förderung des KAHR-Projekts. Diese Forschung wurde vom BMBF unter dem Förderkennzeichen 01LR2102A-M finanziert. Die Autorinnen danken allen Interview-Partner:innen für ihr Engagement und ihre wertvollen Beiträge.

Literatur

Adli M (2017) Stress and the City. Warum Städte uns krank machen. Und warum sie trotzdem gut für uns sind. Bertelsmann, München

Aldrich DP (2012) Building resilience. Social capital in post-disaster recovery. University of Chicago Press, Chicago

Auf der Heide E (1989) Disaster response. Principles of preparation and coordination. Mosby, St. Louis

Barton A (1969) Communities in disaster. A sociological analysis of collective stress situations. Doubleday, Garden City

Baum D (2007) Sozial benachteiligte Quartiere: Der Zusammenhang von räumlicher Segregation und sozialer Exklusion am Beispiel städtischer Problemquartiere. In: Baum D (Hrsg) Die Stadt in der Sozialen Arbeit. Ein Handbuch für soziale und planende Berufe. VS, Wiesbaden, S 136–155

Birkmann J (2007) Risk and vulnerability indicators at different scales. Applicability, usefulness and policy implications. Environ Hazards 70(1):20–31. https://doi.org/10.1016/J.ENVHAZ.2007.04.002

Bohle H-G (2002) Vulnerability. Editorial to the special issue. Geogr Helv 57(1):2–4. https://doi.org/10.5194/gh-57-2-2002

Bohle H-G, Etzold B, Keck M (2009) Resilience as agency. IHDP-Update 2:8–13

Böhme C (2022) Klimaanpassung im Quartier: Herausforderungen, Maßnahmen, Akteure online, Gesundheit und Quartier – Praxis im Gespräch: Klimaanpassung im Quartier. https://hage.de/veranstaltungen/veranstaltungsdokumentationen/09032022-gesundheit-und-quartier-praxis-im-gespraech-klimaanpassung-im-quar-tier/index.php?rex_media_type=hage_download&rex_media_file=praesentation_frau_boehme_klimaanpassung_im_quartier.pdf (Erstellt: 9. März 2022). Zugegriffen: 26. Juli 2024

Bolin R, Bolton P (1986) Race, religion, and ethnicity in disaster recovery. FMHI Publications

Bolin R, Stanford L (1998) The Northridge Earthquake. Routledge

Bonß W (2015) Karriere und sozialwissenschaftliche Potenziale des Resilienzbegriffs. In: Endreß M, Maurer A (Hrsg) Resilienz im Sozialen. Theoretische und empirische Analysen Gemeinsame Tagung der Sektionen „Politische Soziologie“ und „Wirtschaftssoziologie“ der Deutschen Gesellschaft für Soziologie. Springer, Wiesbaden, S 15–32

Browning CR, Wallace D, Feinberg SL, Cagney KA (2006) Neighborhood social processes, physical conditions, and disaster-related mortality: the case of the 1995 chicago heat wave. Am Sociol Rev 71:661–678

Cho A (2014) Post-tsunami recovery and reconstruction: governance issues and implications of the Great East Japan Earthquake. Disasters 38(S2):157–178. https://doi.org/10.1111/disa.12068

Cutter SL (2006) The geography of social vulnerability. Race, class, and catastrophe. http://understandingkatrina.ssrc.org/Cutter/. Zugegriffen: 10. Nov. 2017

Cutter SL, Finch C (2008) Temporal and spatial changes in social vulnerability to natural hazards. Proc Natl Acad Sci U S A 105(7):2301–2306. https://doi.org/10.1073/pnas.0710375105

Dash N, Peacock WG, Morrow BH (1997) And the Poor Get Poorer: A Neglected Black Community. In: Peacock WG, Morrow BH, Gladwin H (Hrsg) Hurricane Andrew. Ethnicity, gender, and the sociology of disasters. Routledge, Taylor & Francis Group, London, New York, S 206–225

Drabek TE, McEntire DA (2003) Emergent phenomena and the sociology of disaster: lessons, trends and opportunities from the research literature. Disaster Prev Manag 12(2):97–112. https://doi.org/10.1108/09653560310474214

DSK – Deutsche Stadt- und Grundstücksentwicklungsgesellschaft (2016) Kupferstadt Stolberg – Durchführung von vorbereitenden Untersuchungen gemäß § 141 BauGB i. V. mit §§ 136–139 BauGB zwecks Festlegung eines Sanierungsgebietes – „Talachse Innenstadt“. Abschlussbericht und Begründung zur Satzung

Dubet F, Lapeyronnie D (1994) Im Aus der Vorstädte. Der Zerfall der demokratischen Gesellschaft. Klett-Cotta, Stuttgart

Farwick A (2007) Soziale Segregation in den Städten – Von der gespaltenen Gesellschaft zur gespaltenen Stadt. In: Baum D (Hrsg) Die Stadt in der Sozialen Arbeit. Ein Handbuch für soziale und planende Berufe. VS, Wiesbaden, S 111–122

Fischer HW (1998) Response to disaster. Fact versus fiction & its perpetuation: the sociology of disaster, 2. Aufl. University Press of America, Lanham

Flade A (2020) Kompendium der Architekturpsychologie. Zur Gestaltung Gebauter Umwelten. Springer, Wiesbaden

Fothergill A (2004) Heads above water. Gender, class, and family in the Grand Forks flood. State University of New York Press, Albany

Fothergill A, Peek LA (2004) Poverty and disasters in the United States: a review of recent sociological findings. Nat Hazards 32(1):89–110. https://doi.org/10.1023/B:NHAZ.0000026792.76181.d9

Galster GC (2011) The mechanism(s) of neighbourhood effects: theory, evidence, and policy implications. In: van Ham M, Manley D, Bailey N (Hrsg) Neighbourhood effects research. New perspectives. Springer, Dordrecht, S 23–56

Gilligan C (2018) Community responses to disaster. Northern Ireland 1969 as a case study. In: Cnaan RA, Milofsky C (Hrsg) Handbook of Community Movements and Local Organizations in the 21st Century. Springer, Cham, S 311–328

Hamedinger A (2002) Sozial-räumliche Polarisierung in Städten: ist das „Quartiersmanagement" eine geeignete stadtplanerische Antwort auf diese Herausforderung? SWS-Rundschau 42(1):122–138

Häußermann H, Siebel W (1987) Neue Urbanität. Suhrkamp, Berlin, Frankfurt am Main

Häußermann H, Siebel W (1995) Dienstleistungsgesellschaften. Suhrkamp, Frankfurt am Main

Heidbrink L, Leggewie C, Welzer H (2010) Von der Natur- zur sozialen Katastrophe. In: Bieber C, Drechsel B, Lang A-K (Hrsg) Kultur im Konflikt. transcript, S 429–432

Hewitt K (2014) Regions of risk. A geographical introduction to disasters. Taylor and Francis, Hoboken

Hornberg C, Maschke J (2017) Soziale Vulnerabilität im Kontext von Umwelt, Gesundheit und sozialer Lage. Umw Mensch – Informationsd 2:43–49

IPCC – Intergovernmental Panel on Climate Change (2023) Climate Change 2022: Impacts, Adaptation and Vulnerability. Contribution of Working Group II to the Sixth Assessment Report of the Intergovernmental Panel on Climate Change. Cambridge University Press, Cambridge, New York

IRI – Institut für Raumforschung & Immobilienwirtschaft (2016) Wohnungsmarktstudie Hagen. Abschlussbericht

IT.NRW – Information und Technik Nordrhein-Westfalen (2023a) Kommunalprofil Hagen. Stadt. Düsseldorf

IT.NRW – Information und Technik Nordrhein-Westfalen (2023b) Kommunalprofil Stolberg (Rhld.). Stadt. Düsseldorf

IWAP – Institut für Wohn- und Architekturpsychologie (2024) Teil 3 – Erholung, Regeneration und Gesundheit [10 teilige Serie – Wie Architektur wirkt]. Graz

Jha AK, Duyne Barenstein JE, Phelps PM, Pittet D, Sena S (2010) Safer homes, stronger communities. A handbook for reconstructing after natural disasters. The World Bank

Kammerbauer M (2014) Asymmetrischer Wiederaufbau in Städten nach Katastrophen. Das Lower Ninth Ward in New Orleans nach Orkan Katrina. Raumforsch Raumordn | Spatial Res Plan. https://doi.org/10.1007/s13147-014-0309-4

Keck M, Sakdapolrak P (2013) What is social resilience? Lessons learned and ways forward. Erdkunde 67(1):5–19. https://doi.org/10.3112/erdkunde.2013.01.02

Kuhlicke C, Pößneck J, Rink D (2024) Wie halten Sie es mit der Resilienz? Kommunale Perspektiven auf ein aktuelles Stadtkonzept. vhw-Schriftenreihe, Bd 47. vhw-Bundesverb. für Wohnen und Stadtentwicklung, Berlin

Lewis J (1999) Development in disaster-prone places. Studies of vulnerability. Intermediate Technology Publ, London

Lorenz DF (2013) The diversity of resilience: contributions from a social science perspective. Nat Hazards 67(1):7–24. https://doi.org/10.1007/s11069-010-9654-y

Medd W, Marvin S (2005) From the politics of urgency to the governance of preparedness: a research agenda on urban vulnerability. J Conting Crisis Manag 13(2):44–49. https://doi.org/10.1111/j.1468-5973.2005.00455.x

Metzler-Amlacher K (Hrsg) (2024) Stress. https://www.psychologin.co.at/stress.html. Zugegriffen: 26. Juli 2024

MHKBD NRW – Ministerium für Heimat, Kommunales, Bau und Digitalisierung des Landes Nordrhein-Westfalen (2025) Daten zu Anträgen auf öffentliche Wiederaufbaumittel in Hagen-Eckesey zum 27. Januar 2025 (Datenset)

MHKBG NRW – Ministerium für Heimat, Kommunales, Bau und Gleichstellung des Landes Nordrhein-Westfalen (2022) Aufbauhilfen für Privathaushalte und Unternehmen der Wohnungswirtschaft. https://www.mhkbg.nrw/aufbauhilfen-fuer-privathaushalte-und-unternehmen-der-wohnungswirtschaft. Zugegriffen: 7. Jan. 2022

Morrow BH (2008) Community resilience. A social justice perspective. Florida International University. CARRI Research Report 4. https://doi.org/10.13140/RG.2.1.1278.9604

O'Brien K, Hayward B, Berkes F (2009) Rethinking social contracts: building resilience in a changing climate. Ecol Soc. https://doi.org/10.5751/ES-03027-140212

Obrist B, Pfeiffer C, Henley R (2010) Multi-layered social resilience. Prog Dev Stud 10(4):283–293. https://doi.org/10.1177/146499340901000402

Park RE, Burgess EW, McKenzie RD (1925) The city. Suggestions for investigation of human behavior in the urban environment. University of Chicago Press, Chicago

Perry RW (2007) What is a disaster? In: Rodríguez H, Quarantelli EL, Dynes RR (Hrsg) Handbook of disaster research. Springer, New York

Roloff S (2019) Städter verarbeiten Stress anders. Interview mit PD Dr. med. Mazda Adli. G + G Gesundh Ges 2019(3) (https://archiv.gg-digital.de/2019/03/staedter-verarbeiten-stress-anders/index.html (abgerufen am 18.09.2024))

Rovai E (1994) The social geography of disaster recovery: differential community response to the north coast earthquakes. Yearb Assoc Pac Coast Geogr 56(1):49–74. https://doi.org/10.1353/pcg.1994.0002

Schott D (2012) Resilienz oder Niedergang? Zur Bedeutung von Naturkatastrophen für Städte in der Neuzeit. In: Wagner U (Hrsg) Stadt- und Stadtverderben 47. Arbeitstagung, Würzburg, 21.–23.11.2008, S 11–32

SINUS Markt- und Sozialforschung (2024) Sinus-Migrantenmilieus. https://www.sinus-institut.de/sinus-milieus/migrantenmilieus. Zugegriffen: 25. Juni 2024

Stadt Hagen (2018) Integriertes Klimaanpassungskonzept Stadt Hagen. https://www.hagen.de/FIRSTspiritWeb/hagen/media/files/fb/fb_69/klima/klimaanpassung/internet_broschklimaanpassungskonz_kpl.pdf. Zugegriffen: 4. Apr. 2022

Stadt Hagen (2020a) HAGENplant 2035. Baustein D: Gesamtstrategie Maßnahmenübersicht

Stadt Hagen (2020b) HAGENplant 2035. Integriertes Stadtentwicklungskonzept

Stadt Hagen (2020c) Bevölkerungsstatistiken des Fachbereichs Öffentliche Sicherheit und Ordnung, Abteilung Statistik und Wahlen 31.12.2020

Stadt Hagen (2023) Zwei Jahre nach dem Hochwasser: Großteil der Schäden behoben, Pressemitteilung vom 11.07.2023. https://www.hagen.de/FIRSTspiritWeb/hagen/de_1/hagen_de/01/0101/010101/PM_509505.jsp?cmcall=true&. Zugegriffen: 18. Aug. 2023

Stadt Stolberg (2023) Wir bauen unsere Stadt wieder auf. https://www.stolberg.de/Verwaltung-und-Politik/Veroeffentlichungen/Aktuelle-Meldungen.htm/Aktuelles/Wir-bauen-unsere-Stadt-wieder-auf.html?#substart. Zugegriffen: 17. Mai 2024

Städteregion Aachen (2022) Sozialberichterstattung StädteRegion Aachen. Kommunalprofil 2022 Stadt Stolberg

Stallings RA, Quarantelli EL (1985) Emergent citizen groups and emergency management. Public Admin Rev 45:93–100

Tierney K (1988) The Whittier Narrows, California Earthquake of October 1, 1987 – Social Aspects. Earthq Spectra 4(1):11–23. https://doi.org/10.1193/1.1585460

Tierney KJ (2007) From the margins to the mainstream? Disaster research at the crossroads. Annu Rev Sociol 33(1):503–525. https://doi.org/10.1146/annurev.soc.33.040406.131743

UNDRR – United Nations Office for Disaster Risk Reduction (2015) Sendai Framework for Disaster Risk Reduction 2015–2030. United Nations, Geneva

Private Hochwasservorsorge in Nordrhein-Westfalen und Rheinland-Pfalz nach der Juli-Flut 2021: Was haben Haushalte umgesetzt und welche Bedarfe bestehen

Helene Meyer, Lisa Dillenardt, Lina Fitz, Guilherme Samprogna Mohor und Georg Johann

J. Birkmann et al. (Hrsg.), *Die Flutkatastrophe 2021*, Sustainable Development Goals (SDG) – Umsetzung in Praxis, Lehre und Entscheidungsprozessen,
https://doi.org/10.1007/978-3-658-48688-4_19

19.1 Einleitung

Das Hochwasser im Juli 2021 in Rheinland-Pfalz (RLP) und Nordrhein-Westfalen (NRW) war das extremste Niederschlagsereignis in Deutschland, gemessen am skalenübergreifenden Extremitätsindex (xWEI) (Lengfeld et al. 2023), mit der höchsten Zahl an Todesopfern seit 1962 (Thieken et al. 2023) und der höchsten Wiederaufbauhilfe von 30 Mrd. EUR (BMI 2022). Nichtsdestotrotz haben seitdem weitere Ereignisse zu Schäden in Milliardenhöhe an Wohngebäuden geführt. Auch wenn die jüngsten Ereignisse nicht von derselben Intensität wie jene vom Juli 2021 waren, verdeutlichen sie erneut, dass Starkregen- und Hochwasserereignisse hohe Schäden verursachen und immer wieder eine erhebliche Herausforderung für die Gesellschaft darstellen. Im Rahmen des integrierten Starkregen- und Hochwasserrisikomanagements hat sich die Erkenntnis durchgesetzt, dass der Umgang mit dem Starkregen- und Hochwasserrisiko eine gesamtgesellschaftliche Aufgabe darstellt, die nicht allein von der öffentlichen Hand getragen werden kann (vgl. Walker et al. 2014). Obgleich die übergeordneten Schutzmaßnahmen der öffentlichen Hand dazu beitragen, das Ausmaß der Schäden zu reduzieren, erweisen sich diese allein als unzureichend, da selbst großzügig dimensionierte Hochwasserschutzanlagen angesichts des fortschreitenden Klimawandels mit seinen zunehmenden Starkregenereignissen (Rosenzweig et al. 2018; Tradowsky et al. 2023) an ihre Grenzen stoßen können. Daher sollten Gebäudeeigentümerinnen und Gebäudeeigentümer Verantwortung im Hochwasserrisikomanagement übernehmen (vgl. Snel et al. 2020; Walker et al. 2014) und umfassende und klimaangepasste Vorsorgemaßnahmen ergreifen. Diese Verpflichtung ist in § 5 des Wasserhaushaltsgesetzes (WHG) verankert.

Durch die Umsetzung vorbeugender Maßnahmen können Gebäudeeigentümerinnen und Gebäudeeigentümer dazu beitragen, Schäden durch Überflutungen zu reduzieren und ihre eigene und damit die gesellschaftliche Resilienz zu erhöhen (vgl. Begg et al. 2017; Bubeck et al. 2012; Kreibich et al. 2011). Diese Maßnahmen müssen privat implementiert und in der Regel finanziert werden. Hierfür gibt es jedoch auch Förderprogramme. So unterstützt beispielsweise die Förderbank Sachsen sowohl die Erstellung von Gutachten als auch die darauf folgende Implementierung von privaten Vorsorgemaßnahmen in Sachsen (SAB, n.y.).

Es existiert eine Vielzahl von Vorsorgemaßnahmen, die sich hinsichtlich ihrer Wirkweise in sechs Gruppen unterteilen lassen, siehe ▫ Abb. 19.1 (DEFRA 2008; Hudson et al. 2014; Kreibich et al. 2011; Lamond et al. 2018; Poussin et al. 2015). Wie und ob Objektschutz für Gebäude möglich ist, hängt von der Beschaffenheit des Gebäudes und der Exposition gegenüber der Überflutungsgefahr ab.

Durch Maßnahmen der Gruppe „Ausweichen“ werden gefährdete Objekte aus der Gefahrenzone entfernt. Dies kann vertikal, z. B. durch die Erhöhung mittels Stelzen erfolgen, oder horizontal durch einen Rückzug aus dem gefährdeten Gebiet. In der Trockenvorsorge wird versucht, Flutwasser vor dem Eindringen in die Gebäude abzuhalten, z. B. durch mobile oder fest installierte Wassersperren. Bis zu einem türhohen Wasserstand ist Trockenvorsorge in der Regel realisierbar und sinnvoll (GDV 2021). Durch höhere Wasserstände kann die Standsicherheit der Gebäude gefährdet sein. Auch sind die Fließgeschwindigkeit und die Möglichkeit von Treibgut, welches an das Haus anprallen kann, wichtige Entscheidungsfaktoren für die Wahl der Vorsorgemaßnahmen. Maßnahmen der Gruppe „Nassvorsorge“ verringern den Schaden von eingedrungenem Flutwasser durch die adäquate Nutzung von gefährdeten Räumen, z. B. durch den Verzicht auf wertvolle und fest installierte Möbel. Von der Gruppe Nassvorsorge lässt sich die Gruppe „Ver-

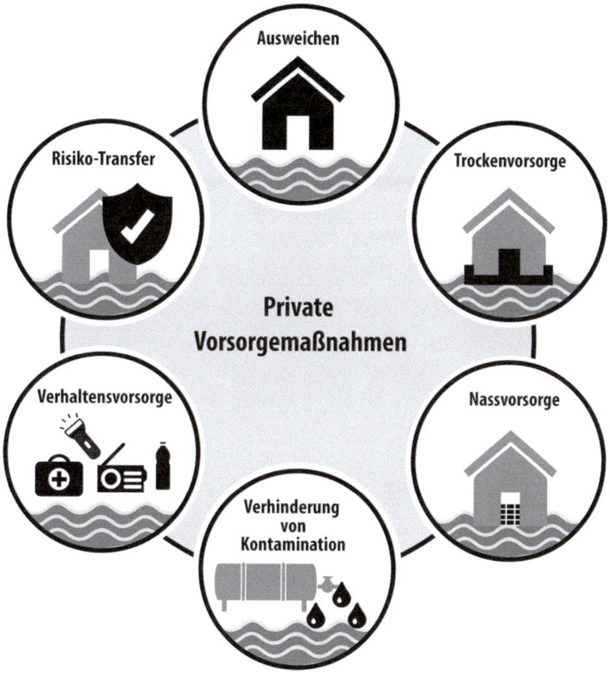

Abb. 19.1 Die sechs Kategorien privater Vorsorgemaßnahmen. Ausweichen, Trockenvorsorge, Nassvorsorge, Verhinderung von Kontamination, Verhaltensvorsorge, Risiko-Transfer

hinderung von Kontamination" differenzieren. Diese Maßnahmen verhindern die Kontamination des Flutwassers durch Gefahrstoffe wie z. B. Heizöl, durch z. B. das Sichern von Öltanks. Neben diesen strukturellen Maßnahmen besteht die Möglichkeit, das Risiko zu transferieren, indem eine Versicherung abgeschlossen wird. Dadurch werden die Schäden nicht reduziert. Jedoch werden die finanziellen Folgen eines Überflutungsereignisses abgefedert, wodurch Betroffene entstandene Schäden besser bewältigen können. Des Weiteren sollte Verhaltensvorsorge z. B. durch das Erstellen eines Notfallplans und das Vorbereiten eines Notfallgepäcks betrieben werden.

Nach den Starkregen- und Hochwasserereignissen 2021 wurde von großen Teilen der politischen Akteure die Gebäude-Pflichtversicherung für Elementarschäden (in der auch Überflutungen eingeschlossen sind) gefordert (Bundesrat 2023). Eine Diskussion über verbesserte Objektschutzmaßnahmen blieb aus. Eine Elementarschadenversicherung vermindert keine Schäden, sondern transferiert diese zu einem Versicherungsunternehmen. Dadurch wird der Schaden auf alle Versicherten und über Zeit und Raum verteilt und somit für den Einzelnen beherrschbar (Surminski und Thieken 2017). Es gibt zudem Hinweise darauf, dass der Abschluss von Versicherungen das Vorsorgeverhalten von Haushalten positiv beeinflussen kann (Surminski und Thieken 2017). Private Hochwasservorsorge mit Objektschutzmaßnahmen hat das große Potenzial, Schäden zu verringern oder gar zu verhindern (DEFRA 2008; Hudson et al. 2014; Kreibich et al. 2011; Lamond et al. 2018; Poussin et al. 2015). Dabei werden auch Leib und Leben geschützt, denn die (Teil-)Zerstörung von Wohngebäuden und Notunterkünften führt zu erhöhten Todeszahlen (Jonkman et al. 2024). Die Anzahl der Todesopfer und die verursachten Schäden der letzten Flutereignisse verdeutlichen, dass eine verbesserte private Vorsorge für die Erhöhung der gesellschaftlichen Resilienz gegenüber Überflutungen notwendig ist. Dieses Kapitel gibt einen Überblick über den Stand der privaten Vorsorge sowie Kenntnisse über

Tab. 19.1 Informationen zu den in Nordrhein-Westfalen und Rheinland-Pfalz durchgeführten Befragungen

Name der Befragung	Durchgeführt durch	Feldzeit	Methodik	Teilnehmende
Haushaltsbefragung	Universität Potsdam	18. Nov–31. Dez 2022	Online-Fragebogen; Haushalte, die im Juli 2021 von Hochwasser betroffen waren	609
Infomobil-Befragung	HKC	1. Mai 2022–30. Apr 2023	Vor-Ort Befragung; Besucherinnen und Besucher am Infomobil; Fragebogen konnte in Papierform oder über einen QR-Code online ausgefüllt werden	773

und Einstellungen zu privaten Vorsorgemaßnahmen in den von der Sturzflut Bernd im Jahr 2021 betroffenen Gebieten.

19.2 Befragungen in den von der Flut 2021 betroffenen Gebieten

Um die Umsetzung von Vorsorgemaßnahmen durch die Bevölkerung vor und nach der Flut 2021 zu untersuchen, wurden zu dieser Thematik zwei Befragungen in Nordrhein-Westfalen und Rheinland-Pfalz durchgeführt (vgl. auch ▶ Kap. 5). An diesen Befragungen nahmen insgesamt 1382 Personen teil (Tab. 19.1). Im Rahmen einer Haushaltsbefragung wurden betroffene Haushalte dazu aufgefordert, einen Online-Fragebogen auszufüllen (Dillenardt und Thieken 2024; Zenker et al. 2024). Insgesamt beantworteten 609 Personen Fragen u. a. zum Stand der eigenen Vorsorge, der Verfügbarkeit von Informationen und zu Meinungen über private Vorsorge und zur Wahrnehmung der Gefahrenlage von Hochwasser allgemein. Das HochwasserKompetenzCentrum e. V. (HKC) organisierte kostenfreie Informations- und Beratungsveranstaltungen, wobei ein eigens dafür konzipiertes Infomobil zum Einsatz kam. Im Rahmen dieser Veranstaltungen vermittelten Expertinnen und Experten Informationen und Empfehlungen zu den Themen Objektschutz und Verhalten (siehe ▶ Kap. 20). Am KAHR-Infomobil wurden 773 Personen in den von der Flut betroffenen Regionen befragt. Von den Befragten waren 64 % von der Flut betroffen, 36 % waren nicht betroffen. Der Fragebogen konnte in gedruckter Form oder durch Scannen eines QR-Codes am Mobiltelefon ausgefüllt werden. Der Fragebogen befasste sich mit den Themen Gefahrenbewusstsein und private Vorsorge.

19.3 Welche Vorsorgemaßnahmen wurden in den betroffenen Gebieten von Haushalten umgesetzt?

19

Im Rahmen der Infomobil-Befragung wurde die Bekanntheit verschiedener privater Vorsorgemaßnahmen der Gruppen Trockenvorsorge, Verhinderung von Kontamination und Risiko-Transfer in retrospektiver Wahrnehmung vor dem Flutereignis untersucht, siehe Abb. 19.2. Den meisten Befragten waren vor allem Maßnahmen der Trockenvorsorge bekannt. 64 % wussten vor dem Flutereignis von mindestens einer Möglichkeit, mit Hilfe stationärer Wassersperren das Eindringen von Flutwasser in das Gebäude zu verhindern.

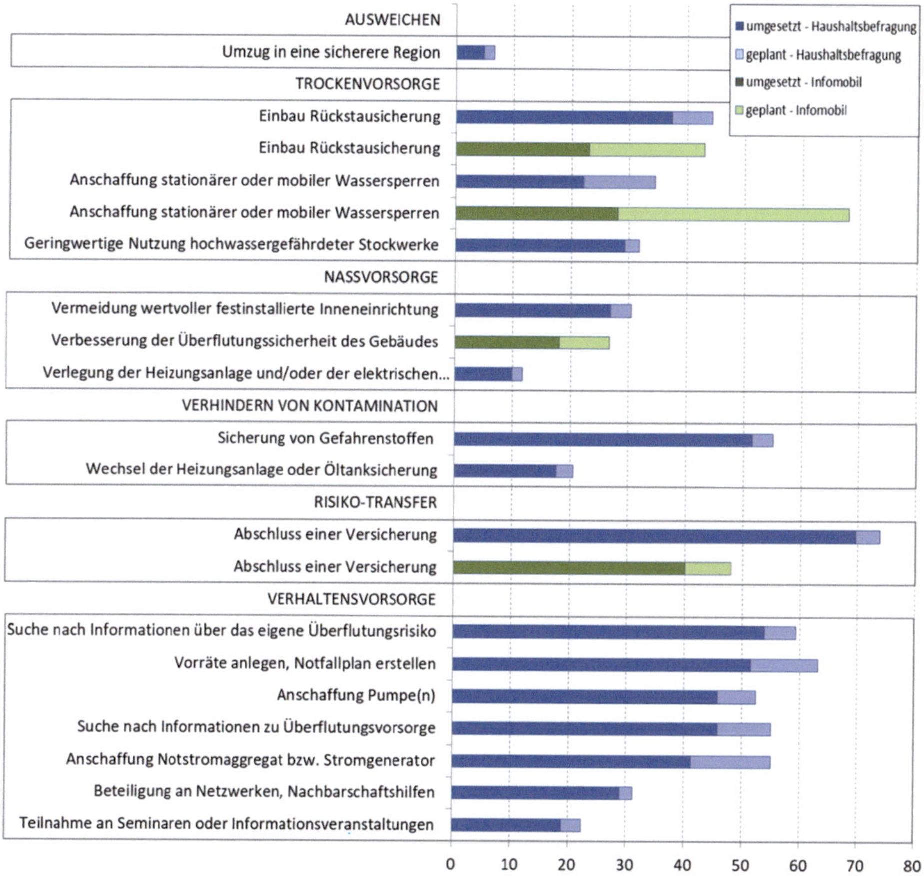

Abb. 19.2 Ergebnisse der Haushaltsbefragung (*blaue Balken*) und der Infomobil-Befragung (*grüne Balken*) zu den Vorsorgemaßnahmen, die von den Befragten bis zum Zeitpunkt der Befragung umgesetzt worden sind (*dunklere Balken*) oder deren Umsetzung geplant ist (*hellere Balken*), Angaben in Prozent (X-Achse). Informationen zu den Befragungen finden sich in Tab. 19.1

Öltanksicherungen zur Verhinderung von Kontamination waren 8 % der Befragten bekannt. Dabei sollte berücksichtigt werden, dass nur noch jeder vierte Haushalt mit Öl heizt (BMWi 2019) und diese Maßnahme daher nur für einzelne Haushalte von Relevanz ist. Neben den genannten Objektschutzmaßnahmen war auch die Möglichkeit, finanzielle Schäden durch einen Versicherungsschutz gegen Überflutungsschäden an Dritte zu transferieren, vielen Befragten nicht bekannt. Elementarschäden müssen als Zusatzbaustein zur bestehenden Wohngebäude- oder Hausratpolice mitversichert werden (GDV 2023). 42 % der Befragten gaben an, nicht von der Notwendigkeit einer zusätzlichen Elementarschadenversicherung gewusst zu haben. Die Ergebnisse zeigen, dass viele Haushalte grundlegende Vorsorgemaßnahmen nicht kennen. Es sind Wissenslücken vorhanden, die dazu führen, dass Maßnahmen nicht umgesetzt werden, so dass kein ausreichender Schutz erreicht wird.

Die Teilnehmenden der Haushaltsbefragung (N = 609) gaben selten an, dass sie nach der Flut den Wohnort gewechselt haben (4 %) oder einen solchen Wegzug planten (2 %).

Sehr viel häufiger wurde von Befragten am Infomobil (N = 773) Trockenvorsorge in Form von stationären oder mobilen Wassersperren nach der Flut umgesetzt (27 %) oder geplant (38 %). In der Haushaltsbefragung wurden vier Maßnahmen der Gruppe Nassvorsorge abgefragt, siehe ◘ Abb. 19.2. Ein Viertel der Befragten gab an, dass sie die gefährdeten Stockwerke geringwertig nutzen und/oder festinstallierte oder teure Inneneinrichtung in diesen Stockwerken vermieden. Unter 5 % der Befragten planten, eine dieser Maßnahmen in Zukunft umzusetzen. Sehr viel seltener (9 % der Befragten) wurden die Heizungsanlage und/oder die elektrische Versorgung in die höheren Stockwerke verlegt und nur 2 % der Befragten planten, diese Maßnahme in Zukunft umzusetzen. Innerhalb der Nassvorsorge gibt es Maßnahmen, die die Schäden durch eine Überflutung minimieren, indem Kontaminationen des Flutwassers verhindert werden. Besonders häufig vermieden Befragte Kontaminationen, indem sie Gefahrstoffe nicht in gefährdeten Bereichen lagern (52 %) und weitere 4 % planten die Umsetzung solcher Maßnahmen. Besonders verheerende Schäden an der Bausubstanz von Gebäuden entstehen, wenn Heizöl das Flutwasser kontaminiert (Kron 2007; Thieken et al. 2005). Aus diesem Grund ist es seit 2018 nicht mehr erlaubt, Heizölverbraucheranlagen in Überschwemmungs- und Risikogebieten zu implementieren (WHG, § 78c). Die Heizungsanlage wurde von 4 % der Befragten vor der erlebten Überflutung gewechselt (bzw. es wurde der Öltank mit einem Überflutungsschutz versehen). Weitere 14 % setzten diese Maßnahme nach der Überflutung um und 3 % planten die Umsetzung.

Der Abschluss von Versicherungen vor oder nach dem Ereignis war die am häufigsten umgesetzte Maßnahme (53 %) aller Befragten (N = 1382). Diese Zahl spiegelt den Versicherungsstatus in gesamt Deutschland wider, welcher bei 52 % im Jahr 2022 lag (GDV 2023). Verhaltensvorsorge wurde besonders häufig nach dem Erleben einer Flut umgesetzt. So gaben nur 1 % im Rahmen der Haushaltsbefragung an, bereits vor der Überflutung an Seminaren und/oder Informationsveranstaltungen zum Thema Starkregen- und Hochwasservorsorge teilgenommen zu haben. Nach der Flut waren es 17 %, weitere 3 % planten eine Teilnahme. Vor dem Ereignis waren 3 % der Befragten Teil von Netzwerken, Nachbarschaftshilfen und Notgemeinschaften, nach der Überflutung waren es 25 %, weitere 2 % planten eine Teilnahme. Hier muss jedoch angemerkt werden, dass aufgrund unserer Datengrundlage nicht untersucht werden konnte, ob es nach der Flut ein größeres Angebot an Seminaren, Infoveranstaltungen, Nachbarschaftshilfen oder Notgemeinschaften gab. 12 % der Befragten hatten sich bereits im Vorhinein vorbereitet, weitere 39 % holten dies nach der Überflutung nach und 12 % planten dies. Eine Möglichkeit vorzusorgen besteht darin, ein Notstromaggregat oder Pumpen zu kaufen (und diese nicht in gefährdeten Stockwerken zu lagern). Vor der Flut hatten 9 % bereits ein Notstromaggregat erworben. Danach besorgten sich 32 % der Befragten ein Notstromaggregat und weitere 14 % planten eine solche Anschaffung. Eine Pumpe hatten vor der Flut bereits 24 % der Haushalte angeschafft. Weitere 22 % schafften sich nach dem Ereignis eine Pumpe an und weitere 7 % planten eine solche Anschaffung.

Neben den Vorsorgemaßnahmen wurden bei der Haushaltsbefragung (N = 609) Meinungen zur privaten Vorsorge abgefragt (◘ Abb. 19.3). Die Mehrheit der Befragten stimmt der Aussage zu, dass sie alles Mögliche tun werden, um ihr Haus vor Überflutungen zu schützen und fühlen sich in der Lage, Vorsorgemaßnahmen umzusetzen. Jedoch waren viele der Befragten auch der Meinung, dass solche Maßnahmen zu teuer seien. Zudem waren viele Befragte der Meinung, dass private Vorsorgemaßnahmen die Schäden eher nicht effektiv reduzieren können. Es ist somit festzuhalten, dass viele Befragte Vorsorgemaßnahmen vor dem Ereignis nicht kannten, diese aber nach dem Ereignis um-

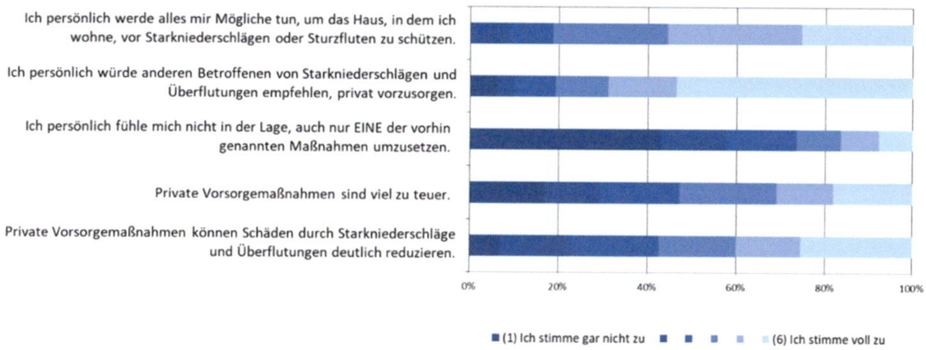

Abb. 19.3 Ergebnisse der Haushaltsbefragung (N = 609) bzgl. der Einstellungen betroffener Haushalte gegenüber privater Vorsorgemaßnahmen. Mehr Informationen zu der Befragung finden sich in Tab. 19.1

setzten oder deren Umsetzung planten. Es ist notwendig, potenziell von Überflutungen betroffene Haushalte über Vorsorgemaßnahmen zu informieren und über die Wirkweise und mögliche Finanzierungsmöglichkeiten aufzuklären.

19.4 Bedarfe für die Umsetzung von Vorsorgemaßnahmen

Für die Umsetzung von Maßnahmen zur Überflutungsvorsorge haben Bürgerinnen und Bürger spezifische Bedarfe. Grundsätzlich zeigt sich ein großer Bedarf in der Informationsvorsorge, welche die Voraussetzung für die Umsetzung von Maßnahmen bildet. Knapp 50 % der Befragten am Infomobil (N = 773), die bei der Implementierung von Vorsorgemaßnahmen auf Probleme stießen, führten diese auf Wissenslücken zurück. Darüber hinaus besteht ein großer Bedarf an finanziellen Ressourcen: So gaben über 20 % der Befragten an, dass sie fehlende finanzielle Mittel in der Überflutungsvorsorge eingeschränkt haben. Die Teilnehmenden gaben an, dass sie sich über Finanzierungslösungen und Möglichkeiten von Förderprogrammen nicht ausreichend informiert gefühlt haben.

Hinweis für Entscheidungsprozesse

Sowohl die Ergebnisse der Infomobil-Befragung (N = 773) als auch die Erfahrungen aus den Beratungssituationen unterstützen die Feststellung, dass ein grundlegender Informationsbedarf im Bereich der Trockenvorsorge besteht. Viele dieser Schutzmaßnahmen sind noch unbekannt.

Ein Drittel der Infomobil-Besuchenden benötigte eine Grundberatung, bei der ein Überblick über alle Maßnahmen gegeben und grundlegendes Wissen vermittelt wurde. Bei dieser Gruppe handelte es sich meist um Personen, die bisher nicht von Überflutungen betroffen, jedoch zur Vorsorge motiviert waren. Um ihren Informationsbedarf zu decken und sich Wissen anzueignen, besuchten sie die Infomobil-Veranstaltungen. Etwa gleich viele Beratungsgespräche entfielen auf Maßnahmen zur Gebäudeabdichtung. Darunter zählen neben Schott-Systemen auch druckdichte Kellerfenster und Schachtabdichtungen.

Bei diesen Beratenen handelte es sich größtenteils um Betroffene, die durch ein Überflutungsereignis die Schwachstellen ihres Gebäudes entdeckt und dadurch einen konkreten Informationsbedarf hatten.

Die Nachfrage nach anderen Maßnahmen ist am Infomobil deutlich geringer. Dies ist darauf zurückzuführen, dass nicht alle Vorsorgemaßnahmen die gleiche Fachexpertise benötigen. Vor allem in Bezug auf Maßnahmen der Trocken- und Nassvorsorge genügt die Erstberatung am Infomobil, um die grundlegenden Fragen zu beantworten. Teilweise ist dann die Weitervermittlung einer spezifischen Fachberatung erforderlich, die im Fall komplexer baulicher Maßnahmen (Ausweichen) und dem Verhindern von Kontaminationen meist durch Fachfirmen erfolgt, die die Maßnahmen auch umsetzen. Bei der Absicherung eines Schadens durch Versicherung (Risiko-Transfer) erfolgt die Informationsweitergabe durch bestehende Ansprechpersonen der Versicherungen.

Zusammenfassung

- Private Vorsorgemaßnahmen können Überflutungsschäden reduzieren oder gar verhindern, müssen jedoch durch die Gebäudeeigentümerinnen und Gebäudeeigentümer selbst implementiert und häufig auch selbst finanziert werden.
- Die Maßnahmen dafür sind vielfältig und können den Gruppen Ausweichen, Trockenvorsorge, Nassvorsorge, Verhinderung von Kontamination, Risiko-Transfer und Verhaltensvorsorge zugeordnet werden.
- Die Infomobil- und Haushaltsbefragungen identifizierten Wissenslücken bei den Bürgerinnen und Bürgern, die vor der Flut in Bezug auf Vorsorgemaßnahmen bestanden. Nach dem Ereignis wurden vor allem Maßnahmen des Risiko-Transfers, weitere Maßnahmen jedoch deutlich weniger umgesetzt. Zukünftig wird mehr Expertise zur Umsetzung benötigt.
- Im Rahmen eines integrierten Starkregen- und Hochwasserrisikomanagements wäre es von Vorteil, wenn Privatpersonen umfassende und klimaangepasste Vorsorgemaßnahmen ergreifen und somit zu einer gesamtgesellschaftlichen Resilienzsteigerung beitragen. Bürgerinnen und Bürger sind oft motiviert, Maßnahmen zu implementieren, um Schäden zu minimieren und zeigen somit Bereitschaft, ihrer Aufgabe gerecht zu werden. Jedoch benötigen sie eine angemessene Unterstützung in Form von Informationen und finanziellen Förderungen.
- Um die Implementierung von privaten Vorsorgemaßnahmen zu fördern, müssen die potenziell Betroffenen durch zielgerichtete Risikokommunikation über ihre Gefahr und Möglichkeiten der Anpassung informiert werden.
- Gerade im Wiederaufbauprozess nach einer Flut sind solche Informationen essentiell, da eine Vielzahl von Maßnahmen umgesetzt werden kann, die Motivation sich zu schützen bei den Betroffenen nach einer Überflutung hoch ist und sie dann besonders häufig Maßnahmen umsetzen, siehe ▶ Kap. 20.

Danksagung Die Autorinnen und Autoren von ▶ Kap. 19 bedanken sich beim Bundesministerium für Bildung und Forschung (BMBF) und beim DLR Projektträger für die Unterstützung und Förderung des KAHR-Projekts. Diese Forschung wurde vom BMBF unter dem Förderkennzeichen 01LR2102A-M finanziert.

Literatur

Begg C, Ueberham M, Masson T, Kuhlicke C (2017) Interactions between citizen responsibilization, flood experience and household resilience: insights from the 2013 flood in Germany. Int J Water Resour Dev 33(4):591–608

Bubeck P, Botzen WJW, Aerts JCJH (2012) A review of risk perceptions and other factors that influence flood mitigation behavior. Risk Analysis 32(9):1481–1495

Bundeministerium für Wirtschaft und Klimaschutz (BMWi) (2019) So heizen die Deutschen. https://www.bmwk-energiewende.de/EWD/Redaktion/Newsletter/2019/10/Meldung/direkt-erfasst_infografik.html

Bundesministerium des Innern und für Heimat (BMI) (2022) Bericht zur Hochwasserkatastrophe 2021: Katastrophenhilfe, Wiederaufbau und Evaluierungsprozesse. Technical report

Bundesrat (2023) Entschließung des Bundesrates „Bundesweite Einführung einer Elementarschaden-Pflichtversicherung", (Vol. Drucksache: 102/23). https://www.bundesrat.de/SharedDocs/TO/1032/to-node.html

DEFRA (2008) Developing the evidence base for flood resistance and resilience. Summary Report. Environment Agency and DEFRA, London

Dillenardt L, Thieken AH (2024) Individual flood risk adaptation in Germany: exploring the role of different types of flooding. EGUshpere [preprint], 1–28

GDV (2021) Baukonstruktive Überflutungsvorsorge. Leitfaden mit Hinweisen für die Wahl geeigneter Bauarten und deren Umsetzung. Gesamtverband der Deutschen Versicherungswirtschaft e. V. [GDV], German Insurance Association. https://shop.vds.de/download/vds-6002/47b06914-3498-4637-901c-283f0babee53

GDV (2023) Datenservice zum Naturgefahrenreport 2023. Gesamtverband der Deutschen Versicherungswirtschaft e. V. [GDV]. German Insurance Association

Hudson P, Botzen WJW, Kreibich H, Bubeck P, Aerts JCJH (2014) Evaluating the effectiveness of flood damage risk reductions by the application of Propensity Score Matching. Nat Hazards Earth Syst Sci 14:1731–1747

Jonkman SN, Curran A, Bouwer LM (2024) Floods have become less deadly: an analysis of global flood fatalities 1975–2022. Nat Hazards 120:6327–6342

Kreibich H, Seifert I, Thieken AH, Lindquist E, Wagner K, Merz B (2011) Recent changes in flood preparedness of private households and businesses in Germany. Reg Environ Change 11(1):59–71

Kron A (2007) Mikroskalige Ermittlung potenzieller Hochwasserschäden zur Gefahren- und Risikoanalyse. Universität Karlsruhe (TH) (Dissertation)

Lamond J, Rose C, Bhattacharya-Mis N, Joseph R (2018) Flood Re and UWE Bristol. In: Evidence review for property flood resilience phase 2 report

Lengfeld K, Voit P, Kaspar F, Heistermann M (2023) Brief communication: On the extremeness of the July 2021 precipitation event in western Germany. Nat Hazards Earth Syst Sci 23(3):1227–1232

Poussin JK, Botzen WJW, Aerts JCJH (2015) Effectiveness of flood damage mitigation measures: empirical evidence from French flood disasters. Glob Environ Chang 31:74–84

Rosenzweig BR, McPhillips L, Chang H, Cheng C, Welty C, Matsler M, Davidson CI (2018) Pluvial flood risk and opportunities for resilience. Wiley Interdiscip Rev Water 5(6):e1302

Snel KA, Witte PA, Hartmann T, Geertman SC (2020) The shifting position of homeowners in flood resilience: from recipients to key-stakeholders. Wires Water 7:e1451

Surminski S, Thieken AH (2017) Promoting flood risk reduction: the role of insurance in Germany and England. Earth's Future 5(10):979–1001

Thieken A, Zenker M-L, Bubeck P (2023) Flood-related fatalities during the flood of July 2021 in North Rhine-Westphalia, Germany: what can be learnt for future flood risk management? J Coast Riverine Flood Risk 2:5

Thieken AH, Müller M, Kreibich H, Merz B (2005) Flood damage and influencing factors: new insights from the August 2002 flood in Germany. Water Resour Res 41(12):W12430

Tradowsky JS, Philip SY, Kreienkamp F et al (2023) Attribution of the heavy rainfall events leading to severe flooding in Western Europe during July 2021. Clim Change 176:90. https://doi.org/10.1007/s10584-023-03502-7

Walker G, Tweed F, Whittle R (2014) A framework for profiling the characteristics of risk governance in natural hazard contexts. Hazards Earth Syst Sci 14:155–164

Zenker ML, Bubeck P, Thieken AH (2024) Always on my mind: indications of post-traumatic stress disorder among those affected by the 2021 flood event in the Ahr valley, Germany. Nat Hazards Earth Syst Sci 24(8):2837–2856

Kommunikation während des Wiederaufbauprozesses

Lisa Dillenardt, Helene Meyer, Philip Bubeck, Lina Fitz und Georg Johann

J. Birkmann et al. (Hrsg.), *Die Flutkatastrophe 2021*, Sustainable Development Goals (SDG) – Umsetzung in Praxis, Lehre und Entscheidungsprozessen,
https://doi.org/10.1007/978-3-658-48688-4_20

20.1 Einleitung

Nach verheerenden Überflutungen wie in Nordrhein-Westfalen und Rheinland-Pfalz in 2021 stehen betroffene Haushalte vor enormen Herausforderungen, die bei der Kommunikation mit der Bevölkerung im Wiederaufbau berücksichtigt werden müssen (siehe ◘ Abb. 20.1; Abunyewah et al. 2018; Balog-Way et al. 2020; Bubeck et al. 2012; Bubeck et al. 2023; Rogers 1983).

Vielen Haushalten sind hohe finanzielle Schäden entstanden. Die Sturzflut Bernd 2021 verursachte in Deutschland insgesamt versicherte Sachschäden von 8,1 Mrd. € (GDV 2022). Allein im Ahrtal waren 42.000 Menschen direkt betroffen und 9000 Häuser wurden schwer beschädigt oder zerstört (Birkmann et al. 2023). Um Gelder der Wiederaufbauhilfe oder Versicherungsleistungen zu erhalten, müssen Haushalte oftmals komplizierte Anträge stellen (siehe auch ► Kap. 18). Zudem hat das Erlebte Auswirkungen auf die mentale Gesundheit der Betroffenen. So zeigten 28 % der Teilnehmenden einer Betroffenenbefragung auch ein Jahr nach der Flut Anzeichen einer posttraumatischen Belastungsstörung (siehe ► Kap. 9). Des Weiteren müssen sich Betroffene auch im Nachgang des Hochwassers vor akuten Gefahren im Katastrophengebiet schützen. Sie sehen sich dem Risiko ausgesetzt, dass Gebäude einsturzgefährdet sein können (siehe ► Kap. 7). Das Flutwasser kann zudem kontaminiert sein (DKKV 2015), wodurch weitere Gefahren, wie z. B. Schimmelbildung in Gebäuden oder eine Tetanusinfektion (Yamaguchi 2023), entstehen können.

Daneben stellt sich die Frage, inwieweit sich ein solches Ereignis wiederholen kann. Eine Grundannahme des integrierten Hochwasserrisikomanagements ist es, dass ein Restrisiko für das Auftreten eines Ereignisses nie ausgeschlossen werden kann, und, dass dessen Intensität und Ereignisstärke vorab nicht bekannt sein wird. Aus diesem Grund wird ein integriertes Hochwasserrisikomanagement als Zyklus beschrieben. Auf ein Hochwasserereignis folgen Phasen der Bewältigung und Nachsorge, welche von einer Phase der

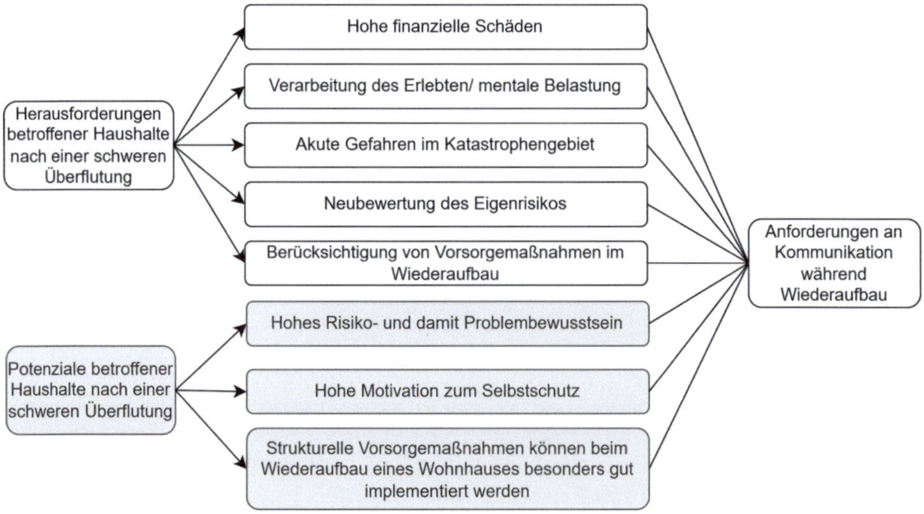

◘ **Abb. 20.1** Herausforderungen betroffener Haushalte nach einer Überflutung

Vorbereitung auf ein eventuelles zukünftiges Ereignis abgelöst wird (UBA 2022). Teil dieser Vorbereitung sollte eine Risikoanalyse sein (DKKV 2015), um aktuelle Erkenntnisse in die Einschätzung zukünftiger Ereignisse einzubeziehen (Vorogushyn et al. 2022). Dieses periodische Vorgehen ist explizit in der Hochwasserrisikomanagement-Richtlinie der Europäischen Union verankert (Directive 2007/60/EC 2007). Diese sieht gemäß Artikel 14 vor, dass die Mitgliedstaaten ihre vorläufige Bewertung des Hochwasserrisikos, ihre Hochwassergefahren- und Hochwasserrisikokarten und ihre Risikomanagementpläne alle sechs Jahre überprüfen.

Auch der Wiederaufbau an sich und wie er gestaltet werden soll muss den Betroffenen kommuniziert werden. Gemäß den Empfehlungen des KAHR-Projektes (Birkmann et al. 2023) sollte ein Wiederaufbau im Sinne eines „Build Back Better" (BBB), in welchem nicht der Status Quo wiederhergestellt wird, sondern die Resilienz gegenüber einem zukünftigen Ereignis durch die Berücksichtigung von Anpassungsmaßnahmen im Wiederaufbau gesteigert wird, angestrebt werden. Im Sinne des BBB sollten Betroffene bereits während des Wiederaufbaus das Risiko von zukünftigen Überflutungsschäden durch die Implementierung von privaten Vorsorgemaßnahmen reduzieren (siehe ► Kap. 19). Viele bauliche Maßnahmen zur privaten Vorsorge können besonders gut während Sanierungs- oder Modernisierungsmaßnahmen implementiert werden (BMWSB 2022). Die Ergebnisse in ► Kap. 19 zeigen auch, dass die Mehrheit der Betroffenen kurz nach einer Überflutung motiviert ist, sich und ihr Eigentum zu schützen und sich zudem in der Lage sieht, Vorsorgemaßnahmen zu implementieren. Jedoch brauchen die Betroffenen Informationen über Vorsorgemaßnahmen. Eine gelungene Kommunikation mit den Betroffenen bei all diesen Aspekten ist für einen resilienzfördernden Wiederaufbau von zentraler Bedeutung und gleichzeitig eine große Herausforderung. Das BBK unterscheidet in seinem Leitfaden zur Risikokommunikation zwischen der Krisenkommunikation während und kurz nach einem Ereignis und der Risikokommunikation als Vorbereitung auf ein zukünftiges Ereignis (BBK und BfR 2022). Dieses Kapitel widmet sich der Kommunikation während des Wiederaufbauprozesses, da an diese besondere Ansprüche bestehen, welche bisher noch nicht konkret diskutiert wurden. Es wird argumentiert, dass sich die Kommunikation während des Wiederaufbaus an der Schnittstelle zwischen der Krisen- und der Risikokommunikation befindet und die Übergänge fließend und nicht klar abzugrenzen sind. So müssen während des Wiederaufbaus Aspekte der Krisenkommunikation fortgesetzt werden. Akute Gefahren in den betroffenen Gebieten müssen weiterhin kommuniziert werden. Auch Informationen zu Hilfsangebote, z. B. zur psychosozialen Betreuung oder administrativen Aspekten müssen längerfristig aufrechterhalten werden. Gleichzeitig brauchen Betroffene aber auch bereits im Wiederaufbau Informationen zu ihrem Risiko und zu möglichen Vorsorgemaßnahmen, wenn sie sich an einem BBB beteiligen sollen.

Anhand von Befragungsdaten wird in diesem Kapitel dargestellt, wie das Risikobewusstsein nach einer Überflutung bei den Betroffenen ausgebildet ist und welche Bedarfe sich daraus für die Risikokommunikation ableiten. Auf Basis der Befragungsdaten werden Bedarfe der Betroffenen an eine gelungene Kommunikation aufgezeigt, sowie die Zufriedenheit der Betroffenen mit der Kommunikation während des Wiederaufbau nach dem Ereignis 2021 ausgewertet. Das Kapitel endet mit dem Vorstellen von Kommunikationstool, welche mit wenig Kosten und Aufwand von Gemeinden genutzt werden können, um mit den Betroffenen zügig in Kontakt zu treten.

20.2 Erhöhtes Risikobewusstsein als Chance für die Kommunikation im Wiederaufbau

(Potenziell) Betroffene müssen über ihr Risiko informiert sein, um die Grenzen des Bevölkerungsschutzes durch den Staat zu verstehen und ihre Eigenverantwortung wahrnehmen zu können (BBK und BfR 2022). In ▶ Kap. 19 wurden private Vorsorgemaßnahmen besprochen, welche teilweise besonders gut während des Wiederaufbaus implementiert werden können. Die Betroffenen müssen hierzu jedoch schon während des Wiederaufbauprozesses Informationen über ihr Risiko und mögliche Anpassungsmaßnahmen erhalten. Daher sollte die Risikovorsorge als auch die Informationsvorsorge bereits in der Phase der Nachsorge bzw. während des Wiederaufbaus starten.

Um Informationsangebote an die Bedarfe der Betroffenen anzupassen, ist es wichtig zu verstehen, ob die Überflutung einen Einfluss auf das Risikobewusstsein der Betroffenen hatte und wie das Risiko nach einer Flut in den betroffenen Gebieten wahrgenommen wird. An einem Infomobil des HochwasserKompetenzCentrums e. V. (HKC) wurden Betroffene zu ihrem Risikobewusstsein (in der retrospektiven Wahrnehmung) vor dem Ereignis und nach dem Ereignis befragt, siehe ◘ Abb. 20.2, linke Darstellung. Im Rahmen einer Haushaltsbefragung wurde gefragt, wie die Gefahr eines zukünftigen Ereignisses hinsichtlich der Wahrscheinlichkeit des Auftretens und der Auswirkungen eingeschätzt werden, siehe ◘ Abb. 20.2 rechte Darstellung. Die Befragungsdaten und ihre Erhebung sind in ▶ Kap. 19 im Detail beschrieben.

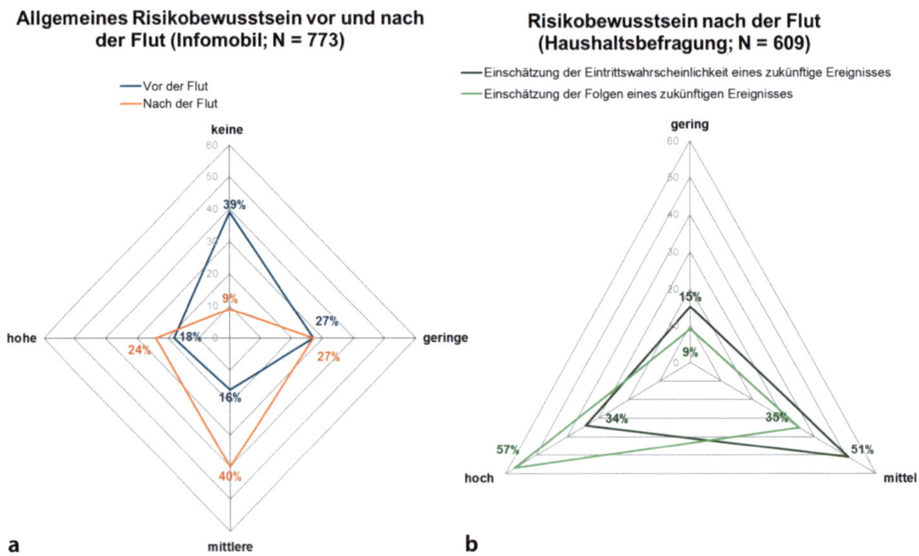

◘ **Abb. 20.2** Ergebnisse der Befragungen hinsichtlich des Risikobewusstseins von Haushalten in den betroffenen Gebieten, Befragungen wurden nach der Sturzflut Bernd durchgeführt. Dargestellt sind die Ergebnisse vom Infomobil (*links*) und von der Haushaltsbefragung (*rechts*) in einem Netzdiagramm. Dabei wird auf den Achsen in Prozent angegeben, wie viele Befragte der jeweiligen Kategorie zustimmen. Die unterschiedlich farbigen Linien beziehen sich auf unterschiedliche Fragen, welche in der Legende des jeweiligen Netzdiagramms benannt werden

Über die Hälfte der Befragten (66 %) am Infomobil gaben an, dass sie vor der Flut kein (39 %) oder ein geringes (27 %) Risikobewusstsein hatten. 16 % der Befragten am Infomobil berichteten von einem mittleren und 18 % von einem hohen Risikobewusstsein vor der Flut. Die Befragten gaben an, dass sich ihr Risikobewusstsein nach der Flut verändert hat. Nach ihrem Risikobewusstsein nach der Flut gefragt, gaben nur noch 9 % an, dass sie keine Gefahr wahrnehmen und 40 % berichteten von einer mittleren und 24 % von einem hohen Risikobewusstsein. Die Gruppe jener, die ein geringes Risikobewusstsein nach der Flut hatten, blieb gleich groß bei 27 % aller Befragten. Im Rahmen der Haushaltsbefragung wurde die Einschätzung eines potenziellen zukünftigen Ereignisses erfragt. Dabei gab die Mehrheit der Befragten (66 %) an, dass sie die Wahrscheinlichkeit für das Auftreten eines zukünftigen Ereignisses als nicht hoch einschätzen. Gleichzeitig gab die Mehrheit der Befragten (57 %) an, dass sie die Auswirkungen eines zukünftigen Ereignisses als schlimm oder sehr schlimm, einschätzen. Das Gesamtbild beider Befragungen zeigt, dass sich das Risikobewusstsein durch das Erleben der Flut deutlich erhöht hat. Diese erhöhte Risiko- und damit Problemwahrnehmung der Bevölkerung nach dem Ereignis kann genutzt werden, um Betroffene mit Informationsangeboten zu erreichen und in einen BBB-Prozess einzubinden.

Gleichzeitig zeigt sich, dass Betroffene nach der Flut die Eintrittswahrscheinlichkeit eines potenziellen zukünftigen Ereignisses als geringer einschätzen als die Auswirkungen eines solchen Ereignisses. Ein Grund für die geringere Einschätzung der Wiederkehrwahrscheinlichkeit kann die Angabe von Wiederkehrintervallen sein, wie sie in Hochwassergefahrenkarten verwendet werden. Laien können diese oftmals nicht richtig interpretieren, wie Studien gezeigt haben (Grounds et al. 2018). Eine Lösung kann sein, zusätzlich die Eintrittswahrscheinlichkeit pro Jahr anzugeben (Grounds et al. 2018; Kellens et al. 2012). Jedoch ist zu beachten, dass nach einem extremen Ereignis wie der Juli-Flut 2021 die bisherigen Wiederkehrperioden hinterfragt werden müssen. Unter Umständen auch unter Einbeziehung weiterer Daten, wie z. B. historischen Ereignissen oder überflutungsfördernder lokaler Besonderheiten wie z. B. die Verklausungen von Brücken (Vorogushyn et al. 2022). Es ist wichtig, dass die Aufklärung der Bevölkerung über ihr Risiko auf den aktuellsten Erkenntnissen beruht und auch bestehende Unsicherheiten kommuniziert werden.

20.3 Bedarfe der Bevölkerung an eine Kommunikation während des Wiederaufbaus

Personen, die motiviert zum Selbstschutz sind, setzen eher Vorsorgemaßnahmen um (Dillenardt et al. 2022). Die Motivation zum Selbstschutz kann gesteigert werden, wenn Menschen eine hohe Selbstwirksamkeit gegenüber der Implementierung von Vorsorgemaßnahmen haben und die Maßnahmen als sinnvoll und nicht zu teuer einschätzen (Dillenardt et al. 2022; Dillenardt et al. 2024; Grothmann und Reusswig 2006). Informationsvorsorge sollte daher kontinuierlich gewährleistet werden, um sicherzustellen, dass sich (potenziell) Betroffene angemessen auf zukünftige Hochwasserereignisse vorbereiten können und hierzu motiviert sind. Doch welcher Informationsbedarf besteht bei den Betroffenen?

Aus dem direkten Kontakt zu den Bürgerinnen und Bürgern am Infomobil ergaben sich konkrete Bedarfe an die Kommunikation während des Wiederaufbaus. Am Infomobil wurden vorrangig Informationen zum Wiederaufbau, der privaten Vorsorge und zu Fördermitteln nachgefragt. Diese Bedarfe bestanden längerfristig nach dem Flutereignis. Es hat sich gezeigt, dass persönliche Beratungsgespräche sehr wichtig sind, da sich die Bedarfe für die Umsetzung von Vorsorgemaßnahmen individuell unterscheiden. Die Gespräche ermöglichten es Interessierten, die Vielfalt der Vorsorgemöglichkeiten kennenzulernen, diese vor Ort zu sehen, auszuprobieren und so ein Verständnis für die Technik zu erlangen. Gleichzeitig wurden sie für Überflutungsrisiken sensibilisiert und zur Vorsorge aktiviert. In den Beratungsgesprächen konnten weitere Bedarfe, beispielsweise Unterstützung bei der konkreten Umsetzung oder Finanzierung, ermittelt werden. Betroffene konnten gezielt mit Informationen zum Wiederaufbau und zu Fördermitteln unterstützt werden. In individuellen Beratungsgesprächen bestand für Betroffene zudem die Möglichkeit einer seelischen Betreuung.

Durch Überflutungsschäden werden Betroffene auf die Schwachstellen ihres Gebäudes aufmerksam gemacht. Die Infomobil-Beratungsgespräche haben gezeigt, dass bei Betroffenen ein besonderer Bedarf nach gebäudespezifischen Informationen bestand, damit zukünftige Schäden minimiert oder gar verhindert werden können. Spezifische Empfehlungen zu privaten Vorsorgemaßnahmen für Gebäude bedürfen einer Gebäudebewertung, die nur durch Expertinnen und Experten vor Ort erfolgen kann. Instrumente zur Gebäudebewertung, wie der Hochwasser-Pass (siehe Info-Box), vereinen den ermittelten Bedarf nach einer Erhöhung des allgemeinen Wissensstandes sowie der Unterstützung bei der Wahl geeigneter Vorsorgemaßnahmen. Gleichzeitig werden geeignete Vorsorgestrategien professionell und ganzheitlich betrachtet und die konkreten Bedarfe bedient. Die Befragung am Infomobil ($N = 773$) ergab, dass 57,3 % der Befragten Interesse am Hochwasser-Pass hatten.

Die Gespräche am Infomobil haben aber auch gezeigt, dass manche Bürgerinnen und Bürger zum Infomobil kamen, sich aber nicht mehr mit dem Thema Hochwasser befassen wollen, weder im Allgemeinen noch mit konkretem Bezug zur Vorsorge. Dies scheint auf den ersten Blick ein Widerspruch. Jedoch schließt sich eine Motivation zum Selbstschutz auf der einen und den vermeintlich entgegengesetzten Einstellungen wie Fatalismus (d. h. einem Ohnmachtsgefühl gegenüber einem unvermeidlichen Schicksal) oder Verleugnung auf der anderen Seite nicht zwangsläufig aus. Vielmehr können eine Motivation zum Selbstschutz und Fatalismus bzw. Verleugnung koexistieren (Dillenardt et al. 2024). In einer Studie von Berghäuser et al. (2023) zeigt sich, dass Betroffene, die Anzeichen von Fatalismus zeigen, gleichzeitig durchaus Vorsorgemaßnahmen umsetzen. Es ist daher wichtig, das Gespräch mit Menschen, die Anzeichen von Fatalismus oder Verleugnung zeigen, zu suchen bzw. aufrechtzuerhalten.

20.4 Kommunikation während des Wiederaufbaus in Nordrhein-Westfalen und Rheinland-Pfalz: Wie zufrieden ist die Bevölkerung und welche Kanäle wurden genutzt?

Die Befragung am Infomobil sowie die Haushaltsbefragung untersuchten die Zufriedenheit der betroffenen Bevölkerung mit der Kommunikation zu verschiedenen Aspekten während des Wiederaufbaus. Darüber hinaus wurde erhoben, wie zufrieden die Befragten

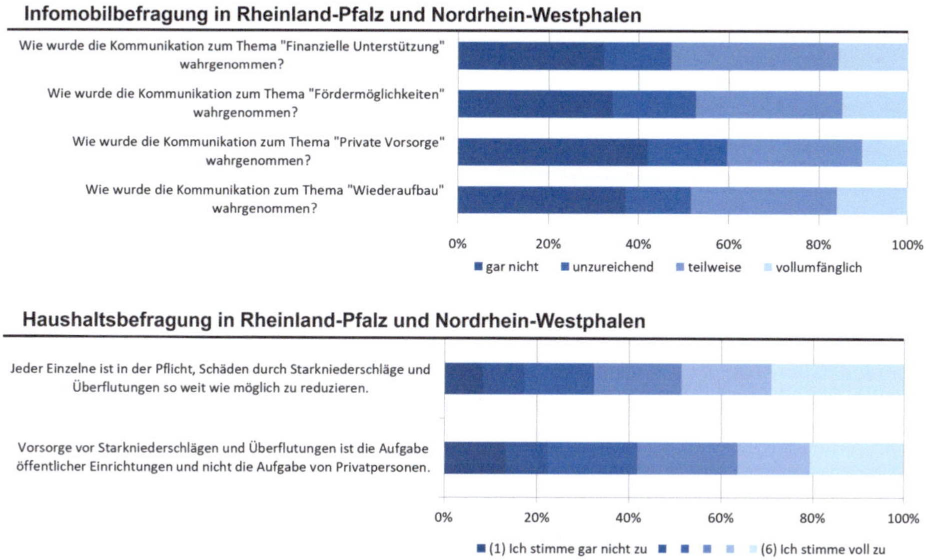

Abb. 20.3 Ergebnisse der Befragungen hinsichtlich der Wahrnehmung der Verfügbarkeit von Informationen nach der Sturzflut Bernd in den betroffenen Gebieten

mit der Risikokommunikation vor dem Ereignis waren und über welche Kanäle Informationen vor der Flut eingeholt wurden.

Nach dem katastrophalen Ereignis 2021 stellten Bund und Länder Wiederaufbauhilfen und Unterstützungen auch für betroffene Haushalte bereit. Um diese in Anspruch nehmen zu können, müssen die Betroffenen informiert sein. Die Befragung am Infomobil ergab (N = 773), dass die Zufriedenheit mit Informationen zur finanziellen Unterstützung zwar am höchsten war, doch waren 40 % der Befragten auch damit unzufrieden. Mit Informationen zum Wiederaufbau (52 %) und Fördermöglichkeiten (53 %) war gut die Hälfte der Befragten am Infomobil unzufrieden. Die größte Unzufriedenheit bestand in Bezug auf Informationen zu Vorsorgemaßnahmen: 60 % der Befragten gaben an, nach der Flut unzureichend oder gar nicht darüber informiert worden zu sein. Diese Einschätzungen finden sich auch in der Haushaltsbefragung (N = 609). Hier gab über die Hälfte der Befragten an, dass es einen Bedarf an Informationen und Beratungsangeboten zur privaten Vorsorge gibt. Beide Befragungen weisen somit auf die Notwendigkeit hin, die Kommunikation von lokalen Informationen über die Gefahrenlage und Vorsorgemaßnahmen während des Wiederaufbauprozesses zu verbessern und zu intensivieren. Eine noch geringere Zufriedenheit ergab sich im Hinblick auf die Kommunikation vor dem Ereignis. Mit 77 % gab der Großteil der Befragten am Infomobil an, sich vor der Flut unzureichend bis gar nicht informiert gefühlt zu haben. Die Haushaltsbefragung (N = 609) ergab, dass knapp 80 % der Befragten sich (eher) nicht gut durch ihre Gemeinden informiert fühlten (Abb. 20.3 unten).

Von Interesse ist zudem, von wem Betroffene die Informationen erhalten. Eine wichtige Rolle sollten dabei die Kommunen spielen. Jedoch gab nur ein eher geringer Anteil der Befragten an, durch die Kommune (5 %) oder Hilfsorganisationen (15 %) über die Überflutungsgefahr und mögliche Vorsorgemaßnahmen informiert worden zu sein. Der geringe Anteil der Kommunen erscheint problematisch, da diesen gemäß vieler Katastrophenschutzgesetze der Länder die Aufgabe zukommt, die Selbsthilfe der Bevölkerung zu fördern (siehe ▶ Abschn. 20.5). Der Großteil der Befragten nutzte Kanäle, wie Internet und Apps (32 %), das Verfolgen von Nachrichten im Fernsehen oder Radio (17 %) sowie Gespräche mit der Familie und Bekannten (15 %) um sich über ihr Risiko und Vorsorge zu informieren. Trotz der Eigeninitiative vieler Betroffener gaben jedoch die meisten (43 %) vor der Flut (in retrospektiver Wahrnehmung) an, überhaupt keine Informationen zur Gefahr erhalten zu haben. Insgesamt wird deutlich, dass die Bevölkerung in der Risikokommunikation überwiegend Informationen über Kanäle wie das Internet oder persönliche Kontakte eingeholt hat. Informationen durch Institutionen, wie z. B. Kommunen oder Hilfsorganisationen, erreichten dabei deutlich weniger und konnten daher weniger zur Risikokommunikation vor der Flut beitragen. Es muss berücksichtigt werden, dass diese Informationen eventuell nicht für alle Befragten zugänglich waren.

20.5 Auf die Kommunikation im Wiederaufbau gut vorbereitet sein

Die Kommunen sehen sich im Wiederaufbau mit einer Vielzahl an Kommunikationsbedarfen seitens der Bevölkerung konfrontiert. Im Sinne des BBB ist es wichtig, dass Informationen einerseits bereits kurz nach dem Ereignis zur Verfügung stehen und andererseits langfristig bereitgestellt werden. Dies kann für Gemeinden aufgrund eigener Betroffenheiten und Überlastung oftmals eine große Herausforderung darstellen. So wurde z. B. das Rathaus der Stadt Stolberg durch das Hochwasser schwer beschädigt und in seiner Funktionsfähigkeit betroffen. Daher stellt sich die Frage, welche Informationsmedien zügig und ohne zu großen Aufwand für die Kommunen zur Verfügung stehen. Hier hat sich herausgestellt, dass Projekte wie KAHR als Schnittstelle zwischen Kommunen, Wissenschaft und Betroffenen agieren und eine Quelle für Informationen sein können. So stellt der Internetauftritt des KAHR-Projektes Informationen in Form von Flyern und Broschüren, Interviews zu Themen wie z. B. Hochwasserschutz sowie weiterführende Links für Gemeinden, betroffene Bürgerinnen und Bürger, Unternehmen und Interessierte bereit.

Die folgenden drei Tools zeigen, wie Gemeinden und Kommunen niederschwellige Kommunikationsangebote im Wiederaufbau anbieten können.

Tools für Kommunikationsangebote im Wiederaufbau

Virtueller Baukasten für Informationsmaterialien

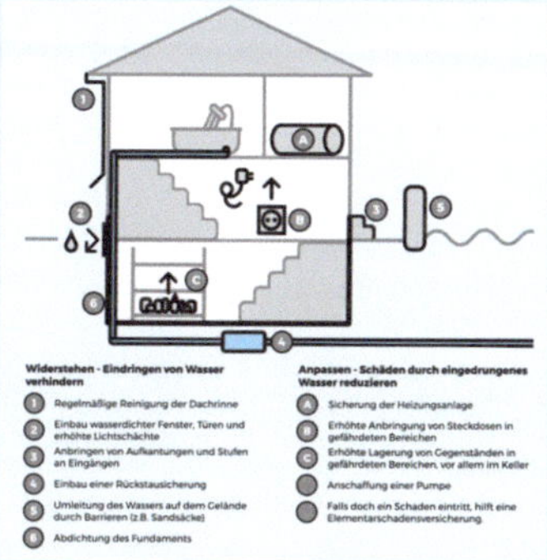

Im Rahmen des Projektes ExTrass wurde ein virtueller Baukasten erarbeitet, mit welchem Gemeinden und andere Akteure die Möglichkeit haben, zwei Informationsflyer für eine erste Kontaktaufnahme mit der Bevölkerung an die lokalen Bedarfe anzupassen. Der Flyer kann auch ohne Anpassung genutzt werden. Die Nutzung ist kostenlos.

Es steht ein Flyer zur „Was tun bei Starkregen" zur Verfügung, welcher über das korrekte Verhalten im Ernstfall informiert, sowie ein Flyer „Wie schütze ich mein Haus richtig", welcher über private Vorsorgemaßnahmen informiert. Beide Flyer verweisen via QR-Code auf weiterführende Informationen.

Beratungsunterstützung durch das Infomobil

Das HKC unterstützt seit 2011 mit Infomobilen die Information der Bürger*innen über Überflutungs- und Verhaltensvorsorge. Die Mobile sind mit Informationsmaterialien und Exponaten ausgestattet, die den Bürger*innen Überflutungsschutz „zum Anfassen" erlebbar machen. So wird ein Dialog eröffnet, der das Thema auf die persönliche Ebene und in die Lebensrealität der Besuchenden bringt. Gleichzeitig bietet es Betroffenen eine Anlaufstelle, die jemanden brauchen, der ihnen zuhört. Das HKC besucht mit den Infomobilen

Veranstaltungen und fährt gezielt in Kommunen im gesamten Bundesgebiet. Die Beratung erfolgt nicht nur in Überflutungsgebieten, sondern auch proaktiv in Regionen, die noch nicht betroffen waren. Es wird der Dialog zwischen Bürger*innen und Expert*innen gestärkt und eine wirkungsvolle Kommunikation ermöglicht.

Hier kann das Infomobil angefragt werden: ▶ https://www.hkc-online.de/de/Infomobile.

Gebäudespezifische Informationen durch Expert*innen (Hochwasser-Pass)

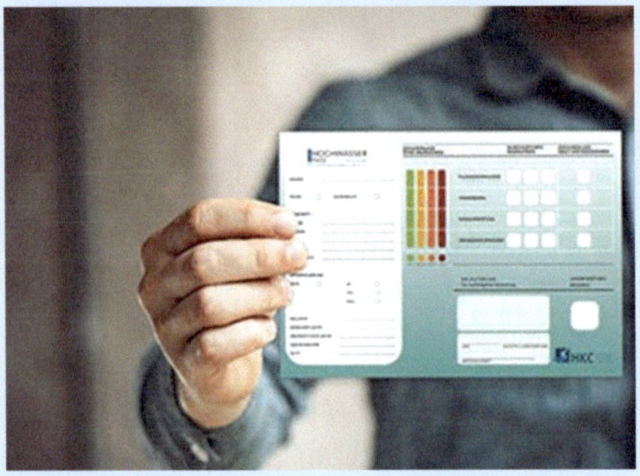

Mit dem Hochwasser-Pass des HKC erhalten Eigentümer*innen eine umfangreiche und professionelle Risikobewertung ihres Gebäudes. Zertifizierte Sachkundige beurteilen die Gefährdung bei einer Vor-Ort-Begehung. Im Dialog wird basierend auf der vorherigen Bewertung gemeinsam ein Schutzkonzept entwickelt, welches auf die individuellen Bedürfnisse ausgerichtet ist. Hochwasser-Pass-Sachkundige empfehlen passgenaue Maßnahmen zur Risikoreduktion, die Betroffene und Gefährdete dazu befähigt, aktiv und an ihre persönliche Situation angepasst zu der Reduktion ihrer Gefährdung beizutragen. Nach der erfolgreichen Umsetzung des Hochwasser-Pass können Eigentümer*innen eine Plakette für ihr Gebäude bekommen. Diese dient nicht nur dem direkten visuellen Nachweis über das Bestehen eines Schutzkonzeptes, sondern auch der Kommunikation der Gefährdung einerseits, aber auch von Lösungsansätzen andererseits. Umsetzende des Hochwasser-Pass nehmen so eine Multiplikator*innen-Rolle ein.

Weiter Informationen zum Hochwasser-Pass sind hier zu finden: ▶ https://www.hochwasser-pass.info

Zusammenfassung

- Der Wiederaufbau nach einem Hochwasserereignis ist mit besonderen Herausforderungen für alle gesellschaftlichen Bereiche verbunden. Anzustreben ist ein Aufbau, der langfristig auch die Resilienz in den betroffenen Gebieten steigert (BBB).
- Die Kommunikation während des Wiederaufbauprozesses befindet sich an der Schnittstelle der Krisenkommunikation und der Risikokommunikation, wobei sich eine klare Abgrenzung nicht durchführen lässt.
- Im Anschluss an das Hochwasserereignis vom Juli 2021 manifestierten sich bei den Betroffenen spezifische Informationsbedürfnisse, die sich auf den Wiederaufbau, private Vorsorgemaßnahmen sowie auf Fördermittel bezogen.
- Während des Wiederaufbaus sind jedoch nicht nur die Haushalte stark betroffen, sondern auch die Kommunen und Gemeinden. Daher ist es wichtig, vorab die Kommunikation im Wiederaufbau zu diskutieren und auszuloten, wie die Kommunikation mit der Bevölkerung zügig, aber mit möglichst wenig Belastungen der Kommunen und Gemeinden möglich gemacht werden könnte.
- Während des Wiederaufbaus können die betroffenen Haushalte zeitnah durch, mit virtuellen Baukästen erstellte, Informationsmaterialien, in den Gemeinden über ein Infomobil oder individuell vor Ort durch eine Risikoberatung im Rahmen des Hochwasserpasses mit Informationen adressiert werden.

Danksagung Die Autorinnen und Autoren von ▶ Kap. 20 bedanken sich beim Bundesministerium für Bildung und Forschung (BMBF) und beim DLR Projektträger für die Unterstützung und Förderung des KAHR-Projekts. Diese Forschung wurde vom BMBF unter dem Förderkennzeichen 01LR2102A-M finanziert.

Literatur

Abunyewah M, Gajendran T, Maund K (2018) Conceptual framework for motivating actions towards disaster preparedness through risk communication. Procedia Eng 212:246–253. https://doi.org/10.1016/j.proeng.2018.01.032

Balog-Way D, McComas K, Besley J (2020) The Evolving Field of Risk Communication. Risk Anal 40(S1):2240–2262. https://doi.org/10.1111/risa.13615

BBK, BfR (2022) Risikokommunikation Bd 1. Bundesamt für Bevölkerungsschutz, Katastrophenhilfe (BBK) Bundesinstitut für Risikobewertung (BfR)

Berghäuser L, Bubeck P, Hudson P, Thieken A (2023) Identifying and characterising individual flood precautionary behaviour dynamics from panel data. Int J Disaster Risk Reduct 94:103835. https://doi.org/10.1016/j.ijdrr.2023.103835

Birkmann J, Schüttrumpf H, Handmer J, Thieken A, Kuhlicke C, Truedinger A, Sauter H, Klopries E, Greiving S, Jamshed A, Merz B, Solecki W, Kirschbauer L (2023) Strengthening resilience in reconstruction after extreme events – Insights from flood affected communities in Germany

BMWSB (2022) Hochwasserschutzfibel Bd 9. Bundesministerium für Wohnen, Stadtentwicklung und Bauwesen

Bubeck P, Botzen WJW, Aerts JCJH (2012) A review of risk perceptions and other factors that influence flood mitigation behavior. Risk Analysis 32(9):1481–1495

Bubeck P, Osberghaus D, Thieken A (2023) Explaining Changes in Threat Appraisal, Coping Appraisal, and Flood Risk-Reducing Behavior Using Panel Data From a Nation-Wide Survey in Germany. Environ Behav 55(4):1–25. https://doi.org/10.1177/00139165231176069

Dillenardt L, Hudson P, Thieken AH (2022) Urban pluvial flood adaptation: Results of a household survey across four German municipalities. J Flood Risk Manag. https://doi.org/10.1111/jfr3.12748

Dillenardt L, Bubeck P, Hudson P, Wutzlrt B, Thieken A (2024) Property-level adaptation to pluvial flooding: An analysis of individual behaviour and risk communication material. Mitig Adapt Strateg Glob Change 29:53

Directive 2007/60/EC of the European Parliament and of the Council of 23 October 2007 on the assessment and management of flood risks

DKKV (2015) Das Hochwasser im Juli 2013. Deutsches Komitee Katastrophenvorsorge e. V. (DKKV), Universität Potsdam. ISBN 978-3-933181-62-6.

GDV (2022) Serviceteil zum Naturgefahrenreport 2022. https://www.gdv.de/resource/blob/105836/117679ac1f31d229d86a1c424fe0aab8/download-serviceteil-naturgefahrenreport-2022-data.pdf. Zugegriffen: 2024

Grothmann T, Reusswig F (2006) People at Risk of Flooding: Why Some Residents Take Precautionary Action While Others Do Not. Nat Hazards 38:101–120.

Grounds MA, LeClerc JE, Joslyn J (2018) Expressing Flood Likelihood: Return Period versus Probability. Weather Clim Soc 10(1):5–17

Kellens W, Zaalber R, De Maeyer P (2012) The Informed Society: An Analysis of the Public's Information-Seeking Behavior Regarding Coastal Flood Risks. Risk Anal 32(8):1369–1381

Rogers RW (1983) Cognitive and Physiological Processes in Fear Appeals and Attitude Change: A Revised Theory of Protection Motivation. In: Cacioppo JT, Petty RE (Hrsg) Social Psychophysiology: A Sourcebook. Guilford, S 153–176

UBA (2022) Hochwasserrisiken umfassend erkennen und handhaben

Vorogushyn S, Apel H, Kemter M, Thieken AH (2022) Analyse der Hochwassergefährdung im Ahrtal unter Berücksichtigung historischer Hochwasser. Hydrol Wasserbewirtsch 66(5):244–254. https://doi.org/10.5675/HyWa_2021.5_2

Yamaguchi J, Kinoshita K (2023) The risk of a new tetanus outbreak due to urban flooding disaster requires vigilance: a narrative review. Acute Med Surg 10(1):e839

Erfahrungen im Wiederaufbau

Bert Droste-Franke, Tanja Nietgen und Michael Kunkis

J. Birkmann et al. (Hrsg.), *Die Flutkatastrophe 2021*, Sustainable Development Goals (SDG) – Umsetzung in Praxis, Lehre und Entscheidungsprozessen,
https://doi.org/10.1007/978-3-658-48688-4_21

21.1 Herausforderungen und Instrumente des Transfers von Forschungsergebnissen in die Praxis

Um Forschungsergebnisse in der Praxis nutzen zu können, reicht es in den meisten Fällen nicht aus, die Ergebnisse vorliegen zu haben und ohne zusätzliche Aufarbeitung weiterzugeben. Abhängig von der Art des Ergebnisses und dem Anwendungsfeld stellt der Transfer in die Praxis einen erheblichen zusätzlichen Aufwand dar, der geeignete Instrumente erfordert und bei dem eine ganze Reihe an Herausforderungen beachtet werden müssen. Idealerweise beginnt der Transfer bereits möglichst frühzeitig, z. B. bei der Projektplanung, jedoch spätestens bei der Erarbeitung der Ergebnisse, sodass diese bestmöglich auf den Anwendungskontext angepasst erarbeitet werden können.

Im Folgenden werden die Herausforderungen des Transfers zunächst aus verschiedenen Blickwinkeln betrachtet. Dazu wird als Erstes die Perspektive der Erarbeitung lösungsorientierten Wissens zur Technikgestaltung aus wissenschaftlicher Expertise heraus eingenommen. Eine weitere Perspektive ergibt sich aus der Erarbeitung von Innovationen aus Wissen. Die Betrachtung der Wirklogik von Projektergebnissen zeigt weitere Herausforderungen auf. Vor dem Hintergrund dieser Herausforderungen wird zusätzlich diskutiert, nach welchen Kriterien Transfermaßnahmen ausgewählt bzw. gestaltet werden können. Die prozessorientierte Darstellung des Wiederaufbaus zeigt außerdem auf, welche Prozesse für eine Umsetzung von Maßnahmen berücksichtigt werden sollten, um möglichst wirksam zu sein.

21.2 Herausforderungen aus Sicht der Erarbeitung lösungsorientierten Wissens in der Technikgestaltung

Erste Herausforderungen werden sichtbar, wenn zunächst die Perspektive der Erarbeitung lösungsorientierten Wissens – wie in der Technikgestaltung bzw. in der Technikfolgenabschätzung – eingenommen wird. Die benötigten Wissenskomponenten zur Technikgestaltung lassen sich sehr gut zerlegen in die drei Komponenten „Orientierungswissen" (oder Zielwissen), „Systemwissen" und „Handlungswissen". Zusammengenommen lässt sich daraus das „Gestaltungswissen" generieren (vgl. Droste-Franke 2018; Grunwald 2004). Orientierungswissen oder Zielwissen beinhaltet dabei vor allem Wissen über wünschenswerte (gesellschaftliche) Ziele und damit vorwiegend normatives Wissen bzw. Bewertungswissen. Systemwissen ist hingegen das Wissen über Zusammenhänge zwischen einer Handlung und ihren Rahmenbedingungen und Auswirkungen. Dieses enthält idealerweise nur deskriptive Komponenten, durch die Notwendigkeit von Abstraktionen oft aber auch implizit normative Elemente. Handlungswissen beinhaltet vor allem Wissen darüber, wie man unter bestimmten Rahmenbedingungen handeln muss, um das gesetzte direkte Ziel des Handelns zu erreichen und damit die Handlung erfolgreich zu gestalten (vgl. Janich 2001). In Summe werden damit alle Wissenskomponenten benötigt, um so zu handeln, dass die Auswirkungen des Handelns zum (gesellschaftlich) gewünschten Ziel bzw. zu einer Verbesserung der Sachverhalte (im gesellschaftlichen Sinne) führen. In der konkreten Anwendung hier gilt es das Wissen so zu erarbeiten und zu kombinieren, dass Maßnahmen zum Wiederaufbau so gestaltet sind, dass mit ihnen gesellschaftlich gewünschte Ziele erreicht werden können.

Die einzelnen Wissenskomponenten sind mit verschiedenen Herausforderungen verbunden, die für den Transfer des Gestaltungswissens in die Praxis berücksichtigt werden müssen. Basierend auf Droste-Franke et al. (2015) ergeben sich unter anderem folgende Aspekte:

Bei der Erzeugung von Gestaltungswissen und, allgemeiner, in der Praxis anwendbarem Wissen ist zu berücksichtigen, dass neben reinem wissenschaftlichem Wissen auch Kontextwissen für den Anwendungsbereich zentral ist, um die Ergebnisse für die Praxis passend zu erstellen. Grundsätzlich erfordert die Aufgabe, generelles wissenschaftliches Wissen für die praktische Gestaltung einzusetzen, eine Anwendung dieses Wissens auf konkrete Kontexte und dafür oftmals eine Ergänzung mit weiterem Wissen, zum Teil auch rein praktischem Wissen bzw. technisches (Wissen und Können zur Bedienung von Geräten), professionelles (bspw. konkrete Rechtsprechungen, Wirkungen und Behandlungen von Krankheiten) bzw. lokal-empirisches Wissen (Erfahrungswissen z. B. zu Wirkungen in realen Umgebungen). Dies steht im Gegensatz zu üblichen disziplinären Untersuchungen, in denen entweder unter Laborbedingungen gearbeitet wird, um Rahmenbedingungen zu kontrollieren, oder die sich spezifisch auf eine konkrete Situation beziehen, aber dadurch nicht oder nur schwer übertragbar sind. Durch Anpassung auf die konkreten Kontexte und die inter- und transdisziplinäre Kombination allgemeiner wissenschaftlicher Erkenntnisse ergeben sich zum Teil Inkonsistenzen, die für die Erarbeitung anwendbarer Ergebnisse und ihrer Interpretation zu beachten sind.

Zusätzlich wird die Relevanz der Ergebnisse gesellschaftlich zum einen auf Basis ihrer Passung für die Anwendung beurteilt. Zum anderen ist die gesellschaftliche Bewertung der mit ihrer Anwendung verursachten Wirkungen und erreichbaren Ziele zentral. Bewertungsgrundlagen sind damit in erster Linie nicht-epistemische, normative Kriterien. Im Gegensatz zur akademischen Forschung, die auf Erkenntnisgewinn abzielt, fließen damit ganz wesentlich gesellschaftlich-normative Aspekte in die Bewertung der Ergebnisse ein.

Vor diesem Hintergrund ist es bei der Erarbeitung von Wissen für die Praxis wichtig, möglichst frühzeitig praktische Expertise in Form von professioneller, technischer und lokal-empirischer Expertise hinzuzuziehen. Zusätzlich sollte den beteiligten wissenschaftlichen Expertinnen und Experten bewusst sein, dass ihre Arbeiten nicht auf zusätzlichen wissenschaftlichen Erkenntnisgewinn abzielen, sondern auf ihre möglichst gute Einsetzbarkeit in der Praxis, hier konkret im Wiederaufbau.

21.3 Herausforderungen aus der Perspektive von Innovationsprozessen

Die Betrachtung von Innovationsprozessen zeigt weitere Herausforderungen, die mit der Erzeugung anwendbaren Wissens für die Praxis verbunden sind. In Innovationsprozesse gehen verschiedene Inputs ein. Diese können unterschieden werden in einen Anteil gefestigten universitären Wissens, Komponenten aus der Grundlagen-, System- und Prozessforschung sowie „geleitetes empirisches“ Wissen, sozusagen Wissen aufgrund intelligenten Ausprobierens (Kline und Rosenberg 1986). Damit verbunden ist, dass eben auch für die Nutzung des Wissens ggf. gewisse Grundkenntnisse notwendig sind. Es muss die Kompetenz auf Seiten des Rezipienten bestehen das Wissen zu verstehen bzw. aufzunehmen, was auch als „Absorptionsfähigkeit“ bezeichnet wird (Alexander 2012; Adams et al.

2006; Pyka 2007; Cohen und Levinthal 1989; Eliasson 1990), und darüber hinaus auch noch anwenden zu können. Diese „Absorptionsfähigkeit" muss im Rahmen des Transfers bewerkstelligt werden.

Des Weiteren ist bei der Entwicklung und Auswahl geeigneter Formate für den Transfer von Wissen in die Praxis zu beachten, dass wissenschaftliches Wissen meist explizites, kodifiziertes Wissen darstellt. Dieses muss oft um sogenanntes nicht-kodifiziertes bzw. implizites Wissen, das nicht explizit dargestellt ist und damit oft direkt vom Wissensträger weitergegeben werden muss, ergänzt werden (s. auch die Diskussion zu Gestaltungswissen oben). Vor allem Handlungswissen bzw. praktisches Wissen enthält oft sogar Komponenten, die nicht explizit gemacht werden können (Polanyi 1966; Foray 2007; Alexander 2012). Im Übergang zur Anwendung muss wissenschaftliches Wissen um passendes implizites Wissen ergänzt werden. Dazu sind zusätzlich entsprechende Entwicklungs- und Transferformate notwendig. Je nach Adressat der Ergebnisse werden außerdem in der Praxis geeignete Austauschformate und Trainings benötigt, um das Wissen, inklusive der impliziten Komponenten, in der Praxis in bereits bestehende Prozesse zu integrieren, zu testen und zudem möglichst langfristig zu festigen. Im Wiederaufbauprozess scheint dieses z. B. in Bezug auf die Vorbereitung auf und das Verhalten in Katastrophensituationen wichtig zu sein. Für den Katastrophenschutz sind neben theoretischen Weiterbildungen entsprechende Übungen zentral.

21.4 Herausforderungen aus Sicht der Wirkungsbetrachtung

Um den Erfolg von Projekten im Vorhinein zu antizipieren, wird oft ein grobes Modell einer Wirklogik der Ergebnisse im gesellschaftlichen Kontext erstellt. Die Ergebnisse aus Projekten (Output) sind dabei der erste Schritt. Erst wenn sie tatsächlich in der Praxis angewendet werden (Outcome), ergibt sich eine Veränderung von gesellschaftlichen Prozessen und Sachverhalten (Impact) (vgl. W. K. Kellogg Foundation 2004). Aufgrund der Unsicherheiten zukünftiger Entwicklungen und vielfältiger Einflüsse nicht vorhersehbarer Rahmenbedingungen sind Outcome und Impact oft nur sehr schwer a priori einzuschätzen. Selbst wenn Systemwissen vorhanden ist, enthält dieses Unsicherheiten. Im Falle komplexer (gesellschaftlicher) Prozesse, die oft auch durch individuelle Entscheidungen geprägt sind, können abstrahierte systemische Betrachtungen oft nur einen Hinweis auf mögliche Entwicklungen geben. Nichtsdestotrotz kann ein kontinuierliches Monitoring der Zielerreichung und ggf. Nachbesserung der Outputs die Erreichung des gewünschten „Outcomes" und „Impacts" unterstützen. Entsprechende Indikatoren für ein solches Monitoring und die Nachsteuerung erfordern jedoch idealerweise Aktivitäten, die deutlich über die reine Erarbeitung des Outputs hinausgehen und teilweise langfristig angelegt sein müssen.

Im Projekt KAHR geht es um Klimaanpassungsmaßnahmen, mit denen Regionen resilient gestaltet werden können. Eine wesentliche Grundlage dazu ist es, eine langfristige Verstetigung und damit den nachhaltigen Transfer in die Praxis zu gewährleisten. Dazu sind Formate und Maßnahmen zu nutzen und ggf. zu entwickeln, mit denen auch ohne permanente Unterstützung von außen eine erfolgreiche Anwendung in der Region gewährleistet werden kann.

21.5 Auswahl und Gestaltung von Transferinstrumenten

Die oben diskutierten Herausforderungen zeigen, dass für die Auswahl bzw. die Gestaltung geeigneter Formate und Instrumente für den Transfer verschiedenste Aspekte berücksichtigt werden müssen. Führt man diese zusammen, so erhält man über Eigenschaften, die die Instrumente aufweisen müssen, eine Auswahl geeigneter Instrumente. Das Vorgehen ist ähnlich eines Trichters, in den oben alle Instrumente hineingegeben werden und aus dem unten die geeigneten Instrumente herauskommen (s. Abb. 21.1).

Verschiedene Filterkriterien können dazu auf die Auswahl der Instrumente angewendet werden:

- Zum einen ist es wichtig nach den Arten der Produkte zu unterscheiden, die im Projekt angestrebt werden: Handelt es sich um reine Kommunikation von Wissen bzw. Information, handelt es sich darum Kompetenzen mit praktischen Inhalten zu vermitteln, ist es letztendlich eine wissenschaftliche Dienstleistung, die mit dem Output bzw. Produkt des Projekts erbracht werden soll oder ist es eine einzelne Veranstaltung?
- Des Weiteren ist die Zielgruppe für die konkrete Gestaltung des Transfers zentral. Hier sind im Projekt Gesellschaft, Verwaltung, Politik, Wirtschaft und Wissenschaft zu unterscheiden.
- Je nachdem welches Ziel erreicht werden bzw. in welchem Bereich Impact erzeugt werden soll, ist des Weiteren der thematische Kontext zu unterscheiden, wie etwa Klimaschutz, Krisenprävention, Risikomanagement und Anwendungsebene.
- Um wirksam zu sein, muss das gewählte Instrument auch auf den Unterstützungsbedarf im Transfer angepasst sein: Ist das Ergebnis verallgemeinerbar, skalierbar und muss es für andere Zielgruppen angepasst werden? Ist es eine Hardware, Software oder Information? Sind vorhandene Netzwerke für die Verbreitung nutzbar oder werden neue Netzwerke benötigt? Und schließlich: Wie hoch ist der Aufwand, die Nutzung zu verstetigen?
- Des Weiteren können generalisierte Impact-Ziele definiert werden, für die jeweils passende Indikatoren für den Impact genutzt werden können. Mögliche Ziele im Kontext

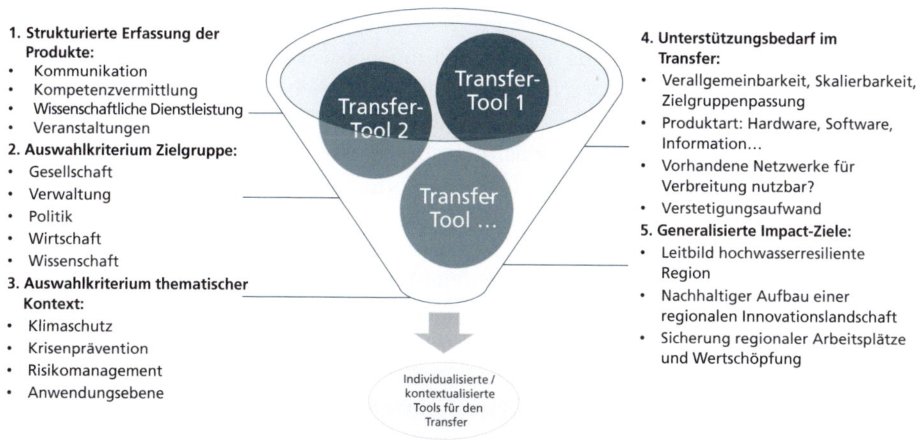

Abb. 21.1 Schematisch dargestellte Vorgehensweise zur Auswahl jeweils passender Transfertools. (Quelle: eigene Darstellung)

des KAHR-Projekts sind etwa: Erreichung des Leitbilds hochwasserresiliente Region, nachhaltiger Aufbau einer regionalen Innovationslandschaft oder Sicherung regionaler Arbeitsplätze und Wertschöpfung.

Anwendung des Filters am Beispiel des Produkts: Kindermaterial – Lernen, verstehen, vorbereitet sein (erstellt vom HochwasserKompetenzCentrum e. V. [HKC])

1. Die vom HKC erstellten Materialien für Kinder fallen in die Produktgruppe „Kompetenzvermittlung", da sie – z. B. im Rahmen des konzipierten Kinderworkshops – eine Interaktion mit der Zielgruppe auslösen.
2. Das Material wendet sich an Kinder (Zielgruppe Gesellschaft) im Grundschulalter.
3. Die für Kinder erstellten Materialien sind thematisch der Krisenprävention zuzuordnen: Kinder erlernen unter pädagogischer Betreuung für sie relevante Aspekte der Überflutungsvorsorge, sie erproben richtiges und schnelles Handeln in Gefahrensituationen. Des Weiteren ist dieses Produkt dem Risikomanagement im Bildungsbereich zuzuordnen: Kinder können mit diesen Materialien im schulischen und außerschulischen Kontext (z. B. Umweltlernangebote) angesprochen werden. Ziel ist es ihnen die entsprechenden Kompetenzen zu vermitteln.
4. Die Kindermaterialien und das damit verbundene Konzept (Kinderworkshop) sind skalierbar, also ohne große Anpassung und in vielen Einrichtungen anwendbar. Es handelt sich somit um ein niedrigschwelliges und nahezu unmittelbar anwendbares Produkt.
5. Das generalisierte Impact-Ziel entspricht dem Leitbild einer hochwasserresilienten Region. Eine Sensibilisierung für dieses Leitbild sollte möglichst frühzeitig erfolgen und Kindern die Möglichkeit gegeben werden eine Steigerung ihrer Selbstwirksamkeit durch erlernte Verhaltensweisen zu erreichen. Die Kinder tragen ihr neues Wissen in ihre Familien, diese sind wiederum elementarer Bestandteil einer resilienten Gesellschaft und somit einer resilienten Region.

21.6 Prozessblick für eine wirkungsorientierte Umsetzung von Maßnahmen

Um den Transfer von Wissen aus dem Projekt möglichst wirkungsvoll zu gestalten, ist eine Prozesssicht auf Maßnahmen im Wiederaufbau hilfreich. Hierbei sind verschiedene Phasen der Gestaltung zu unterscheiden. Grundlage dieser Darstellung bildet die Kategorisierung von Wirkungen für wissenschaftliche Beratungen im Politikbereich. Zum einen wird Wissen vergrößert (deskriptive Dimension), zum anderen werden Meinungen geformt (normative Aspekte) und zum Dritten werden durch sie Aktivitäten initiiert (vgl. Hennen et al. 2004). Nimmt man diese Wirkungsdimension zusammen mit den Dimensionen, in denen die Wirkungen erfolgen (ebd.), vor allem die gesellschaftlichen und politischen Aspekte, und ergänzt diese um regionale und überregionale strukturelle Faktoren, statt technologisch/wissenschaftlicher Aspekte, ergibt sich eine Matrix wie sie in ◘ Abb. 21.2 dargestellt ist (s. z. B. Schöttmer und Droste-Franke 2022). Prozesse aus den Phasen der Status-quo-Erhebung, der Strategieentwicklung und der Umsetzungsplanung werden immer wieder iterativ durchgeführt werden müssen. Die Matrix ermöglicht eine strukturierte Einordnung einzelner Schritte in den gesamten Komplex. Transfer ist hier in verschiedenen Schritten notwendig. Je nachdem welcher Art das Forschungsergebnis ist, ordnet es sich in verschiedene Bereiche der verschiedenartigen Wiederaufbauprozes-

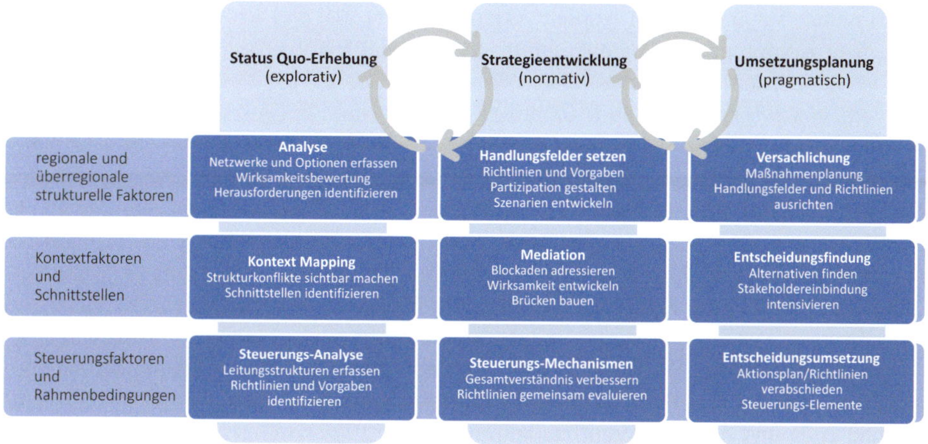

Abb. 21.2 Iteratives Vorgehen und Feedbackschleifen im Transferprozess. (Quelle: Schöttmer und Droste-Franke 2022)

se ein. Die Einordnung der verschiedenen Ergebnisse in die Matrix kann helfen, konkrete Ansatzpunkte für Transferinstrumente zu finden und diese im Hinblick auf ihren Einsatz weiter zu schärfen.

21.7 Das Projektbüro als Baustein eines nachhaltigen Transfers

Grundlegendes Hauptziel des Forschungsprojekts KAHR ist die Bereitstellung wissenschaftlicher Erkenntnisse aus der Forschung für die betroffenen Regionen und Akteure. Somit ist die Auseinandersetzung mit der Fragestellung, wie wissenschaftliches Wissen in die Handlungspraxis überführt und dort nachhaltig angewendet werden kann, zentral. Die Unterstützung des Transferprozesses ist sowohl Voraussetzung zur Steigerung des Impacts öffentlicher Forschungsinvestitionen als auch Grundlage der Unterstützung der Pilotregionen auf ihrem Weg hin zu erhöhter Resilienz gegenüber Hochwasser- und Starkregenereignissen und weiteren (Natur-)Katastrophen.

Die KAHR-Projektbüros als zentrale Vor-Ort-Koordinierungsstellen für Rheinland-Pfalz und Nordrhein-Westfalen sind elementare Bausteine der Transferaktivitäten in die Regionen. Der Transferprozess wird auf vielfältige Art und Weise unterstützt, u. a. durch die Etablierung einer direkten und regelmäßigen Kommunikation mit den zentralen Akteuren im Wiederaufbau aus Politik und Verwaltung. Der enge Austausch und die feste Verankerung im KAHR-Projekt und damit die Möglichkeit Produkte frühzeitig mitzugestalten ermöglicht es, praktisches Wissen und konkrete Erfahrungen aus dem Wiederaufbau direkt in die Erarbeitung der Produkte und in den Transfer einzubringen.

Aufgrund der Größe des Schadensgebietes mit einer Vielzahl betroffener Gemeinden und Akteure sind eine kohärente und transparente Kommunikation sowie die kontinuierliche Erhebung und Systematisierung der Bedarfe der Region essentieller Schlüsselfaktor eines nachhaltigen Transfers.

Die Projektbüros sind zentrale Anlaufstelle und Schnittstelle zwischen der Wissenschaft und der Region und kommunizieren und vernetzen somit in beide Richtungen:

Im Hinblick auf die Region werden Akteure, Netzwerke und Bedarfe identifiziert und geeignete Kommunikationskanäle gefunden und etabliert. Im Hinblick auf die jeweiligen verschiedenen wissenschaftlichen Themen werden für die Projektpartner relevante Akteure, Netzwerke und Bedarfe identifiziert und vermittelt. Dieses Wissen ist ganz wesentlich für die bedarfsgenaue und zielgruppenorientierte Generierung wissenschaftlicher Erkenntnisse und Produkte. Diese werden wiederum den von der Flut 2021 betroffenen Regionen in Form von Informationsveranstaltungen, Leitfäden, direkten Beratungen zur Verfügung gestellt – in enger Abstimmung werden die Fachkompetenz des Projektkonsortiums mit den praktischen Bedarfen verbunden. Diese enge Vernetzung ist Voraussetzung einer optimierten Kohärenz zwischen den Projektaktivitäten und den Bedürfnissen aller Akteure. Für diese Prozesse sind die Projektbüros maßgeblich verantwortlich.

Über diese Aktivitäten hinaus sind die Projektbüros erste Anlaufstelle für die Wissenschaft jenseits des KAHR-Projektkonsortiums mit Fragestellungen die Hochwasserkatastrophe 2021 und den Wiederaufbau betreffend. Die Projektbüros unterstützen mit ihrer regionalen Expertise Wissenschaftlerinnen und Wissenschaftler bei u. a. Rechercheaufgaben, Untersuchungen wissenschaftlicher Fragestellungen, Datenerhebungsprozessen und Exkursionen (▶ Kap. 5). Diese Unterstützung wird gleichermaßen bei wissenschaftlichen Facharbeiten und Abschlussarbeiten wie bei nationalen und internationalen Projekten und Projektverbünden gewährt. Des Weiteren erfolgt über die Projektbüros eine Vernetzung zwischen Wissenschaftlerinnen und Wissenschaftlern, die nicht dem KAHR-Konsortium angehören mit dem KAHR-Projektverbund. Für alle diese Aktivitäten gilt, dass sie sowohl im regionalen als auch im nationalen und internationalen Kontext stattfinden.

Die Projektbüros repräsentieren das Projekt KAHR auf regionaler Ebene im Hinblick auf den Transfer anwendungsnaher Ergebnisse. Im Zuge der Teilnahme an Veranstaltungen werden die Sichtbarkeit des Projekts erhöht, Verständnis für die Projektinhalte geschaffen, lokale Akteure gewonnen und Anwendungsmöglichkeiten erschlossen. Des Weiteren wird im Rahmen dieser Aktivitäten ein regionaler und öffentlicher Zugang zu den Projektinhalten und -partnern etabliert. In Summe stellen die Projektbüros die Grundlage für einen engen Austausch mit den Akteuren vor Ort und einen nachhaltigen Transfer dar. Dabei dient die Kenntnis der oben diskutierten Herausforderungen immer als Grundlage.

21.8 Wissen über die Region als Grundlage passgenauer Ergebnisse

Die KAHR-Projektbüros für Rheinland-Pfalz und Nordrhein-Westfalen begleiten die KAHR-Projektpartner im Vorfeld und bei ihren Vor-Ort-Aktivitäten und liefern wesentliche Kenntnisse über die Region und laufende regionale Aktivitäten sowie über die Interessen und Bedarfe der Akteure vor Ort. Aufgrund der umfangreichen Schadens- und Interessenslage erfolgt die Sichtung und Sammlung aller regional relevanter Informationen die Hochwasserkatastrophe 2021 betreffend in strukturierter Form; diese Informationen fließen in eine zentrale Datenbank ein, die den Projektpartnern uneingeschränkt zur Verfügung steht. Diese Datenbank ist Grundlage einer engen Abstimmung zwischen den Expertinnen und Experten des Projektkonsortiums mit den Bedarfen, Netzwerken und Akteuren der betroffenen Gemeinden und des Landkreises Ahrweiler. Mithilfe dieser Datenbank werden direkte und indirekte Kommunikationsstrukturen

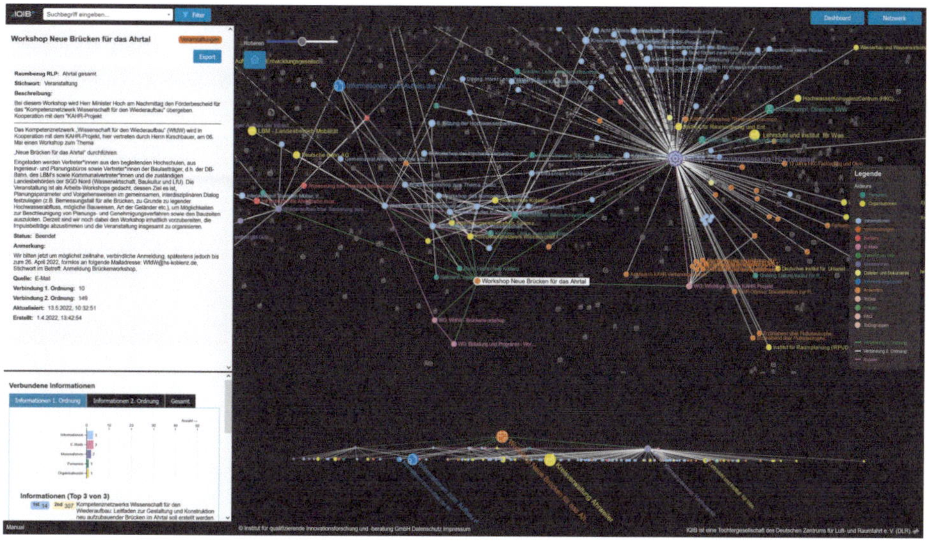

Abb. 21.3 Einbindung von Akteuren und Akteursorganisationen in das gesamte den Wiederaufbau umfassende Netzwerk. (Quelle: Screenshot der Netzwerkdarstellung der KAHR-Wissensdatenbank)

geschaffen, auf deren Grundlage bedarfsorientierte wissenschaftliche Ergebnisse generiert werden können. Auf Basis dieser Vorgehensweise wird sichergestellt, dass sich die wissenschaftlichen Forschungsfragen unmittelbar an den Bedarfen, Netzwerken und Akteuren zum Wiederaufbau der Region orientieren können.

Technische Grundlage ist eine auf Basis von Atlassian (Jira Service Management) entwickelte Wissensdatenbank. In diese werden, unmittelbar nach der Hochwasserkatastrophe 2021 beginnend, Akteure, Akteursorganisationen, Bedarfe und Maßnahmen aufgenommen sowie – für das KAHR-Projekt – relevanten Themenfeldern wie „Baugebiete", „Hochwasserschutz", „Traumabewältigung", „Wirtschaft" zugeordnet. Aufgrund dieser Struktur ist eine direkte und detaillierte Suche – auf Basis stetig aktualisierter Daten – über die für die Region Ahrtal relevanten Aspekte möglich. Grundlage der Daten sind u. a. Veröffentlichungen in der regionalen und überregionalen Presse, regionale Veranstaltungen, themenrelevantes Informationsmaterial und historische Quellen.

Einen weiteren Zugang zu den in der Datenbank hinterlegten Informationen stellen die visualisierten Daten dar (s. Abb. 21.3): Durch die Netzwerkdarstellungen ist z. B. die Einbindung individueller Akteure oder Akteursorganisationen im Gesamtkontext direkt sichtbar. Die interaktive Darstellung ermöglicht es dem Nutzer zudem, sich entlang der Zusammenhänge von Element zu Element durch die Datenbank durchzuarbeiten.

Die auf Grundlage der erhobenen regionalen Bedarfe durch die Projektpartner generierten Ergebnisse und Produkte werden im Rahmen geeigneter Transferformate, wie u. a. Veranstaltungen, Leitfäden und Stellungnahmen der Region zur Verfügung gestellt und sind – als Resultat einer kontinuierlichen und nachhaltigen Anwendung durch die Akteure vor Ort – Grundlage eines wissenschaftlich begleiteten Wiederaufbaus und somit einer resilienten Region.

21.9 KAHR: Produkte und Ergebnisse für die Region

Im Rahmen der Projektlaufzeit von KAHR wurden von den Konsortialpartnern eine Vielzahl von Produkten entwickelt[1]. Losgelöst von ihrer Verschiedenheit haben die Produkte alle einen Aspekt gemein: Im Zuge eines optimalen Transfers der Produkte, die an den Bedarfen und Akteuren der betroffenen Regionen ansetzen, sollen sie zu einem verbesserten Wiederaufbau und somit zu einer in der gesamten Region gestärkten Resilienz beitragen. Die Gesamtheit der für Regionen bestimmten Produkte lässt sich in die folgenden vier Gruppen unterteilen (vgl. auch Abb. 21.4):

- Kommunikation,
- Kompetenzvermittlung,
- Wissenschaftliche Dienstleistungen und
- Veranstaltungen.

Zusätzlich ist in der Abbildung der jeweilige Transferaufwand angegeben: Während viele Produkte, die direkt im geeigneten Format erstellt werden, wie z. B. Leitfäden, auch direkt verwendet werden können, erfordern andere, wie z. B. bestimmte Veranstaltungs- oder Weiterbildungsformate, teilweise einen erheblichen Aufwand, um sie in anderen Kontexten durchführen zu können.

In der Gruppe „Kommunikation" finden sich Produkte, die von KAHR-Partnern entwickelt wurden und sich direkt an eine oder mehrere Zielgruppen – aus den Bereichen Politik, Verwaltung, Zivilgesellschaft, Wissenschaft – wenden. Bei diesen Produkten handelt es sich in der Regel um Bekanntgaben aus dem Konsortium mit der Intention, die Zielgruppe zu informieren. Eine wechselseitige Kommunikation ist bei diesen Produkten i. d. R. nicht vorgesehen. So finden sich hier Informationsmaterialien wie u. a. Flyer (z. B. „Erste Handlungsempfehlungen für von Hochwasser und Starkregen Betroffene", erstellt durch das Institut für Umweltwissenschaften und Geographie [IUG]) und (Fach-)Broschüren (z. B. „So schütze ich mein Haus", erstellt durch das HochwasserKompetenzCen-

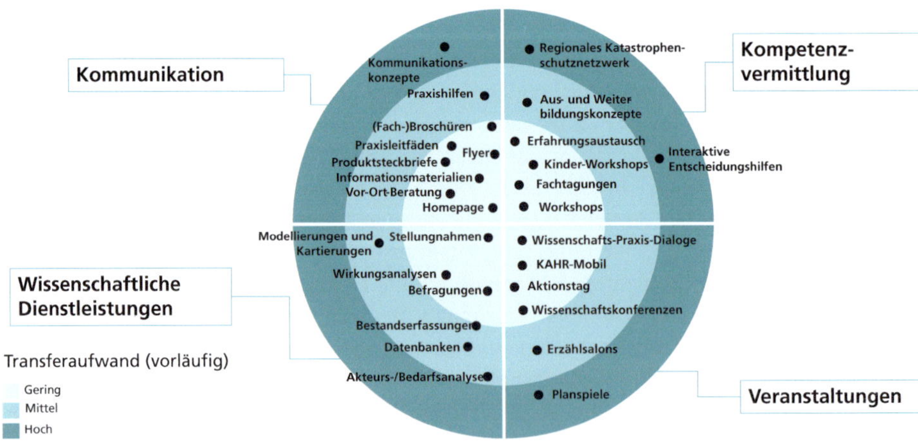

Abb. 21.4 Strukturierung der in KAHR erarbeiteten Produkte nach Produktgruppen und Transferaufwand. (Quelle: eigene Darstellung)

1 Vgl. ▶ https://hochwasser-kahr.de/index.php/de/kahr-produkte.

trum e. V. [HKC]), die Homepage des Projekts („hochwasser-kahr.de") sowie Praxisleitfäden (z. B. „Praxisleitfaden für einen klimaresilienten Wiederaufbau", erstellt durch das gesamte Konsortium unter der Leitung des Deutschen Instituts für Urbanistik [difu]) und Praxishilfen (z. B. Praxishilfe zur Stärkung der Hochwasserresilienz von Sportstätten, erstellt unter der Leitung des Instituts für Raum- und Entwicklungsplanung [IREUS]).

Im Bereich „Kompetenzvermittlung" finden sich Produkte, deren Inhalt über die einseitige Information der Zielgruppe hinausgeht, und die eine Interaktion auslösen. Hier werden wissenschaftliche Erkenntnisse zielgruppengerecht aufbereitet – der Adressat soll dazu befähigt werden, Kompetenzen aufzubauen, um Schadensereignisse besser zu beherrschen als zuvor. Die Vermittlung von Kompetenzen – also Schulung von verschiedenen Zielgruppen – steht hier im Fokus. In diesen Formaten kann teilweise auch implizites Wissen vermittelt werden. So wendet sich u. a. ein Kinder-Workshop (erstellt durch HKC) an Vorschul- und Grundschulkinder, der persönliche Erfahrungsaustausch zwischen ehemals von der Flut betroffenen und 2021 von der Flut betroffenen Regionen („Austausch mit Referenzkommunen zum Wiederaufbau", durchgeführt von difu) an Politik und Verwaltung, während sich Schulungskonzepte (z. B. „Aus- und Weiterbildungskonzept für Einsatzkräfte aus Feuerwehren und weiterer Institutionen" erstellt durch das Fachgebiet Wasserbau und Wasserwirtschaft [FWW] der Rheinland-Pfälzischen Technischen Universität) an die operativen Einsatzkräfte im Rahmen des Hochwasser- und Starkregeneinsatzes wendet.

Die Produkte der Gruppe „Wissenschaftliche Dienstleistungen" wenden sich i. d. R. an die Fachcommunity bzw. direkt an die KAHR-Konsortialpartner sowie weitere Adressaten aus der Zielgruppe „Wissenschaft". Teilweise werden die Produkte allgemeinverständlich für weitere Adressaten aufbereitet. Hier finden sich für das KAHR-Projekt grundlegende Arbeiten wie u. a. die Datenbanken „Schadensdatenbank" sowie „Objektschutzmaßnahmen" – sie fließen unmittelbar in das KAHR-Produkt „Entscheidungsmatrix für Objektschutzmaßnahmen" (interaktive Entscheidungshilfe, Bereich Kompetenzvermittlung) ein (Datenbanken und Entscheidungsmatrix erstellt durch die Hochschule Koblenz). Die Matrix gibt Hinweise auf Maßnahmen, die zum Schutz eines individuellen Gebäudes in Bezug auf Wassereintrag (Hochwasser oder Starkregen) sinnvoll sind. Diese sollen Fachleute und interessierte Laien gleichermaßen nutzen können. Weitere Produkte dieser Gruppe, u. a. „Modellkette Ahr" (▶ Abschn. 4.2) – ein umfassendes hydrologisches-hydraulisches Modell des gesamten Ahr-Einzugsgebiets für die Hochwasserrisikoanalyse – sowie „Wettergenerator" – zur Erstellung langer synthetischer Zeitreihen von realistischen Klimadaten unter Berücksichtigung des Klimawandels – (beide Modellierungen erstellt durch das Helmholtz-Zentrum für Geoforschung [GFZ]) modellieren das Flutereignis 2021 retrospektiv und helfen zukunftsgerichtete Aussagen zu formulieren. Diese Arbeiten fließen in weitere KAHR-Produkte (z. B. „Hydraulisches 2-D-Modell", erstellt durch Institut für Wasserbau und Wasserwirtschaft [IWW] der RWTH Aachen) ein. Ziel ist hier vor allem die Analyse naturwissenschaftlich-systemischer Zusammenhänge. Weitere Arbeiten, wie u. a. Berichte (z. B. „Akteurs- und Bedarfsanalyse", erstellt unter der Leitung des Helmholtz Zentrums für Umweltforschung [UFZ]) unterstützen die Projektpartner bei der pass- und zielgruppengenauen Entwicklung ihrer eigenen Arbeiten.

In der Gruppe „Veranstaltungen" finden sich zahlreiche durch das KAHR-Konsortium entwickelte und durchgeführte Formate, die sich einmalig (z. B. verschiedene Aktionstage) oder wiederholend (z. B. „Wissenschaftskonferenz", „Wissenschaft-Praxis-Dialoge", beide entwickelt und durchgeführt von dem gesamten KAHR-Konsortium) an unterschiedliche Adressaten wenden. Im Mittelpunkt der Veranstaltungen steht einerseits die Information über neue wissenschaftliche und für die Praxis relevante Informationen, andererseits die Bewertung dieser Erkenntnisse durch die Praxis. Diese Formate dienen somit dem direkten Austausch. Deutlich wird diese Vorgehensweise u. a. an dem Produkt „Planspiel: ‚Bauleitplanerische Umsetzung des risikobasierten Planungsansatzes'" (entwickelt und durchgeführt vom Lehrstuhl für Regionalentwicklung und Risikomanagement [RER] der Universität Dortmund). Das Planspiel zielte auf die Simulation einer bauleitplanerischen Umsetzung des risikobasierten Planungsansatzes ab, der die städtebauliche Entwicklung von hochwassergefährdeten Flächen mithilfe einer differenzierten Berücksichtigung von Gefährdungsintensitäten und Vulnerabilitäten behandelt. Im Zuge der aktiven Teilnahme von in Bauleitplanungsverfahren involvierten Mitarbeiterinnen und Mitarbeitern aus Politik und Verwaltung konnten direkt in der Praxis anwendbare Ergebnisse erarbeitet werden.

Zur Systematisierung und transparenten Darstellung der erarbeiteten Produkte liegt für jedes Produkt ein strukturierter Steckbrief vor (erstellt unter der Leitung von Institut für qualifizierende Innovationsforschung und -beratung [IQIB] und IWW). Als Beispiel ist in ◘ Abb. 21.5 der Steckbrief des Planspiels „Bauleitplanerische Umsetzung des risikobasierten Planungsansatzes" dargestellt. Die Steckbriefe geben Auskunft zu den Fragen, mit welchen Arbeiten das KAHR-Projekt die von der Flut 2021 betroffenen Gebiete – sowie weitere Regionen – unterstützt, auf welchem Weg wissenschaftliche Ergebnisse und Produkte in den Wiederaufbauprozess einfließen und nennen die jeweiligen wissenschaftlichen Ansprechpartner mit Kontaktdaten. Die Produktsteckbriefe ermöglichen somit einen direkten und unmittelbaren Zugriff auf die im Rahmen von KAHR erarbeiteten Produkte.

21.10 Fazit

Im Zuge der Projektarbeit hat sich gezeigt, dass in der Praxis eines laufenden Projekts zu den theoretisch beschriebenen Herausforderungen noch weitere praktische Herausforderungen hinzukommen. Ergebnisse können erst dann effektiv transferiert werden, wenn sie auch so vorliegen, dass sie nachhaltig Anwendung in der Praxis finden können. Dazu sind aus den wissenschaftlichen Ergebnissen entsprechende Lösungsansätze herauszuarbeiten, was für die Wissenschaftlerinnen und Wissenschaftler einen zusätzlichen Aufwand im Rahmen der Ergebnisaufbereitung und -darstellung bedeutet. Die Betrachtung der Herausforderungen aus verschiedenen Blickwinkeln helfen entsprechende Transfermaßnahmen zu gestalten und umzusetzen. So wird etwa erst nach der genaueren Betrachtung der administrativen Einbettung und der beteiligten Akteure deutlich, wie umsetzungsnahe Ergebnisse und Instrumente aus der Forschung fruchtbar ergänzend oder unterstützend genutzt werden bzw. zur Verbesserung vorhandener Instrumente in der Praxis einfließen können. In der Regel sind dazu intensive detaillierte Abstimmungen zwischen den Wissenschaftler:innen und Praktikern in verschiedenen Schritten und auf verschiedenen Ebenen notwendig. Mit dem beschriebenen theoretischen Hintergrund konnten in KAHR in den ohnehin schon vorhandenen Prozessen des Wiederaufbaus Anknüpfungspunkte für entsprechende Transfermaßnahmen und Maßgaben für ihre Gestaltung gefunden werden. Die einzelnen Maßnahmen mussten dabei immer sehr spezifisch erarbeitet werden.

Das im Rahmen des KAHR-Projekts erarbeitete Wissen sowie die entwickelten Ansätze für die Produktdarstellung bieten eine wichtige Grundlage für die Aufbereitung der

KAHR – Unser Beitrag für die Regionen

Planspiel „Bauleitplanerische Umsetzung des risikobasierten Planungsansatzes“ (Erftstadt, 25.01.2023)

Das Planspiel zielte auf die Simulation einer bauleitplanerischen Umsetzung des risikobasierten Planungsansatzes ab, der die städtebauliche Entwicklung von hochwassergefährdeten Flächen mithilfe einer differenzierten Berücksichtigung von Gefährdungsintensitäten und Vulnerabilitäten behandelt. Vor diesem Hintergrund wurde die Anwendung und Realisierbarkeit einer hochwasserangepassten, differenzierten Bauleitplanung in dem Stadtteil Liblar (Stadt Erftstadt) erprobt. In diesem Planspiel wurden komplexe Verhandlungs- und Entscheidungsprozesse im Zusammenhang mit gezielten Problemstellungen in einer realitätsnahen Situation simuliert. Wirklichkeitsrelevante Aus- und Folgewirkungen von komplexen Strukturen und Entscheidungsprozessen konnten simuliert und überprüft werden.

Für wen war das Planspiel vorgesehen?

Das Planspiel wendete sich an in Bauleitplanungsverfahren involvierte Mitarbeiterinnen und Mitarbeiter aus Politik und Verwaltung (kommunale Ebene) sowie die Träger öffentlicher Belange. Mitarbeiterinnen und Mitarbeiter aus Regionalplanungsbehörden und Wissenschaft konnten dem Planspiel (beobachtend) beisitzen.

Dokumentation der Ergebnisse und Fragen zu diesem Planspiel

Eine Dokumentation der Veranstaltung (inkl. Präsentationen und Planspielunterlagen) und die Verbreitung an die TeilnehmerInnen erfolgte im August 2023. Vorbereitung, Durchführung und Ergebnisse des Planspiels sind darüber hinaus als Beitrag in Korrespondenz Wasserwirtschaft 2023 (16), Nr. 7, DOI: 10.3243/kwe2023.07.003 sowie in Sustainability 2023, 15, 15340. https://doi.org/10.3390/su152115340 erschienen.

Für Rückfragen wenden Sie sich an	Lehrstuhl Regionalentwicklung und Risikomanagement (RER), Fakultät Raumplanung, TU Dortmund Prof. Dr. Stefan Greiving (stefan.greiving@tu-dortmund.de)
oder an unsere Projektbüros	RLP: kahr-rlp@iqib.de (IQIB, Ahrweiler) NRW: kahr-nrw@iww.rwth-aachen.de (RWTH/IWW, Aachen)

„Das Planspiel ermöglichte den Teilnehmenden, in einem geschützten Raum den neuen risikobasierten Planungsansatz zu testen, zu diskutieren und zu reflektieren.“
Stefan Greiving, RER

Das BMBF-Verbundprojekt KAHR (KlimaAnpasung – Hochwasser – Resilienz) begleitet den Wiederaufbauprozess in NRW und RLP und leistet einen wissenschaftlichen Beitrag zum Hochwasserrisikomanagement nach der Flutkatastrophe, um die betroffenen Regionen resilienter zu gestalten. Weitere Informationen finden Sie unter: www.hochwasser-kahr.de

Abb. 21.5 Steckbrief des Produkts: Planspiel „Bauleitplanerische Umsetzung des risikobasierten Planungsansatzes“

Ergebnisse und ihrer Verbreitung und Anwendung in der Praxis. Da nicht alle Ergebnisse frühzeitig vorlagen und entsprechend noch im Projekt aufbereitet werden konnten bzw. können, ist es allerdings absehbar, dass noch weitere Schritte notwendig sind, um sie in die breite Anwendung bringen zu können. Der Fokus im Projekt lag dabei vor allem in der Anwendung der Ergebnisse vor Ort. Die weitere Nutzung von Produkten in anderen Regionen wurde bereits, national und international, über diverse Veranstaltungen angestoßen. Jedoch werden mit dem fortschreitenden Klimawandel die Themen Starkregen und Hochwasser zum einen zunehmend relevanter und erfahren gleichzeitig breitere Aufmerksamkeit in der Bevölkerung, die weitere gezielte Maßnahmen zum Transfer des KAHR-Wissens und der KAHR-Produkte erfordern.

Hinweis für die Praxis

Aufgrund der Belastung der Akteure in einer von einer Naturkatastrophe getroffenen Region musste der Transfer in die betroffene Region hinein im Projekt sehr niedrigschwellig angelegt sein. D. h., die Wissenschaft musste die ersten Anwendungen in der Region intensiv begleiten und evtl. bei den eigenen Ergebnissen nachsteuern. Die Ergebnisse müssen zudem eine hohe „Transferreife" innehaben und ihre Anwendung muss in der Praxis unproblematisch möglich sein. Dies gilt sowohl für Akteure aus Politik und Verwaltung als auch für Unternehmer:innen und Privatpersonen.

Des Weiteren ist zu berücksichtigen, dass die wissenschaftlichen Ergebnisse Eingang in den Wiederaufbau finden sollen, der Wiederaufbau an sich aber bereits grundsätzlich aus vielen sich gegenseitig beeinflussenden Prozessen besteht und damit sehr komplex ist. Diese Tatsache bedingt eine möglichst passgenaue Entwicklung von Produkten und eine zielgruppenspezifische Ansprache sowie einen intensiven Austausch zur Notwendigkeit der Berücksichtigung neuer Forschungserkenntnisse. Solche Diskussionen können in von Starkregen und Hochwasser betroffenen Regionen konkret und lösungsorientiert geführt werden. Ziel muss es allerdings insgesamt sein, dass sich auch andere, bisher noch nicht von Hochwasser und Starkregen direkt bedrohte aber gefährdete Regionen proaktiv mit geeigneten Schutz- und Vorsorgemaßnahmen auseinandersetzen, die sich beim Auftreten des Ereignisses schadensmindernd auswirken können (z. B. Berücksichtigung der Hochwasserangepasstheit beim Bau neuer Brücken). Trotz aktuell erhöhter Aufmerksamkeit bezüglich Starkregen und Hochwasser ist dies noch nicht flächendeckend der Fall. Der vereinzelte Austausch mit Akteuren außerhalb der Region hat im Projekt gezeigt, dass die KAHR-Ergebnisse dazu sehr gut geeignet sind.

Mit weiterer Aufbereitung und gezieltem Transfer in die Breite können die in KAHR erarbeiteten Ergebnisse wichtige Anstöße geben und als zentrale Grundlagen für Klimaanpassungsmaßnahmen im Bereich Starkregen und Hochwasser national und international dienen. Kombiniert mit der Weiterentwicklung einzelner Produkte kann so eine wesentliche Basis für die resiliente Gestaltung von Regionen in ganz Deutschland – und darüber hinaus – geschaffen werden. Übertragbare Erkenntnisse und Produkte können außerdem zusätzlich direkt zur Bewältigung weiterer Katastrophenfälle und Krisen herangezogen werden.

Zusammenfassung

- Der Transfer von in der Wissenschaft erarbeiteten Ergebnissen, die in Form von anwendungsorientierten Lösungen und Produkten nachhaltig in der Praxis Einsatz finden sollen, stellt einen erheblichen Aufwand dar. Im Idealfall werden die Anforderungen an einen erfolgreichen Transfer bereits im Zuge der Projektplanung, spätestens aber bei der Erarbeitung der Ergebnisse, berücksichtigt.
- Unter anderem ist es für einen erfolgreichen Transfer elementar, frühzeitig praktische Expertise einzubinden. Idealerweise werden die Zielgruppen, für die diese Ergebnisse erstellt werden, bereits in den Generierungsprozess eingebunden.

- Ein wichtiges Instrument, um die von der Flut 2021 betroffenen Regionen im Wiederaufbau in ihren konkreten Bedarfen mit wissenschaftlich erarbeiteten Ergebnissen und Produkten zu unterstützen, sind die KAHR-Projektbüros für Rheinland-Pfalz und Nordrhein-Westfalen, die mit ihrer regionalen Expertise den Transfer in die betroffenen Regionen unterstützen sowie eine genaue Analyse der Bedarfe vor Ort und die systematische Darstellung der Produkte ermöglichen.
- Die erarbeiteten Ansätze können auch für den weiteren Transfer in andere, auch noch nicht betroffene Regionen als Vorbereitung auf extreme Naturkatastrophen genutzt werden.

Danksagung Die Autorinnen und Autoren von ▶ Kap. 21 bedanken sich beim Bundesministerium für Bildung und Forschung (BMBF) und beim DLR Projektträger für die Unterstützung und Förderung des KAHR-Projekts. Diese Forschung wurde vom BMBF unter dem Förderkennzeichen 01LR2102A-M finanziert.

Literatur

Adams R, Bessant J, Phelps R (2006) Innovation management measurement: a review. Int J Management Reviews 8:21–47

Alexander AT (2012) Maximising the outcome of University to Industry Knowledge Transfer. University of Exeter, Exeter (Doctoral thesis)

Cohen WM, Levinthal D (1989) Innovation and learning: the two faces of R&D. Econ J 99:569–596

Droste-Franke B (2018) Instrumente in der Systemanalyse für die Realisierung einer robusten Energieversorgung. In: Decker M, Lindner R, Lingner S, Scherz C, Sotoudeh M (Hrsg) „Grand Challenges" meistern, Der Beitrag der Technikfolgenabschätzung. Reihe: Gesellschaft – Technik – Umwelt, Bd 20. Nomos, S 331–346

Droste-Franke B, Carrier M, Kaiser M, Schreurs M, Weber C, Ziesemer T (2015) Improving Energy Decisions. Towards Better Scientific Policy Advice for a Safe and Secure Future Energy System. Ethics of Science and Technology Assessment, Bd 42. Springer, Heidelberg New York

Eliasson G (1990) The firm as a competent team. J Econ Behav Organ 19:273–298

Foray D (2007) Tacit and codified knowledge. In: Hanusch H, Pyka A (Hrsg) Elgar Companion to Neo-Schumpeterian Economics. Edward Elgar, Cheltenham

Grunwald A (2004) Strategic knowledge for sustainable development: the need for reflexivity and learning at the interface between science and society. Int J Foresight Innov Policy 1(1/2):150–167

Hennen L, Bellucci S, Berloznik R, Cope D, Cruz-Castro L, Karapiparis T, Ladikas M, Klüver L, Menéndes S, Staman J, Stephan S, Szairo T (2004) In: Decker M, Ladikas M (Hrsg) Bridges between Science, Society and Policy. Ethics of Science and Technology Assessment, Bd 22. Springer, Heidelberg

Janich P (2001) Logisch-pragmatische Propädeutik. Ein Grundkurs im philosophischen Reflektieren. Velbrück Wissenschaft, Weilerswist

Kline S, Rosenberg N (1986) An overview of innovation. In: Landau R, Rosenberg N (Hrsg) The positive sum strategy. National Academies Press, Washington, S 275–306

Polanyi M (1966) The Tacit Dimension. Doubleday

Pyka A (2007) Innovation networks. In: Hanusch H, Pyka A (Hrsg) Elgar Companion to Neo-Schumpeterian Economics. Edward Elgar, Cheltenham

Schöttmer S, Droste-Franke B (2022) Allgemeines Vorgehensmodell. In: Runder Tisch Erneuerbare Energien (Homepage): Projekt-Vorschlag (Projektskizze) Nachhaltiger Wiederaufbau und Nutzung regenerativer Energien im Kreis Ahrweiler. https://energiewende-2030.de/wp-content/uploads/2022/05/Projekt-Vorschlag_2.0_korr.pdf. Zugegriffen: 19. Aug. 2024

W. K. Kellogg Foundation (2004) Logic Model Development Guide. W. K. Kellogg Foundation, Battle Creek

21

Anwendung wissenschaftlicher Ergebnisse im Aufbau nach der Flutkatastrophe im Ahrtal 2021

Charlotte Burggraf und Anja Toenneßen

J. Birkmann et al. (Hrsg.), *Die Flutkatastrophe 2021*, Sustainable Development Goals (SDG) – Umsetzung in Praxis, Lehre und Entscheidungsprozessen,
https://doi.org/10.1007/978-3-658-48688-4_22

22.1 Einleitung

Die Flutkatastrophe vom 14./15. Juli 2021 hat verheerende Schäden im gesamten Ahrtal verursacht. 135 Menschen haben ihr Leben verloren (MdI 2022b). Häuser, Straßen, Brücken und jegliche Infrastruktur wurden massiv beschädigt oder gar vollständig zerstört. Der Aufbau stellt das Ahrtal vor große Herausforderungen. Ziel ist es, aus der Katastrophe zu lernen, gewonnene Erkenntnisse zu berücksichtigen und die Chance zu nutzen, resilient und zukunftsgerichtet aufzubauen. Klimaresilientes Planen und Bauen, Hochwasser- und Starkregenvorsorge und Strategien, die die Gefährdung und das Risiko bei zukünftigen Ereignissen minimieren und den Schutz von Leib und Seele erhöhen, sind dabei nur einige Aspekte, die es zu berücksichtigen gilt (Truedinger et al. 2024b).

Bei der Komplexität dieses Aufbauprozesses leistet die Wissenschaft einen wertvollen Beitrag. Es ist stets wichtig, Entscheidungen auf Grundlage von Fakten und fundierten Erkenntnissen treffen zu können. Durch eine enge Verzahnung von Wissenschaft und Praxis können Forschungsvorhaben an den konkreten praktischen Bedürfnissen vor Ort ausgerichtet werden und wissenschaftliche Empfehlungen unmittelbar in Entscheidungsprozesse einfließen. In großen Zukunftsprojekten treffen oft unterschiedliche Interessenslagen aufeinander. Wissenschaftler:innen nehmen hier mit ihrer Expertise als neutrale Akteur:innen eine zentrale Rolle ein und unterstützen so auch die Umsetzung von Projekten. Von großer Bedeutung ist auch die Akzeptanz in der Bevölkerung für Maßnahmen beispielsweise des Hochwasserschutzes, da diese das Orts- und Landschaftsbild, die Infrastruktur und das alltägliche Leben im Ahrtal verändern können. Auch bei der Kommunikation von Konfliktthemen zwischen beteiligten Parteien im Aufbau und der Bevölkerung vor Ort helfen faktenbasierte Aussagen, um eine sachliche Diskussion zu fördern und Emotionalität in Diskussionen zu minimieren.

Es lassen sich neben der Vielzahl an Beteiligungen durch die Verbundpartner zwei übergeordnete Themengebiete festhalten, in denen das KAHR-Projekt mit seinen Empfehlungen, aber auch ganz konkret im Aufbauprozess im Ahrtal maßgeblich mitgewirkt hat: der Aufbau der Infrastruktur und die Hochwasser- und Starkregenvorsorge.

Im Themenfeld des Aufbaus der Infrastruktur können die Gebiete Brücken sowie kritische und sensible Infrastrukturen genannt werden, bei denen ein resilienter Aufbau durch die wissenschaftliche Expertise unterstützt und befördert wurde (s. ▶ Kap. 10 und 16).

Im Bereich der Hochwasser- und Starkregenvorsorge lieferten die Forschungsergebnisse aus dem KAHR-Projekt einen wichtigen Input für den überörtlichen Maßnahmenplan zur Hochwasser- und Starkregenvorsorge im Ahr-Einzugsgebiet und zu Maßnahmen zum Wasserrückhalt im Forst sowie durch eine vertiefte Retentionsraumanalyse zum Wasserrückhalt in der Fläche und zum Wasserrückhalt technischer Art in Form von Hochwasserrückhaltebecken.

22.2 Aufbau der Infrastruktur

22.2.1 Brücken

Im Rahmen der Untersuchungen des Instituts für Wasserbau und Wasserwirtschaft der RWTH Aachen wurden alle Brücken am gesamten Flussverlauf aufgenommen und nach ihren Schäden klassifiziert (Burghardt et al. 2022). Die Verklausung der Brücken trug in großen Teilen dazu bei, dass sich die Flutwelle um bis zu drei Meter erhöhte (s. ► Kap. 10) enorme Kräfte und Fließgeschwindigkeiten entstanden. Insgesamt wurden von den 115 Brücken entlang der Ahr 70 % komplett zerstört oder schwer beschädigt (Burghardt et al. 2022).

Jedes Brückenbauwerk stellt eine potentielle Gefahr bei Hochwasser durch Behinderung des Abflussgeschehens dar. Gemeinsam mit der Wissenschaft wurden daher Empfehlungen für einen hochwasserangepassten Neubau der zerstörten Brücken erarbeitet.

In der Planung von hochwasserangepassten Brücken gilt es, folgende Aspekte zu betrachten: Abflussquerschnitt (vergrößerter Abflussquerschnitt und angepasster Fließquerschnitt sowie mehr Raum für den Fluss), Überbau (Hubbrücke, Klappbrücke und strömungsgünstiger Überbau), Pfeiler (Pfeilerform, Tiefengründung und Erosionsschutz sowie größere Stützweiten) und Bemessung (zusätzliche Bemessungslasten, Verzicht auf Stabbogenbrücken oder Verzicht auf Brücken generell) (Burghardt et al. 2022; ► Kap. 10). Detailliert geht die vierte Empfehlung der „10 Empfehlungen der Wissenschaft" aus dem KAHR-Projekt auf die Hochwassergefahr durch Brücken und die zwingend notwendige hochwasserangepasste Bauweise der Brückenbauwerke ein (KAHR-Verbund 2022).

Auf der Grundlage dieser Erkenntnisse haben die Baulastträger ein Gestaltungshandbuch für die Brücken im Ahrtal erarbeiten lassen, welches die gestaltbestimmenden und für die Lage in einem Überflutungsgebiet technisch relevanten Aspekte berücksichtigt. Die wissenschaftlichen Erkenntnisse haben auch einen Beitrag dazu geleistet, dass schwer beschädigte Brücken unter bestimmten Umständen nicht saniert werden müssen, sondern aus Mitteln des Aufbauhilfefonds durch einen hochwasserresilienten Neubau ersetzt werden können.

22.2.2 Kritische und sensible Infrastruktur

Zu den besonders schutzwürdigen Infrastrukturen zählen neben kritischen Infrastrukturen (Wasser-, Strom-, Wärmeversorgung, Verkehrswege etc.), die eine wichtige Rolle für das staatliche Gemeinwesen darstellen, auch die sensiblen Infrastrukturen (Altenheime, Schulen, Kindergärten), deren Nutzende oftmals nicht selbst in der Lage sind, sich zu schützen, und auf deren Schutz eine hohe soziale Verantwortung liegt (Truedinger et al. 2024b).

Kritische und sensible Infrastrukturen sollten zwingend resilient wiederaufgebaut werden, um eine funktionierende Versorgung und den höchstmöglichen Schutz für vulnerable Bevölkerungsgruppen sicherzustellen (► Kap. 16). Dazu hat das KAHR-Projekt besonders im Bereich hochwasserresiliente Strominfrastruktur und Wiederaufbau einer Schule für Kinder und Jugendliche mit Beeinträchtigungen unterstützt.

22.2.2.1 Hochwasserresiliente Stromversorgung

Mit Hilfe der Westnetz GmbH als größter Verteilnetzbetreiber im Ahreinzugsgebiet konnten Empfehlungen und Erkenntnisse für eine hochwasserresiliente Stromversorgung in Form einer Praxishilfe für Netzbetreibende, Politik und Verwaltung auf kommunaler und regionaler Ebene (insb. Raumplanung und Katastrophenschutz) festgehalten werden.

Da besonders der Ausfall des Stromnetzes weitere Kaskadeneffekte mit sich bringt, ist eine Störung möglichst zu vermeiden. Es lassen sich unterschiedliche Kategorien der Hochwasserresilienz festhalten, in denen jeweils beispielhaft bauliche, technologische oder organisatorische Maßnahmen aufgelistet werden (Truedinger et al. 2024b). Grundsätzlich muss in der Bauvorsorge eine Schutzstrategie gewählt werden, die effektiv gegen Hochwasser schützt. Hier wird differenziert zwischen folgenden Strategien: Ausweichen, Widerstehen und Anpassen. Dabei ist der effektivste Schutz die Bebauung aus hochwassergefährdeten Gebieten zu verlegen und somit dem Hochwasser auszuweichen und der mindeste Schutz das Anpassen von Gebäuden und Infrastruktur (Truedinger et al. 2024b). Die Kategorien der Hochwasserresilienz und die Strategien der baulichen Vorsorge sind mit Beispielen und Umsetzungsmöglichkeiten verdeutlicht, die in der Praxis angewendet werden können und mit Hilfe deren kommunale Planungsstellen, Ingenieurbüros und betroffene Bürgerinnen und Bürger Anregungen erhalten.

22.2.2.2 Levana-Schule

Bei der Flutkatastrophe am 14./15. Juli 2021 wurden in Rheinland-Pfalz 29 Schulen beschädigt, davon 17 im Landkreis Ahrweiler (MdI 2022a). Einige Schulen konnten nach Wochen bzw. Monaten zumindest teilweise wieder genutzt werden, da hauptsächlich das Erdgeschoss betroffen war und die Sanierung im laufenden Betrieb stattfinden konnte. Andere Schulen müssen jedoch aufwendig wiederaufgebaut werden und sind seitdem an alternativen Standorten in Form von Containerbauten in Betrieb (Kreisverwaltung Ahrweiler 2023). Die Betroffenheit der Schulen zeigt, dass diese teils sehr ungünstig im neu festgesetzten Überschwemmungsgebiet („$HQ_{100,neu}$") liegen und zwingend hochwasserangepasst aufgebaut werden müssen. So wurden beispielsweise Fach- und Unterrichtsräume ins 1. Obergeschoss verlegt, um Wasserschäden bei Hochwasserereignissen zu minimieren (Kreisverwaltung Ahrweiler 2023).

Einen besonderen Fall stellt die Levana-Schule in Bad Neuenahr-Ahrweiler dar. Es handelt sich hierbei um eine Einrichtung für Kinder und Jugendliche mit Behinderungen, die eine intensive pädagogische Betreuung brauchen. Die Levana-Schule ist ein Bungalow-Bau und befindet sich auf der rechten Ahrseite auf einem Campusgelände zusammen mit der Don-Bosco-Schule und einer Kindertagesstätte.

Mit der wissenschaftlichen Expertise des KAHR-Projekts konnte durch die Erstellung einer Stellungnahme erreicht werden, dass anstelle einer Sanierung des Bestandsgebäudes die Finanzierung eines Neubaus der Schule an einem anderen Standort außerhalb des Überschwemmungsgebiets mit Mitteln des Aufbauhilfefonds möglich ist (Birkmann et al. 2023). Die zehnte Empfehlung der „10 Empfehlungen der Wissenschaft" aus dem KAHR-Projekt betont ebenfalls, dass neue Schutzstandards und Schutzziele für kritische und sensible Infrastrukturen definiert und Menschen stärker in den Fokus gerückt werden müssen.

Die wissenschaftliche Stellungnahme kommt zu dem Ergebnis, dass trotz deutlich höherer Kosten ein Neubau der Sanierung vorgezogen werden sollte, da die Gesundheit von Leib und Seele ein nicht monetär ausdrückbares, zu schützendes Gut ist. Sie hebt die besondere Exposition und Vulnerabilität, die Evakuierungswege und -möglichkeiten und das Schadensmaß bei kleineren Hochwassern hervor (Birkmann et al. 2023).

Es wird deutlich, dass die Evakuierung – auch durch fehlende Möglichkeiten der vertikalen Evakuierung – durch die notwendige Einzelbetreuung vieler der Kinder sehr zeit- und personalintensiv ist (Birkmann et al. 2023). Hinzu kommt, dass der Ausgang der Schule und somit auch der Evakuierungsweg und Anfahrtsweg für (Kranken-)Transporter Richtung Ahr gerichtet und bereits bei kleinen Hochwasserereignissen nicht mehr nutzbar ist.

Des Weiteren ist es für die Kinder und Jugendlichen der Levana-Schule eine nicht zumutbare psychische Mehrbelastung, wenn ein Hochwasserereignis eintritt und sie die Schule notfallmäßig verlassen müssen. Laut Schulleitung und Psycholog:innen kann das betroffene Kinder weit in ihrer Entwicklung zurückwerfen und bestehende Krankheitsbilder verstärken (Birkmann et al. 2023). Im Herbst 2024 wurde auf politischer Ebene entschieden, die Schule an einen alternativen Standort zu verlagern. Die KAHR-Stellungnahme trug maßgeblich dazu bei, dass weitere Diskussionen mit dem zuständigen Ministerium geführt wurden und weitere Gutachten in Auftrag gegeben wurden.

22.2.3 Sportstätten

Vereinsleben und Sportplätze gehören seit jeher zur Kultur und Identifikation der Gemeinden an der Ahr. Auch unmittelbar nach der Flut 2021 halfen sie bei der Stressbewältigung und wurden Begegnungs- und Bewegungsstätte für Kinder, Jugendliche, Familien und Senior:innen.

Durch die Flutkatastrophe wurden auch viele Sportstätten an der Ahr zerstört. Ein schneller Wiederaufbau dieser Bewegungs- und Begegnungsstätten ist zur Identifikation und Stärkung der sozialen Infrastruktur wichtig (Burggraf et al. 2024). Mit Blick auf die Hochwasservorsorge und die Vermeidung zukünftiger Schäden sowie die Nutzbarkeit auch bei Hochwasser wäre es zielführend, Sportplätze außerhalb des Überschwemmungsgebiets zu errichten. Auf Grund der steilen und schmalen Topographie im Ahrtal ist es jedoch schwierig, alternative Standorte zu finden. Daher ist eine multifunktionale Nutzung von Sportplätzen näher zu betrachten: Sportplätze im Überschwemmungsgebiet können so gestaltet werden, dass sie gleichzeitig auch als Retentionsfläche dienen und mit einem verzögerten und gezielten Abfluss dazu beitragen, Wasser schadfrei abzuleiten und den Wasserrückhalt bei Starkregen zu vergrößern (Burggraf et al. 2024).

Als Vorzeigeprojekt für eine multifunktionale und hochwasserangepasste Sportanlage dient das interkommunale Projekt der Ortsgemeinden Insul, Schuld und Dümpelfeld. Anstelle von drei einzelnen Sportstätten wird eine hochwasserangepasste Sportanlage mit Kunstrasenplatz und zusätzlichem Kleinspielfeld für Basketball, Fußball, Volleyball und Tennis in Insul errichtet (Truedinger et al. 2024a). Das gegossene Granulat des Kleinspielfeldes ist hochwassersicher und bereits in vergleichbaren Anlagen am Rhein verbaut. Der Kunstrasen hingegen wird vollständig verankert, die Ballfangzäune orthogonal zur Ahr sind abbaubar und das Vereinsheim wird auf die gegenüberliegende Platzseite in den Hang versetzt (Truedinger et al. 2024a). Somit wird es zukünftig knapp zwei Meter höher liegen als der Sportplatz.

Eine praxisnahe Handreichung des KAHR-Projekts für Kommunen, Sportplatzbetreiber:innen und Vereine gibt wichtige Informationen und Unterstützung zur multifunktionalen Nutzung von Sportplätzen, bauliche Maßnahmen zur hochwasserangepassten Gestaltung und Strategien zur Notfallvorsorge.

22.3 Hochwasser- und Starkregenvorsorge

Eine wesentliche Erkenntnis für das Hochwasserrisikomanagement im Ahrtal ist die Notwendigkeit einer ganzheitlichen Hochwasser- und Starkregenvorsorge für das Einzugsgebiet der Ahr. In einer landes- und kreisübergreifenden Kooperation wird daher ein überörtlicher Plan mit konkreten Maßnahmen erstellt, die extreme Wassermengen zurückhalten und damit mehrere Städte und Gemeinden übergreifend schützen können. Gegenstand sind sowohl technische Lösungen wie z. B. Rückhaltebecken als auch natürlicher Wasserrückhalt in der Fläche.

In die Planungen fließen Forschungsergebnisse des KAHR-Projekts zum potentiellen Wasserrückhaltevolumen durch natürliche und technische Maßnahmen ein. Schätzungsweise knapp 100 Mio. m^3 Wasser flossen während des Ereignisses 2021 durch das Ahrtal. Rund 30 Mio. m^3 müssten zurückgehalten werden, um die Spitze der Flutwelle zu kappen und den Abfluss auf ein HQ_{100} abzusenken, das einen schadlose Ableitung zum Ziel hat (Wolf et al. 2024). Auch raumplanerische Potenzialbeschränkungen wurden durch die Wissenschaft in den Planungsprozess eingebracht und Tabu- und Konfliktbereiche mit unterschiedlich gewichteter Restriktion bzw. Eignung identifiziert.

Ein wichtiger Baustein der Hochwasser- und Starkregenvorsorge ist die Verbesserung des natürlichen Wasserrückhalts auf land- und forstwirtschaftlichen Flächen. Hochwassermindernde Maßnahmen auf diesen Flächen können nur in einem Schulterschluss mit den Flächenbewirtschaftern erreicht werden. Im Rahmen der Hochwasserpartnerschaft Ahr wurden daher gemeinsam mit den Bewirtschaftern Arbeitsgruppen zu den Themen Forstwirtschaft sowie Landwirtschaft/Weinbau gebildet. Ziel ist es, Rückhaltepotenziale, die der Wald sowie landwirtschaftliche und Weinbau-Flächen bieten, zu ermitteln und Maßnahmen zur Optimierung des Wasserrückhalts zu erarbeiten. Dabei werden die Arbeitsgruppen durch die Expertise der Wissenschaftler:innen aus dem KAHR-Projekt unterstützt.

22.3.1 Forstwirtschaft

Bei der Forstwirtschaft geht es vor allem um einen gesunden, klimaresilienten Wald, der gleichzeitig auch am meisten Wasser zurückhalten kann. Die Naturverjüngung, eine strukturierte Vegetation und gestufte Gewächse sowie eine gezielte Wiederaufforstung des Mischwaldes sind die zentralen Faktoren für einen guten Gesundheitszustand des Forstes. Des Weiteren können sich Abflussspitzen verzögern, wenn Gräben und Wege entschärft und Rigolen gebaut werden und somit das Oberflächenwasser räumlich verteilt und der Abfluss in die Fläche geleitet wird (Ergebnisse der Arbeitsgruppe Forst, 11.10.2023).

Die Probleme an der Ahr sind der oftmals dünne Bodenaufbau, der wenig Infiltration zulässt, die Beschädigung durch den Borkenkäfer und Dürreperioden, Hangrutschungen, Wildverbiss und fehlendes Unterholz sowie Puffergewächse (Berichte der zuständigen Forstämter im Kreis Ahrweiler im Rahmen der Arbeitsgruppe Forst, 11.10.2023). Der lichte Wald mit einer geringen Baumwipfeldichte führt zu vermehrtem Lichteinfall auf den Waldboden, was stärkere Austrocknung und geringe Humusschichten zur Folge hat.

Dass kein Zielkonflikt zwischen Wasserrückhalt, Naturschutz und Forst besteht, begünstigt die Implementierung der Maßnahmen zur Gesunderhaltung des Forstes. Als praxisnahe Handreichung wurden gemeinsam von Wissenschaftler:innen und zuständigen Forstämtern, Förster:innen und Behörden fundamentale Grundsätze zur klimaresilienten und wasserrückhaltenden Bewirtschaftung des Forstes aufgestellt. Dabei arbeitet die Handreichung nicht nur technisch und fachlich Maßnahmen auf, sondern stellt diese auch allgemeinverständlich zur Unterstützung der Kommunikation mit der Bevölkerung dar. Die Kommunikation mit den Bürger:innen ist insofern wichtig, damit das Verständnis für die Wichtigkeit eines gesunden Waldes als Wasserrückhaltefläche und die dafür notwendigen Maßnahmen erhöht wird. Wissenschaftler:innen der RWTH Aachen haben berechnet, dass bis zu ca. 4,5 Mio. m^3 Wasser im Wald des Einzugsgebiets zurückgehalten werden könnten (▶ Kap. 11).

22.3.2 Landwirtschaft und Weinbau

Die Arbeitsgruppe Landwirtschaft und Weinbau beschäftigt sich mit Unterstützung der Wissenschaft mit Möglichkeiten der Landnutzungsänderung und mit dem Zusammenhang der Landnutzungsart zur sogenannten Curve Number (CN). Der Wert der Curve Number stellt das Wasserspeichervermögen dar und berechnet den Direktabfluss des Bodens (s. ▶ Abschn. 11.3.1). Je höher die Curve Number, desto geringer ist die Infiltration im Boden und desto höher ist der Abfluss.

Durch Landnutzungsanpassungen in der Bewirtschaftungsform, die weg von konventionellen Ackerflächen hin zu Mulchsaaten gehen, kann der Abfluss gemindert werden. Ein angepasster Wald kann ebenfalls einen gewissen Teil des Einzugsgebiet-Abflusses zurückhalten (Ergebnisse der Arbeitsgruppe Landwirtschaft und Weinbau, Januar 2024).

Am wenigsten ausschlaggebend ist der Weinbau. Dort ist vor allem die Begrünung zwischen den Reben ausschlaggebend für die Abflussreduktion. Je intensiver begrünt, desto besser der Rückhalt.

22.4 Fazit

Die Beispiele zeigen, dass die Wissenschaft in unterschiedlichen Themengebieten in der Praxis Anwendung findet.

Hinweis für Entscheidungsprozesse

Fundierte und faktenbasierte Ergebnisse können Prozesse verbessern und Aufbaumaßnahmen resilienter gestalten und zudem auch in der Kommunikation zwischen verschiedenen Interessengruppen, der Politik und Bevölkerung helfen. So wird die Akzeptanz für einen nachhaltigen, zukunftsorientierten Aufbau gestärkt und das Bewusstsein für Hochwasser- und Starkregengefahren geschärft.

Die Ergebnisse und Forschungen helfen nicht nur den Prozessen im Aufbau nach der Flutkatastrophe im Ahrtal und Nordrhein-Westfalen, sondern sind auch lokal unabhängig übertragbar, allgemeingültig und für zukünftige Hochwasser- oder Starkregenereignisse nutzbar. Dies zeigt beispielsweise die Schaffung des § 246 c Baugesetzbuch, der in sogenannten Wiederaufbaugebieten nach einem Katastrophenfall in Teilen einen hochwasserangepassten Wiederaufbau und beschleunigte Verfahren für Bauleitpläne ermöglicht.

Für alle von der Flut betroffenen Menschen im Ahrtal und in anderen Regionen ist es wichtig, den Aufbau auch als Chance zu sehen, in Zukunft widerstandsfähiger zu sein. Das Zusammenwirken von Wissenschaft und Praxis trägt stetig dazu bei, einen 1:1 *Wieder*-Aufbau zu vermeiden und ein zukunftsgerichtetes, auch überregional als Modellregion wahrgenommenes Ahrtal zu schaffen, das mit einer resilienten Vorsorge gegen Hochwasser- und Starkregenereignisse gewappnet ist.

22.5 Zusammenfassung

Die Flutkatastrophe vom 14./15. Juli 2021 im Ahrtal hat massive Schäden angerichtet, darunter 135 Todesfälle und die Zerstörung diverser Infrastruktur. Der Aufbau stellt eine Herausforderung dar, deren Bewältigung durch die Anwendung wissenschaftlicher Erkenntnisse unterstützt werden soll. Ziel ist es, resiliente, zukunftsorientierte Lösungen zu entwickeln, die den Schutz und die Vorsorge vor künftigen Hochwasser- und Starkregenereignissen verbessern. Die enge Zusammenarbeit zwischen Wissenschaft und Praxis ist dabei ausschlaggebend, um faktenbasierte Entscheidungen zu treffen und die Akzeptanz der Bevölkerung für Schutzmaßnahmen zu fördern.

Unterschiedliche Handlungsfelder wurden seitens der Wissenschaft untersucht. Wissenschaftliche Empfehlungen für einen hochwasserresilienten Neubau von Brücken umfassen Aspekte wie Abflussquerschnitt, Überbau und Pfeilerform. Ein Gestaltungshandbuch wurde erstellt, das die relevanten technischen Aspekte für Brücken im Überflutungsgebiet berücksichtigt.

Die Wiederherstellung kritischer Infrastrukturen wie Wasser- und Stromversorgung ist unerlässlich. Das KAHR-Projekt arbeitete Empfehlungen für eine hochwasserresiliente Stromversorgung aus, um Kaskadeneffekte durch Stromausfälle zu vermeiden. Ein weiterer Fokus lag auf dem Aufbau von sensiblen Infrastrukturen, wie der Levana-Schule für Kinder mit Behinderungen. Hier wurde durch die Erarbeitung einer wissenschaftlichen Stellungnahme ein Neubau an einem sichereren Standort die Finanzierung sichergestellt, um die Vulnerabilität der Schüler:innen zu verringern und eine Evakuierung zu gewährleisten. Die Schul- und Sportstätten im Überschwemmungsgebiet sollen ebenfalls hochwasserangepasst errichtet werden, um ihre Funktionalität in Zukunft zu sichern.

Eine ganzheitliche Hochwasser- und Starkregenvorsorge ist notwendig, um extreme Wassermengen zu kontrollieren. Das KAHR-Projekt lieferte wichtige Erkenntnisse zur Wasserrückhaltung durch natürliche und technische Maßnahmen, die in den durch eine landes- und kreisübergreifende Kooperation erstellten, überörtlichen Maßnahmenplan zur Hochwasser- und Starkregenvorsorge einfließen. Die Verbesserung des natürlichen Wasserrückhalts in land- und forstwirtschaftlichen Flächen ist ein weiterer Baustein zur Verringerung und Verzögerung des Abflussgeschehens. Empfehlungen zur klimaresilienten Bewirtschaftung des Waldes sowie zur Anpassung der Landnutzung und die Identifizierung von Wasserrückhaltepotenzialen wurden mit entsprechenden Akteuren aus Behörden, der Forst- und Landwirtschaft und dem Weinbau erarbeitet.

Die Wissenschaft spielt eine zentrale Rolle beim Aufbau nach der Flutkatastrophe im Ahrtal. Ihre fundierten Erkenntnisse tragen zur Verbesserung der Resilienz und zur Akzeptanz von Schutzmaßnahmen in der Bevölkerung bei. Die Ergebnisse sind nicht nur lokal im Ahreinzugsgebiet relevant, sondern bieten auch allgemeingültige Ansätze für zukünftige Hochwasserereignisse. Der Aufbau sollte als Chance betrachtet werden, um ein widerstandsfähigeres Ahrtal zu schaffen, das in Zukunft als Modellregion fungiert.

Zusammenfassung

- Der Aufbau nach der Flutkatastrophe im Ahrtal wird durch wissenschaftliche Erkenntnisse unterstützt, um resiliente und zukunftsorientierte Lösungen zu entwickeln, die den Schutz vor künftigen Hochwasser- und Starkregenereignissen verbessern.
- Die enge Zusammenarbeit zwischen Wissenschaft und Praxis ist maßgeblich für faktenbasierte Entscheidungen in unterschiedlichen Themengebieten und erhöht die Akzeptanz in der Bevölkerung.
- Das KAHR-Projekt konnte Konzepte für eine resiliente Stromversorgung und weitere sensible Infrastrukturen entwickeln. Eine umfassende Hochwasser- und Starkregenvorsorge sowie die Verbesserung des natürlichen Wasserrückhalts in der Fläche und des technischen Wasserrückhalts im gesamten Einzugsgebiet sind weitere Themengebiete, die in Zusammenarbeit mit Akteuren vor Ort erarbeitet wurden. Die Erkenntnisse sollen nicht nur lokal, sondern auch als allgemeingültige Ansätze für zukünftige Ereignisse dienen.

Danksagung Die Autorinnen von ▶ Kap. 22 bedanken sich beim Bundesministerium für Bildung und Forschung (BMBF) und beim DLR Projektträger für die Unterstützung und Förderung des KAHR-Projekts. Diese Forschung wurde vom BMBF unter dem Förderkennzeichen 01LR2102A-M finanziert.

Literatur

Birkmann J, Truedinger A, Sauter H, Greiving S (2023) Stellungnahme zur Levana-Schule als besonders sensible Infrastruktur. BMBF-Förderkennzeichen 01LR2102 (Veröffentlichung im Rahmen des Vorhabens KAHR – Klima-Anpassung, Hochwasser und Resilienz (2021–2025))

Burggraf C, Reckhaus Z, Schäfer M, Trüdinger A, Birkmann J, Kirschbauer L (2024) Handreichung Stärkung der Hochwasserresilienz von Außensportanlagen. Veröffentlichung im Rahmen des Vorhabens KAHR – Klima-Anpassung, Hochwasser und Resilienz (2021–2024). BMBF-Förderkennzeichen 01LR2102 (https://www.hochwasser-kahr.de/index.php/de/kahr-produkte)

Burghardt L, Klopries E-M, Wolf S, Schüttrumpf H (2022) Analyse der Schäden an Brückenbauwerken in Folge des Hochwassers 2021 an der Ahr. In: Wasser und Abfall, 11-2022. Springer, Wiesbaden

KAHR-Verbund (2022) 10 Empfehlungen aus Sicht der Wissenschaft (hochwasser-kahr.de, abgerufen am 04.10.2024)

Kreisverwaltung Ahrweiler (2023) Hochwasserkatastrophe 2021 – Sachstandsbericht der Kreisverwaltung, Hochwasserkatastrophe 2021 – Sachstandsbericht der Kreisverwaltung (kreis-ahrweiler.de, abgerufen am 04.10.2024)

Ministerium des Innern und für Sport Rheinland-Pfalz (MdI) (2022a) Der Wiederaufbau in Rheinland-Pfalz nach der Flutkatastrophe vom 14./15. Juli 2021. Entwicklungsagentur Rheinland-Pfalz e. V., Mainz (http://wiederaufbau.rlp.de/de/wiederaufbaurlp/, abgerufen am 04.10.2024)

Ministerium des Innern und für Sport Rheinland-Pfalz (MdI) (2022b) Ein Jahr Wiederaufbau. Entwicklungsagentur Rheinland-Pfalz e. V., Mainz (wiederaufbau.rlp.de/de/wiederaufbau2022 (abgerufen am 30.09.2024))

Truedinger A, Birkmann J, Burggraf C (2024a) Wiederaufbau oder Neuaufbau? Stärkung der Resilienz durch hochwasserangepasste Raumnutzungen und Raumentwicklung – Befunde und Überlegungen am Beispiel des Ahrtals. In: Flächenmanagement und Bodenordnung, Bd 01. Chmielorz, Wiesbaden

Truedinger A, Birkmann J, Fleischhauer M, Greiving S, Kirschbauer L, Schäfer M, Burggraf C, Fabisch M, Jüpner R, Droste-Franke B, Apel H, Arenz, H-T, Johann G, Wolf S, Schüttrumpf H, Dolfus F, Tholl L (2024b) Praxishilfe – Stärkung der Resilienz der Stromversorgung gegenüber Starkregen und Hochwasser: Vorsorge, Anpassung, Wiederaufbau. Veröffentlichung im Rahmen des Vorhabens KAHR – Klima-Anpassung, Hochwasser und Resilienz (2021–2025), BMBF-Förderkennzeichen 01LR2102. https://www.hochwasser-kahr.de/index.php/de/kahr-produkte. Zugegriffen: 30. Sept. 2024

Wolf S, Steudtner F, Buchholz O, Yörük A, Schüttrumpf H (2024) Natürlicher vs. Technischer Hochwasserschutz – Wo sind die Möglichkeiten und Grenzen? Gewässerschutz – Wasser – Abwasser, Bd 257. Springer, Aachen

Synthese: 10 KAHR-Empfehlungen, Wirkungen der Begleitforschung und offene Forschungsfragen

Jörn Birkmann, Holger Schüttrumpf, Alessa Trüdinger,
Stefanie Stenger-Wolf und Benjamin David Ehring

J. Birkmann et al. (Hrsg.), *Die Flutkatastrophe 2021*, Sustainable Development Goals (SDG) – Umsetzung in Praxis, Lehre und Entscheidungsprozessen,
https://doi.org/10.1007/978-3-658-48688-4_23

23.1 Einleitung

Das Forschungsprojekt KAHR hat durch die Flutkatastrophe 2021 betroffene Gemeinden und Regionen in Rheinland-Pfalz und Nordrhein-Westfalen beim Wieder- und Neuaufbau drei Jahre lang wissenschaftlich begleitet. Die Begleitforschung und entsprechende Befunde und Empfehlungen basieren auf Arbeiten zu unterschiedlichen Forschungsgegenständen, Fragestellungen und Anforderungen, die sich auch im Laufe des Projekts weiterentwickelt haben. Gemeinsam ist den Forschungsarbeiten, dass diese alle die übergeordnete Frage des hochwasser- und klimaresilienten Wieder- und Neuaufbaus aufgreifen und thematisieren. Dies reicht von der Erfassung und Analyse von Schadensmustern über die Rolle von Brücken bis hin zu Aspekten der sozialen Vulnerabilität und der kritischen und sensiblen Infrastrukturen. Im Rahmen des Forschungsprojektes wurden Lösungsansätze zu verschiedenen Sustainable Development Goals (SDG) thematisiert und erarbeitet. Hervorzuheben sind hier beispielsweise die Ziele der Gesundheit und des Wohlergehens, der Stärkung von Industrie, Innovation und Infrastruktur, der Reduktion von Ungleichheiten sowie der Förderung nachhaltiger Städte und Gemeinden. Die unterschiedlichen Kapitel dieses Sammelbands zeigen daher auch die Bandbreite der von 13 Partnern im Konsortium durchgeführten Untersuchungen und wissenschaftlichen Beratungen auf. Dieses Kapitel erläutert wichtige Empfehlungen des KAHR-Projekts (die sog. 10 Empfehlungen) und stellt dar, wie diese im Prozess des Wiederaufbaus Berücksichtigung fanden. Zudem werden übergreifende Schlussfolgerungen und offene Forschungsfragen skizziert, die für zukünftige Forschungen wichtige Anregungen und Hinweise bieten können.

Eine Besonderheit des KAHR-Projekts liegt darin, dass die wissenschaftliche Begleitforschung im Vordergrund stand und nicht ein einzelnes Produkt oder technisches System. Die Forschungen innerhalb von KAHR umfassen sehr unterschiedliche Themenfelder und Forschungsobjekte, wobei allen Ansätzen und Produkten gemein ist, dass diese nicht nur eine theoretische Auseinandersetzung mit dem Gegenstand Wiederaufbau, Klimaresilienz und Hochwasservorsorge aufweisen, sondern konkrete Probleme, Fragen und Prozesse in flutgeschädigten Regionen aufgreifen. Im Kern war die Forschung daher wie eine Arbeit am „lebenden Patienten", der noch offene Wunden zu beklagen hatte und gleichzeitig konkrete Fragen im Rahmen des Wieder- und Neuaufbaus beantworten musste. Forschungsgegenstände, die bei der Konzeption des Projekts direkt nach der Flutkatastrophe noch nicht so stark auf dem wissenschaftlichen Radar waren, sind u. a. die Fragen der Rolle der Brücken und kritischen sowie sensiblen Infrastrukturen gewesen (SDG 9 – „Industrie, Innovation und Infrastruktur"). Beispielsweise haben wissenschaftliche Untersuchungen den Einfluss von Brücken auf das Abflussgeschehen des Hochwassers näher untersucht und daraus Empfehlungen für die Weiterentwicklung von Brückenstandards abgeleitet. Auch hat die KAHR-Forschung im Bereich kritischer und sensibler Infrastrukturen nicht nur die allgemeine Bedeutung dieser thematisiert, sondern am konkreten Beispiel der Levana-Schule die Problematik des Wiederaufbaus an gleicher Stelle intensiv diskutiert (► Kap. 16). Die Stellungnahmen aus dem KAHR-Projekt führten auch maßgeblich zu der Entscheidung einen neuen Standort für diese sensible Infrastruktur zu suchen. Hierbei zeigte sich, dass die wissenschaftlichen Befunde nicht nur für die betroffene Gemeinde oder den Kreis, sondern auch für die Landesebene von erheblicher Bedeutung waren, um Mittel des Wiederaufbaus für diese räumliche Verlagerung verfügbar zu machen.

Des Weiteren trafen wissenschaftliche Befunde und Empfehlungen auch auf ein Umfeld, welches durch Spannungen und Konflikte sowie unterschiedliche Lösungsideen geprägt war. Die wissenschaftlichen Daten und Methoden waren in dieser Hinsicht für die Versachlichung der Diskussion hilfreich und boten zudem die Möglichkeit, Akteure zusammen zu bringen und Beratungen neutral und auf wissenschaftlichen Fakten und Empfehlungen beruhend vorzunehmen. Ein zentrales Ziel der Forschungsarbeiten war die Identifikation von Strategien und Ansätzen zur Verbesserung der Hochwasserrisikovorsorge und zur Stärkung der Klimaresilienz im Rahmen des Wieder- und Neuaufbaus. Bereits der Begriff Wiederaufbau wurde kritisch diskutiert, da am Anfang teilweise die Ansicht vertreten wurde, dass der Wiederaufbauprozess einer Schadenskompensation gleichzusetzen ist. Demgegenüber vertritt das KAHR-Projekt – auch in den im Buch dokumentierten Beiträgen – die Auffassung, dass der Wieder- und Neuaufbauprozess deutliche Verbesserungen und neue Standards für mehr Hochwasserschutz und Klimaresilienz umfassen muss, damit die Region für zukünftige Ereignisse besser gewappnet ist und die erheblichen Finanzmittel für den Wiederaufbau in sicherere Strukturen fließen.

Zudem wird in den Beiträgen des Buches deutlich, dass die KAHR-Forschungsergebnisse nicht nur auf die Ebene der betroffenen Kommunen zielen, sondern auch Empfehlungen für übergeordnete Ebenen umfassen, wie z. B. die Landes- oder Bundesebene. Des Weiteren greift das KAHR-Projekt trotz der Fokussierung auf die Flutkatastrophe 2021 in Deutschland auch internationale Konzepte und Strategien, wie das Konzept des Building Back Better des Sendai Rahmenwerks (siehe UNDRR 2015) auf.

Dabei ging es zum einen um Beiträge zur Konkretisierung der Ansätze der Resilienz und des Building Back Betters, und zum anderen um die Identifikation von unterschiedlichen Phasen des Wiederaufbaus und unterschiedlicher Akteure, da die Ansatzpunkte für entsprechende Strategien und Maßnahmen eben auch von diesen Phasen und Akteuren abhängen. Beispielsweise zeigte sich, dass der Wiederaufbauprozess gerade in der zweiten bis vierten Phase – also nach dem Katastrophenmanagement, wenn es um den Wieder- und Neuaufbau von Häusern und Infrastrukturen geht – Anforderungen an eine klimaresiliente Siedlungsentwicklung aufgreifen kann. Das heißt, es lassen sich unterschiedliche Phasen im Wiederaufbauprozess unterscheiden, die gewisse Überlappungen aufweisen, aber dennoch unterschiedliche Handlungsoptionen für die Hochwasserrisikovorsorge und Klimaresilienz bieten (siehe Birkmann et al. 2023a). So waren im direkten Nachgang der Katastrophe Fragen des Katastrophenmanagements, der Bereitstellung von temporären Unterkünften oder der Versorgung mit Strom und Beleuchtungen zentrale Herausforderungen (SDG 13 – „Maßnahmen zum Klimaschutz"). Auch die Beseitigung und Lagerung der erheblichen Mengen an Abfall und Müll aus den betroffenen Orten und zerstörten Häusern war ein erhebliches Problem. In einer weiteren Phase stand die Diskussion über den Wieder- und Neuaufbau von privaten Häusern und öffentlichen Infrastrukturen wie Schulen und Kindergärten im Mittelpunkt. Diese Diskussion dauert auch heute noch an (Stand: Januar 2025). Im dritten Jahr nach der Katastrophe wurden auch weitergehende Konzepte des Hochwasserschutzes und der Resilienzstärkung diskutiert, wie z. B. die Standorte für größere Hochwasserrückhaltebecken (► Abschn. 11.4), allerdings ist man von einer Umsetzung dieser Maßnahmen meist noch weit entfernt. Ein Beispiel dafür, dass Konzepte aber auch schnell nach Katastrophen aufgestellt und Maßnahmen entwickelt sowie auf ihr Potenzial hin untersucht werden können und im Anschluss entsprechend in die Umsetzung gelangen können, ist der von Wasserverband Eifel-Rur und dem Institut für Wasserbau und Wasserwirtschaft der RWTH Aachen entwickelte Masterplan „Hochwasserresiliente Stadt- und Gebietsentwicklung für das Einzugsgebiet von

Inde und Vicht", durch welchen potenzielle Maßnahmen, die den Hochwasserschutz erhöhen und die Hochwasserresilienz der Region stärken, identifiziert werden, sodass das Risiko zukünftiger Hochwasser minimiert wird (s. ▶ Kap. 12). Der Masterplan konzentriert sich dabei auf einen ganzheitlichen Ansatz, der auch über das gesetzlich festgesetzte HQ_{100} hinaus langfristig nachhaltige und widerstandsfähige Strukturen schafft.

Insgesamt dauert der Wiederaufbauprozess insbesondere in Teilen des Ahrtals – aber auch beispielsweise an der Erft und der Inde – deutlich länger als nach der Katastrophe angenommen. Es zeigt sich, dass der Wieder- und Neuaufbau einzelner Ortsteile, aber auch größerer Infrastrukturen, wie Schulen, Brücken und Bahnstrecken, deutlich mehr Zeit in Anspruch nimmt als ein oder zwei Jahre. Zudem ändern sich Daten und Planungsgrundlagen für den Wiederaufbau, beispielsweise bezogen auf genauere oder überarbeitete Hochwassergefahrenkarten, wobei hier auch innovative Ansätze zur Hochwassermodellierung genutzt werden können (s. auch ▶ Kap. 4), sodass Anpassungen im Wiederaufbauprozess erforderlich werden. Ein wichtiger Befund ist daher, dass die wissenschaftliche Begleitforschung nicht nur im ersten Jahr nach einer entsprechenden Katastrophe von Bedeutung ist und wirken kann, sondern wissenschaftliche Erkenntnisse und Empfehlungen auch im weiteren Wieder- und Neuaufbauprozess in den Folgejahren Berücksichtigung finden können. So wurden wissenschaftliche Empfehlungen zur Verlagerung von kritischen und sensiblen Infrastrukturen oder zu hochwasserangepassten Brücken entwickelt, die (heute) drei Jahre nach dem Ereignis in der konkreten Planung und Projektausführung zur Anwendung kommen. Beispielsweise hat die Schulverwaltung die Verlagerung der Levana-Schule, die knapp 100 Kinder mit geistigen und teils auch körperlichen Behinderungen beherbergt, jetzt vom Kreis Ahrweiler beschließen lassen (siehe u. a. Bonner General-Anzeiger 2024), sodass in den kommenden Jahren der Neubau an anderer Stelle erfolgen kann.

Eine zentrale Schlussfolgerung aus der wissenschaftlichen Begleitforschung ist daher, dass durch neue Daten zu verschiedenen Themenfeldern, wie z. B. der Hochwasser- und Schadensmodellierung (s. ▶ Kap. 4 und 7), der mentalen Belastung und Traumatisierung von Menschen (s. ▶ Kap. 9), der besonderen Herausforderungen des Wiederaufbaus in sozial benachteiligten Quartieren (s. ▶ Kap. 18), der Analyse der Verklausung von Brücken (s. ▶ Kap. 10) und den Befunden zum Thema Vorsorgestrategien von betroffenen Haushalten (s. ▶ Kap. 19) sowie Unternehmen (s. ▶ Kap. 8), neue Erkenntnisse in den laufenden Wiederaufbau und die Diskussions- sowie Entscheidungsprozesse einfließen können. In einigen Fällen haben diese Befunde und Empfehlungen auch sehr konkrete Wirkungen in der Praxis gehabt, auf die weiter unten noch eingegangen wird (SDG 3 – „Gesundheit und Wohlergehen", SDG 9 – „Industrie, Innovation und Infrastruktur", SDG 10 – „Weniger Ungleichheiten", SDG 11 „Nachhaltige Städte und Gemeinden", SDG 13 – „Maßnahmen zum Klimaschutz", SDG 15 – „Leben an Land").

Trotz zahlreicher Erfolge der wissenschaftlichen Begleitforschung ist anzumerken, dass die wissenschaftlichen Empfehlungen des KAHR-Projekts auch bei einigen Akteuren auf Kritik gestoßen sind. Beispielsweise wurde die Kritik an der Auslegung der Wiederaufbauförderung von manchen Ministerien kritisch gesehen, weswegen es entsprechende Erörterungsgespräche direkt mit den Ministerien gab. Auch konnte sich das KAHR-Projekt durch seine wissenschaftliche Forschung vor Ort mit Stellungnahmen z. B. im Rahmen von Enquete-Kommissionen der Landtage (siehe u. a. Heidbreder 2023; Schüttrumpf 2022; Thieken 2024) sowie im Bauausschuss des Bundestages (siehe u. a. Birkmann 2022) einbringen.

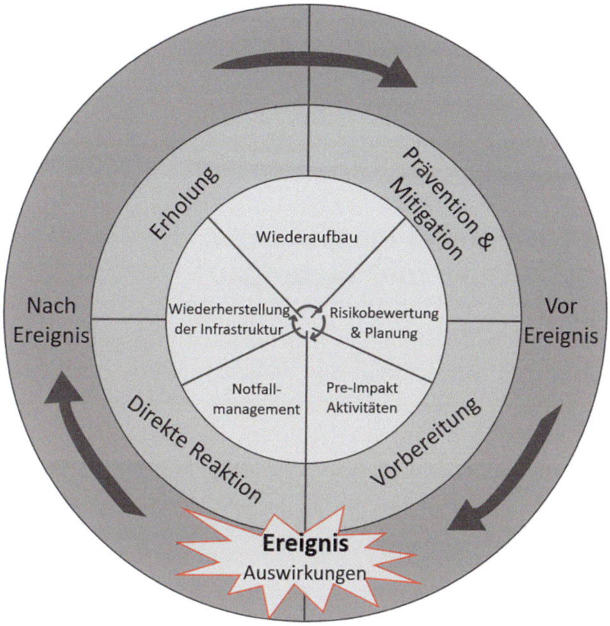

Abb. 23.1 Der Katastrophen-Risiko-Management-Kreislauf. (Eigene Darstellung auf Basis von Lumbroso (2007) und Vanneuville et al. (2011))

Die hohe mediale Aufmerksamkeit, die der Forschung zum Wieder- und Neuaufbau entgegengebracht wurde, hat zudem dazu beigetragen, dass die Stimme der Wissenschaft in der Öffentlichkeit und Politik stärker wahrgenommen wurde. Daher fanden im Zusammenhang der rund fünf Wissenschaft-Praxis-Dialoge des KAHR-Projekts sowie der Synthese-Konferenz auch jedes Mal entsprechende Pressegespräche statt. Die Berichte in den lokalen, überregionalen und nationalen Medien führten u. a. dazu, dass die Einschätzungen der Wissenschaft eine höhere Sichtbarkeit in Politik und Verwaltung erhielten und damit indirekt mehr Wirkung erzielten. Demzufolge ist eine unserer Schlussfolgerungen, dass die Kommunikation wissenschaftlicher Befunde und Empfehlungen über die Medien eine wichtige Bedeutung für die Sensibilisierung und Versachlichung der Diskussion zum Wieder- und Neuaufbau haben kann.

Gerade Wieder- und Neuaufbauprozesse nach Katastrophenprozessen – wie nach der Flutkatastrophe im Sommer 2021 –, bei denen enorme Schäden behoben (siehe Mohor et al. 2024) und Infrastrukturen wiederaufgebaut sowie die sozialen und gesundheitlichen Folgen der Katastrophe über längere Zeiträume (siehe Zenker et al. 2024) bewältigt werden müssen, bieten einen wichtigen Forschungsgegenstand und Ansatzpunkte zur Stärkung der Vorsorge gegenüber zukünftigen Risiken. Bisher wird diesen Phasen im klassischen Risikomanagementkreislauf (siehe u. a. Lumbroso 2007) für die Resilienzbildung noch zu wenig Gewicht beigemessen (siehe Abb. 23.1).

Die Begleitforschungen in KAHR zeigen sehr deutlich, dass die Ansätze und Maßnahmen in der Reaktions- und Erholungsphase wichtige Weichenstellungen für die Vorsorge und Risikoreduktion beinhalten. Ansätze der Prävention von Risiken und zur Stärkung von Resilienz sollten daher auch in den Reaktions- und Erholungsphasen aktiv gestärkt werden. Dies betrifft u. a. Förderinstrumente für den Wiederaufbau oder temporäre Infrastrukturen, aber auch den operativen Hochwasserschutz, der ebenso nicht allein auf die Phase während eines Ereignisses zu beschränken ist, sondern um die Vor- und Nachsorge erweitert werden muss (s. auch ► Kap. 13).

Bereits im ersten Jahr der wissenschaftlichen Begleitung hat das KAHR-Projekt, auch in Kooperation mit weiteren Experten aus den unterschiedlichen Fachperspektiven, gemeinsam „10 Empfehlungen aus Sicht der Wissenschaft zum Thema Wiederaufbau und Zukunftsfähigkeit der flutbetroffenen Regionen" formuliert (siehe Birkmann et al. 2022a), auf die vor Ort bereits häufig zurückgegriffen wurde.

23.2 Empfehlungen für einen klima- und hochwasserresilienten Wieder- und Neuaufbau

Im Folgenden werden die 10 KAHR-Empfehlungen, die früh im Projektverlauf zusammengestellt wurden, skizziert und ausgewählte Empfehlungen bezogen auf ihren Impact diskutiert. Die Empfehlungen spiegeln auch die interdisziplinäre Zusammensetzung des Forschungsteams im KAHR-Projekt wider. Einige Empfehlungen sind zudem im ersten Teil recht allgemein, d. h. es sind zuerst allgemeine, prägnante Empfehlungen aufgeführt, die dann aber durch jeweilige beschreibende Texte weiter konkretisiert werden und vielfach auf spezifische Handlungsoptionen hinweisen.

- **[1] Der Wiederaufbau nach dem Hochwasserereignis 2021 bietet auch eine Chance, einen strategischen Transformationsprozess einzuleiten und Katastrophenresilienz zu stärken.**

Der Wiederaufbau nach der Flutkatastrophe zeigt die Notwendigkeit von Veränderungen auf und ist daher auch eine Gelegenheit, resiliente, hochwasser- und klimaangepasste Raum- und Siedlungsentwicklungskonzepte sowie Managementkonzepte für Flüsse zu konkretisieren, zu fördern und umzusetzen. Auch Synergien zu anderen strategischen Zielen wie dem Klimaschutz sind zu nutzen. Hierunter fällt beispielsweise die Abkehr von Ölheizungen, die sowohl die Schadenshöhe im Hochwasserfall erheblich erhöhen als auch im Sinne des Klimaschutzes eine Veränderung erfordern. Für diese zielgerichtete Transformation sollten Wiederaufbaumittel strategisch mit weiteren Fördermitteln kombiniert werden. Der reine Wiederaufbau der alten Strukturen ist nicht hinreichend (SDG 11 – „Nachhaltige Städte und Gemeinden", SDG 13 – „Maßnahmen zum Klimaschutz").

- **[2] Alle Potenziale der Hochwassermodellierung und Risikoanalyse sollten zur Planung von Schutzstrategien sowie zur Vorbereitung und Warnung Betroffener ausgeschöpft werden.**

In der Wissenschaft werden neue Methoden zur Hochwassermodellierung erfolgreich eingesetzt, deren Detailgrad zwischen Kilometern und Millimetern und deren Berechnungsdauern zwischen Sekundenbruchteilen und Stunden variieren. Diese Methoden sollten auf ihre praktische Einsetzbarkeit kritisch geprüft werden und zielgerichtet in die Erstellung und Überarbeitung von Hochwassergefahren- und -risikokarten, in die Ausweisung hochwassergefährdeter Gebiete, in die Hochwasservorhersage und -warnung sowie in den operativen Katastrophenschutz integriert werden. Extreme Ereignisse können insbesondere dann zur Katastrophe werden, wenn Bevölkerung, Entscheidungsträgerinnen und -träger und Katastrophenschutz überrascht und unvorbereitet mit einer Extremsituation konfrontiert werden. Risikoanalysen sollten u. a. dazu beitragen, solche Überraschungen zu vermeiden und Vorsorge auch in die strategische Entwicklung von Dörfern, Städten und Infrastrukturen frühzeitig zu integrieren. Dies erfordert einen breiteren Ansatz der Risikomodellierung, der sich nicht auf wenige Szenarien beschränkt, sondern auch Worst-Case Szenarien jenseits des HQ_{extrem} umfasst. Es sollte auch vorab abgeschätzt werden, wo sich besonders ungünstige Situationen einstellen könnten, die zum Verlust von Menschenleben führen könnten (SDG 13 – „Maßnahmen zum Klimaschutz").

- **[3] Mehr Raum für den Fluss ist wichtig, dies bedeutet aber nicht nur Siedlungsrückzug – sondern auch angepasste Landnutzungen.**

Flüsse brauchen Raum. Ist dieser Raum nicht vorhanden, so werden Siedlungsflächen, Gewerbe- und Industriegebiete überflutet. Siedlungsrückzug bietet sich vor allem in der Wiederaufbauphase dort an, wo ohnehin Strukturen geschädigt oder zerstört sind. Wichtig ist neben der Gefährdungsintensität (Hochwasser/Starkregen) am betroffenen Standort auch ein Blick auf die Schutzwürdigkeit der betroffenen Strukturen. Einrichtungen, in denen sich besonders verwundbare Menschen aufhalten (z. B. Krankenhäuser, Schulen, Kindergärten, Altersheime), sind hier anders zu behandeln als „normale" Wohnnutzungen oder Gewerbestandorte. Auch Störfallbetriebe, von denen im Schadensfall Sekundäreffekte ausgehen können, sind besonders zu betrachten. Demzufolge müssen die unterschiedlichen Verwundbarkeiten stärker in den Fokus rücken. Auch bietet die Schaffung von multifunktionalen Flächennutzungen Chancen für die Stadtentwicklung im Bereich der Gewässer. So können Parks und Sportplätze im Hochwasser- oder Starkregenfall als Überflutungsflächen dienen, ohne dass hohe Schäden entstehen, und gleichzeitig stehen diese die allermeiste Zeit der Bevölkerung zur Verfügung. Zudem sollten für ausgewählte Flächennutzungen Schutzziele formuliert werden, da die Risiken und Schäden durch Hochwasser und Starkregen auch von der unterschiedlichen Verwundbarkeit und Vorsorge abhängen. Diese Aspekte müssen Bestandteile von Konzepten angepasster Landnutzung werden. Der Wiederaufbau sollte mit der Förderung von Modellansätzen einhergehen (SDG 15 – „Leben an Land").

- **[4] Brücken müssen in Zukunft in der Betrachtung von Hochwassergefahren stärker berücksichtigt werden. Brücken können bei Hochwasserereignissen die Überflutungsgefahr deutlich erhöhen.**

Während des Juli-Hochwassers 2021 sind in einigen Flussgebieten über 50 % der vorhandenen kommunalen, Landes-, Bundes- und Bahnbrücken beschädigt oder zerstört worden, da diese für die enormen Staudrücke und Anpralllasten durch den hohen Anfall an Geschiebe und Treibgut nicht ausgelegt waren. Bei der Neuplanung und zukünftigen Überprüfung von Brückenbauwerken an Fließgewässern sollten vorhandene Bemessungsansätze in Hinblick auf die hydraulische Leistungsfähigkeit des Bauwerks, die erforderliche Gründungsform und -tiefe sowie die statischen Lastfälle erweitert werden, um der erhöhten Belastung durch mitgeführtes Geschiebe und Treibgut gerecht zu werden. Gleichzeitig muss zukünftig die hochwasserverstärkende Wirkung von Brücken durch lokalen Aufstau sowie das Ausbilden von unerwarteten Strömungswegen bei Bauwerksumströmung und die Flutwelle bei Bauwerksversagen in Hochwassergefahren- und -risikokarten sowie der Planung berücksichtigt werden. Vergrößerte oder angepasste Abflussquerschnitte, größere Spannweiten, verstärkter Erosionsschutz, etc. sind Aspekte, die in Zukunft Anwendung finden sollten (SDG 9 – „Industrie, Innovation und Infrastruktur").

- **[5] Die Frühwarnung vor Hochwasserereignissen ist zu stärken.**

Das Hochwasserereignis vom Juli 2021 hat die lebenswichtige Bedeutung effektiver Frühwarnung und vorbereitender Verhaltensvorsorge deutlich sichtbar gemacht. In dieser Hinsicht sind effektive Warnsysteme zu überprüfen und weiterzuentwickeln, sodass diese selbst bei einem kommunalen oder regionalen Stromausfall noch funktionieren. Neuere Entwicklungen erlauben hochaufgelöste, sogenannte impact-basierte Vorhersagen, welche beispielsweise die Vorhersage von Überflutungsflächen und betroffenen Gebäuden ermöglichen. Solche Vorhersagen liefern wichtige Informationen für das Katastrophenmanagement. Die heute bestehenden Wasserstandsvorhersagen an Hochwassermeldepegeln sollten durch solche impact-basierten Vorhersagen ergänzt werden (SDG 13 – „Maßnahmen zum Klimaschutz").

- **[6] Die Signalfunktion von Plänen und Planungen muss gestärkt werden. Starkregengefahren- und -risikokarten müssen öffentlich zugänglich sein.**

Hochwasser und Starkregen müssen gemeinsam gedacht werden! Bisher fehlt die regionalplanerische Darstellung der Vorranggebiete für Hochwasser für die Siedlungsbereiche in Rheinland-Pfalz. Die gleichzeitige Darstellung der HQ_{100} Bereiche sowie betroffener Flächen bei einem HQ_{extrem} in Regional- und Flächennutzungsplänen sowie die Darstellung der Wasserlinie aus dem Hochwasser im Sommer 2021 kann eine wichtige Signalfunktion für andere Fachplanungen haben. Ein Durchgriff aus dem Wasserrecht (§ 78 Wasserhaushaltsgesetz 2009) ist nur für erstmalig baulich in Anspruch genommene Bereiche möglich, nicht aber für Brachflächen oder bereits bebaute Bereiche. Demzufolge ist die Vorsorge in bereits bebauten Bereichen durch das Planungs- und Baurecht sowie durch informelle Instrumente zu stärken. Bei Baugenehmigungen und Umbauten sollten Hauseigentümer automatisch Informationen über Hochwasser- und Starkregengefahren erhalten (SDG 13 – „Maßnahmen zum Klimaschutz").

- **[7] Hochwasser- und klimaresilientes Planen und Bauen muss auf allen Ebenen der räumlichen Planung integriert werden und alle Facetten der Klimawandelauswirkungen berücksichtigen.**

Zur Schadensminimierung bei künftigen Hochwasserereignissen ist es erforderlich, dass alle Ebenen – von der Bundes-, über die Landes- und Regionalebene bis zur kommunalen Bauleitplanung – Resilienz, Klimaanpassung und Hochwasservorsorge in der Siedlungs- und Raumentwicklung berücksichtigen. Dazu sind konkrete Ziele und Grundsätze sowie planerische Darstellungen in den Planwerken erforderlich. Diese können unter anderem konkrete Elemente des Hochwasserflächenmanagements wie das Ausweisen von Überschwemmungsgebieten beinhalten. Für die Stärkung der Resilienz von Kommunen sind die kommunale Bauleitplanung und Flächennutzung unerlässlich, sie sind jedoch allein nicht ausreichend für ein zukunftsfähiges Hochwasserrisikomanagement. Hierzu sind auch vielfach überkommunale Ansätze notwendig. Eine klimaresiliente Planung erfordert zudem die Berücksichtigung zukünftiger Auswirkungen des Klimawandels sowie unterschiedlicher, sich daraus ergebender Phänomene wie Hochwasser- und Dürrephasen. Diesbezüglich spielt die Stärkung der grün-blauen Infrastrukturen eine wichtige Rolle, da hierdurch sowohl Hitze- als auch Hochwasser- und Starkregenvorsorge betrieben werden kann (SDG 11 „Nachhaltige Städte und Gemeinden", SDG 13 – „Maßnahmen zum Klimaschutz").

- **[8] Ein nachhaltiger Wiederaufbau gelingt, wenn Akteure Formen der Zusammenarbeit etablieren und interkommunal zusammenarbeiten. Fördermittel sollten diese Zusammenarbeit stärken.**

Die Etablierung bzw. Stärkung geeigneter Koordinationsstrukturen für die unterschiedlichen Akteure im Wiederaufbau ist innerhalb einer Region sicherzustellen. Aktuelle Strategien des Wiederaufbaus werden vielfach recht kleinräumig gedacht und umgesetzt. Für klimaresiliente Siedlungsstrukturen muss der Wiederaufbau die interkommunale Zusammenarbeit sowie Berücksichtigung des Klimawandels stärken. Einige Kooperationen zwischen Kommunen bestehen und können als Positivbeispiele (z. B. kommunale Zusammenschlüsse, Hochwasserpartnerschaften, Wasserverbände) angeführt werden. Der Fokus ist allerdings oftmals auf einen kleineren Teilraum eines Flusseinzugsgebiets gerichtet. Neben zentralen Koordinierungsstellen sollten adaptive Ansätze gestärkt werden, die es ermöglichen, durch Experimentieren und Lernen innovative Lösungsansätze für den Wiederaufbau zu identifizieren. Zudem ist das Thema des fairen Lastenausgleichs zwischen Gemeinden zu bedenken. Städte, die von einem verbesserten Hochwasserschutz und Wasserrückhalt im Oberlauf oder Umland profitieren, sollten im Gegenzug auch bestimmte Leistungen für die Gemeinden oberstrom erbringen. Die Themen regionale Kooperation sowie fairer Lastenausgleich sollten stärker kommuniziert und im Rahmen der Wiederaufbauförderung sowie weiterer Förderstrukturen (z. B. Dorferneuerung oder Städtebauförderung) unterstützt werden, beispielsweise durch spezifische Fördermittel für die interkommunale Kooperation (z. B. Zusammenlegung von Sportplätzen, um mehr Raum für den Fluss zu schaffen) (SDG 11 „Nachhaltige Städte und Gemeinden", SDG 13 – „Maßnahmen zum Klimaschutz").

- **[9] Intensive Vorbereitung des Katastrophenschutzes und der Wasserwirtschaft auf seltene Hochwasser- und Starkregenereignisse verbessert die Bewältigung dieser Ereignisse.**

Katastrophale Hochwasser- und Starkregenereignisse treten nur selten auf, dann aber mit verheerenden Schäden. Insbesondere die kurze Vorwarnzeit von Starkregenereignissen und die mitunter mühsame Beschaffung eines Lagebildes erschweren dabei die operativen Gefahrenabwehrmaßnahmen und die damit mögliche Reduzierung oder Verhinderung von Schäden. Hier treffen Fachexpertisen und Zuständigkeiten der Wasserwirtschaft und des Katastrophenschutzes zusammen. Eine angemessene und umfassende gemeinsame Vorbereitung von Katastrophenschutz und Wasserwirtschaft kann an dieser Stelle die Schnittstellenarbeit verbessern. Eine zielführende Möglichkeit hierfür ist die Entwicklung und gemeinsame Analyse von Modellszenarien, die konkrete Ereignisauswirkungen abbilden. Die Möglichkeiten der Wasserwirtschaft zur Echtzeitdatenerhebung und -auswertung sollten in der Arbeit des Katastrophenschutzes Berücksichtigung finden. Ebenso wichtig ist die Erarbeitung passgenauer, gemeinsamer Aus- und Fortbildungsangebote für die Einsatzkräfte des Katastrophenschutzes (SDG 13 – „Maßnahmen zum Klimaschutz").

- **[10] Neue Schutzstandards und Schutzziele für Kritische und Sensible Infrastrukturen müssen definiert und das Bewusstsein für ein unvermeidbares Restrisiko muss gestärkt werden.**

Im Rahmen des Wiederaufbaus von Strom- und Kommunikationsinfrastrukturen, Krankenhäusern, Schulen, Kindergärten, Einrichtungen für Menschen mit Behinderung oder Altersheimen, die als sogenannte „Kritische und Sensible Infrastrukturen" bezeichnet werden (siehe Raumordnungsgesetz 2008, [ROG § 2 Abs. 2]), sind höhere Schutzstandards und Schutzziele zu entwickeln. Gerade durch die erheblichen negativen Auswirkungen des Ausfalls und der Schädigung solcher Infrastrukturen – einschließlich von Kaskadeneffekten durch die Funktionsverluste – müssen diese Infrastrukturen und darin befindliche Menschen stärker als bisher in den Fokus rücken. Hierzu sind auch höhere Schutzziele notwendig, die der besonderen Bedeutung dieser Infrastrukturen Rechnung tragen und Extremhochwasser in den Blick nehmen. Bei der Entwicklung von Hochwasservorsorgekonzepten ist es unerlässlich, dass sich Kommunen mit differenzierten Schutzzielen für den Baubestand wie auch den Neubau auseinandersetzen. Diese sollten sich von festen Jährlichkeiten lösen und neben der Gefährdungsintensität auch die Empfindlichkeit der unterschiedlichen Strukturen und ihrer Nutzerinnen und Nutzer berücksichtigen (SDG 11 „Nachhaltige Städte und Gemeinden", SDG 13 – „Maßnahmen zum Klimaschutz").

23.3 Wirkungen der wissenschaftlichen Begleitforschung

Das KAHR-Forschungsprojekt kann bereits nach drei Jahren auf erste dokumentierbare Wirkungen der wissenschaftlichen Begleitforschung verweisen. In der wissenschaftlichen Wirkungsforschung werden verschiedene Stufen der Wirkung unterschieden. Neben der Frage, welche konkreten Aktivitäten und Ergebnisse für die Praxis oder Entscheidungsträger bereitgestellt werden, zeigt ◘ Abb. 23.2, dass in einer ersten Phase die Aufnahme der Forschungsergebnisse erfolgt, beispielsweise in Form der Identifikation spezifischer Zielgruppen oder spezifischer Ansprachen von Akteuren und Zielgruppen. In dieser Hinsicht waren im Rückblick die Wissenschaft-Praxis-Dialoge im KAHR-Projekt ein wichtiges Element, um verschiedene Akteure im Wiederaufbau über die laufenden Befunde und Empfehlungen der wissenschaftlichen Forschung zu informieren und zu sensibilisieren. Eine weitere Stufe umfasst die Nutzung und Umsetzung der wissenschaftlichen Befunde, Daten und Empfehlungen durch Akteure, die auch zu Lernprozessen und

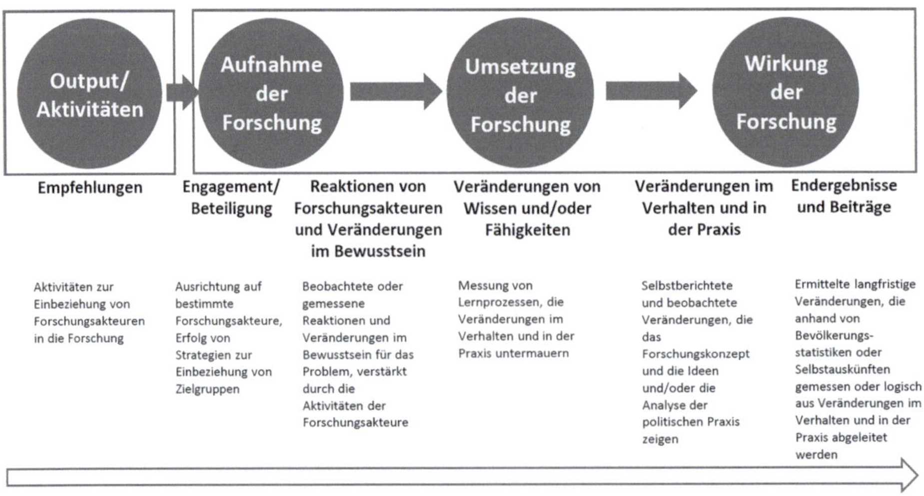

◘ **Abb. 23.2** Phasen der wissenschaftlichen Wirkungsforschung. (Eigene Darstellung auf Basis von Morton (2015))

Veränderungsprozessen führen, wobei die Herausforderungen des Transfers von wissenschaftlichen Ergebnissen in ▶ Kap. 21 genauer betrachtet werden. Die dritte Stufe ist dann die eigentliche Wirkung der Forschung, die vielfach mit Langzeitwirkungen der Forschung verbunden wird. Hier kann das KAHR-Projekt aufgrund seiner bisherigen drei-jährigen Existenz noch nicht auf Langzeitwirkungen verweisen, allerdings sind die auf Basis der KAHR-Forschung getroffenen Änderungen, wie z. B. die Entscheidung zur räumlichen Verlagerung der Levana-Schule oder neue Standards beim hochwasserresilienten Brückenbau, wichtige Anhaltspunkte für die längerfristige Wirkung der Forschungsergebnisse auch über die Projektlaufzeit hinaus.

Da der größte Teil der Wirkungen der zehn wissenschaftlichen KAHR-Empfehlungen noch nicht in der Fachliteratur zu finden ist, basieren die Erkenntnisse zu den ersten Wirkungen des wissenschaftlichen Begleitprojekts auf folgenden Quellen und Methoden:

- Gespräche mit Wiederaufbauagenturen und -gesellschaften,
- Gespräche, Workshops und Planspiele mit betroffenen Gemeinden und Kommunen,
- Gespräche mit den Planungsbehörden auf lokaler und regionaler Ebene,
- Gespräche und Workshops mit Ministerien und Behörden,
- Diskussionen auf wissenschaftlich-politischen Konferenzen, die im Rahmen des KAHR-Projekts durchgeführt wurden,
- Dokumentenanalyse und
- Zeitungsanalyse.

In zahlreichen Fällen bestätigte mehr als eine Quelle die Nutzung ausgewählter Empfehlungen aus den „10 wissenschaftlichen Empfehlungen“, die vom KAHR-Projekt veröffentlicht wurden. Basierend auf dem theoretischen Konzept der „Wirkungskette von Forschung“ nach Morton (2015), ◻ Abb. 23.2 und weiterer Literatur (Pfeifer und Helming 2024; Feiertag 2018) unterscheiden wir „Forschungsaktivitäten“, die Teil des eigenen Forschungsprozesses im KAHR-Projekt (z. B. Beratungen, Workshops, etc.) sind, und „Ergebnisse“, die sich an die Forschungsaktivitäten in Form von entwickelten Dienstleistungen und Produkten anschließen. Hierzu zählen auch die 10 wissenschaftlichen Empfehlungen. Darüber hinaus gibt es Ergebnisse, die sich aus den Aktivitäten und „Outputs“ ergeben. Die Wirkungsforschung befasst sich zudem mit den Wirkungen der Ergebnisse auf verschiedene (Ziel-)Gruppen, wie z. B. die Schaffung eines stärkeren Bewusstseins für die sich verändernden Hochwasser- und Starkregenrisiken im Kontext des Klimawandels. Die abschließenden langfristigen Wirkungen (Impacts) – als eine Art Endphase der Wirkungsweise der Forschung – umfassen und implizieren reale Veränderungen, die durch die Forschungsaktivitäten und -ergebnisse ausgelöst werden. Diese werden oft auf langfristige Veränderungen in Bezug auf gesellschaftliche, wirtschaftliche und ökologische Veränderungen oder Transformationen bezogen (Morton 2015; Feiertag 2018; Pfeifer und Helming 2024). Obschon das KAHR-Projekt aufgrund seiner kurzen Entstehungsgeschichte und Laufzeit (3 Jahre) nicht auf eine längere zeitliche Phase und damit langfristige Wirkungen zurückblicken kann, werden verschiedene Wirkungen der Begleitforschung bereits sichtbar und im Folgenden anhand von drei Fragen systematisiert:

- Wie wurden die Empfehlungen kommuniziert und präsentiert?
- Welche Informationen gibt es darüber, wie diese wissenschaftlichen Empfehlungen aufgegriffen wurden und wie sie sich auf die Wiederaufbauaktivitäten und -politiken in Rheinland-Pfalz und Nordrhein-Westfalen und darüber hinaus auswirken?

- Welche Faktoren behindern oder fördern die Berücksichtigung der wissenschaftlichen Expertise?

23.3.1 Kommunikation und Übernahme der 10 wissenschaftlichen Empfehlungen

Um sicherzustellen, dass die Forschungsergebnisse und die 10 Empfehlungen des KAHR-Projekts von den am Wiederaufbau beteiligten Akteuren wahrgenommen und aufgegriffen wurden, hat das Projekt die Empfehlungen in einem bereits im Jahr 2022 veranstalteten Wissenschaft-Praxis-Dialog zur Diskussion gestellt und dokumentiert. Auch in verschiedenen Workshops mit den am Wiederaufbau beteiligten kommunalen und staatlichen Stellen sowie Präsentationen in der Enquete-Kommission des Landtags von Rheinland-Pfalz sowie in Präsentationen im Bauausschuss des Bundestages wurden die 10 Empfehlungen präsentiert und ausgewählte Empfehlungen näher erläutert. Darüber hinaus waren Interviews in den Medien ein wichtiges Instrument, um die wissenschaftlichen Empfehlungen in der breiteren Öffentlichkeit bekannt zu machen und private sowie staatliche Akteure zu sensibilisieren. Die Kommunikation war dabei nicht auf die Veröffentlichung der Empfehlungen in einer lokalen Zeitung beschränkt, sondern es wurden auch überregionale Medien (ARD, ZDF, Landtagsanhörungen, etc.) genutzt, um Entscheidungsträger auf Landes- und Bundesebene zu sensibilisieren. Die Landes- und Bundesebene definiert wichtige Rahmenbedingungen für den Wiederaufbau, den Hochwasserschutz und die räumliche Planung, sodass sich Innovationen und Änderungen auch in diesen Rahmenbedingungen niederschlagen sollten.

Durch zahlreiche Gespräche mit staatlichen und lokalen Institutionen sowie für den Wiederaufbau zuständigen Behörden und durch Interviews in nationalen und lokalen Medien (mehr als 150 Interviews) wurde die Sichtbarkeit der 10 wissenschaftlichen Empfehlungen des KAHR-Projekts deutlich erhöht. Zudem wurden die Empfehlungen auf der Website des KAHR-Projekts veröffentlicht und sind auch auf anderen Internetseiten, wie z. B. beim Bauausschuss des Bundestags, abrufbar (siehe Birkmann 2022). Zudem wurden einzelne Empfehlungen in den Berichten von Enquete-Kommissionen der Landtage in Rheinland-Pfalz und Nordrhein-Westfalen dokumentiert (Heidbreder 2023; Schüttrumpf 2022; Thieken 2024).

23.3.2 Veränderungen des Wissens und der Entscheidungsgrundlagen mit längerfristigen Folgen

Um die Wirkungen der 10 KAHR-Empfehlungen auf die Veränderungen des Wissensstandes und der Entscheidungsfindung zu erfassen, wurden ausgewählte Empfehlungen näher betrachtet bzw. Informationen darüber eingeholt, inwiefern die Empfehlungen einen praktischen Nutzen für Entscheidungsträger hatten. Da es uns nicht möglich war, die Berücksichtigung aller Empfehlungen bei der Entscheidungsfindung zu ermitteln, konzentriert sich diese Diskussion auf ausgewählte Beispiele, bei denen uns Informationen bekannt sind. Wir untersuchen und dokumentieren daher die Berücksichtigung der Empfehlung Nr. 10, die sich auf kritische und sensible Infrastrukturen bezieht, und die Empfehlung Nr. 4, die sich mit der Rolle von Brücken befasst. Diese zwei Empfehlungen erfassen daher nur einen kleinen Ausschnitt der insgesamt 10 Empfehlungen, aber diese beiden Empfehlungen sind fokussiert und sehr konkret. Darüber hinaus wurden wir von kommunalen und regionalen Akteuren darauf aufmerksam gemacht, dass diese Empfehlungen in die Entscheidungsfindung bei ausgewählten Projekten bereits eingeflossen sind.

Um näher darzulegen, wie die 10 wissenschaftlichen Empfehlungen im Hinblick auf Veränderungen von Wissen und Entscheidungsgrundlagen genutzt wurden und wie dieses Wissen tatsächlich Veränderungen in der Entscheidungsfindung auslöste, geben wir einen kurzen Überblick über Kontexte der Themen, die in den Empfehlungen 4 und 10 thematisiert werden. Danach erörtern wir konkrete Entscheidungen beim Wiederaufbau, die durch die Empfehlungen beeinflusst wurden, und skizzieren damit verbundene längerfristige Wirkungen.

23.3.3 Kritische Infrastrukturen und Wiederaufbau

Durch das Hochwasser 2021 wurden zahlreiche kritische und sensible Infrastrukturen, wie u. a. Wasserversorgungs- und Kläranlagen, schwer beschädigt. Beim Wiederaufbau von Wasserversorgungsanlagen und Kläranlagen wird verlangt, dass diese Anlagen vor einem Hochwasserereignis, das statistisch einmal in 100 Jahren auftritt (also ein HQ_{100}), geschützt werden müssen (aktueller gesetzlicher Standard). Das Hochwasser 2021 hat in drastischer Weise gezeigt, dass ein Schutzniveau für ein HQ_{100} (alt) eigentlich nicht ausreichend ist, da ja enorme Schäden an entsprechenden Infrastrukturen zu beklagen sind. Der anzuwendende gesetzliche Standard ist jedoch der HQ_{100}-Standard, wobei in Rheinland-Pfalz Neuberechnungen des HQ_{100} nach dem Flutereignis 2021 stattfanden. Interessanterweise argumentierten das rheinland-pfälzische Umweltministerium und die Struktur- und Genehmigungsdirektion Nord (SGD Nord) unter Hinzuziehung der KAHR-Empfehlung Nr. 10, dass es sich beim Wiederaufbau der Wasserversorgungsanlage „Wasserwerk Walporzheimer Straße“ um ein zentrales Element der kritischen Infrastruktur handelt und bei der Wasserver- und -entsorgung mit einer Lebensdauer von mehreren Jahrzehnten geplant wird, sodass ein höherer Schutzstandard erforderlich ist. Dieser höhere Schutzstandard ist auch mit Mitteln des Wiederaufbaufonds zu finanzieren, so die Kommunikation zwischen den Behörden.

Eine wichtige Grundlage für die Argumentation in dem entsprechenden Schreiben zwischen den zuständigen Behörden war die wissenschaftliche Empfehlung Nr. 10 des KAHR-Projekts. Die SGD Nord und das Umweltministerium argumentierten, dass auf der Grundlage dieser wissenschaftlichen Erkenntnisse und der wissenschaftlichen Empfehlung des KAHR-Projekts die ausführenden Behörden nicht nur den gesetzlich vorgeschriebenen Hochwasserschutzstandard für ein HQ_{100}-Hochwasserereignis anwenden sollten, sondern eben einen höheren Schutzstandard, da es sich um eine kritische Infrastruktur mit besonderer Schutzwürdigkeit handelt.

Dieser Entscheidungsprozess über den anzuwendenden höheren Schutzstandard beim Wiederaufbau der Wasserversorgungsanlage zeigt zwei wichtige Veränderungen in der Praxis, die zu langfristigen Veränderungen führen. Erstens stehen die KAHR-Empfehlungen den Entscheidungsträgern auf lokaler, regionaler und nationaler Ebene zur Verfügung, sodass sie auf verschiedenen Entscheidungsebenen zur Kenntnis genommen werden. Zweitens bieten die wissenschaftlichen Empfehlungen eine fundierte Argumentationsgrundlage für die Erhöhung des Schutzniveaus und die Aufstockung der Mittel für den

Wiederaufbau der Wasserversorgungsanlage über das gesetzlich vorgeschriebene Maß hinaus. Die langfristige Auswirkung ist ein höheres Schutzniveau für diese Infrastruktur und für ähnliche Infrastrukturen im Rahmen des Wiederaufbaus. Folglich kann von längerfristigen Wirkungen ausgegangen werden.

23.3.4 Sensible Infrastrukturen und vulnerable Gruppen: Beispiel Levana-Schule

Durch das Hochwasser 2021 wurden zahlreiche sensible Infrastrukturen beschädigt, die besonders vulnerable Nutzergruppen beherbergen, welche im Ereignisfall auf die Hilfe Dritter angewiesen sind. Für diese Infrastrukturen ist nur das HQ_{100} rechtsverbindlich als Schutzstandard vorgeschrieben. Des Weiteren gibt es für fast alle teilzerstörten Gebäude Bestandsschutz, weshalb ein Teil der sensiblen und kritischen Infrastrukturen im Prinzip so wieder aufgebaut werden könnte wie vor der Flut. Zu den sensiblen Infrastrukturen entlang der Ahr, die vom Hochwasser 2021 betroffen sind, gehört die Levana-Schule, die in ▸ Kap. 16 intensiv betrachtet wird. Diese Schule für Kinder mit Behinderungen betreut knapp 100 Schülerinnen und Schüler mit geistigen und teils auch körperlichen Behinderungen. Eine erste Risikoanalyse nach dem Hochwasserereignis bestätigte, dass ein Wiederaufbau der Schule an gleicher Stelle möglich wäre (siehe u. a. Rhein-Zeitung 2024). Da die Kreisverwaltung jedoch nicht davon überzeugt war, wurde das KAHR-Projekt angefragt, hier eine eingehendere wissenschaftliche Stellungnahme und Untersuchung zu vollziehen (siehe Truedinger et al. 2023; Birkmann et al. 2023b; Birkmann et al. 2024). Die Schule liegt nur wenige Meter vom Flusslauf der Ahr entfernt. Die Schule selbst ist eingeschossig, d. h. eine vertikale Evakuierung ist nicht möglich, und der Hauptausgang der Schule besteht zur Ahr hin. Im Jahr 2021 wurde die Schule ca. 2 m hoch überflutet – glücklicherweise befanden sich keine Schüler in der Schule, da das Ereignis in der Nacht stattfand. Beim Auftreten des nach dem Hochwasser neu berechneten HQ_{100} werden die Schule und der Schulhof einschließlich der Evakuierungssammelstelle und -wege teilweise bis zu einer Höhe von 1 bis 2 m überflutet – und bei außergewöhnlichen Starkregenereignissen beträgt die Fließgeschwindigkeit des Wassers entlang der Zufahrtsstraße zur Schule bereits bis zu 1 m/s bei einer Wassertiefe von bis zu 30 cm, wobei der Parkplatz, der für die Zufahrt zu Schule gequert werden müsste, sogar bis zu 1 m hoch überflutet wäre (siehe Truedinger et al. 2024).

In einer Stellungnahme zur Levana-Schule im Jahr 2023 (Birkmann et al. 2023b), die auch die KAHR-Empfehlung Nr. 10 aufgreift und konkretisiert, haben wir uns mit diesen Faktoren ausführlich auseinandergesetzt und uns letztlich für die Empfehlung zur Verlagerung der Schule ausgesprochen – denn die Nutzergruppe ist besonders vulnerabel, das Gebäude bietet keine vertikale Evakuierungsmöglichkeit und neben der Überflutung ist auch die Fließgeschwindigkeit im Bereich der Schule an der Ahr nach einer Brücke besonders hoch bzw. stellt eine zusätzliche Gefährdung dar. Auch die schnelle Evakuierung erscheint wenig plausibel, da es sich um Kinder mit geistigen und körperlichen Behinderungen handelt, die nicht selbstständig diesen Prozess vollziehen können. Diese Stellungnahme wurde diskutiert und im September 2024 fasste der Kreistag schließlich den Beschluss, die Kreisschule zu verlagern – auch gerade mit dem Verweis auf unsere wissenschaftliche Stellungnahme und die besondere Bedeutung des Schutzes sensibler Infrastrukturen (siehe Bonner General-Anzeiger 2024). Die wissenschaftliche Stellungnahme war zudem für das Ministerium von hoher Relevanz, um die räumliche Verlagerung auch mit den Mitteln der Wiederaufbauförderung zu finanzieren, da die räumliche Verlagerung Mehrkosten von ca. 10 Mio. € umfassen wird. Aus Sicht der Resilienzforschung sind diese Mittel und Mehrausgaben gerechtfertigt, da es sich um eine besonders schutzwürde Infrastruktur mit sehr vulnerabler Nutzergruppe handelt und mit der räumlichen Verlagerung die betreuten Kinder der Schule, der Schulbetrieb, aber auch das Gebäude und seine Einrichtungen an einen hochwassersichereren Standort verlagert werden.

23.3.5 Hochwasserresiliente Brücken

Die zahlreichen Brücken im Ahrtal haben beim Hochwasser 2021 massiv zu einem Anstieg der Überschwemmungen und Schäden beigetragen. Durch die großen Mengen an Treibgut wie Holz, Wohnwagen oder Tanks ist es an den vielen, alten Steinbrücken zu massiven Verklausungen gekommen, die zu erheblichen Pegelanstiegen von bis zu 2,5 m geführt haben. Faktoren wie die Höhe der Brücke, die Anzahl der Pfeiler, die Bauweise und die Spannweite haben das Ausmaß der Verklausung mit beeinflusst. Dies führte zu erheblichen Kräften und Drücken sowie zu starken Strömungen, die die Brücken umspülten. Die Brücken konnten den hohen Kräften und Strömungsgeschwindigkeiten meist nicht standhalten, sodass 80 der 114 Brücken an der Ahr beschädigt und einige vollständig zerstört wurden. Eine vollständige Analyse der Brückenschäden an Ahr, Inde und Vicht, einschließlich der Ursachen und des Ausmaßes der Schäden, wurde von Burghardt et al. (2022) durchgeführt. Zudem werden Auswirkungen von Brücken auf das Abflussgeschehen 2021 auch in ▶ Kap. 10 näher betrachtet.

Brücken sind nicht nur in normalen Zeiten, sondern auch im Katastrophenfall ein wichtiges Verbindungselement im Verkehrsbereich, sodass ihre Funktionsfähigkeit auch in Extremsituationen gewährleistet sein muss. Es war daher notwendig, aus dem Hochwasserereignis 2021 zu lernen und Empfehlungen für hochwasserresiliente Brücken in Mittelgebirgsregionen zu entwickeln. Die Empfehlungen für hochwasserresiliente Brücken betreffen dabei den Brückenquerschnitt, den Überbau der Brücken, die Pfeiler einschließlich ihrer Fundamente, die Konstruktion der Brücken und die Bauweise. Der Wiederaufbau von Brücken nach einem Hochwasser ist eine ideale Gelegenheit, die Verkehrsinfrastruktur nicht nur an ein modernes Verkehrssystem anzupassen, sondern auch das Risiko von Schäden durch Extremereignisse zu verringern. Beide Aspekte müssen in den vom Hochwasser 2021 betroffenen Gebieten berücksichtigt und umgesetzt werden. Dabei stoßen die Planer jedoch auf eine Reihe von Hindernissen, denn zum Beispiel die alten Steinbrücken in Bogenbauweise prägen das Landschaftsbild einer Tourismusregion und stehen oft unter Denkmalschutz.

Um das Thema „hochwasserresiliente Brücken“ aus der Forschung in die Praxis zu übertragen, wurden mehrere Workshops mit den Planern, den Kommunen, den Landkreisen und den jeweiligen Bundesländern durchgeführt. Ziel war es, nicht nur die Ergebnisse der hydro-numerischen Berechnungen zu präsentieren, sondern auch die Bedeutung von hochwasserresilienten Brücken für eine klimaangepasste Zukunft zu diskutieren und in die Sanierungsplanung einzubringen. Dies hat nun dazu geführt, dass nicht nur das Bewusstsein für hochwasserresiliente Brücken bei den Akteuren vorhanden ist, sondern dass auch erste Planungen in Bad Neuenahr-Ahrweiler diese Empfehlungen aufgreifen und umsetzen. Somit kann ein direkter Einfluss der Forschung auf den Wiederaufbauprozess nachgewiesen werden. Diese Entscheidungen und Veränderungen, die durch die wissenschaftliche Begleitforschung angestoßen und beschleunigt wurden, haben auch mittel- und langfristig nicht nur auf die betroffene Infrastruktur (Schule, Wasserversorgung, Entsorgungseinrichtungen, Brücken) Auswirkungen, sondern auch auf zukünftige Planungen und Wiederaufbaustrategien, so unsere Einschätzung. Demzufolge haben die wissenschaftlichen Befunde und Empfehlungen Änderungen in der Wiederaufbaupolitik und in den Finanzierungsmodi für konkrete Projekte bewirkt.

23.4 Lessons Learned

Im Folgenden werden wichtige Lessons Learned aus der wissenschaftlichen Begleitforschung zusammengefasst. Dabei besteht kein Anspruch auf Vollständigkeit.

- Die wissenschaftliche Begleitforschung nach Katastrophen, wie der Flutkatastrophe in Rheinland-Pfalz und Nordrhein-Westfalen im Sommer 2021, eröffnet nicht nur für die Wissenschaft enorme Einblicke in Risikotreiber, Schadenspotenziale und Lösungsdiskussionen, sondern bietet zudem die Chance, dass wissenschaftliche Daten und Expertise direkt in den Wieder- und Neuaufbauprozess einfließen.
- Katastrophenprozesse, die zu menschlichem Leid und erheblichen Schäden geführt haben, wie z. B. die Flut im Sommer 2021, die u. a. das Ahrtal sowie Orte an der Inde und Vicht verwüstete, oder die Sturzfluten in der Region Valencia in Spanien im Herbst 2024, führen oftmals zu Vertrauensverlusten in Institutionen und zu konfliktiven Wiederaufbauprozessen. In diesen Prozessen des Wieder- und Neuaufbaus kann wissenschaftliche Begleitforschung – gerade in größeren, interdisziplinären Teams – zu einer Versachlichung der Diskussion führen. Wissenschaft kann als neutraler Akteur eigene Empfehlungen und Lösungsvorschläge einbringen.
- Die wissenschaftliche Begleitforschung im Rahmen von KAHR hat gezeigt, dass Veränderungen und Innovationen nicht nur in den von der Katastrophe betroffenen Regionen wichtig sind, sondern auch die Rahmenregelungen – wie Gesetze (BauGB, WHG etc.) oder Wiederaufbauförderungen – von erheblicher Bedeutung sind. Die Frage, ob ein klimaresilienter Wieder- und Neuaufbau möglich ist, hängt demzufolge nicht nur von der Einstellung der Akteure vor Ort ab, sondern auch von den Rahmenbedingungen und der Frage, was die Wiederaufbauförderung auch als Tatbestand finanziert.
- Eine weitere Erkenntnis besteht darin, dass die wissenschaftliche Begleitforschung nach solchen Katastrophen nicht erst am Schluss des Projekts nach z. B. drei Jahren gefragt ist, sondern fortlaufende Analysen, Nutzerbedarfe aus der betroffenen Region oder dem Land, aber auch Stellungnahmen für übergeordnete Ebenen (z. B. Enquete-Kommissionen) in den Wieder- und Neuaufbauprozess eingebracht werden müssen. Damit besteht die Notwendigkeit die Forschungsförderung nicht an einem starren Endprodukt, sondern an dynamisch, sich verändernden Anforderungen und Nachfragen anzupassen.
- Des Weiteren hat sich gezeigt, dass die wissenschaftliche Begleitforschung nach Katastrophen auch eine Vernetzungsfunktion erfüllen muss, die einerseits wissenschaftliche Akteure vernetzt, um Doppelarbeiten in Teilen zu vermeiden, und andererseits auch Akteure aus unterschiedlichen Ebenen und Handlungsfeldern zusammenführt. Hier haben sich neue Kooperationen zwischen Wasserwirtschaft, Raumplanung und Hochwasserschutz (s. auch ▶ Kap. 13) ergeben sowie zwischen verschiedenen Kom-

munen, wobei im Rahmen interkommunaler Kooperationen auch neue Governance-Strukturen und Prozesse, um die Hochwasservorsorge über administrative Grenzen hinweg zu verbessern, entwickelt wurden (s. auch ▶ Kap. 14). Im Wieder- und Neuaufbau bedarf es einer weiteren Stärkung dieser Kooperationen. Bisher laufen Ansätze vielfach noch stark parallel.

- Eine weitere wichtige Erkenntnis ist, dass die Wiederaufbauförderungen deutlich verändert werden müssen, um effektiver und schneller sowie klimaresilienter den Wieder- und Neuaufbau organisieren zu können. Gerade bei erheblichen Schäden und großflächig betroffenen Räumen sollte der Schwerpunkt nicht auf einer Sammlung einzelner geschädigter Objekte liegen, sondern es sollten wie bei städtebaulichen Sanierungsgebieten ganze Stadtquartiere oder Räume definiert werden, die ein Neu- und Umbaukonzept im Rahmen des Wiederaufbaus erfordern. Die Einzelobjektbetrachtung bei Gebäuden oder bestimmten Schadensobjekten bietet teilweise keine hinreichende Grundlage für die Berücksichtigung von funktionalen Zusammenhängen. So sind in Teilen auch nicht-zerstörte oder nicht-geschädigte Infrastrukturen in den Um- und Neuaufbau einzubeziehen, um Klimaresilienz und Hochwasservorsorge fördern zu können.
- Zudem zeigt sich, dass die Normen und Standards im Hochwasserrisikomanagement und im Bereich der risikobasierten und adaptiven räumlichen Planung ebenfalls auf den Prüfstand müssen. Die Vorstellung, dass beispielsweise die Hochwassergefahrenkarten oder Starkregengefahrenkarten im Kontext solcher Extremereignissen und des Klimawandels unverändert bleiben, ist irrig. Hier müssen stärker als bisher Akteure mit Änderungen entsprechender Vorgaben im Wiederaufbau rechnen. Daher sind adaptivere Planungsstrategien, die auch eine Nachjustierung erlauben, gefragt. Unbeschadet dessen gibt es auch Grenzen der Anpassung und Grenzen der Adaptivität von Planung.
- Die wissenschaftliche Begleitforschung zeigt außerdem, dass Schutzziele und Schutzstandards (für den Wieder- und Neuaufbau) stärker differenziert und diskutiert werden sollten. Die bisherige Orientierung an einem Bemessungshochwasser HQ_{100} als Schutzstandard reicht aus unserer Erfahrung nicht aus. Differenziertere Schutzstandards sind notwendig, die auch die unterschiedliche Verwundbarkeit und Schutzwürdigkeit von Landnutzungen und Infrastrukturen berücksichtigen (s. auch ▶ Kap. 15). Ein Bolzplatz mit Rasen braucht nicht dasselbe Schutzniveau wie eine Schule für Kinder mit körperlichen und geistigen Behinderungen. Ein Krankenhaus oder eine Strominfrastruktur (z. B. Umspannwerk) ist anders zu betrachten als ein größerer Parkplatz, der im Falle eines Hochwassers- oder Starkregens nicht mehr aufgesucht werden sollte.
- Abschließend besteht eine wichtige Schlussfolgerung darin, dass der Erfolg oder Misserfolg wissenschaftlicher Begleitforschung auch von der Offenheit und dem Kooperationswillen von betroffenen Kommunen und Ländern abhängig ist. Neben dem Engagement der beteiligten Forschenden ist es von Bedeutung, dass die Adressaten eine gewisse Offenheit gegenüber wissenschaftlichen Institutionen und Akteuren haben. Hier hat im KAHR-Projekt beispielsweise die sehr gute Kooperation mit dem Landkreis Ahrweiler (s. auch ▶ Kap. 22) und dem Land Rheinland-Pfalz erhebliche Synergien und Potenziale geboten. Auch die aktive Unterstützung durch Bundesministerien kann hier sehr hilfreich sein.

23.5 Weitere Forschungsbedarfe und offene Fragen

Auf Basis der intensiven Forschung mit rund 13 Partnerinstitutionen und in Kooperation mit zahlreichen Praxis- und weiteren Forschungsakteuren ergeben sich auch neue Fragen und Forschungsbedarfe. Ohne den Anspruch auf Vollständigkeit erscheinen uns insbesondere die im Folgenden aufgeführten Themen und Fragestellungen besonders wichtig für die weitere wissenschaftliche Forschung und das systematische Lernen aus Krisen- und Wiederaufbauprozessen nach Flutkatastrophen.

So hat es im Zeitraum der drei Jahre nach der Flutkatastrophe deutliche Veränderungen der gesetzlichen Rahmenbedingungen für den Wieder- und Neuaufbau gegeben, wie z. B. durch die Einführung einer Wiederaufbauklausel im Baugesetzbuch (§ 246 c BauGB, s. auch ▶ Kap. 15). Auch gibt es ein neues Klimaanpassungsgesetz (KAnG 2023), das neue Anforderungen an die Risikovorsorge und Anpassung gegenüber Extremereignissen formuliert. Die Anwendung und Wirkung dieser Veränderungen der rechtlichen Rahmenbedingungen sind im Wieder- und Neuaufbau bisher nicht hinreichend erfasst. Vielfach ist zu beobachten, dass bei der Verwendung von Mitteln aus der Wiederaufbauförderung – die mit 30 Mrd. € eine erhebliche Größenordnung umfasst – zwar die sachliche und rechnerische Richtigkeit jeweils genau geprüft wird, aber eine Erfassung der Beispiele, die auch als Best-Practice für andere Wieder- und Neuaufbauprozesse genutzt werden können, vielfach nicht erkennbar ist. D. h. es überwiegt eine haushälterische Berichterstattung über die Verwendung der Wiederaufbaumittel. In dieser Hinsicht ist die Entwicklung einer Methodik zur Ermittlung und Bewertung von Best-Practice-Beispielen noch ein wichtiges zukünftiges Forschungsthema. Auch gibt es noch zahlreiche Maßnahmen und Planungen, die erst jetzt, drei Jahre nach der Katastrophe, in die Detailplanung kommen (z. B. Standorte für Hochwasserrückhaltebecken) und erst in den nächsten Jahren eine praktische Umsetzung erfahren werden.

Im Folgenden werden ausgewählte zukünftige Forschungsbedarfe noch einmal näher erläutert.

23.5.1 Wirkungen gesetzlicher Änderungen für den Wiederaufbau

Ein wichtiges zukünftiges Forschungsfeld ist die Analyse der Wirkungen gesetzlicher Änderungen für einen effektiven, klimaangepassten und schnellen Wieder- und Neuaufbau. Hierbei sollten die ersten Erfahrungen mit der Nutzung der Wiederaufbauklausel § 246c BauGB im Ahrtal und ausgewählten weiteren Regionen erfasst und bewertet werden. Gleichfalls besteht Forschungsbedarf bezogen auf die Frage, inwieweit das KAnG zur Stärkung des Belangs der Klimaresilienz im Kontext von Wiederaufbauprozessen beiträgt und bisher beigetragen hat. Die Forschung in diesem Bereich soll prüfen, inwiefern durch diese gesetzlichen Neuerungen eine bessere Verzahnung von Klimaanpassung und Wiederaufbau erzielt werden kann und ob die gesetzlichen Änderungen auch in der Lage sind, Wieder- und Neuaufbauprozesse sinnvoll zu beschleunigen (mehr Planungssicherheit, Bündelung von Prozessen). Hemmnisse bei der Umsetzung der neuen gesetzlichen Vorgaben in der Praxis sind zu identifizieren, um damit Grundlagen für eine ggf. erforderliche Nachschärfung zu liefern.

23.5.2 Schutz- und Anpassungsziele für den Wieder- und Neuaufbau

Ein weiteres offenes Forschungsfeld ist die Entwicklung von innovativen Ansätzen für die Festlegung von Schutz- und Anpassungszielen für den Wieder- und Neuaufbau. Bisher findet für den Wiederaufbau vielfach das HQ_{100} als Schutzstandard für zukünftige Hochwasser Anwendung. Die räumliche Ausdehnung der Bemessungshochwasser verändert sich allerdings aufgrund des Klimawandels und zudem fehlt bei dieser Betrachtung die Berücksichtigung der unterschiedlichen Schutzwürdigkeit von Landnutzungen, Infrastrukturen und Bevölkerungsgruppen (s. ▶ Kap. 15). Demzufolge sind weitere Forschungen notwendig, die auch die Bewertung der Empfindlichkeit und Schutzwürdigkeit von Nutzungen und Infrastrukturen für die Planung auf regionaler und kommunaler Ebene umfassen. Der Bundesraumordnungsplan Hochwasserschutz und das geplante neue Hochwasserschutzgesetz II des Bundes zeigen die Notwendigkeit einer differenzierteren Betrachtung möglicher betroffener Nutzungen und Bereiche auf, aber es fehlt noch an Methoden, die diese differenzierte Betrachtung der Vulnerabilität und der Risiken für unterschiedliche Landnutzungen und Themenbereiche wie kritische und sensible Infrastrukturen konkretisieren. Zudem sind dabei unterschiedliche Raumkategorien von Bedeutung, beispielsweise sind die aktuellen Schwellenwerte für die kritische Infrastruktur „Krankenhaus", die auch massiv bei den Hochwassern geschädigt wurde, im Ahrtal und eher ländlichen Regionen deutlich unterhalb des in der BSI-Kritisverordnung definierten Schwellenwertes von 30.000 vollstationären Fällen pro Jahr (Verordnung zur Bestimmung Kritischer Infrastrukturen nach dem BSI-Gesetz 2016). Damit müssen auch die Grundlagen für die Abgrenzung von kritischen und sensiblen Infrastrukturen für ländliche Räume auf den Prüfstand für Wieder- und Neuaufbauprozesse nach Flutkatastrophen.

23.5.3 Hochwasserrückhaltebecken und -räume

Aktuell gibt es im Ahrtal, aber auch in anderen flutbetroffenen Regionen in Deutschland Überlegungen, in den Hochwassergebieten von 2021 (z. B. an Ahr, Erft und Inde sowie ihren Nebenläufen) technische Hochwasserrückhalteräume zu planen und zu realisieren. In einem ersten Schritt unterstützte das KAHR-Projekt den Kreis Ahrweiler hierbei mit Untersuchungen zu Landnutzungsanpassungen und der groben Abschätzung des theoretischen Potenzials von Hochwasserrückhaltebecken sowie der Identifikation von Hindernissen für die Umsetzung solcher Becken (vor allem Flächenverfügbarkeit und konkurrierende Nutzungen), wobei deutlich wurde, dass die frühzeitige Flächensicherung entscheidend für eine effektive Hochwasservorsorge ist (s. ▶ Kap. 11). Aus wissenschaftlicher Sicht wäre es nun sinnvoll, insbesondere die aktuelle Planung und Umsetzung von Hochwasserrückhaltebecken im Ahrtal (19 Standorte) wissenschaftlich zu begleiten, da sowohl die weitere Bewertung der Standorte im Rahmen der Strategischen Umweltprüfung (SUP) und Umweltverträglichkeitsprüfung (UVP) von Bedeutung ist, als auch die Frage, wie man Standorte planerisch sichern kann (Landes- und Regionalplanung). Die bauliche Umsetzung wird erst in den nächsten Jahren erfolgen, da u. a. Fragen der Finanzierung und eines Zweckverbandes zu klären sind. Zudem ist zu prüfen, welche Standorte prioritär in Bezug auf die Effektivität und Akzeptanz entwickelt werden sollten. Die systematische Untersuchung und Bewertung größerer Planungs- und Infrastrukturvorhaben, wie technische Hochwasserrückhaltebecken, in Bezug auf ihre Effektivität für den zukünftigen Hochwasserschutz, auch im Kontext des Klimawandels, und die Akzeptanz solcher Vorhaben ist bisher noch nicht hinreichend weit entwickelt.

23.5.4 Aufbereitung von Erfahrungen und Erfahrungsaustausch zwischen Regionen bezogen auf den Wieder- und Neuaufbau sowie die Klimaresilienz

Da auch zahlreiche andere Regionen Schäden durch Hochwasser und Starkregen u. a. im Sommer 2024 verzeichneten und daher dort ebenfalls Wiederaufbauprozesse erforderlich sind, wäre ein stärkerer Erfahrungsaustausch zwischen diesen Regionen mit den Befunden des KAHR-Projekts zielführend. Gerade die Vermittlung wichtiger Lessons Learned aus dem Ahrtal und den flutbetroffenen Regionen in NRW für andere Regionen wäre sinnvoll, beispielsweise bezogen auf Potenziale der Beschleunigung von Wiederaufbauprozessen oder der Ansprache und Information von Bürgern bezüglich eines klimaresilienten Wieder- und Neuaufbaus. Die Formate dieses Erfahrungsaustausches müssen für unterschiedliche Akteure und Regierungsebenen noch entwickelt werden. Landesspezifische Besonderheiten sind dabei zu beachten, wenn es um die Koordinationsprozesse und Beschleunigung von Verfahren zwischen verschiedenen Ebenen (Bund, Land, Kreise, Kommunen) geht. Auch sollten für diesen Erfahrungsaustausch die Wirkungen der wissenschaftlichen Beratung für die Praxis aufbereitet und Empfehlungen für andere Regionen konkretisiert werden. Demzufolge verlangt ein systematischer und effektiver Lern- und Austauschprozess auch die Aufbereitung besonders guter Beispiele für einen hochwasser- und klimaresilienten Wieder- und Neuaufbau in verschiedenen Themenfeldern (Brücken, kritische und sensible Infrastrukturen, Objektschutz, etc.). Neben den Best-Practice-Beispielen könnten auch Erfahrungen dokumentiert werden, die man an anderer Stelle bzw. in anderen Regionen und Wiederaufbauprozessen vermeiden sollte.

Die vier ausgewählten Beispiele zeigen, dass die offenen Forschungsfragen noch weitere wichtige Handlungsfelder und Zukunftsthemen abdecken. Sicherlich gibt es weitere Themenfelder und Forschungsfragen, die ergänzt werden können.

Die in diesem Buch zusammengeführten Beiträge und Ergebnisse der dreijährigen Forschung im KAHR-Projekt bieten dabei einen guten Einblick in die Bedeutung, Praxisrelevanz und Vielschichtigkeit der wissenschaftlichen Begleitforschung nach solchen Katastrophen, die eben nicht nur für die betroffenen Regionen, sondern auch darüber hinaus für zahlreiche Akteure eine hohe Relevanz haben. Auch für zukünftige Wiederaufbauförderungen nach Flutkatastrophen sollten diese Erfahrungen nutzbar gemacht werden. Das vorliegende Buch ist ein Beitrag dazu.

Zusammenfassung

- Ein zentrales Produkt der KAHR-Forschung mit messbarem Impact sind die 10 Empfehlungen für einen klima- und hochwasserresilienten Wieder- und Neuaufbau. Anhand zweier Fallbeispiele im Bereich der kritischen und sensiblen Infrastrukturen kann exemplarisch gezeigt werden, wie die wissenschaftlichen Erkenntnisse und Empfehlungen aus der Begleitforschung in die Praxis des Wieder- und Neuaufbaus nach der Flutkatastrophe eingegangen sind.

- Zudem werden übergreifende Lessons Learned skizziert, die auch auf verschiedenen Beiträgen der KAHR-Forschung aufbauen. Inhaltlich zeichnen sich die zahlreichen Forschungsergebnisse durch eine hohe Bandbreite an inter- und transdisziplinären Ansätzen aus.
- Es gibt jedoch immer noch offene Forschungsfragen und zukünftige Forschungsbedarfe, z. B. in Bezug auf die Wirkung neuer rechtlicher Rahmenbedingungen oder die Begleitung der Planung und Umsetzung neuer Hochwasserrückhaltebecken.

Danksagung Die Autorinnen und Autoren von ▶ Kap. 23 bedanken sich im Namen aller Projektpartner beim Bundesministerium für Bildung und Forschung (BMBF) und beim DLR Projektträger für die Unterstützung und Förderung des KAHR-Projekts, welches vom BMBF unter dem Förderkennzeichen 01LR2102A-M finanziert wurde. Besonderer Dank gilt in diesem Zusammenhang den Mitarbeiterinnen und Mitarbeitern von BMBF und DLR Projektträger für den kollegialen Austausch sowie die vertrauensvolle Zusammenarbeit während der Antragstellung und der gesamten Projektlaufzeit.

Literatur

Birkmann J (2022) Öffentliche Anhörung des Ausschusses für Wohnen, Stadtentwicklung, Bauwesen und Kommunen am 6. Juli 2022 im Deutschen Bundestag. https://www.bundestag.de/dokumente/textarchiv/2022/kw27-pa-bau-flutkatastrophe-901092

Birkmann J, Schüttrumpf H, Trüdinger A, Thieken A, Schüller A, Huck A, Kleber A, Thiebes B, Droste-Franke B, Guse B, Merz B, Brüll C, Kuhlicke C, Stein C, Mudersbach C, Michalski D, Molinari D, Rodriguez-Castro D, Kirsten D, Klopries E, Bachofer F, Steudtner F, Johann G, Lauer H, Luhmann HJ, Arenz HT, Zachert H, Kreibich H, Apel H, Meyer H, Frančik H, Hoppe H, Sauter H, Hasse J, Moßgraber J, Stamm J, Emde K, Laranjeira K, Burghardt L, Kirschbauer L, Madruga de Brito M, Evers M, Disse M, Ravan M, Schaefer M, Garschagen M, Boronowsky M, Paul M, Thiessen N, Koch O, Mahrenholz P, Meier P, Bellanova P, Jüpner R, Reinstädtler S, Vorogushyn S, Schöttmer S, Greiving S, Wolf S, Nietgen T, Hellmund T, Korrmann V, Kron W, Reckhaus Z (2022) 10 Empfehlungen aus Sicht der Wissenschaft zum Thema Wiederaufbau und Zukunftsfähigkeit der flutbetroffenen Regionen. Veröffentlichung im Rahmen des Vorhabens KAHR – Klima-Anpassung, Hochwasser und Resilienz (2021–2025). https://hochwasser-kahr.de/index.php/de/kahr-produkte?view=article&id=83:10-empfehlungen&catid=12

Birkmann J, Schüttrumpf H, Handmer J, Thieken A, Kuhlicke C, Truedinger AJ, Sauter H, Klopries EM, Greiving S, Jamshed A, Merz B, Solecki W, Kirschbauer L (2023a) Strengthening resilience in reconstruction after extreme events – Insights from flood affected communities in Germany. Int J Disaster Risk Reduct. https://doi.org/10.1016/j.ijdrr.2023.103965

Birkmann J, Trüdinger A, Sauter H, Greiving S (2023b) Stellungnahme zur Levana-Schule als besonders sensible Infrastruktur für die Kreisverwaltung, Stuttgart im Mai 2023 (unveröffentlicht)

Birkmann J, Truedinger AJ, Schüttrumpf H (2025) The Impact of Science: Uptake of Scientific Recommendations After Extreme Events – Case Study Floods in 2021 in Germany. J Flood Risk Manag 18(3):e70100. https://doi.org/10.1111/jfr3.70100

Bonner General-Anzeiger (2024) Wir kriegen das Wasser nicht weggezaubert. https://ga.de/region/ahr-und-rhein/bad-neuenahr-ahrweiler/wiederaufbau-der-flut-schutz-gegen-hochwasser-v1_aid-121541797

Burghardt L, Schüttrumpf H, Wolf S, Klopries EM (2022) Analyse der Schäden an Brückenbauwerken in Folge des Hochwassers 2021 an der Ahr. Wasser Abfall 24(11):12–17. https://doi.org/10.1007/s35152-022-1346-x

Feiertag P (2018) Politikrelevantes Wissen in der Raumplanung: Grenzarbeit zwischen Wissenschaft und Politik in Deutschland, den Niederlanden und der Schweiz. Rohn, Lemgo. ISBN 978-3-946319-20-7.

Heidbreder L (2023) Abschlussbericht der Enquete-Kommission 18/1 „Konsequenzen aus der Flutkatastrophe in Rheinland-Pfalz: Erfolgreichen Katastrophenschutz gewährleisten, Klimawandel ernstnehmen und Vorsorgekonzepte weiterentwickeln" („Zukunftsstrategien zur Katastrophenvorsorge"). Landtag von

23

Rheinland-Pfalz, Drucksache 18/8222, Punkt 9. https://dokumente.landtag.rlp.de/landtag/drucksachen/8222-18.pdf

Klimaanpassungsgesetz (KAnG) der Bundesrepublik Deutschland, BGBl. 2023 I Nr. 393 vom 20.12.2023. https://www.recht.bund.de/bgbl/1/2023/393/VO

Lumbroso D (2007) Review report of operational flood management methods and models, FloodSite Consortium, T17-07-01. http://www.floodsite.net/html/partner_area/project_docs/Task17_report_M17_1review_v1_1.pdf

Mohor GS, Sieg T, Koch O, Buhrmann A (2024) Remote sensing-based mapping of structural building damage in the Ahr valley. J Flood Risk Manag. https://doi.org/10.1111/jfr3.12983

Morton S (2015) Progressing research impact assessment: A „contributions" approach. Res Eval 24(4):405–419. https://doi.org/10.1093/reseval/rvv016

Pfeifer L, Helming K (2024) Effective mission-oriented research: A new framework for systemic research impact assessment. Res Eval. https://doi.org/10.1093/reseval/rvae003

Raumordnungsgesetz der Bundesrepublik Deutschland vom 22. Dezember 2008 (BGBl. I S. 2986), das zuletzt durch Artikel 1 des Gesetzes vom 22. März 2023 (BGBl. 2023 I Nr. 88) geändert worden ist. https://www.gesetze-im-internet.de/rog_2008

Rhein-Zeitung (2024) Weg ist frei: Levana-Schule bekommt neuen Standort im Kreis Ahrweiler. https://www.rhein-zeitung.de/lokales/kreis-ahrweiler/weg-ist-frei-levana-schule-bekommt-neuen-standort-im-kreis-ahweiler_arid-2693591.html

Schüttrumpf, H. (2022): Stellungnahme im Anhörverfahren der Enquete-Kommission 18/1 „Zukunftsstrategien zur Katastrophenvorsorge" zum Thema „Technischer Hochwasserschutz, Natürliche Hochwasservorsorge, Erosionsschutz, Wasserrückhalt und Retention" am 11. Oktober 2022. Landtag von Rheinland-Pfalz, EK 18/1-78. Online: https://dokumente.landtag.rlp.de/landtag/vorlagen/1-78-18.pdf&ved=2ahUKEwi5xvGEi-mKAxWQ0QIHHRqKGjYQFnoECA0QAQ&usg=AOvVaw0KFFXwFtaZVC_vrBkP2MN4

Thieken A (2024) Stellungnahme im Anhörverfahren der Enquetekommission II: „Krisen- und Notfallmanagement" – durch die Lehren der Vergangenheit die Zukunft sicher gestalten am 03. Juni 2024. Landtag von Nordrhein-Westfalen. https://www.landtag.nrw.de/portal/WWW/dokumentenarchiv/Dokument/MMST18-1487.pdf&ved=2ahUKEwiLof_hi-mKAxXO0QIHHadkIUAQFnoECBIQAQ&usg=AOvVaw0OZky2wspwxRMuZTApdaSn

Truedinger A, Birkmann J, Fleischhauer M, Ferreira C (2024) Sensitive infrastructures and people with disabilities – Key issues when strenghtening resilience in reconstruction. Egusphere Pre Print. https://doi.org/10.5194/egusphere-2024-1607

Truedinger AJ, Jamshed A, Sauter H, Birkmann J (2023) Adaptation after Extreme Flooding Events: Moving or Staying? The Case of the Ahr Valley in Germany. Sustainability 15(2):1407. https://doi.org/10.3390/su15021407

United Nations Office for Disaster Risk Reduction (2015) Sendai Framework for Disaster Risk Reduction 2015–2030. Veröffentlichung im Rahmen der World Conference on Disaster Risk Reduction, Sendai, 2015 Online, Japan (https://digitallibrary.un.org/record/793460?v=pdf)

Vanneuville W, Kellens W, De Maeyer P, Reniers G, Witlox F (2011) Is 'flood risk management' identical to 'flood disaster management'? IEEE Earthzine 13. See link: https://earthzine.org/is-flood-risk-management-identical-to-flood-disaster-management-3/

Verordnung zur Bestimmung Kritischer Infrastrukturen nach dem BSI-Gesetz (BSI-Kritisverordnung – BSI-KritisV) vom 22. April 2016 (BGBl. I S. 958), die zuletzt durch Artikel 1 der Verordnung vom 29. November 2023 (BGBl. 2023 I Nr. 339) geändert worden ist. https://www.gesetze-im-internet.de/bsi-kritisv/BJNR095800016.html

Wasserhaushaltsgesetz der Bundesrepublik Deutschland vom 31. Juli 2009 (BGBI. I S. 2585), das zuletzt durch Artikel 7 des Gesetzes vom 22. Dezember 2023 (BGBl. 2023 I Nr. 409) geändert worden ist. https://www.gesetze-im-internet.de/whg_2009/BJNR258510009.html

Zenker ML, Bubeck P, Thieken AH (2024) Always on my mind: indications of post-traumatic stress disorder among those affected by the 2021 flood event in the Ahr valley, Germany. Nat Hazards Earth Syst Sci 24:2837–2856. https://doi.org/10.5194/nhess-24-2837-2024